**SOC3**
**Nijole V. Benokraitis**

Executive Editor: Mark Kerr

Acquisitions Editor: Seth Dobrin

Developmental Editor: Kristin Makarewycz and Thomas Finn

Assistant Editor: Mallory Ortberg

Editorial Assistant: Nicole Bator

Media Editor: John Chell

Senior Brand Manager: Liz Rhoden

Senior Market Development Manager: Michelle Williams

Product Development Manager, 4LTR Press: Steven E. Joos

Senior Content Project Manager: Cheri Palmer

Senior Art Director: Caryl Gorska

Manufacturing Planner: Judy Inouye

Rights Acquisitions Specialist: Tom McDonough

Production Service and Composition: Lachina Publishing Services

Photo Researcher: Q2a/Bill Smith

Text Researcher: Pablo D'Stair

Illustrator: Lachina Publishing Services

Text and Cover Designer: Riezebos Holzbaur

Cover Image: David Cowles

Inside Front Cover: © iStockphoto.com/sdominick; © iStockphoto.com/alexsl; © iStockphoto.com/A-Digit

Page i: © iStockphoto.com/CostinT; © iStockphoto.com/photovideostock; © iStockphoto.com/Leontura

Back Cover: © iStockphoto.com/René Mansi

For product information and technology assistance, contact us at
**Cengage Learning Customer & Sales Support, 1-800-354-9706**

For permission to use material from this text or product, submit all requests online at **cengage.com/permissions**
Further permissions questions can be emailed to
**permissionrequest@cengage.com**

Library of Congress Control Number: 2012943491

ISBN-13: 978-1-133-59212-9

ISBN-10: 1-133-59212-0

**Wadsworth**
20 Davis Drive
Belmont, CA 94002-3098
USA

Cengage Learning is a leading provider of customized learning solutions with office locations around the globe, including Singapore, the United Kingdom, Australia, Mexico, Brazil, and Japan. Locate your local office at **www.cengage.com/global**.

Cengage Learning products are represented in Canada by Nelson Education, Ltd.

To learn more about Wadsworth, visit **www.cengage.com/wadsworth**

Purchase any of our products at your local college store or at our preferred online store **www.CengageBrain.com**.

Printed in the United States of America
2 3 4 5 6 7 16 15 14 13

# BRIEF CONTENTS

# CONTENTS

## 1 Thinking Like a Sociologist    1

## 2 Examining Our Social World    21

## 3 Culture    39

# 4 Socialization 61

# 5 Social Interaction and Social Structure 81

# 6 Social Groups, Organizations, and Social Institutions 99

# 7 Deviance, Crime, and the Criminal Justice System   117

# 8 Social Stratification: United States and Global   137

# 9 Gender and Sexuality   157

# 10 Race and Ethnicity 179

# 11 The Economy and Politics 199

# SOC

A sociological imagination can give you more control over your life.

# Thinking Like a Sociologist

Text messaging is associated with the highest risk of car crashes, and headset cell phones aren't much safer than handheld cell phones. In effect, text messaging equates to a driver traveling the length of a football field at 55 miles per hour without looking at the road, and using a cell phone almost triples the risk for being in a car accident (Hanowski 2009). Almost 92 percent of American drivers consider someone talking on a cell phone while driving a serious threat to safety, and 97 percent say the same about texting while driving. However, 69 percent admit using cell phones, and 35 percent send text messages while driving (AAA Foundation for Traffic Safety 2008, 2011).

**sociology** the systematic study of social interaction at a variety of levels.

Why is there such a disconnection between many Americans' attitudes and behavior? This chapter examines these and other questions. Let's begin by looking at what sociology is (and isn't) and how a "sociological imagination" can help us have more control over our lives. We'll then look at how sociologists grapple with complex theoretical issues in explaining human behavior.

## Key Topics

In this chapter, we'll explore the following topics:

1-1 What Is Sociology?

1-2 What Is a Sociological Imagination?

1-3 Why Study Sociology?

1-4 Some Origins of Sociological Theory

1-5 Contemporary Sociological Theories

## 1-1 What Is Sociology?

**S**ociology is the systematic study of social interaction at a variety of levels. *Social interaction* is the process of acting toward and reacting to people around us (see Chapter 5). When sociologists talk about the *systematic* study of social interaction, they mean that social behavior is regular and patterned, and that it takes place between individuals, small groups (such as families), large organizations (such as Apple), and entire societies (such as between the United States and other countries). But, you might protest, "I'm unique."

### 1-1a Are You Unique?

Yes and no. Each of us is unique in the sense that you and I are like no one else on earth. Even identical twins, who have the same physical characteristics and genetic matter, usually differ in personality and interests. One of my colleagues likes to tell the story about his 3-year-old twin girls who received the same doll. One twin chattered that the doll's name was Lori, that she loved Lori and would take good care of her. The second twin muttered, "Her name is Stupid," and flung the doll into a corner.

## What do you think?

Sociology is basically common sense.

| 1 | 2 | 3 | 4 | 5 | 6 | 7 |

strongly agree          strongly disagree

## EVERYBODY KNOWS THAT . . .

**True or False?**

1. The death penalty reduces crime.

2. Opposites attract.

3. People age 65 and older make up the largest group of those who are poor.

4. A majority of Latino families receive public assistance.

5. Divorce rates are higher today than in the past.

6. Among married and unmarried couples, jealousy is a normal trait that shows love.

7. The best way to get an accurate measure of public opinion is to poll as many people as possible.

8. Most child kidnappings are committed by strangers.

*The answers are on the next page.*

© Cengage Learning

**sociological imagination**
the ability to see the relationship between individual experiences and larger social influences.

Despite some differences, identical twins, you, and I are also like other people in many ways. Around the world, we experience grief when a loved one dies, participate in rituals that celebrate marriage or the birth of a child, and want to have healthy and happy lives. Some actions, like terrorist attacks, are unpredictable. For the most part, however, people conform to expected and acceptable behavior. From the time that we get up until we go to bed, we follow a variety of rules and customs about what we eat, how we drive, how we act in different social situations, and how we dress for work, classes, and leisure activities.

So what? you might shrug. Isn't it "obvious" that we dress differently for classes than for job interviews? Isn't all of this just plain old common sense? Before reading further, take the quiz above.

### 1-1b Isn't Sociology Just Common Sense?

No. Sociology ~~is~~ [handwritten note: Sociology **is** more than common sense.] isdom, what we call c

- *Common* [covered by note] ...es into my car, I [covered] conventional wis... [covered] ver the years, tha... [covered] In fact, most driv... [covered] pecially teenagers and those age 70 and older (Insurance Institute for Highway Safety 2010). Thus, *objective* data show that, overall, men are worse drivers than women.

- *Common sense ignores facts.* Because common sense is subjective, there is little room for facts that might be disturbing or challenge cherished beliefs. For example, many Americans are most concerned about street crimes, such as robbery, or violent assaults by strangers. However, FBI and sociological data show that we're much more likely to be assaulted or murdered by someone we know or live with (see Chapters 7 and 12).

- *Commonsense perceptions vary across groups and cultures.* In the United States, it's common sense to date before choosing a marriage mate. In many other societies, however, it's common sense to marry someone that parents and relatives have selected. Thus, commonsense notions about mate selection vary considerably around the world.

- *Much of our common sense is based on myths and misconceptions.* A common myth (false notion about life that ignores evidence to the contrary) is that living together is a good way to find out whether partners will get along after marriage. Generally, however, couples who live together before marriage have higher divorce rates than those who don't (see Chapter 12).

Sociol... [covered by note: Our individual behavior is influenced by social factors such as religion, ethnicity, and politics.] ...d wisdom, examines ... ...ders many points of v... ...nd established way... ...e reasons why a soci... ...y a "sociological ima...

### 1-2 What is a Sociological Imagination?

Acco... [covered] ...ght Mills (191... [handwritten note: sociological imagination → connection b/w personal troubles (biography) & structural issues (public & historical)] ...s influenced by s... ...nicity, and politics. M... ...the relationship betw... ...arger social influences... ...sociological imaginati... ...een personal troubles (biography) and structural (public and historical) issues. Mills noted, for example, that if only some people

# Thinking Like a Sociologist

Text messaging is associated with the highest risk of car crashes, and headset cell phones aren't much safer than handheld cell phones. In effect, text messaging equates to a driver traveling the length of a football field at 55 miles per hour without looking at the road, and using a cell phone almost triples the risk for being in a car accident (Hanowski 2009). Almost 92 percent of American drivers consider someone talking on a cell phone while driving a serious threat to safety, and 97 percent say the same about texting while driving. However, 69 percent admit using cell phones, and 35 percent send text messages while driving (AAA Foundation for Traffic Safety 2008, 2011).

**sociology** the systematic study of social interaction at a variety of levels.

Why is there such a disconnection between many Americans' attitudes and behavior? This chapter examines these and other questions. Let's begin by looking at what sociology is (and isn't) and how a "sociological imagination" can help us have more control over our lives. We'll then look at how sociologists grapple with complex theoretical issues in explaining human behavior.

## Key Topics

In this chapter, we'll explore the following topics:

1-1 What Is Sociology?

1-2 What Is a Sociological Imagination?

1-3 Why Study Sociology?

1-4 Some Origins of Sociological Theory

1-5 Contemporary Sociological Theories

**What do you think?**

Sociology is basically common sense.

1 2 3 4 5 6 7

strongly agree      strongly disagree

## 1-1 What Is Sociology?

**S**ociology is the systematic study of social interaction at a variety of levels. *Social interaction* is the process of acting toward and reacting to people around us (see Chapter 5). When sociologists talk about the *systematic* study of social interaction, they mean that social behavior is regular and patterned, and that it takes place between individuals, small groups (such as families), large organizations (such as Apple), and entire societies (such as between the United States and other countries). But, you might protest, "I'm unique."

### 1-1a Are You Unique?

Yes and no. Each of us is unique in the sense that you and I are like no one else on earth. Even identical twins, who have the same physical characteristics and genetic matter, usually differ in personality and interests. One of my colleagues likes to tell the story about his 3-year-old twin girls who received the same doll. One twin chattered that the doll's name was Lori, that she loved Lori and would take good care of her. The second twin muttered, "Her name is Stupid," and flung the doll into a corner.

**EVERYBODY KNOWS THAT . . .**

**True or False?**

1. The death penalty reduces crime.

2. Opposites attract.

3. People age 65 and older make up the largest group of those who are poor.

4. A majority of Latino families receive public assistance.

5. Divorce rates are higher today than in the past.

6. Among married and unmarried couples, jealousy is a normal trait that shows love.

7. The best way to get an accurate measure of public opinion is to poll as many people as possible.

8. Most child kidnappings are committed by strangers.

*The answers are on the next page.*

---

**sociological imagination**
the ability to see the relationship between individual experiences and larger social influences.

Despite some differences, identical twins, you, and I are also like other people in many ways. Around the world, we experience grief when a loved one dies, participate in rituals that celebrate marriage or the birth of a child, and want to have healthy and happy lives. Some actions, like terrorist attacks, are unpredictable. For the most part, however, people conform to expected and acceptable behavior. From the time that we get up until we go to bed, we follow a variety of rules and customs about what we eat, how we drive, how we act in different social situations, and how we dress for work, classes, and leisure activities.

So what? you might shrug. Isn't it "obvious" that we dress differently for classes than for job interviews? Isn't all of this just plain old common sense? Before reading further, take the quiz above.

## 1-1b Isn't Sociology Just Common Sense?

No. Sociology goes well beyond conventional wisdom, what we call common sense, in several ways:

- *Common sense is subjective.* If a woman crashes into my car, I might conclude, according to the conventional wisdom statements that we've heard over the years, that "all women are terrible drivers." In fact, most drivers involved in crashes are men—especially teenagers and those age 70 and older (Insurance Institute for Highway Safety 2010). Thus, *objective* data show that, overall, men are worse drivers than women.

- *Common sense ignores facts.* Because common sense is subjective, there is little room for facts that might be disturbing or challenge cherished beliefs. For example, many Americans are most concerned about street crimes, such as robbery, or violent assaults by strangers. However, FBI and sociological data show that we're much more likely to be assaulted or murdered by someone we know or live with (see Chapters 7 and 12).

- *Commonsense perceptions vary across groups and cultures.* In the United States, it's common sense to date before choosing a marriage mate. In many other societies, however, it's common sense to marry someone that parents and relatives have selected. Thus, commonsense notions about mate selection vary considerably around the world.

- *Much of our common sense is based on myths and misconceptions.* A common myth (false notion about life that ignores evidence to the contrary) is that living together is a good way to find out whether partners will get along after marriage. Generally, however, couples who live together before marriage have higher divorce rates than those who don't (see Chapter 12).

Sociology, in contrast with conventional wisdom, examines claims and beliefs critically, considers many points of view, and enables us to move beyond established ways of thinking. These are some of the reasons why a sociological perspective, and especially a "sociological imagination," is important.

## 1-2 What Is a Sociological Imagination?

According to sociologist C. Wright Mills (1916–1962), our individual behavior is influenced by social factors such as religion, ethnicity, and politics. Mills (1959) called this ability to see the relationship between individual experiences and larger social influences the **sociological imagination**. The sociological imagination emphasizes the connection between personal troubles (biography) and structural (public and historical) issues. Mills noted, for example, that if only some people

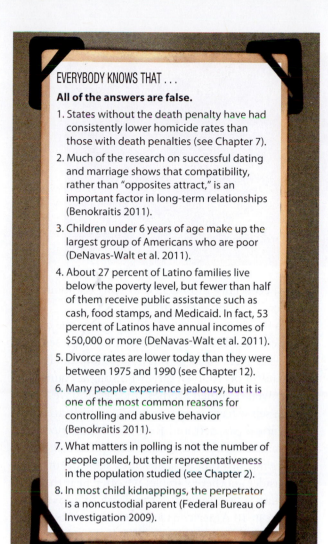

**EVERYBODY KNOWS THAT . . .**

**All of the answers are false.**

1. States without the death penalty have had consistently lower homicide rates than those with death penalties (see Chapter 7).

2. Much of the research on successful dating and marriage shows that compatibility, rather than "opposites attract," is an important factor in long-term relationships (Benokraitis 2011).

3. Children under 6 years of age make up the largest group of Americans who are poor (DeNavas-Walt et al. 2011).

4. About 27 percent of Latino families live below the poverty level, but fewer than half of them receive public assistance such as cash, food stamps, and Medicaid. In fact, 53 percent of Latinos have annual incomes of $50,000 or more (DeNavas-Walt et al. 2011).

5. Divorce rates are lower today than they were between 1975 and 1990 (see Chapter 12).

6. Many people experience jealousy, but it is one of the most common reasons for controlling and abusive behavior (Benokraitis 2011).

7. What matters in polling is not the number of people polled, but their representativeness in the population studied (see Chapter 2).

8. In most child kidnappings, the perpetrator is a noncustodial parent (Federal Bureau of Investigation 2009).

© Cengage Learning

*Microsociology* focuses on the relationships between individuals, whereas *macrosociology* examines social dynamics across a society.

## 1-2a Microsociology: How People Affect Our Everyday Lives

We have many choices in our everyday lives, such as deciding where to shop, what to eat, and whether to buy a car. **Microsociology** is a sociological approach that examines the patterns of individuals' social interaction in specific settings. In most of our relationships, we interact with others on a micro, or "small," level (such as members of a work group discussing who will perform which tasks). These everyday interactions involve what people think, say, or do on a daily basis.

## 1-2b Macrosociology: How Social Structure Affects Our Everyday Lives

**Macrosociology** focuses on large-scale patterns and processes that characterize society as a whole. Macro, or "large," approaches are especially useful in understanding some of the constraints—such as economic

> **microsociology** a sociological approach that examines the patterns of individuals' social interaction in specific settings.
>
> **macrosociology** the study of large-scale patterns and processes that characterize society as a whole.

are unemployed, that's a *personal trouble*. If unemployment is widespread, it's a *public issue*, because economic opportunities have collapsed and the problem requires solutions at the societal rather than at the individual level.

A sociological imagination helps us understand how and why our personal troubles often reflect larger public issues and policies over which we have little, if any, control. For example, Americans have among the lowest voting rates in industrialized countries. Why? It's tempting to conclude that Americans are apathetic or satisfied with the way that government functions (i.e., individual choice). As you'll see in Chapter 11, however, and compared with many other countries, our political system often discourages voting. Examples include the scheduling of elections when most people are at work and transportation problems among many people at lower socioeconomic levels (i.e., structural factors). Thus, a sociological imagination "empowers people to think about themselves, others, and what life is and could be in new and liberating ways" (Dandaneau 2001: 12).

A sociological imagination relies on both micro- and macro-level approaches to understand our social world.

## Marriage Without Love? No Way!

When I ask my students, "Would you marry someone you're not in love with?" most laugh, raise an eyebrow, or stare at me in disbelief. "Of course not!" they exclaim. In fact, the "open" courtship and dating systems common in Western nations, including the United States, are foreign to much of the world. In many African, Asian, Mediterranean, and Middle Eastern countries, marriages are arranged. In these societies, marriages forge bonds between families rather than individuals and preserve family continuity along religious and socioeconomic lines. Love is not a prerequisite for marriage in societies that value the intergenerational and community relations of a kin group rather than an individual's choices (see Chapters 9 and 12).

© Cengage Learning

The Wire was a popular television show that aired on HBO between 2002 and 2008. It depicted the lives of poor people and blue-collar workers in Baltimore, Maryland. Many faculty still include The Wire in their lectures or require their students to watch selected episodes (Bennett 2010). For sociologists, The Wire illustrates the connection between macro-level structural constraints (such as poverty and unemployment) and micro-level individual behavior (such as dropping out of school, dealing drugs, and engaging in corrupt political practices).

© AF archive/ Alamy

forces, laws, and social and public policies—that limit many of our personal options on the micro level.

Microsociology and macrosociology differ conceptually, but they're interrelated. Consider divorce. On a micro level, sociologists might analyze the everyday interactions that fuel marital tension and unhappiness, leading to divorce. On a macro level, sociologists might look at how economic factors—such as job loss, home foreclosures, and high credit card debts—affect divorce rates. Examining micro, macro, and micro–macro forces is one of the reasons why sociology is a powerful tool in understanding (and changing) our behavior and society at large (Ritzer 1992).

## 1-3 Why Study Sociology?

Sociology offers explanations that can greatly improve the quality of your everyday life. These explanations can influence choices that range from your personal decisions to expanding your career opportunities.

### 1-3a Making Informed Decisions

Knowing some sociology can help us make more informed decisions. In 1982, psychologist Carol Gilligan published an influential book, *In a Different Voice*, which maintained that adolescent girls face a devastating drop in self-regard that boys don't experience. A decade later, clinical psychologist Mary Pipher's (1994) best seller, *Reviving Ophelia*, contended similarly that teenage girls experience a decline in self-esteem from which many never recover. Both books, publicized by the popular press, generated considerable anxiety among

© Suzanne Tucker/Shutterstock

parents, especially mothers, who worried about their daughters' emotional health.

Was the distress justified? No, because the conclusions were based on very small and nonrepresentative groups—Gilligan's on a private girls' school and Pipher's on a handful of troubled girls who sought counseling. In fact, well-designed studies since then have shown that the self-esteem scores of boys and girls are virtually identical (Barnett and Rivers 2004). Nonetheless, many academics and the mainstream press continue to promote the idea that girls have low self-esteem.

Consider another example. We often hear that grief counseling is essential after the death of a loved one. In fact, 4 in 10 Americans are better off without such counseling. Grief is normal, and most people work through their losses on their own, whereas counseling sometimes prolongs the feelings of depression and anxiety (Stroebe et al. 2000).

### 1-3b Understanding Diversity

The racial and ethnic composition of the United States is becoming more diverse. By 2020, 64 percent of the U.S. population will be white, down from 76 percent in 1990 and 86 percent in 1950 (U.S. Census Bureau 2012). As you'll see in later chapters, this racial/ethnic shift has already affected interpersonal relationships, as well as education, politics, religion, and other spheres of social life.

## Sociology in Your Life

How useful is your Intro Soc course? You're probably taking it because you "have to," but you'll enjoy the course and find it useful in the future. Dr. Steven Steele, who teaches at Anne Arundel Community College in Maryland, explains why applied sociology is important. Go to Sociology CourseMate at CengageBrain.com to access this site.

Recognizing and understand diversity is one of the central themes in sociology. Our gender, social class, marital status, ethnicity, sexual orientation, and age—among other factors—shape our beliefs, behavior, and experiences. If, for example, you are a white middle-class male who attends a private college, your experiences are very different from those of a female Vietnamese immigrant who is struggling to pay expenses at a community college.

Increasingly, nations around the world are intertwined through political and economic ties. What happens in other societies often has a direct or indirect impact on contemporary U.S. life. Decisions in oil-producing countries, for example, affect gas prices, spur the development of hybrid cars that are less dependent on oil, and stimulate research on alternative sources of energy.

## 1-3c Shaping Social and Public Policies and Practices

Sociology is valuable in applied, clinical, and policy settings because many jobs require understanding society and research, and applying theoretical perspectives to create social change. According to a director of a research institute that focuses on patient–provider relationships in health care, for example, sociology increased her contributions: "I can look at problems of concern to the National Institutes of Health and say 'here is a different way to solve this problem'" (Nyseth et al. 2011: 48).

## 1-3d Thinking Critically

We develop a sociological imagination not only when we understand and can apply the concepts, but also when we can think, speak, and write critically. Much of our thinking and decision making is often impulsive and emotional. In contrast, critical thinking enhances our knowledge and problem solving (Paul and Elder 2007).

Critical sociological thinking goes further

because we begin to understand how our individual lives, choices, and troubles are shaped by larger forces such as race, gender, social class, and social institutions, including the economy, politics, and education (Eckstein et al. 1995; Grauerholz and Bouma-Holtrop 2003). *Table 1.1* summarizes some of the basic elements of critical sociological thinking.

## 1-3e Expanding Your Career Opportunities

A degree in sociology is a springboard for entering many jobs and professions. A national survey of sociology majors found that 30 percent were in administrative support or management positions, 27 percent were employed in social service and counseling, and more than 10 percent were in sales and marketing occupations (Spalter-Roth and Van Vooren 2008).

What specific skills do sociology majors learn that are useful in their jobs? Some of the most important skills are being better able to work with people (71 percent), to organize information (69 percent), to use a computer (64 percent), to write reports that nonsociologists understand (61 percent), and to interpret research findings (56 percent) (Van Vooren and Spalter-Roth 2010). In other cases, students major in sociology because they see it as a broad liberal arts foundation for professions such as law, education, medicine, and social work.

Even if you don't major in sociology, developing your sociological imagination can bring a depth and breadth of understanding to your job. Sociology courses help you learn to think abstractly and critically, formulate problems, ask incisive questions, search for answers in the most reliable and up-to-date sources, organize material, and improve your oral presentations.

---

### TABLE 1.1

# What Is Critical Sociological Thinking?

Critical sociological thinking requires a combination of skills. Some of the basic elements include the ability to:

- rely on reason rather than emotion
- ask questions, avoid snap judgments, and examine popular and unpopular beliefs
- recognize one's own and others' assumptions, prejudices, and points of view
- remain open to alternative explanations and theories
- require and examine competing evidence (see Chapter 2)
- understand how public issues affect or cause many private troubles

© Cengage Learning

© Sira Anamwong/Shutterstock

# Sociology and Other Social Sciences: *What's the Difference?*

How would different social scientists study the same phenomenon, such as homelessness? Criminologists might examine whether crime rates are higher among homeless people than in the general population. Economists might measure the financial impact of programs for the homeless. Political scientists might study whether and how government officials respond to homelessness. Psychologists might be interested in how homelessness affects individuals' emotional and mental health. Social workers are most likely to try to provide needed services such as food, shelter, medical care, and jobs. Sociologists have been most interested in examining homelessness across gender, age, and social class, and explaining how this social problem devastates families and communities.

According to sociologist Herbert Gans (2005), sociologists "study everything." There are currently 43 different subfields in sociology, and the number continues to increase, because sociologists' interests range across many areas.

> **theory** a set of statements that explains why a phenomenon occurs.

## 1-4 Some Origins of Sociological Theory

**Theories = Tools**

During college, most of my classmates and I avoided taking theory courses (in all majors) as long as possible. "This stuff is boring, boring, boring," we'd grump, "and has nothing to do with the real world." However, theorizing is, in fact, part of our everyday lives. Every time you try to explain why your family and friends behave as they do, for example, you're theorizing.

As people struggle to understand human behavior, they develop theories. A **theory** is a set of statements that explains why a phenomenon occurs. Theories produce knowledge, guide our research, help us analyze our findings, and, ideally, offer solutions for social problems.

Sociologist James White (2005: 170–171) describes theories as "tools" that don't profess to know "the truth" but "may need replacing" over time as our understanding of society becomes more sophisticated. Like hardware tools that change over the years, theories

Mitchell Funk/Photographer's Choice/Getty Images

© The Art Gallery Collection/Alamy

Father of Sociology—Auguste Comte

evolve over time because of cultural and technological transformations. As you'll see shortly, for example, sociological theories about behavior changed considerably after the rise of feminist scholarship during the late 1960s.

Sociological theories didn't emerge overnight. Nineteenth-century thinkers grappled with some of the same questions that sociologists try to answer today: Why do people behave as they do? What holds society together? What pulls it apart? Of the many early sociological theorists, some of the most influential were Auguste Comte, Harriet Martineau, Émile Durkheim, Karl Marx, Max Weber, Jane Addams, and W. E. B. Du Bois.

## 1-4a Auguste Comte

Auguste Comte (pronounced oh-gust KONT; 1798–1857) coined the term *sociology* and is often described as the "father of sociology." Comte maintained that the study of society must be **empirical**. That is, information should be based on observations, experiments, or other data collection rather than on ideology, religion, intuition, or conventional wisdom.

He saw sociology as the scientific study of two aspects of society: social statics and social dynamics. *Social statics* investigates how principles of social order explain a particular society, as well as the interconnections between institutions. *Social dynamics* explores how individuals and societies change over time. Comte's emphasis on social order and change within and across societies is still useful today because many sociologists examine the relationships between education and politics (social statics), as well as how such interconnections change over time (social dynamics).

## 1-4b Harriet Martineau

Harriet Martineau (1802–1876), an English author, published several dozen books on a wide range of topics in social science, politics, literature, and history. Her translation and condensation of Auguste Comte's difficult material for popular consumption was largely responsible for the dissemination of Comte's work. "We might, say, then, that sociology had parents of both sexes" (Adams and Sydie 2001: 32). She emphasized the importance of systematic data collection through observation and interviews, and an objective analysis of data to explain events and behavior, and she published the first sociology research methods textbook.

Martineau, a feminist and strong opponent of slavery, denounced many aspects of capitalism as alienating and degrading, and criticized dangerous workplaces that often resulted in injury and death. Martineau promoted improving women's position in the workforce through education, nondiscriminatory employment, and training programs. She advocated women's

**empirical** information that is based on observations, experiments, or other data collection rather than on ideology, religion, intuition, or conventional wisdom.

Spencer Arnold/Getty Images

Harriet Martineau

**social facts** aspects of social life, external to the individual, that can be measured.

**social solidarity** social cohesiveness and harmony.

**division of labor** an interdependence of different tasks and occupations, characteristic of industrialized societies, that produces social unity and facilitates change.

admission into medical schools and emphasized issues such as infant care, the rights of the aged, the prevention of suicide, and other social problems (Hoecker-Drysdale 1992).

After a long tour of the United States, Martineau described American women as being socialized to be subservient and dependent rather than equal marriage partners. She also criticized American and European religious institutions for expecting women to be pious and passive rather than educating them in philosophy and politics. Most scholars, including sociologists, ridiculed and dismissed her ideas as too radical.

## 1-4c Émile Durkheim

Émile Durkheim (1858–1917), a French sociologist and writer, agreed with Comte that societies are characterized by unity and cohesion because its members are bound together by common interests and attitudes. According to Durkheim, however, Comte did not show that sociology could be scientific. Like Comte, Durkheim ignored Martineau's contributions on the importance of systematic and objective data collection (Adams and Sydie 2001).

### Social Facts

To be scientific, Durkheim maintained, sociology must study **social facts**—aspects of social life, external to the individual, that can be measured. Sociologists can determine *material facts* by examining demographic characteristics such as age, place of residence, and population size. They can gauge *nonmaterial facts*, such as communication processes, by observing everyday behavior and how people relate to each other (see Chapters 3 to 6). For contemporary sociologists, social

© The Art Gallery Collection/Alamy

Émile Durkheim

facts also include collecting and analyzing data on *social currents*, such as collective behavior and social movements (see Chapter 16).

### Division of Labor

One of Durkheim's central questions was how people can be autonomous and individualistic while being integrated in society. **Social solidarity**, or social cohesiveness and harmony, according to Durkheim, is maintained by a **division of labor**—an interdependence of different tasks and occupations, characteristic of industrialized societies, that produces social unity and facilitates change.

As the division of labor becomes more specialized, people become increasingly more dependent on others for specific goods and services. Today, for example, many couples who are getting married often contract "experts" such as photographers, florists, deejays, caterers, bartenders, travel agents (for the honeymoon), and even a "wedding planner."

### Social Integration

Durkheim was one of the first pioneering theorists to use data to test a theory. In his classic study *Suicide*, Durkheim (1897) relied on extensive data collection to test his theory that suicide is associated with social integration. He concluded that people who experience meaningful social relationships in families, social groups, and communities are less likely to commit suicide than those who feel alone, helpless, or hopeless. Thus, many seemingly isolated individual acts, including suicide, are often the result of structural arrangements, such as weak social ties.

Are Durkheim's findings on social integration dated? No. We typically read about the high suicide rates of teens. As *Figure 1.1* shows, however, suicide rates are much higher at later ages. As Durkheim found, men still have higher suicide rates than women across all age groups, especially those age 75 and older, and across all racial and ethnic groups. White males age 85 and older have the highest suicide rates (Kent 2010).

Durkheim's concept of social integration is useful in explaining the high suicide rates of older men. The rates may reflect being widowed and feeling alone, a sense of hopelessness because of terminal illnesses, and not being "connected" to support systems that women tend to develop throughout their lives. Women are more likely than men to have close friends (especially other women), to maintain close ties with family members, and to join community groups that provide assistance during troubling times

FIGURE 1.1

## U.S. Suicide Rates, by Sex and Age

Suicide deaths per 100,000 population

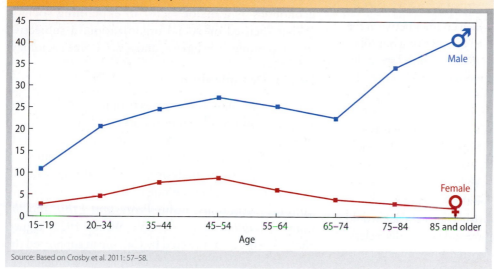

Source: Based on Crosby et al. 2011: 57–58.

(American Association of Suicidology 2009; see also Chapters 9 and 12).

## 1-4d Karl Marx

Karl Marx (1818–1883), a German social philosopher, is often described as the most influential social scientist who ever lived. Marx, like Comte and Durkheim, tried to explain the changes that were taking place in society during the Industrial Revolution.

The Industrial Revolution began in England around 1780 and spread throughout Western Europe and the United States during the nineteenth century. A number of technological inventions—such as the spinning wheel, the steam engine, and large weaving looms—enabled the development of large-scale manufacturing and mining industries over a relatively short period. The extensive mechanization shifted agricultural and home-based work to factories in cities. As masses of people migrated from small farms to factories to find jobs, urbanization and capitalism grew rapidly.

### Capitalism

Unlike his predecessors and contemporaries, Marx (1867/1967, 1964) maintained that economic issues produce divisiveness rather than social solidarity. For Marx, the most important social change was the development of **capitalism**, an economic system in which the ownership of the means of production—such as land, factories, large sums of money, and

machines—is in private hands. As a result, Marx saw industrial society as composed of three social classes:

- *capitalists*—the ruling elite who own the means of producing wealth (such as factories)
- *petit bourgeoisie*—small business owners and owner workers who still have their own means of production but might end up in the proletariat because they are driven out by competition or their businesses fail
- *proletariat*—the masses of workers who depend on wages to survive, who have few resources, and who make up the working class

**capitalism** an economic system in which the ownership of the means of production—such as land, factories, large sums of money, and machines—is in private hands.

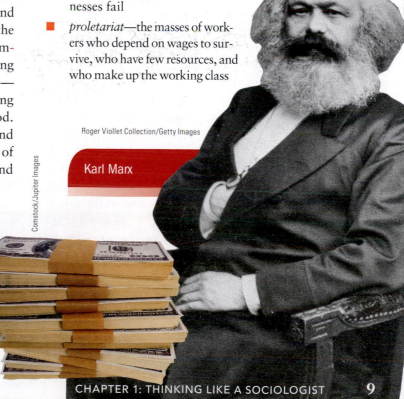

Roger Viollet Collection/Getty Images

Comstock/Jupiter Images

Karl Marx

## Class Conflict

Marx believed that society is divided into the haves (capitalists) and the have-nots (proletariat). For Marx, capitalism is a class system in which conflict between the classes is commonplace and society is anything but cohesive. Instead, class antagonisms revolve around struggles between the capitalists, who increase their profits by exploiting workers, and workers, who resist but give in because they depend on capitalists for jobs.

Marx argued that there is a close relationship between inequality, social conflict, and social class. He maintained that history is a series of class struggles between capitalists and workers. As wealth became more concentrated in the hands of a few capitalists, he predicted, the ranks of an increasingly dissatisfied proletariat would swell, leading to bloody revolution and eventually a classless society. The Occupy Wall Street movement shows that thousands of Americans are very unhappy about the growing inequality between the haves and the have-nots, but, unlike some countries in the Middle East, there hasn't been a "bloody revolution" in the United States.

### Alienation

In industrial capitalist systems, Marx (1844/1964) contended, **alienation**—the feeling of separation from one's group or society—is common across all social classes. Workers feel alienated because they don't own or control either the means of production or the product. Because meaningful labor is what makes us human, Marx maintained, our workplace has alienated us "from the essence of our humanness." In plain English, instead of collaborating, a capitalistic society encourages competition, backstabbing, and "looking out for number one."

According to Marx, capitalists are also alienated. They regard goods and services as important simply because they are sources of profit. Capitalists don't care who buys or sells their products, how the workers feel about the products they make, or whether buyers value the products. The major focus, for capitalists, is on increasing profits as much as possible rather than feeling "connected" to the products or services they sell. Every year, for example, companies must recall cars, pharmaceutical items, toys, and food products that cause injuries, illness, or death.

© iofoto/Shutterstock

## 1-4e Max Weber

Max Weber (pronounced VAY-ber; 1864–1920) was a German sociologist, economist, legal scholar, historian, and politician. Unlike Marx's emphasis on economics as a major factor in explaining society, Weber focused on social organization, a subjective understanding of behavior, and a value-free sociology.

### Social Organization

For Weber, economic factors were important, but ideas, religious values, ideologies, and charismatic leaders were just as crucial in shaping and changing societies. He maintained that a complete understanding of society requires an analysis of the social organization and interrelationships among economic, political, and cultural institutions. In his *Protestant Ethic and the Spirit of Capitalism*, for example, Weber (1920) argued that the self-denial fostered by Calvinism supported the rise of capitalism and shaped many of our current values about working hard (see Chapters 3 and 6).

### Subjective Understanding

Weber posited that an understanding of society requires a "subjective" understanding of behavior. Such understanding, or *verstehen* (pronounced fer-SHTAY-en), requires knowing how people perceive the world in which they live. Weber described two types of *verstehen*. In *direct observational understanding*, the social scientist observes a person's facial expressions, gestures, and listens to his/her words. In *explanatory understanding*, the social scientist tries to grasp the intention and context of the behavior.

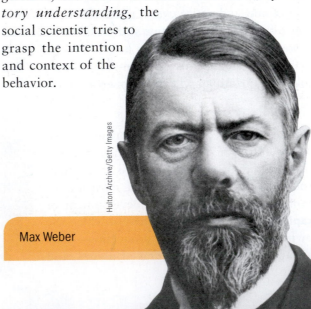

Hulton Archive/Getty Images

Max Weber

# Is It Possible to Be a Value-Free Sociologist?

**M**ax Weber was concerned about popular professors who took political positions that pleased many of their students. He felt that these professors were behaving improperly because science, including sociology, must be "value free." Faculty must set their personal values aside to make a contribution to society. According to one sociologist who agrees with Weber, sociology's weakness is its tendency toward moralism and ideology:

*Many people become sociologists out of an impulse to reform society, fight injustice, and help people. Those sentiments are noble, but unless they are tempered by skepticism, discipline, and scientific detachment, they can be destructive. Especially when you are morally outraged and burning with a desire for action, you need to be cautious (Massey 2007: B12).*

There is considerable disagreement on whether sociologists can *really* be value free. Some argue that being value free is a myth because it's impossible for a scholar's attitudes and opinions to be totally divorced from her or his scholarship (Gouldner 1962). Many sociologists, after all, do research on topics that they consider significant and about which they have strong views. If you talk to your sociology instructor, for example, you'll probably find that she or he often teaches and does research on topics that are intensely and personally interesting.

> *Can* sociologists be value free —especially when they have strong feelings about many societal issues? *Should they be?*

Other sociologists maintain that one's values should be passionately partisan, frame research issues, and have an impact on improving society (Feagin 2001). That is, sociologists should not apologize for being subjective in their teaching and research.

Can sociologists be value free—especially when they have strong feelings about many societal issues? Should they be?

If a person bursts into tears (direct observational understanding), the observer knows what the person may be feeling (anger, sorrow, and so on). An explanatory understanding goes a step further by spelling out the reason for the behavior (rejection by a loved one, frustration when your computer crashes, humiliation if a boss yells at you in public).

> **value free** separating one's personal values, opinions, ideology, and beliefs from scientific research.

## Value-Free Sociology

One of Weber's most lasting and controversial views was the notion that sociologists must be as objective, or "value free," as possible in analyzing society. A researcher who is **value free** is one who separates her or his personal values, opinions, ideology, and beliefs from scientific research.

During Weber's time, the government and other organizations demanded that university faculty teach the "right" ideas. Weber encouraged everyone to be involved as citizens, but he maintained that educators and scholars should be as dispassionate as possible about political and ideological positions. The task of the teacher, Weber argued, was to provide students with knowledge and scientific experience, not to "imprint" the teacher's personal political views and value judgments (Gerth and Mills 1946). The box "Is It Possible to Be a Value-Free Sociologist?" examines this issue further.

## 1-4f Jane Addams

Jane Addams (1860–1935) was a social worker who cofounded Hull House, one of the first settlement houses in Chicago that served as a community center for the neighborhood poor. An active reformer throughout her life, Jane Addams was a leader in the women's suffrage movement and, in 1931, was the first American woman to be awarded the Nobel Peace Prize for her advocacy of negotiating, rather than waging war, to settle disputes.

Sociologist Mary Jo Deegan (1986) describes Jane Addams as "the greatest woman sociologist of her day." She was ignored by her colleagues at the University of Chicago (the first sociology department established in the United States in 1892), however, because discrimination against women sociologists was "rampant" (p. 8).

Despite such discrimination, Addams published articles in numerous popular and scholarly journals, as well as many books on the everyday life of urban neighborhoods, especially the effects of social

Jane Addams with a child at Hull House.

Wallace Kirkland/Time Life Pictures/Getty Images

believed that the race problem was one of ignorance and wanted to provide a "cure" for prejudice and discrimination. Such cures included promoting black political power and civil rights and providing blacks with a higher education rather than funneling them into technical schools.

These and other writings were unpopular at a time when Booker T. Washington, a well-known black educator, encouraged black people to be patient instead of demanding equal rights. As a result, Du Bois was dismissed as a radical by his contemporaries but rediscovered by a new generation of black scholars during the 1970s and 1980s. Among his many contributions, Du Bois examined the oppressive effects of race and class, advocated women's rights, and played a key role in reshaping black–white relations in America (Du Bois 1986; Lewis 1993).

All of the early thinkers agreed that people are transformed by each other's actions, social patterns, and historical changes. They and other scholars shaped contemporary sociological theories.

disorganization and immigration. Much of her work contributed to symbolic interactionism, an emerging school of thought. One of Addams' greatest intellectual legacies was her emphasis on applying knowledge to everyday problems. Her pioneering work in criminology included ecological maps of Chicago that were later credited to men (Moyer 2003).

### 1-4g W. E. B. Du Bois

W. E. B. Du Bois (pronounced Do-BOICE; 1868–1963) was a prominent black sociologist, writer, editor, social reformer, and passionate orator. The author of almost two dozen books on Africans and black Americans, Du Bois spent most of his life responding to the critics and detractors of black life. He was the first African American to receive a Ph.D. from Harvard University, but once remarked, "I was in Harvard but not of it."

Du Bois helped found the National Association for the Advancement of Colored People (NAACP) and became editor of its journal, *Crisis*. The problem of the twentieth century, he wrote, is the problem of the color line. Du Bois

W. E. B. DuBois

## 1-5 Contemporary Sociological Theories

How one defines "contemporary sociological theory" is somewhat arbitrary. The mid-twentieth century is a good starting point because "the late 1950s and 1960s have, in historical hindsight, been regarded as significant years of momentous changes in the social and cultural life of most Western societies" (Adams and Sydie 2001: 479). Some of the sociological perspectives had earlier origins, but all matured during this period.

Sociologists typically use more than one theory to explain behavior. The theories view our social world somewhat differently, but all of them analyze why society is organized the way it is and why we behave as we do. Four of the most influential

Hulton Archive/Getty Images

theoretical perspectives are functionalism, conflict theory, feminist theories, and symbolic interactionism.

# 1-5a Functionalism

**Functionalism** (also known as *structural functionalism*) maintains that society is a complex system of interdependent parts that work together to ensure a society's survival. Much of contemporary functionalism grew out of the work of Auguste Comte and Émile Durkheim, both of whom believed that human behavior is a result of social structures that promote order and integration in society.

One of their contemporaries, English philosopher Herbert Spencer (1820–1903), used an organic analogy to explain the evolution of societies. To survive, Spencer (1862/1901) wrote, our vital organs—like the heart, lungs, kidneys, liver, and so on—must function together. Similarly, the parts of a society, like the parts of a body, work together to maintain the whole structure.

## Society Is a Social System

Prominent American sociologists, especially Talcott Parsons (1902–1979) and Robert K. Merton (1910–2003), developed the earlier ideas of structure and function. For these and other functionalists, a society is a system that is composed of major institutions such as government, religion, the economy, education, medicine, and the family.

Each institution or other social group has *structures*, or organized units, that are connected to each other and within which behavior occurs. Education structures such as colleges, for instance, are not only organized internally in terms of who does what and when, but depend on other structures such as government (to provide funding), business (to produce textbooks and construct buildings), and medicine (to ensure that students, staff, and faculty are healthy).

## Functions and Dysfunctions

Each structure fulfills certain *functions*, or purposes and activities, to meet different needs that contribute to a society's stability and survival (Merton 1938). The purpose of education, for instance, is to transmit knowledge to the young, to teach them to be good citizens, and to prepare them for jobs (see Chapter 11).

**Dysfunctions** are social patterns that have a negative impact on a group or society. When one part of society isn't working, it affects other parts by creating conflict, divisiveness, and social problems. Consider religion. In the United States, the Catholic Church's

stance on issues such as not ordaining women to be priests and denouncing abortion and homosexuality has produced a rift between those who espouse the Pope's edicts and those who question them. In other countries, religious intolerance has led to wars and terrorist attacks (see Chapter 13).

## Manifest and Latent Functions

There are two kinds of functions. **Manifest functions** are intended and recognized; they are present

© Simone van den Berg/Shutterstock

Sociologists typically use more than one theory to explain behavior and why society is organized the way it is.

and clearly evident. **Latent functions** are unintended and unrecognized; they are present but not immediately obvious. Consider the marriage ceremony. The primary manifest function of the marriage ceremony is to publicize the formation of a new family unit and to legitimize sexual intercourse and childbirth (even though both might occur outside of marriage). Its latent functions include communicating a "hands-off" message to suitors, providing the new couple with household goods and products through bridal showers and wedding gifts, and redefining family boundaries to include in-laws or stepfamily members.

### Critical Evaluation

Functionalism is useful in seeing the "big picture" of interrelated structures and functions. According to some critics, however, functionalism is so focused on order and stability that it often ignores social change. For example, functionalists typically see high divorce rates as dysfunctional, signaling the disintegration of the family, but tend to ignore positive effects such as allowing people to leave unhappy marriages.

A second criticism is that functionalism often glosses over the inequality that a handful of powerful people create and maintain. Some critics have also charged that functionalism views society narrowly through white, male, middle-class lenses. According to some feminist scholars, for example, "functionalism tends to support a white middle-class family model emphasizing the economic activities of the male household head and domestic activities of his female subordinate" while ignoring nontraditional families such as single-parent households (Lindsey 2005: 6).

© iStockphoto.com/Rhienna Cutler

## 1-5b Conflict Theory

In contrast to functionalism, which emphasizes order, stability, cohesion, and consensus, **conflict theory** examines how and why groups disagree, struggle over power, and compete for scarce resources (such as property, wealth, and prestige). Conflict theorists see disagreement and the resulting changes in society as natural, inevitable, and even desirable.

### Sources of Conflict

The conflict perspective has a long history. As you saw earlier, Karl Marx predicted that conflict would result from widespread economic inequality, and W. E. B. DuBois criticized U.S. society for its ongoing and divisive racial discrimination. Since the 1960s, as you'll see in later chapters, many sociologists—especially feminist and minority scholars—have emphasized that the key sources of economic inequity in any society include race, ethnicity, gender, age, and sexual orientation.

Conflict theorists agree with functionalists that many societal arrangements are functional. But, conflict theorists ask, who benefits? And who loses? When corporations merge, workers in lower-end jobs are often laid off while the salaries and benefits of corporate executives soar and the value of stocks (usually held by higher social classes) increase. Thus, mergers might be functional for those at the upper end of the socioeconomic ladder, but dysfunctional for those on the lower rungs.

### Social Inequality

Unlike functionalists, conflict theorists see society not as cooperative and harmonious, but as a system of widespread inequality. For conflict theorists, there is a continuous tension between the haves and the have-nots, most of whom are children, women, minorities, people with low incomes, and the poor.

Many conflict theorists focus on how those in power—typically, wealthy white Anglo-Saxon Protestant males (WASPs)—dominate political and economic decision making in U.S. society. This group controls a

Among other *manifest functions*, schools transmit knowledge and prepare children for adult economic roles. Among their *latent functions*, schools provide many matchmaking opportunities. What are some other examples of education's manifest and latent functions?

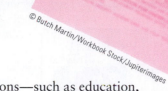
© Butch Martin/Workbook Stock/Jupiterimages

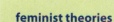

variety of institutions—such as education, criminal justice, and the media—and passes laws that benefit primarily small groups of people like themselves (see Chapters 8 and 11).

## Critical Evaluation

Conflict theory explains how societies create and cope with disagreements. However, some have criticized conflict theorists for overemphasizing competition and coercion at the expense of order and stability. Inequality exists and struggles over scarce resources occur, critics agree, but conflict theorists often ignore cooperation and harmony. Voters, for example, can boot dominant white males out of office and replace them with women and minority group members. Critics also point out that the have-nots can increase their power through negotiation, bargaining, lawsuits, and strikes.

Some critics also believe that conflict theory presents a negative view of human nature and neglects the importance of love and self-sacrifice, which are essential to family and other personal relationships. Because conflict perspectives examine institutional rather than personal choices and constraints, they don't give us insights into everyday individual behavior.

## 1-5c Feminist Theories

Rebecca West, a British journalist and novelist, once said, "I myself have never been able to find out precisely what feminism is; I only know that people call me a feminist whenever I express sentiments that differentiate me from a doormat." Feminist scholars agree with West and conflict theorists that much of society is characterized by tension and struggle between groups.

**Feminist theories** go a step further because they explain women's social, economic, and political inequality. Feminist theorists maintain that women often suffer injustice primarily because of their gender rather than because of personal inadequacies such as low educational levels or not caring about success. Feminist scholars assert that people should be treated fairly and equally regardless not only of their sex but also of other characteristics such as their race, ethnicity, national origin, age, religion, class, sexual

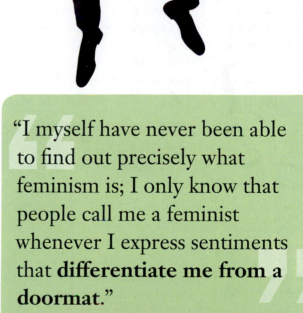

"I myself have never been able to find out precisely what feminism is; I only know that people call me a feminist whenever I express sentiments that **differentiate me from a doormat**."

—Rebecca West, British journalist

orientation, or disability. They emphasize the importance of freeing women from traditionally oppressive expectations, constraints, roles, and behavior.

FIGURE 1.2

## Women Should Have the Same Rights as Men. Well, Maybe Not Always . . .

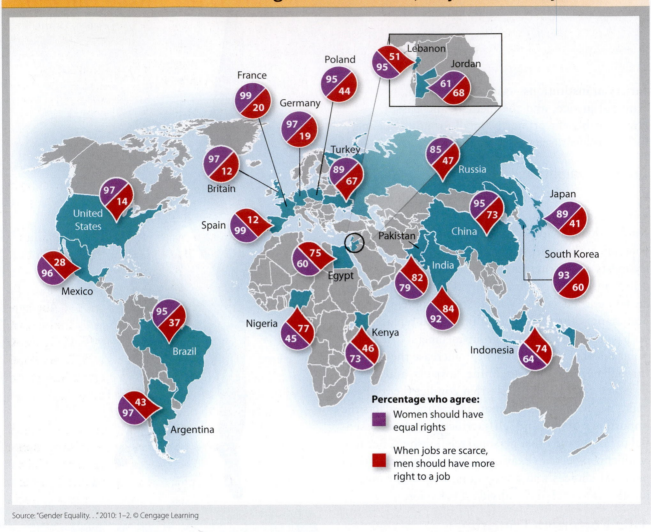

**Percentage who agree:**

- Women should have equal rights
- When jobs are scarce, men should have more right to a job

Source: "Gender Equality. . ." 2010: 1–2. © Cengage Learning

### Focusing on Gender

Many feminist scholars contend that women have historically been excluded from most sociological analyses (Smith 1987). Before the 1960s women's movement in the United States, very few sociologists published anything about gender roles, women's sexuality, fathers, or intimate partner violence. According to sociologist Myra Ferree (2005: B10), during the 1970s, "the Harvard social-science library could fit all its books on gender inequalities onto a single half-shelf." Since then, because of feminist scholars, many researchers—both women and men—now routinely include sex as an important research variable on both micro and macro levels.

Globally, except for some predominantly Muslim countries, solid majorities of both women and men support gender equality and agree that women should be able to work outside the home. When jobs are scarce, however, many women and men believe that men should be given preferential treatment (see *Figure 1.2*). Thus, even equal rights proponents place a higher priority on men's economic rights.

## Sociology in Your Life

Are you a feminist? Think about why or why not. Then, take a short online quiz to learn more by going to Sociology CourseMate at CengageBrain.com.

## Listening to Many Voices

Feminist scholars contend that gender inequality is central to *all* behavior, ranging from everyday interactions to political and economic institutions, but feminist theories encompass many perspectives. For example, *liberal feminism* emphasizes social and legal reform to create equal opportunities for women. *Radical feminism* sees male dominance in social institutions (such as the economy and politics) as the major cause of women's inequality. *Global feminism* focuses on how the intersection of gender with race, social class, and colonization has exploited women in the developing world (see Lengermann and Niebrugge-Brantley 1992). Most of us are feminists because we endorse equal opportunities for women and men in the economy, politics, education, and other institutions.

## Critical Evaluation

Feminist scholars have challenged employment discrimination, particularly practices that routinely exclude women who are not part of the "old boy network" (Wenneras and Wold 1997). One criticism, however, is that many feminists are part of an "old girl network" that has not always welcomed different points of view from black, Asian American, American Indian, Muslim, Latina, lesbian, working-class, and disabled women (Almeida 1994; Lynn and Todoroff 1995; Jackson 1998).

A second criticism is that feminist perspectives tend to downplay social class inequality by focusing on low-income and minority women but not their male counterparts. Thus, some contend, feminist theories are not as gender balanced as they claim. Some critics, including some feminists, also question whether feminist scholars have lost their bearings by focusing on personal issues such as greater sexual freedom rather than broader social issues, particularly wage inequality (Rowe-Finkbeiner 2004; Chesler 2006).

> "Sometimes the best man for the job isn't."
>
> —Author Unknown

# 1-5d Symbolic Interactionism

**Symbolic interactionism** (sometimes called *interactionism*) is a micro-level perspective that examines individuals' everyday behavior through the communication of knowledge, ideas, beliefs, and attitudes. Whereas functionalists, conflict theorists, and some feminist theories emphasize structures and large (macro) systems, symbolic interactionists focus on *process* and keep the *person* at the center of their analysis.

There have been many influential symbolic interactionists whom we'll cover in later chapters. In brief, George Herbert Mead's (1863–1931) proposal that the human mind and self arise in the process of social communication became the foundation of the symbolic interaction schools of thought in sociology and social psychology. Herbert Blumer (1900–1987) coined the term *symbolic interactionism* in 1937, developed Mead's ideas, and emphasized that people interpret or "define" each other's actions instead of merely reacting to them, especially through symbols.

Erving Goffman (1922–1982) contributed significantly to these earlier theories by examining human interaction in everyday situations ranging from jobs to funerals. Among his other contributions, Goffman used "dramaturgical analysis" to compare everyday social interaction to a theatrical presentation (see Chapter 5).

## Constructing Meaning

Our actions are based on **social interaction** in the sense that people take each other into account in their own behavior. Thus, we act differently in different social settings and continuously adjust our behavior, including our body language, as we interact (Goffman 1959; Blumer 1969). A woman's interactions with her husband are different from those with her children. And she will interact still differently when she is teaching, talking to a colleague in the hall, or addressing an audience of colleagues at a professional conference.

For symbolic interactionists, society is *socially constructed* through human interpretation (O'Brien and Kollock 2001). That is, meanings are not inherent but are created and modified through interaction with others. For example, a daughter who has batting practice with her dad will probably interpret her

**symbolic interactionism (*interactionism*)**
a micro-level perspective that examines individuals' everyday behavior through the communication of knowledge, ideas, beliefs, and attitudes.

**social interaction**
a process in which people take each other into account in their own behavior.

For many people, a diamond, especially in an engagement ring, signifies love and commitment. For others, diamonds represent Western exploitation of poor people in Africa who are paid next to nothing for their backbreaking labor in mining these stones.

## 1 symbol ≠ 1 meaning

The American flag is one symbol, but it has different meanings for different groups.

father's behavior as loving and involved. In contrast, she will see batting practice with her baseball coach as less personal and more goal oriented. In this sense, our interpretations of even the same behavior, such as batting practice, vary across situations and depend on the people with whom we interact.

### Symbols and Shared Meanings

Symbolic interactionism looks at subjective, interpersonal meanings and at the ways in which we interact with and influence each other by communicating through *symbols*—words, gestures, or pictures that stand for something and that can have different meanings for different individuals.

After the 9/11 terrorist attacks, many Americans displayed the flag on buildings, bridges, homes, and cars to show their solidarity and pride in the United States. In contrast, some groups in the Middle East burned the U.S. flag to show their contempt for American culture and policies. Thus, symbols are powerful forms of communication that show how people feel and interpret a situation.

For us to interact effectively, our symbols must have *shared meanings,* or agreed on definitions. One of the most important of these shared meanings is the *definition of the situation,* or the way we perceive reality and react to it. Relationships often end, for example, because partners view emotional closeness differently ("We broke up because Tom wanted sex. I wanted an emotional connection."). We typically learn our definitions of the situation through interaction

with *significant others*—especially parents, friends, relatives, and teachers—who play an important role in our socialization (as you'll see in Chapters 4 and 5).

### Critical Evaluation

Unlike other theorists, symbolic interactionists show how people play an active role in shaping their lives on a micro level. One of the most common criticisms, however, is that symbolic interactionism overlooks the widespread impact of macro-level factors such as economic forces, social movements, and public policies on our everyday behavior and relationships. When the U.S. economy began to plunge in late 2007, for example, more than 75 percent of job losses were among men. Their unemployment and ensuing financial problems created considerable interpersonal conflict among married and unmarried couples (Whelan 2009). Symbolic interactionism rarely considers such macro-level changes in explaining everyday behavior.

A related criticism is that interactionists sometimes have an optimistic and unrealistic view of people's everyday choices. Most of us enjoy little flexibility in our daily lives because deeply embedded social arrangements and practices benefit those in power. For example, people are usually powerless when corporations transfer many jobs overseas or cut the pension funds of retired employees.

Some also believe that interaction theory is flawed because it ignores the irrational and unconscious aspects of human behavior (LaRossa and Reitzes 1993). People don't always consider the meaning of their actions or behave as reflectively as interactionists assume. Instead, we often act impulsively or say hurtful things without weighing the consequences of our actions or words.

For a summary of all these perspectives, see *Table 1.2*.

## TABLE 1.2

# Leading Contemporary Perspectives in Sociology

| THEORETICAL PERSPECTIVE | FUNCTIONALIST | CONFLICT | FEMINIST | SYMBOLIC INTERACTIONIST |
|---|---|---|---|---|
| **Level of Analysis** | Macro | Macro | Macro and Micro | Micro |
| **Key Points** | • Society is composed of interrelated, mutually dependent parts.<br>• Structures and functions maintain a society's or group's stability, cohesion, and continuity.<br>• Dysfunctional activities that threaten a society's or group's survival are controlled or eliminated. | • Life is a continuous struggle between the haves and the have-nots.<br>• People compete for limited resources that are controlled by a small number of powerful groups.<br>• Society is based on inequality in terms of ethnicity, race, social class, and sex. | • Women experience widespread inequality in society because, as a group, they have little power.<br>• Sex, ethnicity, race, age, sexual orientation, and social class—rather than a person's intelligence and ability—explain many of our social interactions and lack of access to resources.<br>• Social change is possible only if we change our institutional structures and our day-to-day interactions. | • People act on the basis of the meaning they attribute to others. Meaning grows out of the social interaction that we have with others.<br>• People continuously reinterpret and reevaluate their knowledge and information in their everyday encounters. |
| **Key Questions** | • What holds society together? How does it work?<br>• What is the structure of society?<br>• What functions does society perform?<br>• How do structures and functions contribute to social stability? | • How are resources distributed in a society?<br>• Who benefits when resources are limited? Who loses?<br>• How do those in power protect their privileges?<br>• When does conflict lead to social change? | • Do men and women experience social situations in the same way?<br>• How does our everyday behavior reflect our sex, social class, age, race, ethnicity, sexual orientation, and other factors?<br>• How do macro structures (such as the economy and the political system) shape our opportunities?<br>• How can we change current structures through social activism? | • How does social interaction influence our behavior?<br>• How do social interactions change across situations and between people?<br>• Why does our behavior change because of our beliefs, attitudes, values, and roles?<br>• How is "right" and "wrong" behavior defined, interpreted, reinforced, or discouraged? |
| **Example** | • A college education increases one's job opportunities and income. | • Most low-income families can't afford to pay for a college education. | • Gender affects decisions about a major and which college to attend. | • College students succeed or fail based on their degree of academic engagement. |

Every day, we are inundated with unscientific, false, and misleading information.

# Examining Our Social World

Spring break is all about beer fests, wet T-shirt contests, frolicking on the beach, and hooking up, right? Maybe not. A recent survey found that 70 percent of college students stay home with their parents, and 84 percent of those who throng to vacation spots report consuming alcohol in moderation (The Nielsen Company 2008).

**social research** systematic study of human behavior.

If you suspect that these numbers are too high or too low and wonder how the survey was done, you're thinking like a researcher, the focus of this chapter. We'll begin by looking at what sociologists mean by social research and how it affects our everyday lives, then examine the scientific method, explore some of the major data collection methods, and discuss how ethical guidelines shape sociological research.

## 2-1 Doing Sociology: What Is Social Research?

**S**ocial research is a systematic study of human behavior. The process requires curiosity and imagination, but also knowing the research rules and procedures that describe and explain why people behave as they do.

You'll recall that sociologists debate whether the discipline can or should be value free, especially in teaching (see Chapter 1). When it comes to research, they agree that the selection of a topic may be subjective (e.g., examining immigration because of personal experiences or family background), but a researcher should be objective in collecting, analyzing, and interpreting the data. That is, private beliefs, biases, and value judgments shouldn't bias the research process (Gray et al. 2007).

## 2-2 Why Is Sociological Research Important in Our Everyday Lives?

**H**ow do we know what we know? Much of our knowledge is based on *tradition*, a handing down of statements, beliefs, and customs from generation to generation ("The groom's parents should pay for the wedding rehearsal dinner" or "Flying the American flag at home shows one's patriotism"). Another common source of knowledge is *authority*, a socially accepted source of information that includes experts, parents, government officials, police,

## Key Topics

In this chapter, we'll explore the following topics:

**2-1** Doing Sociology: What Is Social Research?

**2-2** Why Is Sociological Research Important in Our Everyday Lives?

**2-3** The Scientific Method

**2-4** Some Major Data Collection Methods

**2-5** Ethics, Politics, and Sociological Research

## What do you think?

Trust your instincts, not research studies.

1  2  3  4  5  6  7
strongly agree          strongly disagree

judges, and religious leaders ("My mom says that . . ." or "According to the American Heart Association . . .").

Knowledge based on tradition and authority simplifies our lives because it provides us with basic rules about socially and legally acceptable behavior. Often, however, the information is misleading or downright incorrect. Suppose a 2-year-old throws a temper tantrum at a family barbecue. One adult comments, "What that kid needs is a smack on the behind." Another person immediately disagrees: "All kids go through this stage. Just ignore it."

Who's right? Much research shows that neither ignoring a problem nor inflicting physical punishment (such as spanking) stops a toddler's bad behavior. Instead, according to many researchers, most young children's misbehavior can be curbed by having simple rules, being consistent in disciplining misbehavior, praising good behavior, and setting a good example for how to act (Benokraitis 2011).

## "Facebook Causes 20 Percent of Today's Divorces" What?!

The founder and self-described leader of the United Kingdom's online divorce site (Divorce-Online) sent out a press release titled "Facebook Is Bad for Your Marriage—Research Finds," and claimed that Facebook causes 20 percent of today's divorces. News media around the world ran stories about this press release with headlines such as "Facebook to Blame for Divorce Boom." You'll see in Chapter 12 that there's no divorce boom, so where did the 20 percent number come from?

In 2009, the managing director of Divorce-Online scanned its online divorce petition database for the use of the word Facebook, and found 989 instances in about 5,000 petitions. Divorce-Online never said that the petitions were only those filed by members of the American Academy of Matrimonial Lawyers, who comprise a very small percentage of all divorce attorneys. Two years later, numerous Internet sites and blogs were still spreading the fiction that "Facebook Causes Divorce" (see Bialik 2011). In reality, as you'll see in Chapter 12, there are a number of interrelated macro- and micro-level reasons for divorce; there is no single "cause," much less a website.

In contrast with knowledge based on tradition and authority, sociological research is important in our everyday lives for several reasons:

1. **It creates new knowledge that helps us understand social life.** Regarding the economy, for example, sociological research has shown that workplace diversity that includes women and minorities leads to better products, services, and higher company profits (Berrey 2009). In effect, then, employment discrimination—whether intentional or not—reduces a company's success.

2. **It exposes myths.** According to many newspapers and television shows, suicide rates are highest during the end of the Christmas holidays. In fact, suicide rates are lowest in December and highest in the spring and fall (but the reasons for these peaks are unclear). Another myth is that more women are victims of domestic violence on Super Bowl Sunday than on any other day of the year, presumably because men become intoxicated and abusive. In fact, intimate partner violence is common throughout the year and doesn't spike on Super Bowl Sunday (Mikkelson and Mikkelson 2005; Romer 2011).

3. **It helps explain *why* people behave as they do.** A recent study by the Insurance Institute for Highway Safety, an industry group that covers the costs of auto accidents, predicted that older drivers, particularly those age 70 and older, would be more likely than younger drivers to have fatal crashes, but found the opposite (Braitman et al. 2011). The researchers couldn't explain why their prediction turned out to be false. Sociologists posit that older Americans are less likely to have fatal crashes because many avoid driving at night or during bad weather, and they are much less likely than younger drivers to use cell phones or text while driving (Halsey 2010).

4. **It influences social policies.** In the past, many psychologists attributed school bullying to students with emotional and behavioral problems. Recently, however, sociologists have found that bullying is due largely to jockeying for social status rather than to mental or emotional problems (Faris and Felmlee 2011). That is, the most aggressive bullies are trying to eliminate their rivals from becoming members of a higher-ranking group. Because of these findings, some educators have enlisted the support of students at the top of the high school social ladder to prevent or decrease bullying (Parker-Pope 2011).

5. **It sharpens critical thinking skills that affect our everyday lives.** Many Americans, especially women, rely on talk shows for information on a number of topics. During 2009 alone, Oprah Winfrey featured and applauded guests who maintained, among other

things, that children contract autism from the measles, mumps, and rubella (MMR) vaccinations they receive as babies, that fortune cards can help people diagnose their illnesses, and that people can wish away cancer (Kosova and Wingert 2009)—*all of which are false.* Such misinformation can be dangerous. Because of the "MMR vaccinations can cause autism" scare, about 30 percent of U.S. parents are hesitant to vaccinate their children (Kennedy et al. 2011). Partly because of such fears, by mid-2011, the United States was experiencing the largest increase in measles cases since 1996. For every 1,000 children who get measles, one or two die (McCauley and Chenowith 2011; see also Gibson 2012 for other recent examples of "celebrity bogus science").

Oprah Winfrey regularly featured guests, such as actress Suzanne Somers (pictured here), who dismissed scientific findings about health and endorsed treatments such as taking up to 60 vitamin pills every day and wishing cancer away.

AP Photo/Jennifer Graylock

## Sociology in Your Life

Are you as an American more likely to be healthy or sick than people in other industrialized nations? The World Health Organization (WHO), a premier source of information, provides answers to such questions. See their Data and Statistics link by going to Sociology CourseMate at CengageBrain.com.

## 2-3 The Scientific Method

To discover patterns that explain behavior, sociologists rely on the **scientific method**. The steps involved include careful data collection, exact measurement, accurate recording and analysis of the findings, thoughtful interpretation of results, and, when appropriate, generalization of the findings to a larger group. Before collecting any data, however, social scientists must grapple with a number of research-related issues. Let's begin with concepts, variables, and hypotheses.

### 2-3a Concepts, Variables, and Hypotheses

A basic element of the scientific method is having a **concept**—an abstract idea, mental image, or general notion that represents some aspect of our social life. Some examples of concepts are "blood pressure," "climate change," and "marriage."

Because concepts are abstract and may vary among individuals and cultures, scientists rely on variables to measure (*operationalize*) concepts. A **variable** is a characteristic that can change in value or magnitude under different conditions. Variables can be attitudes, behaviors, or traits such as ethnicity, age, and social class. An **independent variable** is a characteristic that has an effect on the **dependent variable**, the outcome. A **control variable** is a characteristic that is constant and unchanged during the research process.

Scientists can simply ask a research question ("Why do people get a divorce?"), but they usually begin with a **hypothesis**—a statement of the expected relationship between two or more variables—such as "Unemployment increases divorce rates." In this example, "unemployment" is the independent variable and "divorce rate" is the dependent variable.

Researchers might also use control variables, such as education, to explain the relationship between unemployment and divorce. For example, people with at least a college degree generally have lower divorce rates than those with lower educational levels because the former are less likely to experience long periods of unemployment, which is a major source of marital stress and conflict (see Chapter 12).

**scientific method** a research process that includes careful data collection, exact measurement, accurate recording and analysis of the findings, thoughtful interpretation of results, and, when appropriate, generalization of the findings to a larger group.

**concept** an abstract idea, mental image, or general notion that represents some aspect of our social life.

**variable** a characteristic that can change in value or magnitude under different conditions.

**independent variable** a characteristic that has an effect on the dependent variable.

**dependent variable** the outcome, which may be affected by the independent variable.

**control variable** a characteristic that is constant and unchanged during the research process.

**hypothesis** a statement of the expected relationship between two or more variables.

**reliability** the consistency with which the same measure produces similar results time after time.

**validity** the degree to which a measure is accurate and really measures what it claims to measure.

**deductive reasoning** an inquiry process that begins with a theory, prediction, or general principle that is then tested through data collection.

**inductive reasoning** an inquiry process that begins with a specific observation, followed by data collection, a conclusion about patterns or regularities, and the formulation of hypotheses that can lead to theory construction.

theory
+ hypothesis
+ variables
+ data
+ explanation of results
—————————————————
= foundation of scientific method

Image Source/Jupiter Images

### 2-3b Reliability and Validity

After asking a research question or stating one or more hypotheses, how do scientists measure the variables? Sociologists are always concerned about reliability and validity. **Reliability** is the *consistency* with which the same measure produces similar results time after time. If, for example, you ask "How old are you?" on two subsequent days and a respondent gives two different answers, such as 25 and 30, there's either something wrong with the question or the respondent is lying. Respondents might lie, but scientists must make sure that their measures are as reliable as possible.

**Validity** is the degree to which a measure is accurate and *really* measures what it claims to measure. Consider student course evaluations. The measures of a "good" professor often include items such as "The instructor is interesting" and "The instructor is fair." Because we don't know what students mean by "interesting" and "fair," how accurate are such measures in differentiating between "good" and "bad" professors?

### 2-3c Deductive and Inductive Reasoning

Deduction and induction are two different but equally valuable approaches in examining the relationship between variables. Generally, **deductive reasoning** begins with a theory, prediction, or general principle that is then tested through data collection. An alternative mode of inquiry, **inductive reasoning**, begins with specific observations, followed by data collection, a conclusion about patterns or regularities, and the formulation of hypotheses that can lead to theory construction (see *Figure 2.1*).

Taking a deductive approach, you might decide to test a theory of academic success using the following hypothesis: "Students who study in groups perform better on exams than those who study alone." You would collect the data, ultimately confirming or rejecting your hypotheses (or theory). Alternatively, you might notice that your classmates who participate in study groups seem to get higher grades on exams than those who study alone. Using an inductive approach, you would collect the data systematically and formulate hypotheses (or suggest a theory) that could then be tested deductively. Most social science research involves both inductive and deductive reasoning.

### 2-3d Sampling

Early in the research process, sociologists decide what sampling procedures to use. Ideally, researchers would

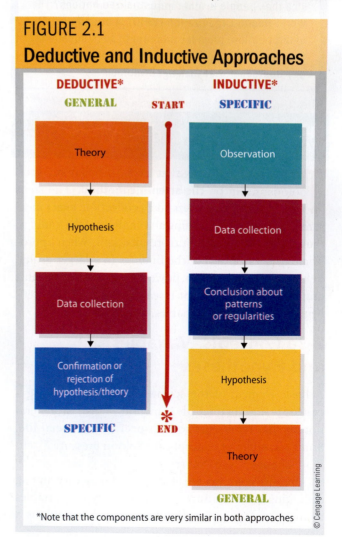

**FIGURE 2.1**
**Deductive and Inductive Approaches**

DEDUCTIVE*  START  INDUCTIVE*
GENERAL  SPECIFIC

Theory → Hypothesis → Data collection → Confirmation or rejection of hypothesis/theory
SPECIFIC ... END

Observation → Data collection → Conclusion about patterns or regularities → Hypothesis → Theory
GENERAL

*Note that the components are very similar in both approaches

© Cengage Learning

Are *American Idol* voters an example of a probability or nonprobability sample of the show's fans?

S_bukley/Shutterstock.com

like to study all the units of the population in which they are interested—say, all adolescents who use drugs. A **population** is any well-defined group of people (or things) about which researchers want to know something. Obtaining information about and from populations is problematic, however. The population may be so large that it would be too expensive and time consuming to conduct the research. In other cases— such as all adolescents who use drugs—it's impossible even to identify the population.

As a result, researchers typically select a **sample**, a group of people (or things) that is representative of the population they wish to study. In obtaining a sample, researchers decide whether to use probability or nonprobability sampling. A **probability sample** is one in which each person (or thing, such as an email address) has an equal chance of being selected because the selection process is *random*. The most desirable characteristic of a probability sample is that the results can be generalized to the larger population because all the people (or things) have had an equal chance of being selected.

A **nonprobability sample** is a sample for which there is little or no attempt to get a representative cross section of the population. Instead, researchers use sampling criteria such as convenience or the availability of respondents or information. Nonprobability samples are very useful when sociologists are exploring a new topic or want to get insights on how people feel about a particular topic before launching a larger study (Babbie 2002).

Television news programs, newsmagazines, and entertainment shows often provide a toll-free number, a texting number, or a Web address and encourage viewers to vote on an issue (such as whether U.S. troops should leave Afghanistan). How representative are these voters of the general population? And how many enthusiasts skew the results by voting more than once?

But, you might think, if as many as 100,000 people respond, doesn't such a large number indicate what most people think? No. Because the respondents are self-selected and don't comprise a random sample, they are not representative of a population.

Stockbyte/Getty Images

## 2-3e The Research Process: The Basics

Hypotheses construction, establishing reliability and validity, using deductive and inductive reasoning, and sampling are some of the preliminary and often most challenging of the seven steps in the research process, illustrated in *Figure 2.2*. This figure outlines the scientific method using a deductive approach that begins with an idea and ends with writing up (and sometimes publishing) the results.

1. **Choose a topic to study.** Good research is generally guided by theory. The topic can be general or very specific. Some sociologists begin with a new question or idea; others extend or refine previous research findings. A topic can generate new information, replicate a previous study, or propose an intervention (such as implementing a new foster homes program).

2. **Summarize the related research.** In what is often called a *literature review*, a sociologist summarizes the pertinent research, shows how her or his topic is related to previous and current research, and indicates

how the study will extend the body of knowledge. If the research is applied, a sociologist also explains how the proposed service or program will improve people's lives.

**3. Formulate a hypothesis or ask a research question.** A sociologist next states a hypothesis or asks a research question. In either case, she or he has to be sure that the measures of the variables are as reliable and valid as possible.

4. **Describe the data collection method(s).** A sociologist describes which method or combination of methods (sometimes called *methodology, procedure,* or *research design*) is best for testing the hypothesis or answering the research question. As you'll see shortly, each data collection method has strengths and weaknesses that the researcher considers before deciding which method is the most appropriate. This step also describes sampling, the sample size, and the respondents' characteristics.

5. **Collect the data.** The actual data collection might rely on fieldwork, surveys, experiments, or existing sources of information such as Census Bureau data.

6. **Present the findings.** After coding (tabulating the results) and running statistical tests, a sociologist presents the findings as clearly as possible.

7. **Analyze and explain the results.** After analyzing the data, a sociologist explains why the findings are important. This can be done in many ways. The researcher might show how the results provide new information, enrich our understanding of behavior or attitudes that researchers have examined previously, or refine existing theories or research approaches.

In drawing conclusions about the study, sociologists typically discuss the study's implications. For instance, does a study of juvenile arrests suggest that new policies should be implemented, that existing ones should be changed, or that current police practices may be affecting the arrest rates? That is, the researcher answers the question "So what?" by showing the importance and usefulness of the study.

## 2-3f Qualitative and Quantitative Approaches

In **qualitative research**, sociologists examine nonnumerical material that they then interpret. In a study of grandfathers who were raising their grandchildren, for example, the researcher tape-recorded in-depth interviews and

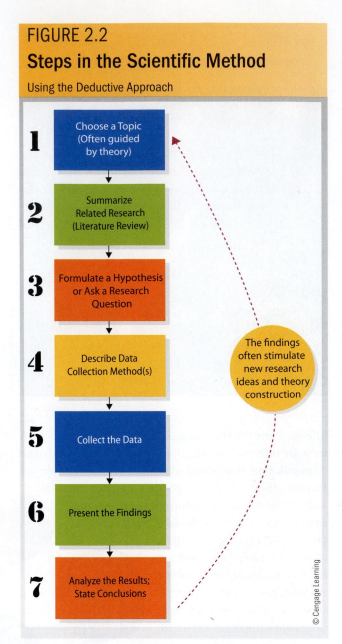

**FIGURE 2.2**

**Steps in the Scientific Method**

Using the Deductive Approach

1. Choose a Topic (Often guided by theory)
2. Summarize Related Research (Literature Review)
3. Formulate a Hypothesis or Ask a Research Question
4. Describe Data Collection Method(s)
5. Collect the Data
6. Present the Findings
7. Analyze the Results; State Conclusions

The findings often stimulate new research ideas and theory construction

© Cengage Learning

then analyzed the responses to questions about financial issues and daily parenting tasks (Bullock 2005).

In **quantitative research**, sociologists focus on a numerical analysis of people's responses or specific characteristics, studying a wide range of attitudes, behaviors, and traits (such as homeowners versus renters). In one national probability study, for example, the researchers surveyed almost 7,000 respondents to understand the influence of grandparents who live with their children and grandchildren (Dunifon and Kowaleski-Jones 2007).

Which approach should a researcher use? It depends on her or his purpose. Quantitative data provide valuable information on characteristics such as national college graduation rates. Qualitative data, in contrast, yield in-depth descriptions of why some college students drop out whereas others graduate. In many studies, sociologists use both approaches.

## 2-3g Correlation and Causation

Sociologists have found that some of the best predictors of divorce include marrying at a young age, having children before marriage, and experiencing domestic violence (see Chapter 12). This doesn't mean, however, that these and other variables *cause* divorce. Instead, divorce is more *likely* or *probable* in certain situations (when the marital partners are teenagers, for example).

Because nothing in life (except death) is certain, sociologists talk about *correlation*—the strength of the relationship between variables. They rarely use the term *cause* when interpreting their results because they can't *prove* that there's a cause-and-effect relationship. Instead, sociologists "can only *suggest*, or at most *indicate*" a relationship between variables (Glenn 2001). Thus, a researcher would say that marrying at a young age is "more likely" to result in divorce, is "associated (or correlated)" with divorce, or "contributes to" rather than "causes" divorce.

Consider another example of the difference between correlation and causation. In 2011, one of the most publicized media stories claimed that cell phones can cause brain cancer. The stories were especially alarming because they were based on a press release by the highly respected International Agency for Research on Cancer. Actually, the scientists stated that there wasn't any evidence of a causal relationship between wireless phone usage and brain cancer. They concluded, instead, that "additional research should be conducted into the long-term, heavy use of mobile phones" (Gaudin 2011: 2).

# 2-4 Some Major Data Collection Methods

During the research process, sociologists typically use one or more of the following data collection methods: surveys, secondary analysis of existing data, field research, content analysis, experiments, and evaluation research. Because each method has strengths and weaknesses, researchers must consider which will provide the most accurate information given time and budget constraints.

## 2-4a Surveys

Many sociologists use **surveys** to systematically collect data from respondents using questionnaires, face-to-face or telephone interviews, or a combination of these methods. Questionnaires can be mailed, used during an interview, or self-administered (such as student course evaluations). *Random sample surveys* are preferred because the results can be generalized to a larger population, but it's often difficult to determine the population from which a random sample can be drawn.

Telephone interviews are popular because they're a relatively inexpensive way to collect data. Researchers can obtain representative samples through *random digit dialing*, which involves selecting area codes and exchanges (the next three numbers) followed by four random digits. In the procedure called *computer-assisted telephone interviewing* (CATI), the interviewer uses a computer to select random telephone numbers, reads the questions to the respondent from a computer screen, and then enters the replies in precoded spaces, saving time and expense by not having to reenter the data after the interview.

Two types of electronic surveys are becoming increasingly popular. The first type is a survey sent via email, either as text in the body of the message or as an attachment. The second type is the more familiar survey that is posted on a website. Electronic surveys are cost effective and can represent a broad spectrum of the general population. In addition, Web surveys provide respondents with visual material, including videos, to look at and respond to (Keeter 2009, 2010).

### Advantages

Surveys are usually inexpensive, simple to administer, and have a fast turnaround. Because the results are anonymous, respondents are generally willing to answer questions on sensitive topics such as income, sexual behavior, and drug usage (Hamby and Finkelhor 2001).

Researchers often simplify surveys to increase response rates. During the 2010 census, for example, the Census Bureau used only a short form and slogans such as "10 Questions, 10 Minutes" to encourage people to mail back the forms. And, for the first time, the Census Bureau sent bilingual questionnaires to predominantly

> **surveys** a systematic method for collecting data from respondents, including questionnaires, face-to-face or telephone interviews, or a combination.

Gerald Bernard/Shutterstock.com

Spanish-speaking neighborhoods. Doing so generated higher response rates, reduced the need for expensive follow-up interviews, and saved millions of dollars (Groves and Vitrano 2011; Mather et al. 2011).

Face-to-face interviews have high response rates (sometimes up to 99 percent) compared with other data collection techniques because they involve personal contact. In-depth interviews can provide rich detail about the respondent's social world and vivid descriptions of personal experiences, such as the emotional reactions to having a baby or losing a job. People are more likely to discuss such sensitive issues in an interview than via a mailed questionnaire or a phone or electronic survey (Babbie 2013).

### Disadvantages

A major limitation of surveys that use mailed questionnaires is a low response rate, often only about 10 percent (Gray et al. 2007). If the questions are unclear, complicated, or seen as offensive, a respondent may simply throw the questionnaire away. Another concern is a *social desirability bias*—the tendency of respondents to give the answer that they think they "should" give or that will cast them in a favorable light (Cooperman 2010). For example, the proportion of Americans who say they voted in a given election is always much higher than the actual number of votes cast. Respondents also underreport behaviors perceived negatively (such as using alcohol or illicit drugs, or having multiple sexual partners), but exaggerate attendance at religious services, watching the evening news, and washing their hands after using a public restroom (Prior 2009; Radwin 2009; Zezima 2010).

Another problem with surveys is that people sometimes lie. A study at two large public universities found that a third of the students admitted being dishonest in end-of-semester course evaluations. Some fibbed to make their instructors look good, but most lied to "punish" professors they didn't like, especially when they received lower grades than they thought they deserved (Clayson and Haley 2011).

For telephone interviews, many researchers still rely heavily on landlines. More than 24

percent of U.S. households use only wireless telephones, not landlines, however. Because cell-only households are more likely than the general population to be young, black or Latino, renters, and low-income, telephone surveys often overlook these groups, which can skew a study's results (Cohn 2010; Blumberg et al. 2011).

Public opinion polls usually use surveys. Asking a few basic questions about a survey, such as those in *Table 2.1*, will help you evaluate its credibility.

"How long have you been dead? Do you have any complaints about your treatment here? Do you have any suggestions?"

Does my having only a cell phone decrease the possibility that I'll be contacted for a survey? The Pew Research Center provides an "Ask the Expert" site that answers this and other questions about surveys. Go to Sociology CourseMate at CengageBrain.com to access this site.

## 2-4b Secondary Analysis of Existing Data

Sociologists also rely heavily on **secondary analysis**, an unobtrusive form of data collection that examines information that has been collected by someone else. The data may be historical materials (such as court proceedings), personal documents (such as letters and diaries), public records (such as state archives on births, marriages, and deaths), or official statistics (such as health information from the Centers for Disease Control and Prevention).

### Advantages

In most cases, secondary analysis is convenient and inexpensive. Census Bureau data on topics such as income and employment are readily available at public libraries, college and university libraries, and online. Also, many academic institutions buy disks of national, regional, and local data that can be used for faculty or student research, or for class-related projects.

In secondary analysis, the data are *longitudinal* (collected at two or more points in time from the same or different samples of respondents) or *cross-sectional* (collected at one point in time). *Figure 2.3* shows a change over time in the attitudes of Americans toward homosexuality; this is an example of a longitudinal study. If the researchers had collected data for different groups at only one point in time (2011, 2000, or 1995), it would have been a cross-sectional study. Cross-sectional studies provide valuable information, but longitudinal studies are especially useful in examining trends in behavior or attitudes; a researcher can compare similar populations across different years or follow a particular group of people over time.

### Disadvantages

Existing data sources may not have the information a researcher needs. In 2000, the Census Bureau changed how it counted race and ethnicity to include the growing numbers of recent immigrants and mixed-race Americans. Doing so increased the accuracy of gauging racial and ethnic diversity but created 63 categories of possible racial-ethnic combinations. As a result, today's numbers are different from those before 2000, making comparisons across time on race and ethnicity problematic (Saulny 2011).

It may be difficult to gain access to historical materials because the documents are deteriorating, housed in only a few libraries in the country, or held in private collections. Also, existing data sources don't always include information the researcher is looking for. If you wanted to find out why some mothers deliberately kill their infants, for example, you'd find very little, if any, national data (Meyer and Oberman 2001). Consequently, you'd have to rely on studies with small and nonrepresentative samples or collect such data yourself.

## 2-4c Field Research

In **field research**, sociologists collect data by systematically observing people in their

**secondary analysis** examination of data that have been collected by someone else.

**field research** data collected by systematically observing people in their natural surroundings.

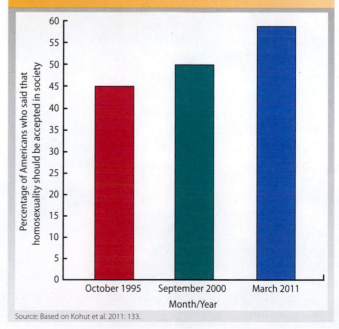

### FIGURE 2.3
### Opposition to Homosexuality Has Declined

Percentage of Americans who said that homosexuality should be accepted in society

| Month/Year | Percentage |
| --- | --- |
| October 1995 | 45 |
| September 2000 | 50 |
| March 2011 | ~59 |

Source: Based on Kohut et al. 2011: 133.

Svetlana Lukienko/Shutterstock.com

natural surroundings. In *participant observation*, researchers interact with the people they are studying; they may or may not reveal their identities as researchers. If you recorded interaction patterns between students and professors during your classes, you would be engaging in participant observation.

In *nonparticipant observation*, researchers study phenomena without being part of the situation. For example, child psychologists, clinicians, and sociologists often study young children in classrooms through one-way mirrors.

Some field research studies are short term (such as observing whether and how parents discipline their unruly children in grocery stores during a few weeks or months). Others, often called *ethnologies*, require a considerable amount of time in the field. For example, Sudhir Venkatesh (2008), while a graduate student at the University of Chicago, spent more than 6 years studying the culture and members of the Black Kings, a crack-selling gang in Chicago's inner city.

Observational studies are usually highly structured and carefully designed projects in which data are recorded and then converted to quantitative summaries. These studies may examine complex communication patterns, measure the frequency of acts (such as the number of head nods or angry statements), or note the duration of a particular behavior (such as the length of eye contact) (Stillars 1991). Thus, observational studies are much more complex and sophisticated than they appear to the general public. Researchers sometimes combine participant and nonparticipant observation.

## Advantages

Unlike secondary analysis and most surveys, field research provides an in-depth understanding of attitudes and behavior. Because observation usually doesn't disrupt the natural surroundings, the researcher doesn't directly influence the subjects. Field research is also more flexible than some other methods. The researcher can modify the research design, for example, by deciding to interview (rather than just observe) key people after the research has started.

In *Nickel and Dimed*, a study based on participant observation, journalist Barbara Ehrenreich (2001) worked at several low-income jobs (such as waitressing in Florida and clerking in a Walmart in Minnesota) to find out whether low-wage American workers could survive on their earnings. Ehrenreich's conclusion that people can't get by on a $7-per-hour job echoed what sociologists had been saying for decades. Because Ehrenreich is a well-known journalist, however, *Nickel and Dimed* has become a much cited example of field research.

## Disadvantages

Observation can be expensive if a researcher needs elaborate recording equipment, must travel far or often, or has to live in a different society or community for an extended period. Researchers who study other cultures must often learn a new language, a time-consuming task.

Doing fieldwork in a country that's wracked by war may be dangerous for both researchers and respondents. In a study of the major causes of death before and after the U.S. invasion of Iraq, for example, a team of American researchers feared the possibility of being killed or abducted. They also had difficulty recruiting translators and convincing subjects to participate in a study conducted by Americans (Guterman 2005).

A field researcher may encounter other barriers in collecting the desired data. Homeless and battered women's shelters, for example, are usually—and understandably—wary of researchers' intruding on their residents' privacy.

© Ianni Dimitrov/Alamy

Field researchers study a variety of topics, including how customers, cocktail waitresses, bouncers, and paid consultants often engage in deception, hustling, and bribes to help people hook up with someone at a nightclub (see Grazian 2008 for a description of this study).

## FIGURE 2.4

# How Do Views of Female and Male Babies Differ?

As these cards illustrate, girls and boys are viewed differently from the time they're born. In these and other birth announcements, girls, but not boys, are typically described as "sweet" or "precious." Also, the images usually show boys as active but girls as passive.

© Andrius Benokraitis

**content analysis** a data collection method that systematically examines some form of communication.

## 2-4d Content Analysis

**Content analysis** is a data collection method that systematically examines some form of communication. This is an unobtrusive approach that a researcher can apply to almost any form of written or oral communication: speeches, TV programs, newspaper articles, advertisements, office emails, songs, diaries, advice columns, poems, or Facebook chatter, to mention just a few.

The researcher develops categories for coding the material, sorts and analyzes the content of the data in terms of frequency, intensity, or other characteristics, and draws conclusions about the results. For example, look at the two greeting cards in *Figure 2.4*. What messages do they send about gender roles? These are only two cards, but we could do a content analysis of all such cards at a particular store or of a large sample of cards from a number of stores or online sites to determine whether such cards reinforce stereotypical gender role expectations.

Sociologists have used content analysis to examine a number of topics, such as the depictions of minority groups in sociology college textbooks, media coverage of heavy metal and rap music, images of women and men in advertisements and music videos, and changes in child-rearing advice in popular parenting magazines (Binder 1993; Shaw-Taylor and Benokraitis 1995; Larson and Hickman 2004; Rutherford 2009; Wallis 2011).

AP Photo/Mel Evans

Counting the homeless is an ongoing research problem. Communities must provide accurate numbers to qualify for federal funds, but census takers have difficulty getting such counts because safety rules prohibit them from entering private property (such as warehouses) or dark alleys. In one approach, the YWCA in Trenton, New Jersey, gave free haircuts at a fair for the homeless to attract the city's homeless people to a place where they could be counted (Jonsson 2007). Is such field research unethical because it violates people's privacy? Or is it innovative in helping a community get more federal funding for the homeless?

## Advantages

A major advantage of content analysis is that it is usually inexpensive and often less time consuming than other data collection methods, especially field research. If, for example, you wanted to examine the content of television commercials that target football fans, you wouldn't need fancy equipment, a travel budget, or a research staff.

A second advantage is that researchers can correct coding errors fairly easily by redoing the work. This is not the case with surveys. If you mail a questionnaire with poorly constructed items, it's too late to change anything.

A third advantage is that content analysis is unobtrusive. Because researchers aren't dealing with human subjects, they don't need permission to do the research and don't have to worry about influencing the respondents' attitudes or behavior.

A fourth advantage is that, in content analysis, researchers can obtain specific data over time. In one study, the researchers examined the amount and intensity of violence in children's animated movies that were released between 1938 and 1999 (Yokota and Thompson 2000). It would be very difficult, using most other data collection methods, to collect such information about or analyze a phenomenon over a 62-year period.

## Disadvantages

Content analysis can be very labor intensive, especially if a project is ambitious. In the research on children's animated movies, for instance, it took several years to code one or more of the major characters' words, expressions, and actions. A related disadvantage is that the

coding may be subjective. Having several researchers on a project can increase coding objectivity, but many content analyses are performed by only one researcher.

Finally, content analysis often reflects social class bias. Because most books, articles, speeches, films, and so forth are produced by people in upper socioeconomic levels, content analysis rarely captures the behavior or attitudes of working-class people and the poor. Even when documents created by lower-class individuals or groups are available, it's difficult to determine whether the coding reflects a researcher's social class prejudices.

## 2-4e Experiments

An **experiment** is a carefully controlled artificial situation that allows researchers to manipulate variables and measure the effects. The classic experimental design includes two equal-size groups that are very similar on characteristics such as sex, age, ethnicity or race, and education.

In the **experimental group**, the participants are exposed to the independent variable. In the **control group**, they aren't. Before the experiment, the researcher measures the dependent variable in both groups using a *pretest*. After the experimental group is exposed to the independent variable, the researcher measures both groups again using a *posttest*. If the researcher finds a difference in the measures of the dependent variable, she or he assumes that the independent variable is having an effect on the dependent variable.

In a recent study, the researchers at a large public university divided their introductory sociology classes into experimental and control groups. Both groups had the same readings, lectures, instructor, films, and classroom activities. In the experimental group, however, the professor incorporated a comedic clip, ranging from 1 to 6 minutes, toward the end of the class. The purpose of the clips (e.g., from

Are many sociologists, comedians, and television satirists (such as Jon Stewart, pictured here) similar in encouraging critical thinking about social life?

AP Photo/Jason DeCrow

## An Experiment That Went Awry

In a well-known 1973 experiment, psychologist Philip Zimbardo created a mock prison, using 21 undergraduate volunteers as prisoners and guards. The experiment was stopped after 6 days because some of the participants experienced intense negative reactions, and the study raised several ethical questions. For more information about this study, go to Stanford Prison Experiment, at www.prisonexp.org. We'll examine this study in more detail in Chapter 6.

Jon Stewart's *The Daily Show* and *The Colbert Report*) was to spark group discussion and active learning that demonstrated sociological thinking and concepts such as "alienation," "social class," and "conflict theory."

Compared with the control groups, the experimental groups (who viewed the clips) had higher scores and final grades, lower rates of withdrawal, and higher completion of online discussion questions. In addition, many of the students in the experimental groups began to bring in their own clips that illustrated sociological concepts. The researchers concluded that comedy is a useful tool in exploring sociological concepts because it resonates with students, and that "there is much to learn about social life by exploring what we laugh at and why" (Bingham and Hernandez 2009: 350).

### Advantages

A major strength of experiments is that they come closer than any other data collection method to suggest a possible cause-and-effect relationship. Experiments are usually less expensive and time consuming than other data collection techniques (especially large surveys and multiyear field research), and there's often no need to purchase special equipment. A related strength is that because researchers recruit students or other volunteers (as in medical research), participants are usually readily available and don't expect much, if any, monetary compensation.

A third advantage is that experiments can be replicated many times with different participants. Such replication strengthens the researchers' confidence in the validity and reliability of the measures and the study's results. For example, doctors stopped prescribing hormone pills for menopausal women because better-designed experiments that replicated earlier studies

found that the pills increased the risk for breast cancer and strokes (Ioannidis 2005).

### Disadvantages

A major disadvantage of laboratory experiments is that "the conditions so carefully controlled for the purpose of establishing causality may be so atypical or artificial that *it becomes difficult to generalize from them to the outside world*" (Gray et al. 2007: 280, emphasis added). Especially in laboratory settings, participants know that they're being studied and might behave very differently than if they were in a natural setting (Babbie 2002).

A second drawback is the possibility that the conclusions drawn from experiments may not be accurate. Among other problems, attrition among the participants may be high, the members of experimental and control groups may communicate with each other about what's going on and behave differently as a result, and the presence of a researcher may affect the participants' behavior (Cook and Campbell 1979; see also Chapter 6 on the Hawthorne effect).

Because of these and other limitations of experiments, consumers are often confused and frustrated by contradictory findings and advice. For example, different experiments have concluded that caffeine can help relieve headaches or lead to chronic headaches, can inhibit memory or improve it, and can increase or reduce depression (Bartlett 2011). Such conflicting findings sometimes lead people to say "If you don't like or believe the latest health advice, wait a few months and you'll hear what you want." The point is that the results of many experiments are speculative because they typically rely on small, nonrepresentative, self-selected samples, and there are often few control variables.

A third disadvantage is that experiments aren't suitable for studying large groups of people, a major focus in sociology. As you saw in Chapter 1, many sociologists examine macro-level issues (such as employment trends) that can't be observed in a laboratory or other experimental setting.

## 2-4f Evaluation Research

**Evaluation research**—which uses all of the standard data collection methods described so far—assesses the effectiveness of social programs in both the public and private sectors. Many government and nonprofit agencies provide services that affect people both directly and indirectly. Examples include housing, teenage pregnancy prevention, work training, and drug rehabilitation programs. Because local and state government budgets have been cut since the early 1980s, social service agencies often rely on evaluation research to streamline their services and programs.

In evaluation research, sociologists use a variety of data collection methods, such as examination of an organization's records (secondary analysis and content analysis), surveys to gauge employee and client satisfaction, and interviews with the staff and program recipients. Unlike the other data collection methods we've looked at, evaluation research is *applied* because it compares a program's achievements with its goals (Weiss 1998). The research findings are generally used to improve the program's efficiency and effectiveness.

### Advantages

Evaluation research examines actual efforts to deal with social problems such as homelessness and poverty. If the researchers rely only on secondary analysis, the data collection costs are usually low. The findings of evaluation research can be valuable to program directors or agency heads in showing discrepancies between the original objectives and the program's actual functioning and accomplishments (Card et al. 1994).

### Disadvantages

Practitioners rarely welcome the results of evaluation research if a sociologist concludes that a particular program isn't working. For example, since its inception in 1983, the DARE (Drug Abuse Resistance Education)

Evaluation research can help an organization streamline its programs and services. Doing so, however, may mean some employees' being laid off, because federal and state budgets for social programs have been shrinking.

program—which relied on trained volunteers from local police departments to address elementary school children on the dangers of drug use—has been popular with schools, police departments, parents, and politicians across the country. When social scientists evaluated DARE in 1994, 1998, and 1999, however, they concluded that there was no significant difference in the drug use of students who had completed the DARE curriculum and those who had not. When DARE funding was threatened, its promoters dismissed the evaluation research results as "voodoo science" (Miller 2001).

The various research methods described in this chapter are summarized in *Table 2.2*. Often, researchers use a combination of methods because "many view the social world as a multi-faceted and multi-layered reality that reveals itself only in part with any single method" (Jacobs 2005: 4). Despite a researcher's commitment to objectivity, ethical debates and politically charged disagreements can affect research.

## 2-5 Ethics, Politics, and Sociological Research

Scientists are always concerned about providing the most valid and reliable data. In conducting their research, what ethical and political dilemmas do sociologists (and other social scientists) encounter?

## TABLE 2.2

## Some Data Collection Methods in Sociological Research

| METHOD | EXAMPLE | ADVANTAGES | DISADVANTAGES |
|---|---|---|---|
| Surveys | Sending questionnaires and/or interviewing students on why they succeeded in college or dropped out | Questionnaires are fairly inexpensive and simple to administer; interviews have high response rates; findings are often generalizable | Mailed questionnaires may have low response rates; respondents tend to be self-selected; interviews are usually expensive |
| Secondary analysis | Using data from the National Center for Education Statistics (or similar organizations) to examine why students drop out of college | Usually accessible, convenient and inexpensive; often longitudinal and historical | Information may be incomplete; some documents may be inaccessible; some data can't be collected over time |
| Field research | Observing classroom participation and other activities of first-year college students with high and low grade-point averages (GPAs) | Flexible; offers deeper understanding of social behavior; usually inexpensive | Difficult to quantify and to maintain observer/subject boundaries; the observer may be biased or judgmental; findings are not generalizable |
| Content analysis | Comparing the transcripts of college graduates and dropouts on variables such as gender, race/ethnicity, and social class | Usually inexpensive; can recode errors easily; unobtrusive; permits comparisons over time | Can be labor intensive; coding is often subjective (and may be distorted); may reflect social class bias |
| Experiments | Providing tutors to some students with low GPAs to find out if such resources increase college graduation rates | Usually inexpensive; plentiful supply of subjects; can be replicated | Subjects aren't representative of a larger population; the laboratory setting is artificial; findings can't be generalized |
| Evaluation research | Examining student records; interviewing administrators, faculty, and students; observing students in a variety of settings (such as classroom and extracurricular activities); and using surveys to determine students' employment and family responsibilities | Usually inexpensive; valuable in real-life applications | Often political; findings might be rejected |

© Cengage Learning

## 2-5a Ethical Research

Because so much research relies on human participants, the federal government, university institutional review boards (IRBs), and many professional organizations have formulated codes of ethics to protect participants. Regardless of the discipline or the research methods used, all ethical standards have at least three golden rules.

■ First, *do no harm* by causing participants physical, psychological, or emotional pain.

■ Second, the researcher must get the participant's *informed consent* to be in a study. This includes the participant's knowing what the study is about and how the results will be used. Sociologists can use deception (such as not revealing that they are researchers) if doing so doesn't harm the participants, if the research has been approved by an IRB, and if

the researcher explains the purpose of the study to participants at the end of the research (American Sociological Association 1999: 14).

■ Third, researchers must always protect a participant's *confidentiality*, even if the participant has broken a law that she or he tells about to the researcher. *Table 2.3* lists some of the basic ethical principles to be followed in conducting sociological research.

## 2-5b Scientific Dishonesty

Some disciplines are more susceptible to ethical violations than others. Medical researchers, especially, have been accused of considerable scientific misconduct. Some of the alleged violations have included the following: changing research results to please the corporation (usually tobacco or pharmaceutical companies) that

## TABLE 2.3

# Some Basic Principles of Ethical Sociological Research

In its most recent update, the American Sociological Association (1999) has reinforced its previous ethical codes and guidelines for researchers. Researchers must:

- Obtain all subjects' consent to participate in the research and their permission to quote from their responses and comments.

- Not exploit subjects or research assistants involved in the research for personal gain.

- Never harm, humiliate, abuse, or coerce the participants in their studies, either physically or psychologically.

- Honor all guarantees to participants of privacy, anonymity, and confidentiality.

- Use the highest methodological standards and be as accurate as possible.

- Describe the limitations and shortcomings of the research in their published reports.

- Identify the sponsors who funded the research.

- Acknowledge the contributions of research assistants (usually underpaid and overworked graduate students) who participate in the research project.

Stephen Stickler/Image Source

© Cengage Learning

sponsored the research; being paid by companies to deliver speeches to health practitioners that endorse specific drugs, even if the medications don't reduce health problems; allowing drug manufacturers to ghostwrite their articles (and even draft textbooks) that are published in prestigious medical journals; and falsifying data (Basken 2011b; Blumenstyk 2009; Johnson 2010; Ornstein and Weber 2011; Project on Government Oversight 2010; Shamoo and Bricker 2011).

In the social sciences, some data collection methods are more prone to ethical violations than others. Surveys, secondary analysis, and content analysis are less vulnerable than field research and experiments because the researchers typically don't interact directly with participants, affect them, or become personally involved with the respondents. In contrast, experiments and field research can raise ethical questions because of deception or influencing the participants' attitudes or behavior.

All scholars face emerging ethical questions when they do research using social networks and other online environments. IRBs lack experience with Web research and sometimes unintentionally approve social science research that might violate ethics guidelines.

## 2-5c Political, Religious, and Community Pressure on Researchers

Throughout the tenure of the George W. Bush administration, many government scientists complained that political officials routinely stifled research results that were unpopular with business and industry. President Obama promised to put science ahead of politics. His new rules require universities to disclose financial ties between their researchers and companies, but some critics claim that the new standards don't go far enough in ensuring the integrity of medical research (Baskin 2011a). Some federal biologists contend, for example, that they are still pressured by high-ranking politicians, who don't want to antagonize corporations, not to report the negative effects of overgrazing on federal land or raise concerns about the impact of oil spills on the environment (Hamburger and Geiger 2010).

Social science research is especially likely to be challenged when it focuses on sensitive social, moral, and political issues. Research on teenage sexual behavior is valuable because it provides information that public health agencies and schools can circulate about sexually transmitted diseases (such as HIV) and contraception. Nonetheless, many local jurisdictions have refused to let social scientists study adolescent sexual behavior. Some parents believe that such research violates student privacy and might make a school district look bad (e.g., if a study reports a high incidence of drug use or sexual activity) (Kempner et al. 2005).

Some religious groups, school administrators, and politicians have also opposed such studies because they believe that the research undermines traditional family values or makes deviant behavior seem normal. As a result, many social scientists don't do research on some of the most controversial issues because there is no funding or because a study would be opposed by a variety of groups that get considerable media attention (Carey 2004).

## 2-5d Do People Believe Scientific Findings?

Not always. For example, 87 percent of scientists say that humans and other living things have evolved over time, but only 32 percent of Americans agree. And although there has been no scientific evidence of paranormal events, 29 percent of Americans say they have been in touch with the dead, and 18 percent report having seen or been in the presence of ghosts (Allen and Auxier 2010).

Why do many people reject scientific findings? One reason may be that when the media publicize the relatively

Ricardo Watson UPI Photo Service/Newscom

In 2010, Susan Reverby, a college professor, found research that showed that U.S. doctors had deliberately infected 700 Guatemalans—primarily prison inmates, mental patients, and soldiers—with venereal diseases between 1946 and 1948 to test the effectiveness of penicillin. John C. Cutler, the public health doctor who led the experiment, also played an important role in a study conducted between 1932 and 1972 in Tuskegee, Alabama, that studied the progression of syphilis among poor, rural black men. The men were deliberately left untreated for decades.

rare occurrences of medical misconduct, readers generalize the few cases to all research and become more skeptical about all scientific studies. Second, research findings often challenge personal attitudes and beliefs that people cherish, such as believing that global warming is a myth (Kahan et al. 2011; Sheppard 2011). Third, many Americans, especially minorities, are suspicious of the scientific community. Between 1906 and 1932, for example, medical researchers purposely infected unknowing subjects with fatal venereal diseases to test the effectiveness of penicillin. The subjects were almost always black men, prison inmates, mental patients, or soldiers (McNeil 2010; Sanburn 2010). Finally, many Americans know little about the scientific method, including the strengths and limitations of the research process.

## Study Tools

Ready to study? Go to **Sociology CourseMate** at **www.cengagebrain.com** to complete practice quizzes, review flashcards, watch videos, and more.

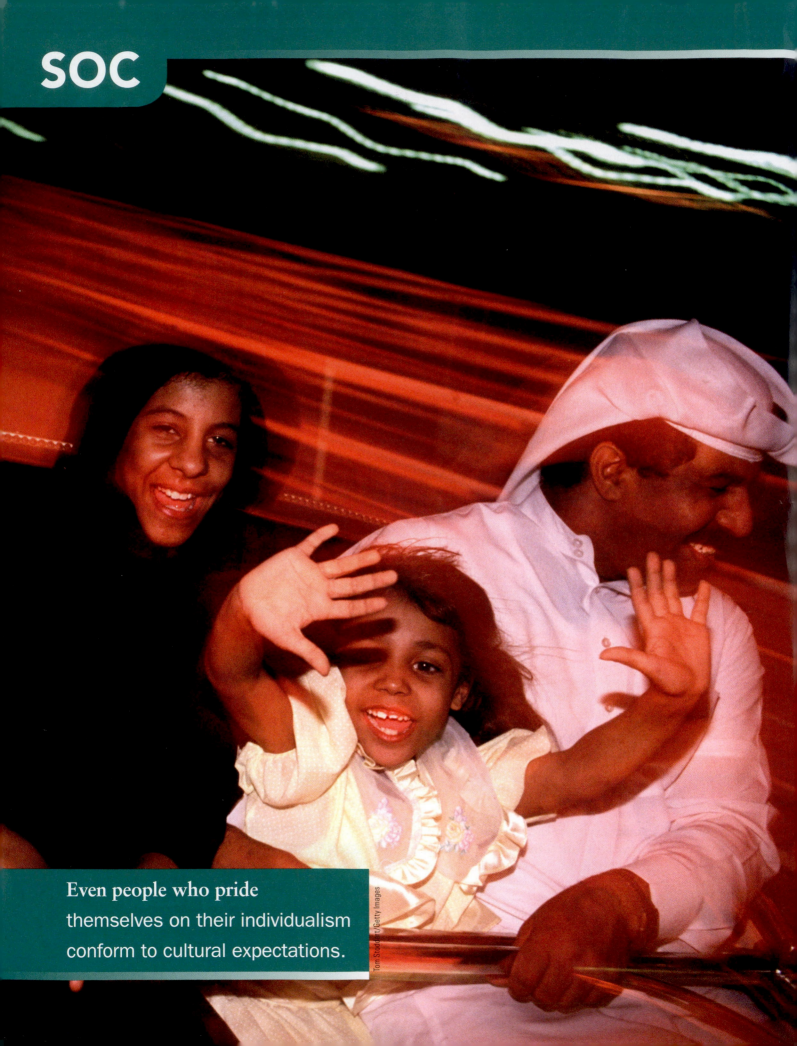

SOC

Even people who pride themselves on their individualism conform to cultural expectations.

Tom Stoddart/Getty Images

# Culture

Once when I returned a set of exams, a student who was unhappy with his grade blurted out an obscenity. A voice from the back of the classroom snapped, "You ain't got no culture, man!" The remark implied that proper classroom behavior doesn't include cursing. Is it true that people who use vulgar language "ain't got no culture"?

**culture** the learned and shared behaviors, beliefs, attitudes, values, and material objects that characterize a particular group or society.

**society** a group of people who have lived and worked together long enough to become an organized population and to think of themselves as a social unit.

## Key Topics

In this chapter, we'll explore the following topics:

## 3-1 Culture and Society

As popularly used, *culture* often means appreciating the finer things in life, such as Shakespeare's sonnets, the opera, and using civil language. In contrast, sociologists use the term in a much broader sense: **Culture** refers to the learned and shared behaviors, beliefs, attitudes, values, and material objects that characterize a particular group or society. Thus, culture shapes a people's total way of life.

Most human behavior is not random or haphazard. Among other things, culture influences what you eat; how you were raised and will raise your own children; if, when, and whom you will marry; how you make and spend money; and what you read. Even people who pride themselves on their individualism conform to most cultural rules. The next time you're in class, for example, count how many students are *not* wearing jeans, T-shirts, sweatshirts, or sneakers—clothes that are the prevalent uniform of adolescents, college students, and many adults in U.S. society.

A **society** is a group of people who have lived and worked together long enough to become an organized population and to think of themselves as a social unit (Linton 1936). Every society has a culture that guides people's interactions and behaviors. Society and culture are mutually dependent; neither can exist without the other. Because of this interdependence, social scientists sometimes use the terms *culture* and *society* interchangeably.

### 3-1a Some Characteristics of Culture

All human societies, despite their diversity, share some cultural characteristics and functions (Murdock 1940). We don't see culture directly, but it shapes our attitudes and behaviors.

1. **Culture is learned.** Culture is not innate but learned, and it shapes how we think, feel, and behave. If a child is born in one region of the world but raised in another, she or he will learn the customs, attitudes, and beliefs of the adopted culture.

## What do you think?

I shape my culture; my culture doesn't shape me.

| 1 | 2 | 3 | 4 | 5 | 6 | 7 |

strongly agree    strongly disagree

**2. Culture is transmitted from one generation to the next.** We learn many customs, habits, and attitudes informally through interactions with parents, relatives, and friends, and from the media. We also learn culture formally in settings such as schools, workplaces, and community organizations. Whether our learning is formal or informal, culture is cumulative because each generation transmits cultural information to the next one.

**3. Culture is shared.** Culture brings members of a society together. We have a sense of belonging because we share similar beliefs, values, and attitudes about what is right and wrong. Imagine the chaos if we did what we wanted (such as physically assaulting an annoying neighbor) or if we couldn't make numerous daily assumptions about other people's behavior (such as stopping at red traffic lights).

**4. Culture is adaptive and always changing.** Culture changes over time. New generations discard technological aspects of culture that are no longer practical, such as replacing typewriters with personal computers. Attitudes can also change over time. Compared with several generations ago, for example, many Americans now feel that premarital sex is acceptable or at least not wrong (see *Figure 3.1*).

Culture reflects who we are, but remember that it's people who create culture. As a result, culture changes as people adapt to their surroundings. Since the attacks on September 11, 2001, for example, many people worldwide, including Americans, have become accustomed to greater surveillance by the government. We don't complain at airports when we have to pass through a metal detector, when our baggage is X-rayed or searched, when items are confiscated, and, increasingly, when airports use full-body scanners. Thus, as the dangers in

© World History Archive/Alamy

iStockphoto.com/Jill Fromer

society have grown, people have passed laws or implemented rules that have both increased their security and decreased their privacy.

## 3-1b Material and Nonmaterial Culture

Cultures that people construct are both material and nonmaterial (Ogburn 1922). **Material culture** consists of the tangible objects that members of a society make, use, and share. These creations include diverse products such as buildings, tools, music, weapons, jewelry, religious objects, and cell phones. **Nonmaterial culture** includes the shared set of meanings that people in a

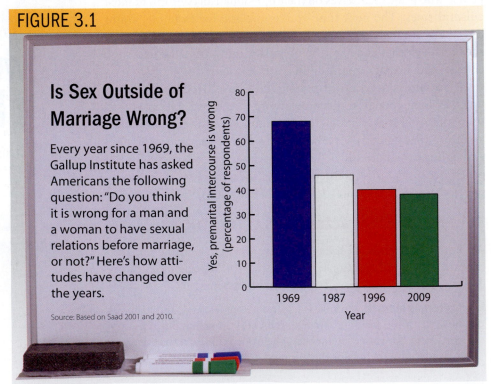

### FIGURE 3.1

**Is Sex Outside of Marriage Wrong?**

Every year since 1969, the Gallup Institute has asked Americans the following question: "Do you think it is wrong for a man and a woman to have sexual relations before marriage, or not?" Here's how attitudes have changed over the years.

Source: Based on Saad 2001 and 2010.

*Yes, premarital intercourse is wrong (percentage of respondents)* vs *Year*: 1969 ≈ 68, 1987 ≈ 46, 1996 ≈ 40, 2009 ≈ 38

Ryan McVay/Getty Images

society use to interpret and understand the world. Symbols, values, beliefs, sanctions, customs, and rules of behavior are examples of nonmaterial culture.

Material and nonmaterial culture influence each other. The automobile, for example, has changed every society that has adopted it. Among other things, cars have provided privacy during hookups and dating, transported passengers relatively inexpensively, and generated new laws (such as not drinking and driving). Frequently, however, as you'll see shortly, there is a "cultural lag" because material culture (invention of the automobile and the cell phone) changes faster than nonmaterial culture (pollution and texting while driving laws).

# 3-2 The Building Blocks of Culture

Consider the following research findings:

- In a survey of 202 corporate hiring managers, 57 percent said that qualified but unattractive candidates are likely to have a harder time landing a job (Bennett 2010).

- On average, good-looking men earn 5 percent more per hour and attractive women earn 4 percent more per hour than their average-looking counterparts (Hamermesh and Parker 2003).

- College students who view their instructors, especially their male professors, as good looking give them better course evaluations, which, in turn, can generate more economic rewards, such as merit increases and promotions, for those professors (Engemann and Owyang 2005).

Such findings contradict the proverb "Beauty is in the eye of the beholder" (that someone or something is beautiful if the viewer perceives it so). How can we explain this contradiction? Why do many Americans agree on who is beautiful, average, or unattractive? Are the less attractive people victims of bias? Or do good-looking people develop self-confidence and social skills that enhance their economic opportunities?

To answer such questions, we must understand the building blocks of culture, especially symbols, language, values, and norms.

## 3-2a Symbols

A **symbol** is anything that stands for something else and has a particular meaning for people who share a culture. In most societies, for example, a handshake communicates friendship or courtesy, a wedding ring signals that a person isn't a potential dating partner, and a siren denotes an emergency. People influence each other through the use of symbols. A smile and a frown communicate different information and elicit different responses. Through symbols, we engage in *symbolic interaction* (see Chapter 1).

### Symbols Take Many Forms

Written words are the most common symbols, but we also communicate by tattooing our bodies, getting breast implants, and purchasing goods and services that we believe will increase our social status. And gestures (such as raised fists, hugs, and stares) convey important messages about people's feelings and attitudes.

### Symbols Distinguish One Culture from Another

In Islamic societies, many women wear a head scarf (*hijab* or *hejab*) that covers their hair and neck, a veil that hides part of the face from just below the eyes (*nijab*), or a garment that covers the entire body (*chador* or *burqa*). Whatever form it takes, the veil is just a piece of cloth, but it means different things to people across and within cultures.

For non-Muslims, especially Westerners, veils are often symbols of women's repression by men and religious zealotry; they conjure up images of terrorism and connote hiding something. For many Muslims,

Comstock/Jupiterimages

The U.S. Flag Code describes the flag as a "living thing" and forbids its display on clothing, bedding, drapery, or in advertising (Luckey 2008). Are these people disrespecting a cherished symbol? Or expressing their patriotism?

**language** a system of shared symbols that enables people to communicate with one another.

on the other hand, veiling symbolizes religious commitment, women's modesty, and an Islamic identity. Other Muslims—including those who live in the Middle East or have emigrated to Western countries—view most veiling, especially *nijabs* and *burqas,* not as a religious rule but as a custom that discourages face-to-face interaction (Murphy 2009).

## Symbols Can Unify or Divide a Society

Symbols usually unify people culturally. About 75 percent of Americans say that they display the nation's flag in places such as their home, workplace, car, or clothing ("75%—A Nation of Flag Wavers" 2011). Every Fourth of July, many Americans celebrate the anniversary of gaining independence from Britain with parades, firecrackers, barbecues, and numerous speeches by local and national politicians. All of this symbolic behavior signifies freedom and democracy. For this reason, many immigrants purposely choose July 4th to be naturalized.

Symbols can also be divisive. For example, some white Southerners fly the Confederate flag because they see it as a proud emblem of their Southern heritage. Others have abandoned the flag because it is used by racist groups, such as the Ku Klux Klan, to symbolize slavery and white domination of African Americans.

## Symbols Can Change Over Time

Symbols communicate information that varies across societies and may change over time. In 1986, for example, the International Red Cross changed its name to

Left: The two emblems of the International Movement of the Red Cross and the Red Crescent. Right: The proposed red diamond emblem.

the International Red Cross and the Red Crescent Movement to encompass a number of Arab branches, and adopted a crescent emblem in addition to the well-known cross. Israel wanted to use a red Star of David, rejecting the cross as a Christian symbol and the crescent as an Islamic one. Red Cross officials offered a red diamond as the new shape, but some countries rejected the diamond because it represents bloody conflicts in many African countries that mine diamonds (Whitelaw 2000). The red cross and crescent continue to be the organization's emblems until the issue is resolved.

## 3-2b Language

Perhaps the most powerful of all human symbols is **language**, a system of shared symbols that enables people to communicate with one another. Essentially, language is an invention of human thought that communities of people have endowed with meaning. In every society, children begin to grasp the essential structure of their language at a very early age and without any instruction. Babbling rapidly leads to uttering words and combinations of words. The average child knows approximately 900 words by the age of 2 and 8,000 words by age 6 (Hetherington et al. 2006).

Anthropological linguist Edward Sapir (1929) and later his student, Benjamin Whorf (1956), conducted comparative research on a wide variety of languages. They theorized that thoughts and behavior are determined by language. Their theory, known as the *Sapir-Whorf hypothesis,* posited that language not only is a tool for interpersonal communication but also provides people with a framework for interpreting social reality and the world around them. Thus, as languages vary, so do interpretations of social reality.

Subsequent research found little evidence to support the *Sapir-Whorf hypothesis.* Some critics maintained that language affects, but does not *determine* culture. Others argued that language influences some thinking processes, but not all (e.g., Carroll 1956; Berlin and Kay 1969; Heider and Olivier 1972; Schlesinger 1991). Still, the *Sapir-Whorf hypothesis* has generated considerable research across disciplines such as philosophy, sociology, and psychology in exploring the relationship between language, thought, and culture.

### Why Language Is Important

Language makes us human: It helps us understand our everyday experiences, conveys our ideas, communicates information, and influences other people's attitudes and behavior. Language directs our thinking, controls our actions, shapes our expression of emotions, and

AP Photo/Salvatore Di Nolfi/FABRICE COFFRINI/AFP/Getty Images gives us a sense of belonging to a group.

Language can also spark anger and conflict. Recognizing this connection, two high schools in Connecticut started to fine students up to $103 for cursing. The schools' officials hoped to decrease the fights that erupted when students used obscenities and vulgar language (Llana 2005). Some middle school students, believing that civility will discourage meanness and bullying, have founded a nationwide No Cussing Club (Severson 2011). The relationship between language, thought, and behavior are especially evident when we consider gender and ethnicity.

## Language and Gender

Language has a profound influence on how we think about and act toward women and men. Those who adhere to traditional language usage contend that nouns such as *businessman*, *chairman*, *mailman*, and *mankind* and pronouns such as *he* refer to both women and men, and that women who object to such usage are too sensitive. Suppose, however, that all of your professors used only *she*, *her*, and *women* when referring to all people. Would the men in the class feel excluded?

Many professors routinely use phrases such as "Okay, guys, in class today . . ." and no one objects. One of my colleagues illustrates the linkage between language and gender, and how it affects our thinking, by saying "Okay, gals, in class today . . ." "I always get a reaction of gaping mouths, laughs, and bewildered looks," he says. The students react differently to "guys" and "gals" because we have internalized male terms (such as *guys*, *policemen*, and *maintenance man*) as normal and acceptable.

# Wanted: Someone Who Can Translate Federal Gibberish Into Plain English

In Sweden, unlike in the United States, it's almost impossible to find a high-level government document that people can't understand. To increase the readability of federal publications, in 2010 President Obama signed the Plain Writing Act, which requires federal documents to be rewritten in plain English. Here's a "before" and "after" example from the Department of Health and Human Services Handbook that assures access to essential health care for low-income HIV/AIDS patients:

- **Before**: Title I of the CARE Act creates a program of formula and supplemental competitive grants to help metropolitan areas with 2,000 or more reported AIDS cases meet emergency care needs of low-income HIV patients. Title II of the Ryan White Act provides formula grants to States and territories for operation of HIV service consortia in the localities most affected by the epidemic, provision of home and community-based care, continuation of insurance coverage for persons with HIV infection, and treatments that prolong life and prevent serious deterioration of health. Up to 10 percent of the funds for this program can be used to support Special Projects of National Significance.
- **After**: Low income people living with HIV/AIDS gain, literally, years, through the advanced drug treatments and ongoing care supported by HRSA's Ryan White Comprehensive AIDS Resources Emergency (CARE) Act.

In all cultures, language is supposed to help people communicate. However, the Plain Writing Act lacks enforcement, and many federal agencies, including the Internal Revenue Service, are exempted. Does a government have more control if its citizens don't understand the government's rules and regulations?

Sources: Based on Health Resources and Services Administration (2010), and "Government Resolves . . ." (2011).

Rubberball/Mike Kemp/Jupiter Images

$*#@%**#@

## Language, Race, and Ethnicity

Words—written and spoken—create and reinforce both positive and negative images about race and ethnicity. Someone might receive a *black mark*, and a *white lie* isn't really a lie. We *blackball* someone, *blacken* someone's reputation, denounce *blackguards* (villains), and prosecute people who participate in the *black market*. In contrast, a *white knight* rescues people in distress, the *white hope* brings glory to a group, and the good guys wear *white hats*.

Racist or ethnic slurs, labels, and stereotypes demean and stigmatize people. Derogatory ethnic words abound: *honky*, *hebe*, *kike*, *spic*, *chink*, *jap*, *polack*, *wetback*, and many others. Self-ascribed racial epithets are as harmful as those imposed by outsiders. When Italians refer to themselves as *dagos* or African Americans call each other *nigger*, they tacitly accept stereotypes about themselves and legitimize the general usage of such derogatory ethnic labels (Attinasi 1994).

Through language, children learn about their cultural heritage and develop a sense of group identity. In many Latino and other immigrant families, the native language begins to disappear after just one generation, resulting in communication problems between Americanized children and their non-English-speaking kin in the homeland. To counter the loss of Spanish and other languages among their children, many bilingual parents have instituted weekend schools where children learn to speak their native language (Tobar 2009; see also Chapter 10).

### Language and Social Change

Language is dynamic and changes over time. A new edition of the New American Bible, an annual best seller, has changed the word "booty," which now has a sexual connotation, to "spoils of war." "Cereal," which many think of as breakfast food, is now "grain" to reference loads of wheat (*Huffington Post* 2011).

U.S. English is composed of hundreds of thousands of words borrowed from many countries and from groups that were in the Americas before the colonists arrived (Carney 1997). *Table 3.1*

provides a few examples of English words borrowed from other languages.

In response to cultural and technological changes, our vocabulary now includes *sexting*, *ebook*, *tweet*, *staycation*, and *unfriend*, among other new words. And to some people's dismay and others' delight, some writers are now substituting *partner* for the traditional *spouse*, *wife*, or *husband*.

Language is sometimes slow to catch up with cultural changes. For example, how does one refer to a romantic partner, especially in social situations? "Significant other" can mean a parent, sibling, or close friend; "partner" or "companion" doesn't sound quite right; "lover" isn't acceptable in most social circles; "girlfriend" or "boyfriend" seems inappropriate unless you're a teenager; and "the person I'm seeing" might elicit unwanted questions about your relationship.

## 3-2c Values

**Values** are the standards by which members of a particular culture define what is good or bad, moral or immoral, proper or improper, desirable or undesirable, beautiful or ugly. They are widely shared within a society and provide *general guidelines* for everyday behavior rather than specific rules for concrete situations. For example, when faculty members catch students plagiarizing, students often plead innocence and blame the instructor ("*You* never told us *exactly* what *you* mean by plagiarism.").

Jiri Hera/Shutterstock.com/Ispace/Shutterstock.com

---

### TABLE 3.1

## U.S. English Is a Mixed Salad

Here are a few example of English words borrowed from other languages, showing our mixture of heritages. What words would you add to the list?

| | |
|---|---|
| ☐ Africa: apartheid, Kwanzaa, safari | ☐ Spain: anchovy, bizarre |
| ☐ Alaska and Siberia: husky, igloo, kayak | ☐ Thailand: Siamese |
| ☐ Bangladesh: bungalow, dinghy | ☐ Turkey: baklava, caviar, kebob |
| ☐ Hungary: coach, goulash, paprika | ☐ France: bacon, police, ballet |
| ☐ India: bandanna, cheetah, shampoo | ☐ Japan: geisha, judo, sushi |
| ☐ Iran and Afghanistan: bazaar, caravan, tiger | ☐ Norway: iceberg, rig, walrus |
| ☐ Israel: kosher, rabbi, Sabbath | ☐ Mexico: avocado, chocolate, coyote |
| ☐ Italy: fresco, spaghetti, piano | ☐ Germany: strudel, vitamin, sauerkraut, kindergarten |

© Cengage Learning

## Major U.S. Values

Sociologist Robin Williams (1970: 452–500) has identified a number of core U.S. values. All are central to the American way of life because they are widespread, have endured over time, and reflect many people's intense feelings.

1.  **Achievement and success.** U.S. culture stresses personal achievement, especially occupational success. Many Americans believe that if they work hard, apply themselves, and save their money, they will be successful in the future.

2.  **Activity and work.** Americans want to "make things happen." They respect people who are focused and disciplined in their jobs, and assume that hard work will be rewarded. Journalists and others often praise those who work past their retirement age.

3.  **Humanitarianism.** U.S. society emphasizes concern for others, helpfulness, kindness, and offering comfort and support. During natural disasters—such as earthquakes, floods, fires, and famines—at home and abroad, many Americans are enormously generous (see Chapter 5).

4.  **Efficiency and practicality.** Americans emphasize technological innovation, up-to-dateness, practicality, and getting things done. Many American colleges, in fact, now tout their programs or courses as being practical instead of emphasizing knowledge and intellectual growth.

5.  **Progress.** Americans focus on the future rather than the present or the past. The next time you walk down the aisle of a grocery store, note how many products are "new," "improved," and "better than ever."

6.  **Material comfort.** Americans consider it normal to want new products and services. Many work hard to pay for fancy new cars, large homes, and dream vacations (even if they can't afford them). Americans never have enough gadgets and always need more stuff. When the stuff accumulates, we buy stuff to stuff it into.

7.  **Freedom and equality.** Countless U.S. documents affirm freedom of speech, freedom of the press, and freedom of worship. U.S. laws also tell Americans that they have been "created equal" and have the same legal rights regardless of race, ethnicity, sex, religion, disability, age, or social class.

8.  **Conformity.** Most people don't want to be labeled as "strange," "peculiar," or "different." They conform because they want to be accepted, to get social approval from those they respect, and to be hired or promoted. Striving for success often means controlling one's impulses and biting one's tongue.

9.  **Democracy.** Democracy provides the average U.S. citizen with equal political rights. Democracy emphasizes equality, freedom, and faith in the people rather than giving power to a monarch, dictator, or emperor.

10. **Individualism.** American culture sets a high value on each person's development. Thus, we often encourage children to be independent, creative, self-directed, self-motivated, and spontaneous.

Williams acknowledged troublesome patterns because these core American values, as you may have noticed, are sometimes contradictory and don't always mesh with reality. For example, Americans value equality but are comfortable with enormous gaps in wealth and power, and continue to discriminate against people because of their ethnicity, race, sexual orientation, sex, or age. We proclaim that we respect individualism but reward people who conform to a group's expectations. And we say that we value responsibility but often blame popular culture rather than parenting for children's bad behavior.

### Values Vary Across Cultures and Change Over Time

As you'll see shortly, cultural values can change as a result of technological advances, immigration, and contact with outsiders. For example, the Japanese parliament recently passed a law making love of country a compulsory part of school curricula. The lawmakers and their numerous supporters hope that teaching patriotism will counteract the American-style emphasis on individualism and self-expression that they believe has undermined Japanese values of cooperation, self-discipline, responsibility, and respect for others (Wallace 2006).

In the United States, surveys of first-year college students show a shift in values. Between 1968 and

© Jaros/Shutterstock

## Sociology in Your Life

Does censorship protect the members of a society? Or limit their freedom of speech and expression? See the books that some groups have succeeded in removing from libraries and schools during the early part of the twenty-first century by going to Sociology CourseMate at CengageBrain.com.

2006, for example, developing a meaningful philosophy of life plummeted in importance, whereas being rich became substantially more important (Pryor et al. 2007). Consistent with our humanitarian values, however, a majority (69 percent) of the students surveyed in 2009 said that "helping others who are in difficulty" is essential or very important, but women were more likely to feel this way (see *Figure 3.2*).

## 3-2d Norms

Values are general standards, whereas **norms** are a society's specific rules of right and wrong behavior. Norms tell us what we *should*, *ought*, and *must* do, as well as what we *should not*, *ought not*, and *must not* do: don't talk in church, stand in line, and so on.

Norms are specific expectations shared by the members of the society at large or by members of particular groups within a society. For example, a female professor's students sometimes comment on her clothes in their course evaluations and want her to "look like a professor." In contrast, her husband, also a professor, "has yet to hear a single student comment about his wardrobe." So, she changes her businesslike outfits every day while her husband usually wears khaki pants (or jeans) and a polo shirt day in and day out (Johnston 2005). Thus, "her" and "his" norms differ.

Here are some general characteristics of norms:

- Most are *unwritten*, passed down orally from generation to generation (use the good dishes and tablecloth on special occasions).

- They are *instrumental* because they serve a specific purpose (get rid of garbage because it attracts roaches and rats).

- Some are *explicit* (save your money "for a rainy day"), whereas others are *implicit* (be respectful at a wake or funeral).

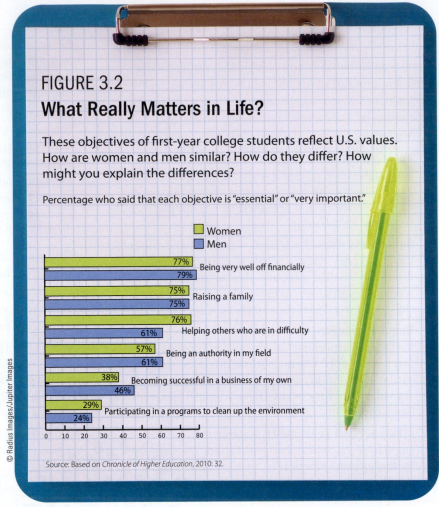

## FIGURE 3.2
## What Really Matters in Life?

These objectives of first-year college students reflect U.S. values. How are women and men similar? How do they differ? How might you explain the differences?

Percentage who said that each objective is "essential" or "very important."

Legend: Women (green), Men (blue)

- Being very well off financially: Women 77%, Men 79%
- Raising a family: Women 75%, Men 75%
- Helping others who are in difficulty: Women 76%, Men 61%
- Being an authority in my field: Women 57%, Men 61%
- Becoming successful in a business of my own: Women 38%, Men 46%
- Participating in a programs to clean up the environment: Women 29%, Men 24%

(scale: 0 10 20 30 40 50 60 70 80)

Source: Based on *Chronicle of Higher Education*, 2010: 32.

© Radius Images/Jupiter Images

- They *change* over time (it's now more acceptable to have a child out of wedlock but much less acceptable to smoke or drink alcohol while pregnant).

- Most are *conditional* because they apply in specific situations (slipping out of your smelly shoes may be fine at home but not on an airplane).

- Because they are situational, norms can be *rigid* ("You *must* turn in your term paper at the beginning of class on May 5.") or *flexible* ("You can have another week to finish the paper.").

Norms organize and regulate our behavior. We may not like many of the rules, but they make our everyday lives more orderly and predictable.

Sociologists differentiate three types of norms—folkways, mores, and laws—that vary because some rules are more important than others. As a result, a society punishes some wrongdoers more severely than others.

### Folkways

**Folkways** are norms that members of a society (or a group within a society) see as not being critical for a society's survival and that, consequently, are not severely punished when violated. Etiquette rules are good examples of folkways: Cover your mouth when you cough or sneeze, say "please" and "thank you," and knock before entering someone's office.

We automatically follow a variety of everyday conventions and customs because we've internalized them since birth. We often don't realize that we conform to norms until someone violates one, such as picking one's nose in public or talking loudly on a cell phone.

Folkways vary from one country to another. Punctuality is important in many Western countries but not in much of Latin America and the Mediterranean. Japan is notable for its gift giving among businesspeople, but this may be viewed as bribery in Western countries. Austrians, as well as Germans and the Swiss, consider chewing gum in public vulgar. There are also many differences in table manners: Europeans keep their hands above the table at all times, not in their laps, and Koreans frown on sniffling or blowing one's nose at the table (Axtell et al. 1997).

Folkways can change in response to macro-level changes. The growth of technology has changed many campus folkways. In the past, a syllabus was typically a few pages long, outlining course requirements and deadlines for exams and papers. Now, many syllabi—some as long as 20 pages—look more like legal documents. They describe, often in great detail, a variety of rules on Internet plagiarism, unacceptable online sources for papers and projects, and laptop and cell phone use during class, even forbidding videos of professors that may appear on YouTube (Wasley 2008).

## Mores

**Mores** (pronounced "MORE-ayz") are norms that members of a society consider very important because they maintain moral and ethical behavior. According to U.S. cultural mores, one must be sexually faithful to one's spouse or sexual partner, must be loyal to one's country, and must not kill another person (except during war or in self-defense).

Note that folkways emphasize *ought to* behavior whereas mores define *must* behavior. The Ten Commandments are a good example of religious mores. Other mores include ethical guidelines (don't cheat, don't lie), expectations about behavior (don't use illicit drugs), and rules about sexual partners (don't have sexual intercourse with children or family members).

Mores, like folkways, can change. For example, until about the late 1970s, 80 percent of U.S. babies born out of wedlock were given up for adoption. This

A hoarder? Or hobbyist?

rate has dropped to about 2 or 3 percent today because most unwed mothers, who are no longer stigmatized for having out-of-wedlock babies, keep their infants (Benokraitis 2011).

One of the most powerful types of mores are **taboos**, strong prohibitions of any act that is considered to be extremely offensive and forbidden because of social customs, religious or moral beliefs, or laws. Cannibalism, incest, and extramarital sex are examples of tabooed behavior. Taboos define and support cultural norms of moral behavior, but there are many violations. For instance, almost 90 percent of Americans say that having an extramarital affair is morally wrong, but over a lifetime, 21 percent of men and 12 percent of women admit to having had extramarital sex (see Benokraitis 2011 for a summary of these studies).

**mores** norms that members of a society consider very important because they maintain moral and ethical behavior.

**taboos** strong prohibitions of any act that is considered to be extremely offensive and forbidden because of social customs, religious or moral beliefs, or laws.

**laws** formal rules for behavior that are defined by a political authority that has the power to punish violators.

## Laws

The most rigid norms are **laws**, formal rules for behavior that are defined by a political authority that has the power to punish violators. Unlike folkways and most mores, laws are deliberate, formal, "precisely specified in written texts," and "enforced by a specialized bureaucracy," usually police and courts (Hechter and Opp 2001: xi).

Laws change over time. Sixteen states now allow the sale of marijuana for medical use, and others are considering such legislation. Also, every state has legalized gambling such as a lottery, casinos, and racetrack betting (Bland 2010). It's not clear whether

A AND E/THE KOBAL COLLECTION

U.S. laws about marijuana usage and gambling have changed because of changing mores, states' enhancing their budgets by taxing marijuana and gambling, casinos' providing jobs for the unemployed, or a combination of these reasons.

Laws also vary across societies. Judges in Iran can order public floggings of young men accused of drinking alcohol, distributing Western music, or being alone with women who are not their relatives. Singapore, where chewing gum is now legal after a 12-year ban, requires citizens who want to chew gum to submit their names and ID cards to the government (Anderson 2001; Knickerbocker 2005).

## Sanctions

Most people conform to norms because of **sanctions**, rewards for good or appropriate behavior and/or penalties for bad or inappropriate behavior. Children learn norms through both *positive* sanctions (praise, hugs, smiles, new toys) and *negative* sanctions (frowning, scolding, spanking, withdrawing love) (see Chapter 4).

Negative sanctions vary in the degree of punishment. When we violate folkways, the sanctions are relatively mild: gossip, ridicule, exclusion from a group. If you don't bathe or brush your teeth, you may not be invited to parties because others will see you as crude, but not sinful or evil. The sanctions for violating mores and some laws can be severe: loss of employment, expulsion from college, whipping, torture, banishment, imprisonment, and even execution. The sanctions are usually harsh because the unacceptable behavior threatens the moral foundations of a society. The public floggings in Iran punish offenders for their "immorality" and send a warning to others to follow the rules.

Sanctions aren't always consistent, despite universally held norms. In the United States, someone who can hire a good attorney might receive a lighter penalty for a serious crime. In some cultures, young girls and women may be lashed for engaging in premarital sex, but men aren't punished at all for the same behavior, even though it violates Islamic law (Seager 2009). Also, laws are sometimes enforced selectively or not at all. In the United States and elsewhere, for example, offenders who violate laws against discrimination, sexual harassment, and domestic violence are rarely prosecuted or punished (see Chapters 7 and 12).

At the 2009 MTV Video Music Awards, country singer Taylor Swift received an award for best female video. Rapper Kanye West stormed the stage, grabbed Swift's microphone, and told the audience that the award should have been given to Beyoncé, a popular singer and actress. Which types of norms did West violate—folkways, mores, or laws?

# 3-3 Some Cultural Similarities

You've seen that there's considerable diversity across societies in symbols, language, values, and norms. There are also some striking similarities across and within cultures because of cultural universals, ideal versus real culture, ethnocentrism, and cultural relativism.

## 3-3a Cultural Universals

**Cultural universals** are customs and practices that are common to all societies. Anthropologist George Murdock and his associates studied hundreds of societies and compiled a list of 88 categories that they found among all cultures (see *Table 3.2* for some examples).

There are many cultural universals, but specific behaviors vary across cultures, from one group to another in the same society, and over time. For example, all societies have food taboos, but specifics about what people ought and ought not to eat differ across societies. About 75 percent of the world's people eat insects. In Thailand, locusts, crickets, silkworms, grasshoppers, ants, and other insects are a big part of the diet (Stolley and Taphaneeyapan 2002). Vietnamese restaurants offer cat on the menu, and many poor people in rural China eat dog meat because they can't

TABLE 3.2

## Some Cultural Universals

| | | | |
|---|---|---|---|
| Athletics | Games | Kin terminology | Property rights |
| Cooking | Gift giving | Language | Religious rituals |
| Courtship | Greetings | Magic | Sexual restrictions |
| Dancing | Hairstyles | Marriage | Status differentiation |
| Division of labor | Housing | Medicine | |
| Food taboos | Inheritance rules | Music | |

Source: Based on Murdock 1945.

**ideal culture** the beliefs, values, and norms that people in a society say they hold or follow.

**real culture** the actual everyday behavior of people in a society.

**ethnocentrism** the belief that one's culture and way of life are superior to those of other groups.

afford poultry, pork, and beef. Increasingly, however, when individual incomes increase, many people treat their dogs as pets and family members rather than as nutritional food sources (Masis 2010; Wan 2011).

### 3-3b Ideal Versus Real Culture

The **ideal culture** of a society comprises the beliefs, values, and norms that people say they hold or follow. In every culture, however, these standards differ from the society's **real culture**, or people's actual everyday behavior. For example, Americans say that they love and cherish their children, but every year hundreds of thousands of children experience abuse and neglect on a daily basis. Indeed, 82 percent of people who abuse their children are parents (U.S. Department of Health and Human Services 2011; see also Chapter 12). Thus, our ideal culture and our actual behavior are often inconsistent.

AP Photo/Dario Lopez-Mills

### 3-3c Ethnocentrism and Cultural Relativism

**Ethnocentrism** is the belief that one's culture and way of life are superior to those of other groups. Because people internalize their culture and tend to see their way of life as the best and the most natural, they often dismiss groups with differing attitudes and behavior as inferior, wrong, or backward. How many of you, for example, are repelled at the thought of eating insects, cats, or dogs?

Some of my black students argue that it's impossible for African Americans to be ethnocentric because they suffer so much prejudice and discrimination. *Any* group can be ethnocentric, however (Rose 1997). An immigrant from Nigeria who assumes that all native-born African Americans are lazy and criminal is just as ethnocentric as a native-born African American who assumes that all Nigerians are arrogant and "uppity."

Ethnocentrism can be functional. Pride in one's country promotes loyalty and cultural unity. When children learn their country's national anthem and customs, they have a sense of belonging. Ethnocentrism also reinforces conformity and maintains stability. Members of a society become committed to their particular values and customs, and transmit them to the next generation. As a result, life is (generally) orderly and predictable.

About 75 percent of the world's people consume insects, which are high in protein, vitamins, and fiber, and usually low in fat. More than two out of every three American adults are overweight or obese, the highest percentage on the planet (Flegal et al. 2010). Would we be healthier if we ate insects instead of gulping down hamburgers and french fries?

**cultural relativism** the belief that no culture is better than another and that a culture should be judged by its own standards.

**subculture** a group of people whose distinctive ways of thinking, feeling, and acting differ somewhat from those of the larger society.

Ethnocentrism has its benefits, but it's usually dysfunctional because viewing others as inferior generates hatred, discrimination, and conflict. Many recent wars, such as those in the former Yugoslavia and Rwanda and the ongoing battles between Palestinians and Israelis reflect religious, ethnic, or political intolerance toward others (see Chapter 13). Thus, ethnocentrism discourages intergroup understanding and cooperation.

The opposite of ethnocentrism is **cultural relativism**, a belief that no culture is better than another and that a culture should be judged by its own standards. Most Japanese mothers stay home with their children, whereas many American mothers are employed outside the home. Is one practice better than another? No. Because Japanese fathers are expected to be the breadwinners, it's common for many Japanese women to be homemakers. In the United States, in contrast, some mothers choose to stay home, but economic recessions and the loss of many high-paying jobs have catapulted many women into the job market to help support their families (see Chapters 10, 11, and 12). Thus, Japanese and American parenting may be different, but one culture isn't better or worse than the other.

During a recent trip to India, First Lady Michelle Obama chose an ensemble by designers who use traditional Indian fabrics (Givhan 2010). How was her wardrobe decision an example of cultural relativism?

JIM WATSON/AFP/Getty Images

# 3-4 Some Cultural Variations

There is considerable cultural variation not only *across* societies but also *within* the same society. *Subcultures* and *countercultures* account for some of the complexity within a society.

## 3-4a Subcultures

A **subculture** is a group of people whose distinctive ways of thinking, feeling, and acting differ somewhat from those of the larger society. A subculture is part of the larger, dominant culture but has its own particular values, beliefs, perspectives, lifestyle, or language. Members of subcultures often live in the same neighborhoods, associate with each other, have close personal relationships, and marry others who are similar to themselves.

Whether we realize it or not, most of us are members of numerous subcultures. Subcultures reflect a variety of characteristics, interests, or activities:

- *Ethnicity* (Irish, Mexican, Vietnamese)
- *Religion* (evangelical Christians, atheists, Mormons)
- *Politics* (Maine Republicans, Southern Democrats, independents)
- *Sex and gender* (heterosexual, lesbian, transgendered)
- *Age* (older widows, middle schoolers, college students)
- *Occupation* (surgeons, teachers, prostitutes, truck drivers)
- *Music and art* (jazz aficionados, country music buffs, art lovers)
- *Social class* (billionaires, working poor, middle class)
- *Recreation* (mountain bikers, poker players, dancers)

Some subcultures retreat from the dominant culture to preserve their beliefs and values. The Amish, for example, have created self-sustaining economic units, travel locally by horse and buggy, conduct religious services in their homes, make their own clothes, and generally shun modern conveniences such as electricity and phones. Despite their self-imposed isolation, the Amish have been affected by the dominant U.S. culture. As farming became less self-sustaining, many began small businesses that produce quilts, leather products, and baked goods for tourists. Others have found steady wages in assembling recreational vehicles, making furniture, and in construction (Boak 2009; Mertens 2009).

To fit in, members of most subcultures adapt to the larger society but maintain some of their traditional customs. Among Ghanaian immigrants, funerals are lavish celebrations rather than places of sadness. The funerals are "all-night affairs with open bars and window-rattling music." As in Ghana, the guests need not know the deceased or even the family. Many people attend simply to meet other Ghanaians and to "cut loose on the dance floor" (Dolnick 2011: A1).

In many instances, subcultures arise because of technological or other societal changes. After the emergence of the Internet, for example, subcultures arose that identified themselves as hackers, techies, or computer geeks.

## 3-4b Countercultures

Unlike a subculture, a **counterculture** deliberately opposes and consciously rejects some of the basic beliefs, values, and norms of the dominant culture. Countercultures usually emerge when people believe they can't achieve their goals within the existing society. As a result, such groups develop values and practices that run counter to those of the dominant society. Some countercultures are small and informal, but others, such as religious militants, have millions of members and are highly organized (see Chapter 11).

Most countercultures don't engage in illegal activities. Some, however, are violent and extremist, such as the 1,002 active hate groups across the United States (see *Figure 3.3*) that intimidate ethnic groups and gays. Some skinheads have murdered gays; antigovernment militia adherents bombed a federal building in Oklahoma, killing dozens of adults and children; and antiabortion advocates have murdered physicians and bombed abortion clinics. About two-thirds of terrorist acts in the United States are conducted by non-Islamic extremists. The FBI is particularly concerned by this violence because it's performed by individuals who are not affiliated with any countercultural movement and are, therefore, hard to detect (Masters 2011).

## 3-4c Multiculturalism

**Multiculturalism** (sometimes called **cultural pluralism**) refers to the coexistence of several cultures in the same geographic area, without one culture dominating another. Many applaud multiculturalism because it encourages intercultural dialogue (e.g., U.S. schools offering programs and courses in African American, Latino, Arabic, and Asian studies). Supporters hope that emphasizing multiculturalism—especially in schools and the workplace—will decrease ethnocentrism, racism, sexism, and other forms of discrimination.

Despite its benefits, not everyone is enthusiastic about multiculturalism. Not learning the language of the country where one lives and works, for instance, can be isolating and create on-the-job miscommunication, tension, and conflict. Some also believe that

> **counterculture** a group of people who deliberately oppose and consciously reject some of the basic beliefs, values, and norms of the dominant culture.

> **multiculturalism (cultural pluralism)** the coexistence of several cultures in the same geographic area, without one culture dominating another.

*The Simpsons* was once considered a countercultural icon. It's now been on the air for more than two decades and is the longest running sitcom in the history of American television. Can *The Simpsons* still be considered countercultural, or is it now part of the dominant culture?

20TH CENTURY FOX / THE KOBAL COLLECTION / GROENING, MATT

## FIGURE 3.3
## Active Hate Groups in the United States: 2010

Source: Southern Poverty Law Center, http://www.splcenter.org/get-informed/hate-map (accessed July 20, 2011). Used with permission.

multiculturalism can destroy a country's national traditions, heritage, and identity because ethnic and religious subcultures may not support the dominant culture's values and beliefs. For these reasons, some European political leaders have rejected multiculturalism, especially in integrating Muslims, as a realistic goal (Marquand 2011).

### 3-4d Culture Shock

People who travel to other countries often experience **culture shock**—a sense of confusion, uncertainty, disorientation, or anxiety that accompanies exposure to an unfamiliar way of life or environment. Familiar cues about how to behave are missing or have a different meaning. Culture shock affects some people more than others, but the most stressful changes involve differences of food, clothes, punctuality, ideas about what offends people, language, the general pace of life, and a lack of privacy (Spradley and Phillips 1972; Pedersen 1995).

To some degree, everyone is *culture bound* because we've internalized cultural norms and values. For example, an American journalist who works in Mexico City says that perpetual lateness is common: A child's birthday party may start 2 or 3 hours late, a wedding may begin an hour after the announced time, and interviews with top Mexico City officials may be up to 2 hours late or the official never arrives. These norms are changing because of the growth of Mexico's global economy, but the more relaxed attitude toward time in Mexico City (and elsewhere) can be a culture shock to someone from the United States who has grown up in a "clock-obsessed" culture (Ellingwood 2009).

## 3-5 Popular Culture

Sociologists use the term *high culture* to describe the cultural expression of a society's elite or highest social classes. Examples include opera, ballet, paintings, and classical music. In contrast, **popular culture** refers to beliefs, practices, activities, and products that are widely shared within a population in everyday life.

Popular culture includes television, music, magazines, radio, advertising, sports, hobbies, fads, fashions, and movies, as well as the food we eat, the gossip we share, and the jokes we pass along to others. People produce and consume popular culture: They are not simply passive recipients but influence popular culture by what they buy, how they spend their leisure time, and how they express themselves.

### 3-5a The Impact of Popular Culture

Popular culture can have both positive and negative effects on our everyday lives. Many people don't believe everything they read or see on television or online but weigh the merit and credibility of much of the content. Most of us are highly influenced, nonetheless, by a popular culture that is largely controlled and manipulated by newspapers and magazines, television, movies, music, and ads (see Chapter 4). These **mass media**, or forms of communication designed to reach large numbers of people, have enormous power in shaping public perceptions and opinions. Let's look at a few examples.

# Does Pop Culture Make Kids Fat?

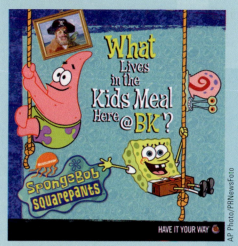

AP Photo/PRNewsFoto

The prevalence of obesity among American children ages 2 to 19 years has more than tripled, increasing from 5 percent in the 1960s to almost 17 percent by 2008 (Ogden and Carroll 2010). There are many reasons for the increase, but physicians and researchers lay much of the blame on popular culture, especially the advertising industry. Much of the advertising, particularly on television, uses cartoon characters such as SpongeBob SquarePants and Scooby-Doo to sell sugary cereals, cookies, candy, and other high-calorie snacks. In contrast, many European countries forbid advertising on children's television programs. After considerable pressure from the federal government, the nation's largest food companies have agreed to curb advertising. Some experts are skeptical, however, because similar efforts in the past to self-regulate haven't been very successful ("Firms Suggest Limits . . ." 2011; "Food Companies Propose . . ." 2011).

## Television

The top-rated television programs are what some critics call "trash TV" (such as sitcoms or reality shows like *Jersey Shore*), in which "the producers see nothing wrong with glorifying drunken idiocy and moral buffoonery in every episode" (Goldberg 2009). A bigger complaint is that the raw language (such as "bitch" and "douche"), violence, and sexual content that formerly were restricted to the 10 P.M. hour have migrated to earlier time slots when young children are watching (Wyatt 2009; see also Parents Television Council 2010).

## Advertising and Commercials

The average American is exposed to 3,000 ads per day and views at least 40,000 commercials on television per year (Kilbourne 1999; Strasburger et al. 2006). We are constantly deluged with advertising in newspapers and magazines, on television and radio, in movie theaters, on billboards and the sides of buildings, on public transportation, and on the Internet.

Many of my students, who claim that they "don't pay any attention to ads," come to class wearing branded apparel: Budweiser caps, Abercrombie shirts, Nike footwear. Advertising affects our self-image and self-esteem. As you'll see in later chapters, many women and men spend considerable time and money to meet ideal standards of femininity and masculinity—generated by ads—that are unrealistic and impossible to attain.

Considerable mass media content is basically marketing. Much of the content on television morning

MTV/THE KOBAL COLLECTION

If, as some critics contend, *Jersey Shore* corrupts our values and norms, why is this television show so popular?

shows and MTV has become "a kind of sophisticated infomercial" (O'Donnell 2007: 30). For example, a third of the content on morning shows is essentially selling something (a book, music, a movie, or another television program) that the corporation owns. One of the most lucrative alliances is between Hollywood and toy manufacturers (see Chapter 4). And, according to some scholars, the U.S. mass media have expanded cultural imperialism abroad.

### 3-5b Cultural Imperialism

In **cultural imperialism**, the cultural values and products of one society influence or dominate those of another. Many countries complain that U.S. cultural imperialism displaces authentic local culture and results in cultural loss.

The United States established the Internet, a global media network that uses mainly English and is heavily saturated with American advertising and popular culture. Iran's government has recently denounced Batman, Spider-Man, and Harry Potter toys as a form of "cultural invasion" that challenges the country's conservative and religious values. The curvaceous and often scantily clad Barbie dolls with peroxide-blond hair have been especially singled out as "destructive culturally and a social danger." However, many of the girls who watch foreign television and (illegal) satellite TV want the dolls (Peterson 2008: 4).

American fast-food restaurants are popular in many Asian countries. Because of greater competition from local fast-food chains, American companies now offer a large number of items geared specifically to Asian tastes (Chu 2010). At one of China's Pizza Huts, for example, the menu includes shrimp pizza, fried squid, and green tea ice cream. In Latin America, McDonald's sells "McMollettes," English muffins with refried beans, cheese, and salsa (York 2012).

## 3-6 Cultural Change and Technology

This section examines why cultures persist, and how and why they change. And what happens when technology changes faster than cultural values, laws, and attitudes?

### 3-6a Cultural Persistence: Why Cultures Are Stable

In many ways, culture is a conservative force. As you saw earlier, values, norms, and language are transmitted from generation to generation. Such **cultural integration**, or the consistency of various aspects of society, promotes order and stability. Even when new behaviors and beliefs emerge, they commonly adapt to existing ones. Recent immigrants, for example, may speak their native language at home and celebrate their own holy days, but they are

expected to gradually absorb the host country's values, obey its civil and criminal laws, and adopt its national language. Life would be chaotic and unpredictable without such cultural integration.

### 3-6b Cultural Dynamics: Why Cultures Change

Cultural stability is important, but all societies change over time. Some of the major reasons for cultural change include diffusion, invention and innovation, discovery, and external pressures.

#### Diffusion

A culture may change because of *diffusion,* the process through which components of culture spread from one society to another. Such borrowing may have occurred so long ago that the members of a society consider their culture to be entirely their own creation. However, anthropologist Ralph Linton (1964) has estimated that 90 percent of the elements of any culture are a result of diffusion (see *Figure 3.4*).

Diffusion can be direct and interpersonal, occurring through trade, tourism, immigration, intermarriage, or the invasion of one country by another. Diffusion can also be indirect and largely impersonal, as in the Internet transmissions that zip around the world.

## Sociology in Your Life

Is social networking a global phenomenon? And what factors affect Internet usage? Go to Sociology CourseMate at CengageBrain.com to find out.

FIGURE 3.4
## The 100 Percent American

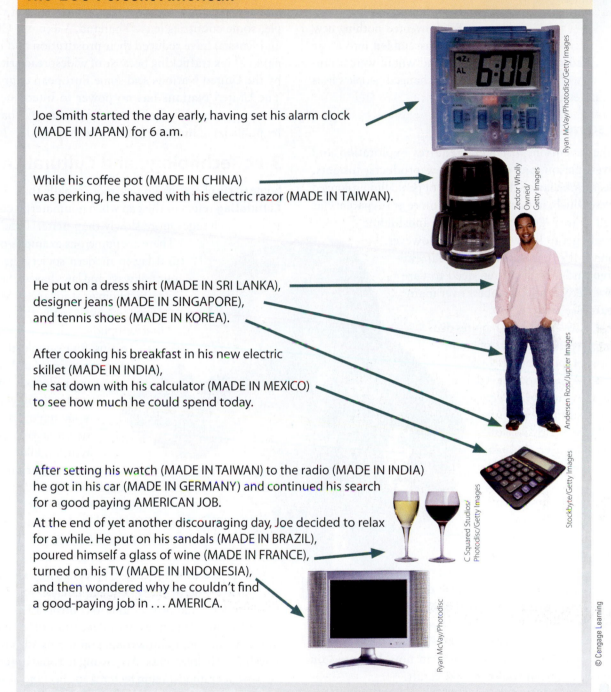

Joe Smith started the day early, having set his alarm clock (MADE IN JAPAN) for 6 a.m.

While his coffee pot (MADE IN CHINA) was perking, he shaved with his electric razor (MADE IN TAIWAN).

He put on a dress shirt (MADE IN SRI LANKA), designer jeans (MADE IN SINGAPORE), and tennis shoes (MADE IN KOREA).

After cooking his breakfast in his new electric skillet (MADE IN INDIA), he sat down with his calculator (MADE IN MEXICO) to see how much he could spend today.

After setting his watch (MADE IN TAIWAN) to the radio (MADE IN INDIA) he got in his car (MADE IN GERMANY) and continued his search for a good paying AMERICAN JOB.

At the end of yet another discouraging day, Joe decided to relax for a while. He put on his sandals (MADE IN BRAZIL), poured himself a glass of wine (MADE IN FRANCE), turned on his TV (MADE IN INDONESIA), and then wondered why he couldn't find a good-paying job in . . . AMERICA.

Ryan McVay/Photodisc/Getty Images

Zedcor Wholly Owned/Getty Images

Andersen Ross/Jupiter Images

Stockbyte/Getty Images

C Squared Studios/Photodisc/Getty Images

Ryan McVay/Photodisc

© Cengage Learning

## Invention and Innovation

Cultures change because people are continually finding new ways of doing things. *Invention*, the process of creating new things, brought about products such as toothpaste (invented in 3000 B.C.), eyeglasses (262 A.D.), flushable toilets (the sixteenth century), fax machines (1843—that's right, invented in 1843!), credit cards (1920s), computer mouses (1964), Post-It notes (1980), and DVDs (1995).

*Innovation*—turning inventions into mass-market products—also sparks cultural changes. An innovator

Tim Robberts/Stone/Getty Images

is someone determined to market an invention, even if it's someone else's good idea. For example, Henry Ford invented nothing new but "assembled into a car the discoveries of other men behind whom were centuries of work," an innovation that changed people's lives (Evans et al. 2006: 465).

## Discovery

Like invention, *discovery* requires exploration and investigation, and results in new products, insights, ideas, or behavior. The discovery of penicillin prolonged lives, which, in turn, meant that more grandparents (as well as great-grandparents) and grandchildren would get to know each other. However, longer life spans also mean that children and grandchildren need to care for elderly family members over many years (see Chapter 12).

Discovery usually requires dedicated work and years of commitment, but some discoveries occur by chance, called the *serendipity effect*. For example, George de Mestral, a Swiss electrical engineer, was hiking through the woods and was annoyed by burrs that clung to his clothing. Why were they so difficult to remove? A closer examination showed that the burrs had hooklike arms that locked into the open weave of his clothes. The discovery led de Mestral to invent a hook-and-loop fastener. His invention, Velcro—derived from the French words *velour* (velvet) and *crochet* (hooks)—can now be found on everything from wallets to spacecraft.

© Henrik Lehnerer/Shutterstock

## External Pressures

External pressure for cultural change can take various forms. In its most direct form—war, conquest, or colonization—the dominant group uses force or the threat of force to bring about cultural change in other groups. When the Soviet Union invaded and took over many small countries (such as Lithuania, Latvia, Estonia, Ukraine, and Armenia) after World War II, it forbade citizens to speak their native languages, banned traditions and customs, and turned churches into warehouses.

Pressures for change can also be indirect. For example, some countries (e.g., Thailand, Vietnam, China, and Russia) have reduced their prostitution and international sex trafficking because of widespread criticism by the United Nations and some European countries. The United Nations has no power to intervene in a country's internal affairs but can embarrass nations by publicizing human rights violations (Farley 2001).

## 3-6c Technology and Cultural Lag

Some parts of culture change more rapidly than others. **Cultural lag** refers to the gap when nonmaterial culture changes more slowly than material culture.

There are numerous examples of cultural lag in modern society, because our folkways and laws haven't kept up with technological advances. Martin Cooper, 82, who made the world's first cell phone call in 1973, complains, for example, that cell phones have unleashed more rudeness: "What can be so urgent that you have to look down at your phone in the middle of a dinner conversation with people who matter to you? That's the equivalent of if a friend or loved one was talking to you, and in the middle of their sentence you pick up a newspaper and start reading it . . ." (Greene 2011). And as you'll see in several later chapters, technology is moving faster than laws to regulate Internet intellectual property and privacy and to control online crime.

Cultural lag often creates uncertainty, ambiguity about what's right and wrong, conflicting values, and a feeling of helplessness. According to some observers, we both fear and worship technology, become obsessed with gadgets (like computers) even though they take up much of our time, don't deal with ethical issues raised by biotechnology (such as the implications of cloning embryos), and rely on technology as quick fixes: "We want to believe that any given solution is only a purchase away" (Naisbitt et al. 1999: 3). Cultural lags have always existed and will continue to do so in the future. They create anxiety, but also offer opportunities to enhance culture.

## TABLE 3.3

# Sociological Explanations of Culture

| THEORETICAL PERSPECTIVE | FUNCTIONALIST | CONFLICT | FEMINIST | SYMBOLIC INTERACTIONIST |
|---|---|---|---|---|
| **Level of Analysis** | **Macro** | **Macro** | **Macro and Micro** | **Micro** |
| **Key Points** | • Similar beliefs bind people together and create stability.<br>• Sharing core values unifies a society and promotes cultural solidarity. | • Culture benefits some groups at the expense of others.<br>• As powerful economic monopolies increase worldwide, the rich get richer and the rest of us get poorer. | • Women and men often experience culture differently.<br>• Cultural values and norms can increase inequality because of sex, race/ethnicity, and social class. | • Cultural symbols forge identities (that change over time).<br>• Culture (such as norms and values) helps people merge into a society despite their differences. |
| **Examples** | • Speaking the same language (English in the United States) binds people together because they can communicate with one another, express their feelings, and influence one another's attitudes and behaviors. | • Much of the English language reinforces negative images about gender ("slut"), race ("honky"), ethnicity ("jap"), and age ("old geezer") that create inequality and foster ethnocentrism. | • Using male language (such as "Congress-man," "fireman," and "chairman") conveys the idea that men are superior to and dominant over women, even when women have the same jobs. | • People can change the language they create as they interact with others. Many Americans now use "police officer" instead of "policeman," and "single person" instead of "bachelor" or "old maid." |

# 3-7 Sociological Perspectives on Culture

**W**hat is the role of culture in modern society? And how does culture help us understand ourselves and the world around us? Functionalist, conflict, feminist, and interactionist scholars offer different answers to these and other questions, but all provide important insights about culture. *Table 3.3* summarizes these perspectives.

## 3-7a Functionalism

Functionalists focus on society as a system of interrelated parts (see Chapter 1). Similarly, in their analysis of culture, functionalists emphasize the social bonds that attach people to society.

### Key Points

For functionalists, culture is a cement that binds society. As you saw earlier, norms and values shape our lives, provide guideposts for our everyday behavior, and promote cultural integration and societal stability.

Especially in countries such as the United States that have high immigration rates, cultural norms and values help newcomers adjust to the host society.

For functionalists, all societies have similar strategies for meeting human needs. These cultural universals, such as religious rituals, may play out differently in different societies, but all known societies have religious rituals.

Functionalists also note that culture can be dysfunctional. For example, when subcultures such as the Amish refuse to immunize their children, some diseases can surge in the community (Brown 2005). Also, various countercultures (such as paramilitary groups) can create chaos by bombing federal buildings and killing or injuring hundreds of people.

### Critical Evaluation

Functionalism is important in showing that shared norms and values create solidarity and stability in a culture. In emphasizing culture's role in meeting people's daily needs, however, functionalism often overlooks diversity and social change. For example, a number of influential functionalists have proposed that immigration should be restricted because it dilutes

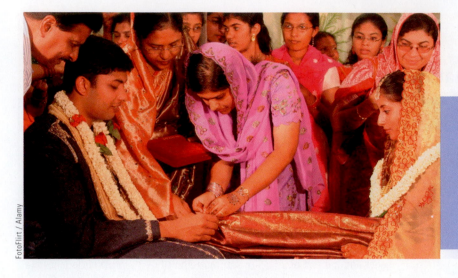

Muslims in the Kashmir region of India surround a groom as he goes to his in-laws' house to fetch his bride. For functionalists, such customs and rituals reinforce cultural identity and bind people with similar beliefs.

shared U.S. values, overlooking the many contributions that newcomers make to society (see Chapter 10).

## 3-7b Conflict Theory

Unlike functionalists, conflict theorists argue that culture can generate considerable inequality instead of unify society. Because the rich and powerful determine economic, political, educational, and legal policies for their own benefit and control the mass media, the average American has little power to change the culture—whether it's low wages, unpopular wars, or corporate corruption (see Chapters 8 and 11).

### Key Points

Conflict theorists maintain that many cultural values and norms benefit some members of society more than others. We are taught to value hard work and success, for example, but who benefits from the average worker's efforts? As you'll see in Chapter 11, U.S. taxpayers paid billions of dollars to bail out faltering financial industries because the top executives made bad decisions based largely on greed. After the bailouts, the corporate executives and many of their staff used the money to give themselves higher salaries and bonuses than ever before. Thus, those at the top of the socioeconomic ladder can benefit by violating mores, including honesty, that other Americans are expected to observe.

Conflict theorists also point out that technology benefits primarily the rich. For example, the C-Leg is a prosthetic leg that allows users to go down slopes or staircases, to run, or to stroll on a hilly path. Because the C-Leg costs about $50,000 and is typically not covered by medical insurance, only affluent people can take advantage of this technology (Austen 2002; see also Chapter 14).

### Critical Evaluation

According to some critics, conflict theorists place too much emphasis on societal discord and downplay a culture's benefits. For instance, cultural values integrate members of a society and decrease divisiveness.

According to conflict theorists, mass media conglomerates control popular culture and promote the purchase of goods and services that most people don't need and can't afford (such as big screen TVs, the most recent smartphones, and designer clothes). Critics counter that people aren't mindless puppets but can make choices, including not living on credit. Thus, conglomerates shouldn't be blamed for people's irrational behavior.

## 3-7c Feminist Theories

Feminist scholars, who use both macro and micro approaches, agree with conflict theorists that material culture, in particular, creates considerable inequality, but they focus on gender differences. Feminist scholars are also more likely than other theorists to examine multicultural variations across groups.

### Key Points

Gender affects our cultural experiences. When media portrayals of women are absent or stereotypical, we get a distorted view of reality. Feminist scholars also emphasize that subcultures—for example, female students or single mothers—may experience culture differently than their male counterparts do because they typically have fewer resources, such as income and power (see Chapters 9 and 11).

Unequal access to resources often results in women having fewer choices than men, and living under laws and customs that subordinate women. In the United States, for example, an out-of-wedlock birth typically impoverishes women but not men. In Japan, where rape is often kept hush-hush, police and judges rarely take the offense seriously (Makino 2009). And in many Islamic societies, men dictate the appropriateness of women's attire and behavior (Heath 2008; Zahedi 2008).

### Critical Evaluation

Feminist analyses expand our understanding of cultural components that other theoretical perspectives ignore or gloss over. Like conflict theorists, however, feminist theorists often emphasize discord rather than how culture integrates women and men into society. Another weakness is that feminist scholars tend to downplay social class inequality by focusing on the cultural experiences of low-income and minority women but not of their male counterparts.

## 3-7d Symbolic Interactionism

Unlike functionalists and conflict theorists, symbolic interactionists examine culture through micro lenses. They are most interested in understanding how people create, maintain, and modify culture.

### Key Points

Interactionists explain how culture influences our everyday lives. Language, you'll recall, shapes our views and behavior. People within and across societies create a variety of symbols that may change over time. Technology has also changed many people's communication patterns. Some complain, for example, that texting dominates our lives. A much viewed YouTube clip shows the bride pulling out her cell phone from under her veil and texting with one hand as she walks down the aisle with her father (Pflaumer 2011).

Symbolic interactionists also note that our values and norms, like other components of culture, aren't

Is the L sign for "loser" an effective sanction? Have you ever used this symbol? Or had it directed at you?

RTimages/Shutterstock.com

superimposed by some unknown external force. Instead, as people construct their perception of reality, they create, change, and reinterpret values and norms through interaction with others. Peer pressure, for example, is often effective in discouraging inappropriate language or behavior.

### Critical Evaluation

Micro approaches are useful in understanding what culture means to people and how these meanings differ across societies. However, symbolic interactionists don't offer a systematic framework that explains exactly how people create and shape culture or develop shared meanings of reality. Why, for instance, are some of us less polite than others even though we share the same cultural values and norms? Another weakness is that interactionists don't address the linkages between culture and institutions. For example, it's important to recognize how language bonds people together, but interactionists say little about how organized groups (such as the English-only movement) try to maintain control over language to promote their beliefs.

### Study Tools

Ready to study? Go to **Sociology CourseMate** at **www.cengagebrain.com** to complete practice quizzes, review flashcards, watch videos, and more.

SOC

Have you ever thought about how you became the person you are today?

# Socialization

Two of MTV's most popular reality shows are *16 and Pregnant* and *Teen Mom*. The producers say that the goal of both programs is to decrease teen pregnancy by showing the harmful effects of not using protection or birth control and the resulting struggles in raising a baby. Critics argue that the shows glamorize teen pregnancy and motherhood, make the teen moms instant celebrities, and ignore the long-term negative impact on the mothers' and children's healthy social development (Thompson 2010). Both sides are talking about **socialization**, the lifelong process of social interaction in which the individual acquires a social identity and ways of thinking, feeling, and acting that are essential for effective participation in a society.

> **socialization** the lifelong process of social interaction in which the individual acquires a social identity and ways of thinking, feeling, and acting that are essential for effective participation in a society.

## Key Topics

In this chapter, we'll explore the following topics:

## 4-1 Socialization: Its Purpose and Importance

Socialization is critical in all societies. To understand why, let's begin by looking at the purpose of socialization.

### 4-1a What Is the Purpose of Socialization?

Socialization—from childhood to old age—can be relatively smooth or very bumpy, depending on factors such as age, sex, race/ethnicity, and social class. Generally, however, socialization has four key functions that range from providing us with a social identity to transmitting culture to the next generation.

#### Socialization Establishes Our Social Identity

Have you ever thought about how you became the person you are today? Sociology professors sometimes ask their students to give 20 answers to the question "Who am I?" (Kuhn and McPartland 1954). How would you respond? You would probably include a variety of descriptions such as college student, single or married, female or male, and son or daughter. Your answers would show a sense of your *self* (a concept we'll examine shortly). You are who you are largely because of socialization.

#### Socialization Teaches Us Role Taking

Why do you act differently in class than with your friends? Because we play different roles in different settings. A *role* is the behavior expected of a person in a particular social position (see Chapter 5). The way we interact with a parent is typically very different from the way we talk to an employer, a child, or a professor. We all learn appropriate roles through the socialization process.

### What do you think?

I'm always the same person, no matter where I am or who's around.

1 2 3 4 5 6 7

strongly agree · strongly disagree

## Socialization Controls Our Behavior

In learning appropriate roles, we absorb values and a variety of rules about how we should (and shouldn't) interact in everyday situations. If we follow the rules, we're usually rewarded or at least accepted. If we break the rules, we may be punished. By teaching us to conform to societal expectations, socialization controls our behavior and makes life more orderly and predictable.

We conform because we've internalized societal norms and values. **Internalization** is the process of learning cultural behaviors and expectations so deeply that we accept them without question. Obeying laws, paying bills on time, and respecting teachers are examples of internalized behaviors.

## Socialization Transmits Culture to the Next Generation

Each generation passes its culture on to the next generation. The culture that is transmitted, as you saw in Chapter 3, includes language, beliefs, values, norms, and symbols.

## 4-1b Why Is Socialization Important?

Social isolation can be devastating. An example of its negative effects is Genie. When Genie was 20 months old, her father decided she was "retarded," locked her away in a back room with the curtains drawn and the door shut, and put her in a wire mesh cage. She had almost no opportunity to overhear any conversation between others in the house. Genie's father frequently beat her with a wooden stick and never spoke to her. When she was discovered in 1970 at age 13, a psychiatrist described Genie as "unsocialized, primitive, and hardly human." Except for high-pitched whimpers, she never spoke. Genie had little bowel control, experienced rages, rubbed her face frantically, and tried to hurt herself. After living in a rehabilitation ward and a foster home, she learned to eat normally, was toilet-trained, and gradually developed a vocabulary, but her language use never progressed beyond that of a 3- or 4-year-old (Curtiss 1977).

| TABLE 4.1 | |
|---|---|
| **The Nature–Nurture Debate** | |
| **Nature** | Human development is . . . |
| | Innate |
| | Biological, physiological |
| | Due largely to heredity |
| | Fairly fixed |
| **Nurture** | Human development is . . . |
| | Learned |
| | Psychological, social, cultural |
| | Due largely to environment |
| | Fairly changeable |

© Cengage Learning

The research on children who are isolated (such as Genie) or institutionalized (such as orphans) demonstrates that socialization is critical to our development. Talking, eating with utensils, and controlling our bowel movements do not come naturally. Instead, we learn to do all of these things beginning in infancy. When children are deprived of interaction with other people, they don't develop the characteristics that most of us see as normal and human.

# 4-2 Nature and Nurture

Biologists focus on the role of heredity (or genetics) in human development. In contrast, most social scientists, including sociologists, underscore the importance of learning, socialization, and culture. This difference of opinion is often called the *nature–nurture debate* (see *Table 4.1*). Many researchers believe that heredity and environment intersect in shaping the developing person, but there is still a tendency to emphasize either nature or nurture.

© Angela Waye/Shutterstock

## 4-2a How Important Is Nature?

Some scientists propose that biological factors—especially the brain—play an important role in our development. For example, some research suggests that when activity in the prefrontal cortex region of the brain increases, people tend to be more anxious, irritable, angry, and unpleasant (Vedantam 2002). It's not clear, however, why many people have high brain activity yet remain calm, friendly, and pleasant.

## How Biology Affects Behavior

John Money, a highly respected medical psychologist at Johns Hopkins University Hospital, published numerous articles and books in which he maintained that gender identity is not firm at birth but is determined as much by culture and nurture as by genes and hormones (Money and Ehrhardt 1972). In 1963, twin boys were being circumcised when the penis of one of the infants, David Reimer, was accidentally burned off. Encouraged by John Money, the parents agreed to raise David as "Brenda." The child's testicles were removed, and surgery to construct a vagina was planned. Money reported that the twins were growing into happy, well-adjusted children, setting a precedent for sex reassignment as the standard treatment for 15,000 newborns with similarly injured genitals (Colapinto 1997, 2001).

In the mid-1990s, a biologist and a psychiatrist followed up on Brenda's progress and concluded that the sex reassignment had not been successful. Almost from the beginning, Brenda refused to be treated like a girl. When her mother dressed her in frilly clothes as a toddler, Brenda tried to rip them off. She preferred to play with boys and stereotypical boys' toys such as machine guns. People in the community said that she "looks like a boy, talks like a boy." Brenda had no friends, and no one would play with her (Diamond and Sigmundson 1997).

When she was 14, Brenda rebelled and stopped living as a girl: She refused to wear dresses, urinated standing up, turned down vaginal surgery, and decided she would either commit suicide or live as a male. When his father finally told David the true story of his birth and sex change, David recalls that "all of a sudden everything clicked. For the first time things made sense and I understood who and what I was" (Diamond and Sigmundson 1997: 300).

David had a mastectomy (breast removal surgery) at the age of 14 and underwent several operations to reconstruct a penis. At age 25, he married an older woman and adopted her three children. At age 38, he committed suicide. Most suicides have multiple motives, but some speculated that David committed suicide because of the "physical and mental torments he suffered in childhood that haunted him the rest of his life" (Colapinto 2004: 96). David's experience suggests to some scientists that nature outweighs nurture in shaping a person's gender identity.

## Sociology in Your Life

For more on the Harlow experiment, go to Sociology CourseMate at CengageBrain.com.

**EMOTIONAL ATTACHMENT**

Nina Leen/Time Life Pictures/Getty Images

## Harlow Studies

In the early 1960s, psychologists Margaret and Harry Harlow (1962) conducted several studies on infant monkeys. In one group, a "mother" made of terry cloth provided no food, while a "mother" made of wire did so through an attached baby bottle containing milk. In another group, the cloth mother provided food but the wire mother didn't. Regardless of which mother provided milk, when both groups of monkeys were frightened, they clung to the cloth mother. The Harlows concluded that physical contact and comfort were more important to the infant monkeys than nourishment. Since then, some sociologists have cited the Harlow studies to argue that emotional attachment may be even more critical than food for human infants. Do you see any problems with sociologists' generalizing the results of animal studies to humans?

© Cengage Learning

*David Reimer was raised as a girl, Brenda, until he was 14.*

REUTERS/Landov

## Sociobiology and Socialization

**Sociobiology** is a theoretical perspective that applies biological principles to explain the behavior of animals, including human beings. Sociobiologists argue, for example, that evolution and genes (nature) can explain why men are generally more aggressive than women. To ensure that their genes will be passed on to offspring, males (human and animal) have to prevail over their rivals. The intense competition often includes aggression and violence (Barash 2002).

Critics reject such sociobiological explanations of male violence. If men were innately aggressive, they would be equally violent across all societies. This isn't the case. The proportion of women who have ever suffered physical violence by a male partner varies considerably: 80 percent in Vietnam, 61 percent in Peru, 20 percent in the United States, and 13 percent in Japan (Chelala 2002; Rennison 2003; World Health Organization 2005). Such variations, sociobiology critics argue, reflect cultural norms and practices and other environmental factors (nurture) rather than biology or genetics (nature) (Chesney-Lind and Pasko 2004).

## 4-2b How Important Is Nurture?

Many sociologists agree that nature affects human development. They maintain, however, that nurture is more significant than nature because socialization and culture shape even biological inputs.

### How Behavior Affects Biology

Much research suggests that environment (nurture) influences children's genetic makeup (nature). For example, birth defects associated with prenatal alcohol exposure can occur in the first 3 to 8 weeks of pregnancy, before a woman even knows she's pregnant. A woman's single drinking "binge"—lasting four hours or more—can permanently damage an unborn child's brain. A man who drinks heavily may have genetically damaged sperm that also leads to birth defects. Thus, alcohol abuse (an environmental factor) by either a woman or a man can have a devastating, irreversible, and lifelong negative impact on a child (Twitchell et al. 2001; Hetherington et al. 2006).

In a growing field known as "fetal origins," scientists claim that the nine months of pregnancy constitute the most consequential period of our lives, permanently influencing the wiring of the brain, the functioning of organs such as the heart and liver, and behavior. That is, the kind and quantity of nutrition you received in the womb; the pollutants, drugs, and infections you were exposed to during gestation; your parents' health and eating habits; and your mother's stress level during pregnancy—all these factors have lasting effects in infancy, childhood, and adulthood (Begley 2010; Paul 2010).

Adult behavior can influence a child's biological makeup in other ways. Physical, psychological, or sexual abuse can impair the developing brain during childhood, and these changes can trigger disorders such as depression in adulthood (Teicher 2000). Thus, childhood mistreatment can blunt biological development and lead to behavioral and emotional problems.

WilleeCole/Shutterstock.com

Margaret Mead (1901–1978) conducted a number of field studies in the Pacific. Her work, especially her demonstration that gender roles and child-rearing practices differ in various cultures, helped to break down stereotypes about what is natural (or innate).

## Culture and Socialization

Sociologists often point to cross-cultural data to illustrate the importance of nurture. In a well-known study, anthropologist Margaret Mead (1935) observed three tribes—the Arapesh, Mundugumor, and Tchambuli—who lived close to one another in New Guinea.

She found three variations of gender roles. Among the Arapesh, both men and women nurtured their children. The men were cooperative and sensitive, and rarely engaged in warfare. The Mundugumor were just the opposite: Both men and women were competitive and aggressive, neither parent showed much tenderness toward their offspring, and both parents often used physical punishment to discipline their children. The Tchambuli reversed Western gender roles. The women were the economic providers; the men took care of the children, sat around chatting, and spent a lot of time decorating themselves for tribal festivities. Mead concluded that attributes long considered either masculine (such as being aggressive) or feminine (such as being emotional) are culturally, not biologically, determined.

## 4-2c Is the Nature–Nurture Debate Becoming Obsolete in Sociology?

You and your cousins, Erik and Joelle, share the same genes, but you're in college whereas they dropped out of high school. Why? A growing number of sociologists—who are calling for a "genetically informed sociology"—maintain that the nature *or* nurture debate is becoming outdated in understanding socialization processes and outcomes (Ledger 2009). Ignoring the impact of genetics, they argue, leads to an incomplete understanding of behavior because genes shape our lives and could help explain why there is so much variation across families and other groups.

People have at least 52 characteristics—such as aggression, leadership traits, and cognitive ability—that are partially inherited (Freese 2008). Our social environment can enhance or dampen these genetic characteristics. For example,

- People who have inherited a genetic predisposition for alcoholism vary depending on other social factors, especially environment. As income and education levels decrease, for example, alcohol usage goes up, and as childhood deprivation and the number of "daily hassles" decreases, so does alcoholism (Pescosolido et al. 2008). Thus, the environment can affect genetic predispositions for alcoholism.

- Children who are genetically predisposed to obesity don't always become overweight if parents discourage overeating and encourage physical recreational activities (Martin 2008).

Sociologists who study the relationship between genetics (nature) and the environment (nurture) admit that their samples are small. They maintain, however, that research that combines both aspects enhances our understanding of how social factors affect genetic predispositions (see Schnittker 2008 and Shanahan et al. 2010).

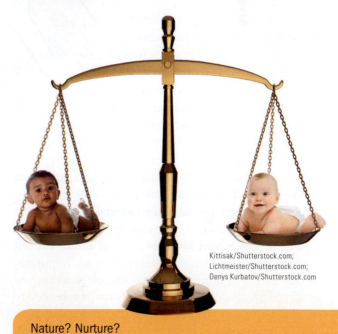

Kittisak/Shutterstock.com;
Lichtmeister/Shutterstock.com;
Denys Kurbatov/Shutterstock.com

Nature? Nurture?

**social learning theories**
maintain that people learn new attitudes, beliefs, and behaviors through social interaction, especially during childhood.

Functionalism provides a foundation for understanding the purposes of socialization described at the beginning of this chapter. However, functionalists don't tell us *how* socialization works on a micro level. There are several well-known psychological and psychosocial theories of human development. In sociology, two influential approaches that explain how socialization works on a micro level are *social learning theories* and *symbolic interaction theories* (*Table 4.2* summarizes these perspectives). Let's begin with social learning theories.

### 4-3a Social Learning Theories

The central notion of **social learning theories** is that people learn new attitudes, beliefs, and behaviors through social interaction, especially during childhood. We learn how to behave both directly and indirectly, for example, through observation and reinforcement. We can also learn how to act without actually performing a behavior.

#### Direct and Indirect Learning

Our socialization is usually *direct*. By being rewarded or punished, we learn many behaviors through reinforcement. A little girl who puts on her mother's makeup may be told she's cute, but her brother will be scolded ("Boys don't wear makeup!"). If a child is caught lying, she or he will be punished through a variety of sanctions, such as not being able to play with friends, not being allowed to watch television or use the computer, or doing extra household chores.

Socialization also involves *indirect* reinforcement through *modeling* (imitating people who are important in our lives). In one study, for example, researchers observed preschoolers ages 2 to 6 as they pretended to shop for a visiting friend of their Barbie or Ken doll. The pretend store stocked 133 miniature items, including meat, fruit, vegetables, snacks, cigarettes, beer, and wine. Overall, 28 percent of the children bought cigarettes, and 61 percent bought alcohol. The children whose parents smoked were almost four times more likely to buy cigarettes. Those whose parents drank at least monthly were three times more likely to buy alcohol. The researchers concluded that observing parental behavior may influence children to view smoking and drinking as appropriate and to adopt such practices later in life (Dalton et al. 2005).

#### Learning and Performing

Social learning theorists also distinguish between *learning* and *performing* behavior. Children and adults can learn to do something through observation, but they don't always imitate the behavior. For example, children may see their friends cheat in school but not do so themselves. Adults, similarly, may see their coworkers steal office supplies but buy their own.

Social learning theorists maintain that we behave in certain ways because of past rewards and punishments, modeling, and observation. We behave as we do, then, because our society teaches us what's appropriate and inappropriate (Bandura and Walters 1963; Mischel 1966; Lynn 1969).

#### Critical Evaluation

Social learning theories help us understand why we behave as we do, but much of the emphasis is on early socialization rather than on what occurs throughout life. Social learning theories also don't explain why reinforcement and modeling affect some children more than others, especially those in the same family. There may be personality differences, but if learning is as effective as social learning theorists maintain, siblings' attitudes and behavior should be more similar than different. This is often not the case, however, even with identical twins.

lenetstan/Shutterstock.com

## Sociology in Your Life

Albert Bandura is one of the originators of social learning theory. He conducted what's become known as the Bobo Doll experiment, in which children imitated adult aggression. Go to Sociology CourseMate at CengageBrain.com to watch a short clip from this experiment.

## TABLE 4.2

# Key Elements of Socialization Theories

| SOCIAL LEARNING THEORIES | SYMBOLIC INTERACTION THEORIES |
|---|---|
| • Social interaction is important in learning appropriate and inappropriate behavior. | • The self emerges through social interaction with significant others. |
| • Socialization relies on direct and indirect reinforcement. | • Socialization includes role taking and controlling the impression we give to others. |
| *Example:* Children learn how to behave when they are scolded or praised for specific behaviors. | *Example:* Children who are praised are more likely to develop a strong self-image than those who are always criticized. |

© Cengage Learning

A third weakness is that most social learning theories ignore factors such as birth order, which brings different advantages and disadvantages. Except for affluent families, larger families have fewer resources for the second, third, and later children. The first child may be disciplined more but may also be encouraged to pursue interests that could lead to higher education and a prestigious career. In contrast, later-born children may enjoy less parental attention and be less likely to receive financial support to pursue a college education (Zajonc and Markus 1975; Conley 2004).

## 4-3b Symbolic Interaction Theories

Symbolic interaction theories have had a major impact in explaining social development. Sociologists Charles Horton Cooley (1864–1929), George Herbert Mead (1863–1931), and Erving Goffman (1922–1982) were especially influential in showing how social interaction shapes socialization.

### Charles Horton Cooley: Emergence of the Self and the Looking-Glass Self

Newborn infants lack a sense of **self**, an awareness of one's social identity. Gradually, they begin to differentiate themselves from their environment and develop a sense of self.

After carefully observing the development of his young daughter, Charles Horton Cooley (1909/1983) concluded that children acquire a sense of who they are through their interactions with others, especially by imagining how others view them. The sense of self, then, is not innate but emerges out of social relationships. Cooley called this social self the *reflected self*, or the **looking-glass self**, a self-image based on how we think others see us. He proposed that the looking-glass self develops in an ongoing process of three phases:

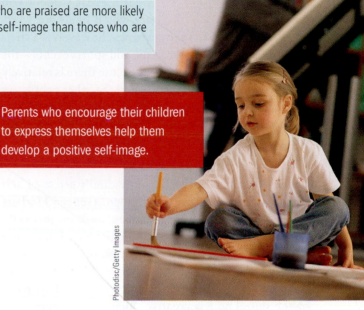

Parents who encourage their children to express themselves help them develop a positive self-image.

Photodisc/Getty Images

- *Phase 1: Perception.* We imagine how we appear to other people and how they *perceive* us ("She thinks I'm attractive" or "I bet he thinks I'm fat").

- *Phase 2: Interpretation of the perception.* We imagine how others *judge* us ("She's impressed with me" or "He's disgusted with the way I look").

- *Phase 3: Response.* We experience *self-feelings* based on what we regard to be others' judgments of us. If we think others see us in a favorable light, we may feel proud, happy, or self-confident ("I'm terrific"). If we think others see us in a negative light, we may feel angry, embarrassed, or insecure ("I'm pathetic").

Our interpretation of others' perceptions (phase 2) may be totally wrong. The looking-glass self, remember, refers to how we *think* others see us rather than how they *actually* see us. However, our perceptions of other people's views—whether we're right or wrong—affect our self-image, which then affects our behavior.

Cooley focused on how children acquire a sense of who they are through their interactions with others, but he noted that the process of forming a looking-glass self doesn't end in childhood. Instead, our self-concept may change over time because we reimagine ourselves as we think others see us—attractive or ugly, interesting or boring, intelligent or stupid, graceful or awkward, and so on.

Sometimes we're aware of the process shaping our looking-glass self. Consider your own self-concept: If others keep telling you what a great student you are, you'll likely see yourself as intelligent. In many settings, however—such as public places, large classrooms, and sports events—the looking-glass self is irrelevant because there is relatively little interpersonal interaction and the other people are strangers or aren't very significant in our lives.

## George Herbert Mead: Development of the Self and Role Taking

Cooley described how an individual's sense of self emerges, but not how it develops. George Herbert Mead, one of Cooley's colleagues, took up this task. For Mead (1934), the most critical social interaction occurs in the family, the foundation of socialization.

Our self develops, according to Mead, when we learn to differentiate the *me* from the *I*—two parts of the self. The *I* is creative, imaginative, impulsive, spontaneous, nonconformist, self-centered, and sometimes unpredictable. The *me* that has been successfully socialized is aware of the attitudes of others, has self-control, and has internalized social roles. Instead of impulsively and selfishly grabbing another child's toys, for example, as the *I* would do, the *me* asks for permission to use someone's toys and shares them with others.

For Mead, the *me* forms as children engage in **role taking**, learning to take the perspective of others. Children gradually acquire this ability early in the socialization process through three sequential stages:

1. **Preparatory stage (roughly birth to 2 years).** An infant doesn't distinguish between the self and others. The *I* is dominant, while the *me* is forming in the background. In this stage, children learn through imitation. They may mimic daddy's shaving or mommy's angry tone of voice without really understanding the parent's behavior. In this exploratory stage, children

engage in behavior that they rarely associate with words or symbols, but they begin to understand cause and effect (e.g., crying leads to being picked up). Gradually, as the child begins to recognize others' reactions (to form a looking-glass self), he or she develops a self.

2. **Play stage (roughly 2 to 6 years).** The child begins to use language and to understand that words (such as *dog* and *cat*) have a shared cultural meaning. Through play, children begin to learn role taking in two ways. First, they emulate the words and behavior of **significant others**, the people who are important in one's life, such as parents (or other primary caregivers), siblings, and grandparents. The child learns that he or she has a self that is distinct from that of others, that others behave in many different ways, and that others expect her or him to behave in specific ways. In other words, the child learns social norms (see Chapter 3).

In the play stage, the child moves beyond imitation and acts out imagined roles ("I'll be the mommy and you be the daddy."). The play stage involves relatively simple role taking because the child plays one role at a time but doesn't yet understand the relationships between roles. This stage is crucial, according to Mead, because the child *is learning to take the role of the other*. For the first time, the child tries to imagine how others behave or feel. The *me* grows stronger because the child is concerned about the judgments of significant others.

Also in the play stage, children experience **anticipatory socialization**, the process of learning how to perform a role they don't occupy. By playing "mommy" or "daddy," children prepare themselves for eventually becoming parents. Anticipatory socialization continues in later years, for example, when expectant parents attend childbirth classes, job seekers practice their skills in mock interviews, and many high school students visit campuses and attend orientation meetings to prepare for college life.

3. **Game stage (roughly 6 years and older).** This stage involves acquiring the ability to understand connections between roles. The child must "not only take the role of the other . . . but must assume the various roles of all participants in the game, and govern his action accordingly" (Mead 1964: 285).

Mead used baseball to illustrate this stage. In baseball (or other organized games and activities), the child plays one role at a time (such as batter) but understands and anticipates the actions of other players (pitcher, shortstop, runners on bases) on both teams. To successfully participate in a game of baseball, the individual must be

© OtnaYdur/Shutterstock

able to anticipate the roles of others. The same is true of participating in society. Thus, as children grow older and interact with a wider range of people, they learn to anticipate, respond to, and fulfill a variety of social roles.

The game stage enables the child to understand and take the role of the **generalized other**, people who don't have close ties to a child but who influence her or his internalization of society's norms and values. The generalized other thus exerts control over the *I* and ensures some predictability in life. For sociologists, the development of the generalized other is a central feature of socialization because the *me* becomes an integral part of the self (see *Figure 4.1*).

Even after the generalized other has developed, the *me* never fully controls the *I*, even in adulthood. We sometimes break rules or act impulsively, even though we know better. When we are frustrated, not feeling well, or angry, we may lash out against other people even though they haven't done anything wrong.

### Erving Goffman: Staging the Self in Everyday Life

Cooley and Mead described how the self and role taking emerge and develop during early socialization. Erving Goffman (1959, 1969) extended these analyses by showing that we interact differently in different settings throughout adulthood. Goffman proposed that social life mirrors the theater because we are like actors: We engage in "role performances," want to influence an "audience," and can have considerable control over the image that we project while we're "on stage." (These concepts are explored in greater detail in Chapter 5.)

In a process that Goffman called **impression management**, we provide information and cues to others to present ourselves in a favorable light while downplaying or concealing our less appealing characteristics. Being successful in this presentation of the self requires managing three types of expressive resources. First, we try to control the *setting*—the physical space, or "scene," where the interaction takes place. In the classroom, "a professor may use items such as chalk, lecture notes, computers, videos, and desks to

facilitate a class and show that he or she is an excellent teacher" (Sandstrom et al. 2006: 105).

A second expressive resource is controlling *appearance*, such as clothing and titles that convey information about our social status. When physicians or professors use the title "Doctor," for example, they are telling the "audience" that they expect respect.

The third expressive resource is *manner*—the mood or style of behavior we display that sends important messages to the audience. For example, faculty members regularly manage their manner when interacting with students. When you email a professor a question about your research project, the response is usually inviting ("I'll be happy to discuss your project") even though the professor may actually be annoyed ("Your project has nothing to do with this course; stop wasting my time"). Sometimes, no matter how much we try to manage setting, appearance, and manner, we slip up in our role performances. This leads to discomfort and embarrassment for everyone involved, but the audience is typically tactful:

Fotoline/Shutterstock.com

## FIGURE 4.1
## Mead's Three Stages in Developing a Sense of Self

**STAGE 1:** **Preparatory Stage** (under age 2)

No distinction between self and others; the child is self-centered and self-absorbed

Learns through observation

**STAGE 2:** **Play Stage** (aged 2 to about 6)

Distinguishes between self and others

Imitates significant others (usually parents)

Learns role taking, assuming one role at a time, in "let's pretend" and other play that teaches **anticipatory socialization**

**STAGE 3:** **Game Stage** (aged 6 and older)

Understands and anticipates multiple roles

Connects to societal roles through the **generalized other**

© Cengage Learning

© Helga Chirkova/Shutterstock

*Imagine that you are giving a presentation in one of your classes and, just as you make a key point, a large droplet of saliva lands near someone sitting in the front row. Although several of your classmates cannot help but notice it, none of them are likely to shout out, "Hey, you just about spit on someone in the first row!" Instead, they will probably act as if nothing unusual or embarrassing took place (Sandstrom et al. 2006: 109).*

As we move from one situation to another, according to Goffman, we maintain self-control, for example, by avoiding emotional outbursts and altering our facial expressions and verbal tones. Thus, all of us engage in impression management practically every day.

## Critical Evaluation

Symbolic interactionists have provided major insights about socialization, particularly of young children, and have shown that fitting into our social world is a complex process that continues through adulthood. Like other theories, however, symbolic interactionism has its limitations. For example, it's not clear why some children have a positive looking-glass self and are successful later in life even when the cues are consistently negative, as happens with some children who grow up in abusive homes or attend low-quality schools.

Some scholars have also criticized the vagueness of essential concepts such as *self, me,* and *I*. Because the concepts are imprecise, it's difficult, if not impossible, to measure them. Some have also challenged Mead's claim that children automatically pass through play and game stages as they get older. Instead, some critics contend, the extent of socialization depends on a child's social context, such as whether parents and other primary

caregivers are actively involved in the child's upbringing and provide enriching interaction. Even then, some critics contend, some children never develop the role of the generalized other (see Ritzer 1992).

Others have questioned the value of the concept of the generalized other in understanding early socialization processes. Because children interact in many social contexts (home, preschool, play groups), they don't simply assume the role of one generalized other, but many. That is, as children get older, they may have several **reference groups**—groups of people who shape an individual's self-image, behavior, values, and attitudes in different contexts (Merton and Rossi 1950; Shibutani 1986).

Another weakness is that interactionists credit people with more free will than they have. Individuals don't always have the power or ability to affect others' reactions or their own lives. For example, impression management is less common among lower than higher socioeconomic groups, whose members have internalized such skills to be successful in jobs. Also, how people react to us may reflect characteristics (e.g., gender, age, or race/ethnicity) that we can't change (Powers 2004).

A major criticism of Cooley, Mead, Goffman, and other interactionists is that they tend to downplay or ignore macro-level factors that affect our development (Ritzer 1992). As the next two sections show, institutions such as the family, education, economy, and popular culture have a major impact on socialization. Let's begin by looking at socialization agents that play important roles in our lives.

# 4-4 Primary Socialization Agents

B y fretting, crying, and whining, infants teach adults when to feed, change, and pick them up. Babies are actively engaged in the socialization process, but the family, peer groups, teachers, and the media are some of the primary **agents of socialization**—the individuals, groups, or institutions that teach us what we need to know to participate effectively in society.

© Gelpi/Shutterstock

## 4-4a Family

Parents, siblings, grandparents, and other family members play a critical role in our socialization. Parents, however, are the first and most influential socialization agents.

## TABLE 4.3

# Parenting Styles

| PARENTING STYLE | CHARACTERISTICS | EXAMPLE |
|---|---|---|
| Authoritarian | Very demanding, controlling, punitive | "You can't borrow the car because I said so." |
| Authoritative | Demanding, controlling, warm, supportive | "You can borrow the car, but be home by curfew." |
| Permissive | Not demanding, warm, indulgent, set few rules | "Borrow the car whenever you want." |
| Uninvolved | Neither supportive nor controlling | "I don't care what you do; I'm busy." |

© Cengage Learning

## How Parents Socialize Children

The purpose of socialization is to enable a child to regulate her or his own behavior and to make responsible decisions. In teaching their children social rules and roles, parents rely on several of the learning techniques, such as reinforcement, discussed earlier in this chapter. Parents also manage many aspects of the environment that influence a child's social development: They often choose the neighborhood (which often determines what school a child attends), decorate the child's room in a masculine or feminine style, provide her or him with particular toys and books, and arrange enriching social events and other activities (such as sports, art, and music).

## Parenting Styles

Parents have four common parenting styles that affect socialization (see *Table 4.3*). In *authoritarian parenting*, parents tend to be harsh, unresponsive, and rigid, using their power to control a child's behavior. *Authoritative parenting* is warm, responsive, and involved but unobtrusive. Parents set reasonable limits and expect appropriately mature behavior from their children. *Permissive parenting* is lax: Parents set few rules but are usually warm and responsive. *Uninvolved parenting* is indifferent and neglectful. Parents focus on their own needs rather than those of the children, spend little time interacting with the children, and know little about their interests or whereabouts (Baumrind 1968, 1989; Maccoby and Martin 1983; Aunola and Nurmi 2005).

Healthy child development is most likely in authoritative homes because parents are consistent in combining warmth, monitoring, and discipline. Authoritative parenting tends to produce children who are self-reliant, achievement oriented, and successful in school. Adolescents in households with authoritarian, permissive, or uninvolved parents tend to have poorer psychosocial development, lower school grades, lower self-reliance, higher levels of delinquent behavior, and are more likely to be swayed by harmful peer pressure (e.g., to use drugs and alcohol) (Dorius et al. 2004; Eisenberg et al. 2005; Hillaker et al. 2008).

Such findings vary by social class and race/ethnicity, however. For many Latino and Asian immigrants, authoritarian parenting produces positive outcomes, such as academic success. This parenting style is also effective in safeguarding children who are growing up in low-income neighborhoods with high rates of crime and drug peddling (Brody et al. 2002; Pong et al. 2005).

AP Photo/Yonhap, Shin Young-kuyn

During the last decade, numerous summer camps have sprung up in South Korea where children ages 7 to 19 live in military-style barracks, undergo drills and exercise from dawn to dusk, eat very simple meals, and have no access to computers, television, or cell phones. Most are sent by parents "who realize that their pampered offspring need more discipline to become better students and grow into conscientious adults" (Glionna 2009: A1). Some of the kids' offenses have included getting a low grade on an important test, accidentally breaking a window in the family's apartment, and most commonly, talking back to their parents.

### Siblings

Siblings, like parents, play important roles in socialization. Children can bully and abuse their (usually younger) brothers and sisters through name calling, ridiculing, destroying personal possessions, and even physical or sexual abuse. Such behavior can leave lasting emotional scars that lower a child's self-esteem, and physical abuse sends the message that violence is acceptable in resolving disputes (Gelles 1997; Wiehe 1997). Several studies have also found that smoking, drinking, and marijuana use by older siblings increase the probability that younger siblings will also engage in these behaviors (Duncan et al. 2005; Altonji et al. 2010).

Siblings can also have a positive impact by helping their younger brothers and sisters with their homework and protecting them from neighborhood and school bullies. During adolescence and adulthood, siblings often act as confidantes and as role models on how to get along with people, and they often offer financial and emotional assistance during stressful times (McCoy et al. 2002; Whiteman et al. 2011).

## 4-4b Play, Peer Groups, and Friends

As you saw earlier, Mead believed that play and games are important in children's development. They help us to become more skilled in using language and symbols,

Grandparents often provide stability in family relationships and a continuity of family rituals and values. When a teacher asked her class of 8-year-olds what they liked best about a grandparent, the essays included comments such as "They don't say 'Hurry up'" and "When they take us for a walk, they slow down for things like pretty leaves and caterpillars."

Dmitriy Shironosov/Shutterstock.com

learn role playing, and internalize roles that we don't necessarily enact (the generalized other). A **peer group** consists of people who are similar in age, social status, and interests. All of us are members of peer groups, but such groups are especially influential until about our mid-20s. After that, coworkers, spouses, children, and close friends are usually more important than peer groups in our everyday lives.

### Play and Its Functions

Play serves several important functions. First, it promotes cognitive development. Whether it's doing simple five-piece puzzles or tackling complex video games such as *Sim City* and *Roller Coaster Tycoon*, play encourages children to think, formulate strategies, and budget and manage resources. From an early age, however, play is generally gender typed. In 2010, the top-selling toys for girls were dolls, animal toys (e.g., Zhu Zhu pets), and cooking sets. The top-selling toys for boys were video games, ride-on toys (e.g., pedal cars), Hot Wheels, Lego, and Transformers (Rattray 2010).

Second, play—especially when it's structured—keeps children out of trouble and enhances their social development. For example, children who devote more of their free time to structured and supervised activities—such as hobbies and sports—rather than just hanging out with their friends, perform better academically, are emotionally better adjusted, and have fewer problems at school and at home. In effect, sports and hobbies provide children with constructive ways to channel their energy and intelligence (McHale 2001).

Third, play can strengthen peer relationships. Beginning in elementary school, few things are more important to most children than being accepted by their peers. Even if children aren't popular, belonging to a friendship group enhances their psychological well-being and ability to cope with stress (Rubin et al. 1998; Scarlett et al. 2005).

### Peers and Friends

Peer influence usually increases as children get older. Especially during the early teen years, friends often reinforce desirable behavior or skills in ways that enhance a child's self-image ("Wow, you're really good in math!"). Thus, to apply Cooley's concept, peers can help each other develop a positive looking-glass self.

Peers also serve as positive role models. Children acquire a wide array of information and knowledge by observing their peers. Even during the first days of school, children learn to imitate their peers at standing in line, raising their hands in class, and being quiet while the teacher is speaking. Among teens and young adults who are lesbian, gay, or bisexual, heterosexual friends can be especially supportive in accepting one's homosexuality and disclosing it to family (Shilo and Savaya 2011).

Not all peer or friend influence is positive, however. A nationally representative study of seventh to twelfth graders found that having a best friend who engages in sexual intercourse, is truant, and uses tobacco and marijuana increases the probability of imitating such behavior (Card and Giuliano 2011). A study of college students at the U.S. Air Force Academy found that students who are the least physically fit have similar friends. Poor physical fitness among friends increases the likelihood of adopting or maintaining an unhealthy diet and developing health problems (Carrell et al. 2010).

## 4-4c Teachers and Schools

Oprah Winfrey often praised her fourth-grade teacher, who recognized her abilities, encouraged her to read, and inspired her to excel academically. Like family and peer groups, teachers and schools affect our socialization.

### The School's Role in Socialization

By the time children are 4 or 5, school fills an increasingly large portion of their lives. The primary purpose of the school is to instruct children and enhance their cognitive development. Schools don't simply transmit knowledge; they also teach children to think about the world in different ways. Because of the emphasis on multiculturalism, for example, children often learn about other societies and customs. Even outside of classes, schools affect children's daily activities through homework assignments and participation in clubs and other extracurricular activities.

Because many parents are employed, schools have had to devote more time and resources to topics—such as sex education and drug abuse prevention—that were once the sole responsibility of families. In many ways, then, schools play an increasingly important role in socialization.

### Teachers' Impact on Children's Development

Teachers are among the most important socialization agents. From kindergarten through high school, teachers play numerous roles in the classroom—instructor, role model, evaluator, moral guide, and disciplinarian,

to name just a few. Once children enter school, their relationships with their teachers are important for academic success. Kindergarten and elementary school teachers' reports of behavioral problems (such as unexcused absences) and poor grades often follow students into middle school (Hamre and Pianta 2001).

## 4-4d Popular Culture and the Media

Because of iPads, smartphones, YouTube, and social networking sites such as Facebook, young people are rarely out of the reach of the electronic media. How does such technology affect socialization?

### Electronic Media

The American Academy of Pediatrics (2001: 424) advises parents to avoid television entirely for children younger than 2. The Academy also counsels parents to limit the viewing time of elementary school children to no more than 2 hours a day to encourage more interactive activities "that will promote proper brain

Class field trips enhance children's socialization. Among other benefits, the field trips expand learning beyond the classroom, enrich students' understanding of subject matter, and encourage them to think about occupational choices and careers that they haven't considered.

© Neil McAllister/Alamy

Siede Preis/Photodisc/Getty Images

TABLE 4.4

# How Do Electronic Media Affect Children?

| AMONG ALL 8- TO 18-YEAR-OLDS, PERCENTAGE WHO SAID THAT THEY ARE. . . | HEAVY USERS (MORE THAN 16 HOURS/DAY) | MODERATE USERS (3–16 HOURS/DAY) | LIGHT USERS (LESS THAN 3 HOURS/DAY) |
|---|---|---|---|
| Get good grades (A's and B's) | 51 | 65 | 66 |
| Get fair/poor grades (C's or lower) | 47 | 31 | 23 |
| Have been happy at school this year | 72 | 81 | 82 |
| Are often bored | 60 | 53 | 48 |
| Get into trouble a lot | 33 | 21 | 16 |
| Are often sad or unhappy | 32 | 23 | 22 |

Source: Based on Rideout et al. 2010: 4.

development, such as talking, playing, singing, and reading together." Still, 68 percent of children under age 2 view 2 to 3 hours of television daily, and 20 percent have a television in their bedroom, as do one-third of 3- to 6-year-olds (Garrison and Christakis 2005; Vandewater et al. 2007).

The average young American now spends practically every waking minute—except for the time in school—using a smartphone, computer, television, or other electronic device. In 2009, those ages 8 to 18 spent 7.5 hours a day engaged with some type of electronic media, which is more time than most adults spend in a full-time job. Generally, youths who spend more time with media have lower grades and lower levels of personal contentment (see *Table 4.4*). These findings are similar for girls and boys across all categories of age, race/ethnicity, parents' social class, and single- and two-parent households (Rideout et al. 2010).

High media usage can decrease academic success, but two-thirds of parents are especially concerned that the media contribute to young people's violent behavior (Borzekowski and Robinson 2005; Rideout 2007). Are such concerns justified? Many scholars agree that electronic media have negative aspects but contend that there is still no evidence that television viewing *causes* violence. For example, violent crimes in society have decreased despite the increased depiction of violence on television and other electronic media (Powell et al. 2007; Sternheimer 2007).

There is also a growing consensus that violent video games make violence seem normal. Playing violent video games such as *Grand Theft Auto IV*, *School Shooter*, and *God of War 4* can increase a person's aggressive

thoughts, feelings, and behavior both in laboratory settings and in real life. Violent video games also encourage male-to-female aggression because much of the violence is directed at women (Anderson et al. 2003; Carnagey and Anderson 2005).

Still, it's not clear why violent video games affect people differently. Many young males enjoy playing such video games, for example, but aren't any more aggressive, vicious, or destructive than those who aren't video game enthusiasts (Williams and Skoric 2005; Kutner and Olson 2008).

## Advertising

Increasingly, advertisers are targeting children as early as possible. A new form of advertising called *advergaming* combines free online games with advertising. Advergaming is growing rapidly. Sites such as those operated by Nestlé (www.wonka.com), Mattel (Barbie.com), and M&Ms (mms.com/us/fungames/games) attract millions of young children and provide marketers with an inexpensive way to "draw attention to their brand in a playful way, and for an extended period of time" (Moore 2006: 5).

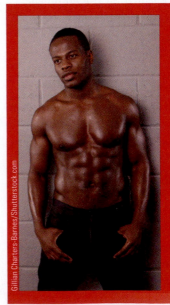

Gillian Charters-Barnes/Shutterstock.com

Many men's health and fitness magazines routinely feature models who have undergone several months of extreme regimens, including starvation and dehydration, to tighten their skin and make their muscles "pop." The magazines also use camera and lighting tricks and Photoshop to project an idealized image of hypermasculinity that, in reality, is impossible to attain (Christina 2011; see also Ricciardelli et al. 2010).

# What? Adolescents Can Buy Ultraviolent Video Games, but Not a Magazine with an Image of a Nude Woman?!

In 2011, the Supreme Court ruled (7–2 in *Brown v. Entertainment Merchants Association*) that video games, even ultraviolent ones, that are sold to minors are protected by the First Amendment's guarantee of free speech. The majority said that none of the scientific studies *prove* that violent games *cause* minors to act aggressively, and that parents, not the government, should decide what is appropriate for their children (Schiesel 2011; Walls 2011).

Hot Property/Shutterstock

Not surprisingly, the video game industry was jubilant over the decision. Many parents and lawmakers, on the other hand, agreed with one of the dissenting justices that it makes no sense to forbid selling a 13-year-old boy a magazine with an image of a nude woman, but not an interactive video game in which the same boy "actively . . . binds and gags the woman, then tortures and kills her" (Barnes 2011: A1).

© Cengage Learning

---

In the print media, young people see 45 percent more beer ads and 27 percent more ads for hard liquor in teen magazines than adults do in their magazines (Strasburger et al. 2006; see also Jernigan 2010). Girls ages 11 to 14 are subjected to about 500 advertisements a day on the Internet, billboards, and magazines in which the majority of models are "nipped, tucked, and airbrushed to perfection" (Bennett 2009: 43).

What effect do such ads have on girls' and women's self-image? About 43 percent of 6- to 9-year-old American girls use lipstick or lip gloss, 38 percent use hairstyling products, and 12 percent use other cosmetics. Eight- to 12-year-old girls spend more than $40 million a month on beauty products (Bennett 2009).

Many women, especially white women, are unhappy with their bodies. An analysis of 77 recent studies of women's media images concluded that there is a strong association between exposure to media depicting ultrathin actresses and models, many women's dissatisfaction with their bodies, and their likelihood of engaging in unhealthy eating behaviors such as excessive dieting (Grabe et al. 2008).

Millions of others turn to cosmetic surgery. Women have 92 percent of all cosmetic procedures, and the number undergoing surgery has increased 155 percent since 1997. The most common cosmetic surgery is breast augmentation. In 2010 alone, nearly 319, 000 women and girls age 18 and younger had breast implants, whereas penis implants aren't even mentioned (American Society for Aesthetic Plastic Surgery 2011).

© vetkit/Shutterstock

## Sociology in Your Life

Watch the transformation of an ordinary girl into a billboard model through cosmetics and digital manipulation. Go to Sociology CourseMate at CengageBrain.com to access this site.

## 4-5 Socialization Throughout Life

Biological aging comes naturally, but social aging is a different matter. As we progress through the life course—from infancy to death—we are expected to learn culturally approved norms, values, and roles.

### THAT'S NOT ALL . . .

There are many agents of socialization other than family, play and peer groups, teachers and schools, and the media. If you think about the last few years, which groups or organizations have influenced who you are? For example, what about the military, a company you've worked for, college clubs, religious organizations, support groups, athletic memberships, community groups, or others? All are important socialization agents throughout our lives.

Until 1938, young children made up a large segment of the U.S. labor force, especially in factories. They worked 6 days a week, 12 or more hours a day, received very low wages, and were often injured because of dangerous equipment.

## 4-5a Infancy

Some scientists describe healthy infants' brains as "small computers" because of their enormous capacity for learning (Gopnik et al. 2001: 142). Normal babies are quick learners, but many proud parents believe that their children can be brilliant. As a result, since 1997, Americans have spent $200 million a year on "Baby Einstein" products such as flash cards and educational software for children as young as 6 months ("Your baby will learn the numbers 1–20!" according to some ads). Other parents have played classical music all day or placed DVDs next to the crib to teach the baby French or German.

Some researchers have found that such products are doing more harm than good. Babies 8 to 16 months old who are exposed to the DVDs learn fewer words than those whose parents talk to them, tell them stories, and use a rich vocabulary (Zimmerman et al. 2007). In 2009, when lawyers threatened a class action lawsuit for deceptive advertising because the "Baby Einstein" products (owned by the Walt Disney Company) didn't increase an infant's intelligence, parents were offered a refund (Lewin 2009).

## 4-5b Childhood

According to French historian Philippe Ariès (1962), childhood—viewed as a distinct stage of development—is a fairly recent phenomenon. In the U.S. colonies, child labor was nearly universal. At the beginning of the nineteenth century, American children began to spend more time playing than working, and adolescence became a new stage of life without adult responsibilities. Parents began for the first time to celebrate children's birthdays, to recognize children's individuality by giving them names that were different from their own, and to show a marked decline in the use of corporal punishment (Degler 1981; Demos 1986). At the turn of the twentieth century, U.S. children were required to attend school until age 16, and in 1938, the federal government passed its first law forbidding child labor.

Most American children today enjoy happy and healthy lives, but many don't. About 1.7 million children younger than 18 (more than 2 percent of all children) have an incarcerated parent, a number that has increased by 77 percent since 1991. Another 429,000 are in the foster care system. Almost 13 percent of all children live in households where a parent or other adult uses, manufactures, or distributes illicit drugs. And in 2010, there were 754,000 substantiated cases of children's mistreatment that included neglect and physical and sexual abuse (Christian 2009; U.S. Department of Health and Human Services 2011; U.S. Department of Justice 2011). Thus, for millions of U.S. children, the socialization process is shaky at best.

## 4-5c Adolescence

Almost 90 percent of U.S. parents say that, compared with the time when they were growing up, it's harder to monitor their 12- to 17-year-olds, especially when both parents are employed (National Center on Addiction and Substance Abuse 2008). Nearly half of such parents believe that they don't spend enough time with their children, but 75 percent of children ages 8 to 17 say that working parents have a positive effect on the quality of their home life (Conlin 2007). It seems, then, that many employed parents are doing a better job in raising their kids than they think.

Parental expectations play an important role in adolescent development. For example, families with high educational aspirations for their offspring provide more out-of-school learning opportunities, the children have more positive attitudes toward school, and they are more likely to attain a four-year college degree (Child Trends 2010).

Academic achievement varies by social class, as you'll see in Chapter 13, but being raised in a low-income family with unschooled parents doesn't predict educational failure. A study conducted in 27 countries

Lewis Hine/Archive Farms/Getty Images

© Howard McWilliam

over 20 years concluded that academic success is strongly influenced by a "family scholarly culture—the way of life in homes where books are numerous, esteemed, read, and enjoyed." The family scholarly culture had a powerful impact on children's education in rich and poor nations, across all economic and political systems, in families with modest incomes, and was transmitted from generation to generation (Evans et al. 2010).

High school counselors typically applaud parental involvement in their children's lives, but they also dread "helicopter parents" who hover over their kids, micromanaging every aspect of their lives. Examples include verbally attacking teachers over their adolescents' low grades, demanding that their child be moved to another class before the school year has even begun, and showing up in the guidance counselor's office with college applications that they have filled out for their children (Krache 2008). Helicopter parenting diminishes teens' sense of self and autonomy, their ability to develop decision-making skills, and their capacity to become responsible problem solvers.

## 4-5d Adulthood

Socialization continues throughout adulthood, the period roughly between ages 21 and 65. Most adults adopt a series of new roles that may include work, marriage, parenthood, divorce, remarriage, buying a house, and experiencing the death of a child, parent, or grandparent. We'll examine these transitions in later chapters. Two of the most important roles in adulthood are work

© Anest/Shutterstock

and parenthood, but let's begin with young adults who are reluctant to leave their parents' nest.

### The Crowded Empty Nest

During the 1960s and 1970s, sociologists almost always included the "empty nest" in describing the family life course. This is the stage in which parents, typically in their 50s, find themselves alone after their children have married, gone to college, or found jobs and moved out.

Today, however, young adults are living at home longer than was generally true in the past. The terms *boomerang children* and *boomerang generation* are used to refer to young adults moving back in with their parents, but many people in their 30s, 40s, and older—often with a spouse and children in tow—are moving back into their parents' home. In 2011, 40 percent of all adults age 18 and older were living with a family member, other relative, or a friend, up from 28 percent in 2007 before the Great Recession began (DeNavas Walt et al. 2011; Mykata and Macartney 2011).

Macro-level factors motivate many people to stay at home or move back. A major reason for this "doubling up," as the Census Bureau calls it, is economic, particularly unemployment. Student loans, low wages, divorce, and credit card debt have made it harder for young middle-class adults to enjoy the lifestyle that their parents provided. And the transition to adulthood gets tougher the lower people are on the social class ladder. Until the late 1980s, it was possible for a high school graduate to achieve a middle-class standard of living. Today, even a college degree doesn't guarantee economic success (Settersten and Ray 2010).

Individual and macro-level variables intersect in explaining the delay of many Americans' transition to adulthood, especially to living on their own. Many doting parents are willing to provide financial support for adult children who spend years searching for jobs they like rather than those that provide a living. This isn't surprising as many college commencement speakers tell the graduates, however unrealistically, to "follow *your* passion, chart *your* own course, march to the beat of *your* own drummer, follow *your* dreams and find *yourself*," and pursue happiness and joy (Brooks 2011: A23). Other young adults, as you'll see in Chapter 12, are delaying marriage and don't feel a need to establish their own homes.

## FIGURE 4.2
## Yearly Cost of Raising Children, 2010

In 2010, middle-income families—those earning $57,660 to $99,730 a year—spent almost $12,000 *per year* for a child under 2. This amount does not include the costs of prenatal care or delivery.

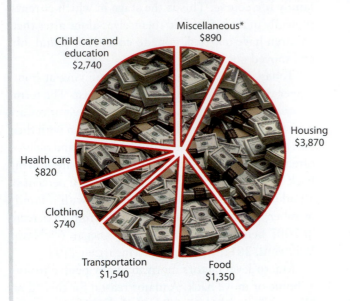

Miscellaneous*
$890

Child care and education
$2,740

Housing
$3,870

Health care
$820

Clothing
$740

Transportation
$1,540

Food
$1,350

$11,950

© Don Farrall/Photodisc/Getty Images/Creatas/Jupiter Images

*Includes personal care items, entertainment, and reading materials.
Source: Based on Lino, 2011, Table 1.

something." Even if we remain in the same job for many years, we must often acquire new skills, especially technological ones, or risk being laid off. As U.S. companies have transferred more jobs overseas, many workers have been laid off. Being laid off can be stressful at any age. In the case of midlife men (generally defined as those ages 45 to 60), such changes are especially traumatic: They must learn new job-hunting skills and often adjust to as much as a 70 percent decrease in their earnings. Instead of enjoying the security of a job in their midlife years, these men must undergo occupational socialization that assaults their self-image as good employees and family providers.

### Parenting Roles

Like workplace roles, parenting does *not* come naturally. Most first-time parents muddle through by trial and error. Family sociologists often point out that we get more training for driving a car than for marriage and parenting. For example, most couples don't realize that raising children is expensive. Middle-income couples, with an average income of $78,700 a year, spend about 16 percent of their earnings on a child during the first 2 years (see *Figure 4.2*). Child-rearing costs are much higher for single parents and especially for low-income families if a child is disabled, chronically ill, or needs specialized care that welfare benefits don't cover (Lino 2011).

Arguments over finances and child-rearing strategies are two of the major reasons for divorce (see Chapter 12). All in all, adjusting to marital and parental roles during adulthood requires considerable patience, effort, and work.

### 4-5e Later Life

Socialization continues in later life. Because we live longer, we may spend 20 percent of our adult life in retirement. When retired people are unhappy, it's usually because of health or income problems rather than the loss of the worker role (see Chapter 14). If retirement benefits don't keep up with inflation or if a retiree isn't covered by a pension plan, older people can plunge into poverty soon after retirement (Toder 2005).

Still others enjoy the comforts of the coddling parental nest. According to one of my male students in his late 20s, for example, "My mom enjoys cooking, cleaning my room, and just having me around. I don't pitch in for any of the expenses, but we get along great because she doesn't hassle me about my comings and goings" (Benokraitis 2011: 335). Perhaps most important, the stigma traditionally linked to young adults' living at home has faded because the practice is widespread enough "to be considered socially acceptable rather than an indicator of the youth's personal failure" (Danziger 2008: F8).

### Work Roles

The average American holds at least nine jobs from ages 18 to 34 alone (U.S. Department of Labor 2002). Such job changing means that our occupational socialization is an ongoing process. Learning work roles is difficult because each job has different demands and expectations. Even when there is formal job training, we must also learn the subtle rules that are implicit in many job settings. A supervisor who says, proudly and loudly, "I have an open door policy. Come in and chat whenever you want," may *really* mean "I have an open door policy if you're not going to complain about

Comedian Woody Allen once quipped, "It's not that I'm afraid to die. I just don't want to be there when it happens." Older people who are in poor health and experience continuous pain sometimes welcome death. Many others reenter the labor force or volunteer in various organizations, are often involved with their grandchildren, and forge new relationships after widowhood (see Chapter 12). Thus, even in the later years, many people continue to learn new roles.

## 4-6 Resocialization

Socialization isn't always predictable. Sometimes people undergo **resocialization**, the process of unlearning old ways of doing things and adopting new attitudes, values, norms, and behavior. Much resocialization takes place in **total institutions**—such as military boot camps, mental hospitals, prisons, concentration camps, and some religious orders—where people are isolated from the rest of society, stripped of their former identities, and required to conform to new rules and behavior (Goffman 1961).

Some resocialization is voluntary, as when an American wife moves to the Middle East with her Iranian husband and follows local customs about veiling, women's submissive roles, and staying out of public life. Other examples of voluntary resocialization include entering a religious order, seeking treatment in a drug abuse rehabilitation facility, or serving in the military.

Even when resocialization is voluntary, the process can be long and difficult. In divorce, people may experience a sense of relief and liberation, but must also cope with emotional distress, financial changes, anxiety about dating again, and child custody disputes (see Chapter 12). Soldiers who return from a year's deployment to Iraq or Afghanistan, even if they haven't suffered physical injuries, describe coming home and living a normal life as very difficult. They must reconnect with their children, adjust to spouses who have become more independent, and deal with posttraumatic stress disorders that include depression and nightmares (Dao 2011).

Resocialization can also be involuntary, as when children are sent to a foster home or a juvenile detention camp, or when people are imprisoned. Whether built in rural or urban areas, prisons are physically separated from the rest of society, have high fences, barred windows, armed corrections officers at the entrance, and the staff has almost complete control over the prisoner. Many guards use *degradation ceremonies*, humiliating rituals that publicly stigmatize people (Garfinkel 1956). The inmates are strip-searched, deloused, and fingerprinted; their heads are shaved; they are issued a uniform; they are given a serial number; and they are told what to do and when to do it.

Prisoners have almost no privacy, limited access to family and friends, and little communication with the outside world. The purpose of these practices and restrictions is to destroy any sense of individualism, autonomy, and past identity, and to produce a more compliant person (Goffman 1961; see also Chapter 7).

As this chapter shows, socialization is a powerful force in shaping who we are. Socialization doesn't produce robots, however, because people are creative, adapt to new environments, and change as they interact with others.

**resocialization** the process of unlearning old ways of doing things and adopting new attitudes, values, norms, and behavior.

**total institutions** places where people are isolated from the rest of society, stripped of their former identities, and required to conform to new rules and behavior.

### Study Tools

Ready to study? Go to **Sociology CourseMate** at **www.cengagebrain.com** to complete practice quizzes, review flashcards, watch videos, and more.

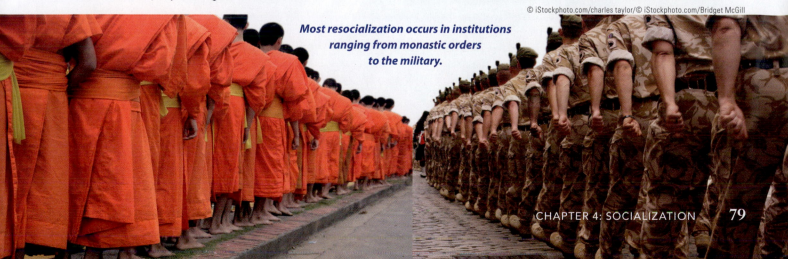

*Most resocialization occurs in institutions ranging from monastic orders to the military.*

Social structure
provides a guide for everyday
living.

# Social Interaction and Social Structure

In mid-2011, a Boeing 767 took off from an airport in Washington, D.C., bound for Ghana, Africa. Shortly after the flight was airborne, a passenger reclined his seat, which was close to the lap of the man sitting behind him. There was a smack to the head, angry words, and a fist fight broke out. A flight attendant and another passenger jumped in between. The pilot returned to the airport to determine the scope of the problem (Halsey 2011: A1; see also Grossman 2011).

**social interaction** the process by which we act toward and react to people around us.

**social structure** an organized pattern of behavior that governs people's relationships.

This incident of air rage is an example of **social interaction**, the process by which we act toward and react to people around us. The scuffle began with a typical passenger annoyance, escalated into an aggressive nonverbal reaction (a smack to the head), burst into an angry argument and fight, and quickly involved others—a third passenger, the airline attendant, and the pilot. This incident illustrates four critical components of social interaction (Maines 2001; Schwalbe 2001):

- It is central to all human social activity. Social interaction includes both nonverbal behavior (such as a physical smack) and words (such as an argument).

- People respond differently during social interaction, depending on what they think is at stake for them ("I'm not going to put up with this guy all the way to Ghana.").

- People influence each other's behavior through social interaction ("No one can hit me on the head and get away with it.").

- Elements of social structure, such as rules about proper behavior, affect all social interaction but can produce different personal outcomes (e.g., most airline passengers tolerate cramped seats quietly whereas others explode).

Almost all of us conform to a cultural social structure that shapes our roles, status, social interaction, and nonverbal and online communication and that governs people's relationships.

## Key Topics

In this chapter, we'll explore the following topics:

**5-1** Social Structure

**5-2** Status

**5-3** Role

**5-4** Explaining Social Interaction

**5-5** Nonverbal Communication

**5-6** Online Interaction

## What do you think?

I spend more time texting and on Facebook than I do talking to my family.

| 1 | 2 | 3 | 4 | 5 | 6 | 7 |

strongly agree     strongly disagree

## 5-1 Social Structure

**S**ocial structure is an organized pattern of behavior that governs people's relationships (Smelser 1988). Because social structure guides our actions, it gives us the feeling that

## FIGURE 5.1

## Is This Your Status Set?

Being a college student is only one of your current statuses. What other statuses comprise your status set?

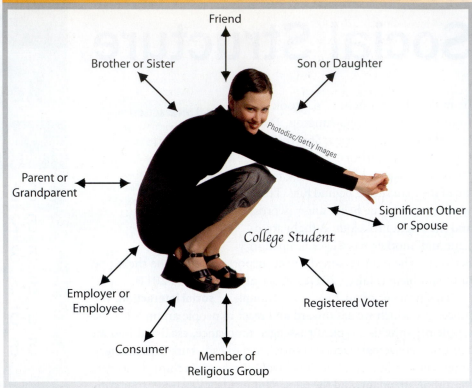

Friend

Brother or Sister

Son or Daughter

Parent or Grandparent

Photodisc/Getty Images

College Student

Significant Other or Spouse

Employer or Employee

Registered Voter

Consumer

Member of Religious Group

Photodisc/Getty Images

life is orderly and predictable rather than haphazard or random. We're often not aware of the impact of social structure until we violate cultural rules, formal or informal, that dictate our daily behavior. People may sometimes resent the impact of social structure because it limits their personal choices. But like it or not, social structure and social interaction—whether in class, at work, or in an airplane— shape our daily lives.

Every society has a social structure that encompasses statuses, roles, groups, organizations, and institutions (Smelser 1988). We'll examine groups, organizations, and institutions in later chapters. Let's take a closer look here at statuses and roles, two building blocks of social structure.

# 5-2  Status

For most people, the word *status* signifies prestige: An executive, for example, has more status than a receptionist, and a physician has a higher status than a nurse. For sociologists, **status** refers to a social position that a person occupies in a society (Linton 1936). Thus, executive, secretary, physician, and nurse are all social statuses. Other examples of statuses are student, professor, musician, voter, sister, parent, police officer, and friend.

Sociologists don't assume that one position is more important than another. A mother, for example, is not more important than a father, and an adult is not more important than a child. Instead, all statuses are significant because they determine social identity, or who we are.

## 5-2a Status Set

Every person has many statuses (see *Figure 5.1*). Together, they form her or his **status set**, a collection of social statuses that a person occupies at a given time (Merton 1968). Dionne, one of my students, is female, African American, 42 years old, divorced, mother of two, daughter, aunt, Baptist, voter, bank supervisor, volunteer at a soup kitchen, and country music fan. All of these socially defined positions (and others) make up Dionne's status set.

Status sets change throughout the life course. Because she will graduate next year and is considering remarrying and starting an after-school program, Dionne will add at least three more statuses to her status set and will also lose the statuses of divorced and college student. As Dionne ages, she will continue to gain new statuses (grandmother, retiree) and lose others (supervisor at a bank, or wife, if she is widowed).

Statuses are *relational*, or complementary, because they are connected to other statuses: A *husband* has a *wife*, a *real-estate agent* has *customers*, and a *teacher* has *students*. No matter how many statuses you occupy, every status is linked to that of one or more other people. These connections between statuses influence our behavior and relationships.

## 5-2b Ascribed and Achieved Status

Status sets include both ascribed and achieved statuses. An **ascribed status** is a social position that a person is born into. We can't control, change, or choose our ascribed statuses, which include sex (male or female), age, race, ethnicity, and family relationships. Your ascribed statuses, for example, might include *male*, *Latino*, and *brother*. Some argue that sex isn't really an ascribed status because people can have sex-change operations. However, a sex-change operation doesn't change the fact that someone was born a male or a female (except in a minority of cases, as you'll see in Chapter 9).

An **achieved status**, in contrast, is a social position that a person attains through personal effort or assumes voluntarily. Your achieved statuses might include college graduate, mother, and employee. Unlike our ascribed statuses, our achieved statuses can be controlled and changed. We have no choice about being a son or daughter (an ascribed status), but we have an option to become a parent (an achieved status).

Students sometimes think that religion and social class are ascribed rather than achieved statuses. It's true that someone may be born into a family that practices a certain religion or one that is poor, middle class, or wealthy. Because we can change these statuses through our own actions, however, neither religion nor social class is an ascribed status. A Catholic might convert to Judaism (or vice versa), and thousands of Americans born into poor or working-class families have become millionaires (see Chapter 8).

## 5-2c Master Status

An ascribed or achieved status can be a **master status** that determines a person's identity (Hughes 1945; Becker 1963). In most societies—including the United States—one's sex, age, physical ability, and race are master statuses because they are very visible and often shape a person's entire life. Occupations are frequently a master status because they denote social class and may override an ascribed status such as one's sex ("She's a chemical engineer; she must be smart.").

A master status can be positive or negative. Many people admire Bill Gates, the founder of Microsoft, because he's a billionaire, a positive master status.

Master statuses can also be negative. Instead of getting to know a person in a wheelchair, for example, we might stigmatize her or him as somehow imperfect and react to the disability rather than the person's accomplishments.

## 5-2d Status Inconsistency

Because we hold many statuses, some clash. **Status inconsistency** refers to the conflict that arises from occupying social positions that are ranked differently. Examples include a computer programmer who works as a bartender or a skilled welder who stocks shelves at Walmart because neither can find a better job in a weak economy. We'll cover status inconsistency in more detail in Chapter 7. For now, you should be aware that you occupy many statuses, and some of them may clash now and in the future.

**ascribed status** a social position that a person is born into.

**achieved status** a social position that a person attains through personal effort or assumes voluntarily.

**master status** an ascribed or achieved status that determines a person's identity.

**status inconsistency** the conflict that arises from occupying social positions that are ranked differently.

AP Photo/David Stluka

Leslie Visser graduated from Boston College with honors, was an accomplished athlete, and has been a successful sports analyst, journalist, and sportscaster for more than 25 years. When describing Visser's achievement, writers typically focus on her gender: "The first woman to cover Monday Night Football" or "The first woman enshrined into the Pro Football Hall of Fame." Does Visser's master status diminish her achieved status?

## 5-3 Role

**role** the behavior expected of a person who has a particular status.

**role performance** the actual behavior of a person who occupies a status.

**role set** the different roles attached to a single status.

Each status is associated with one or more roles. A **role** is the behavior expected of a person who has a particular status. We *occupy* a status but *play* a role. In this sense, a role is the dynamic aspect of a status (Linton 1936).

College student is a status, but the role of a college student requires many *formal* behaviors such as going to class, reading, thinking, completing weekly assignments, writing papers, and taking exams. *Informal* behaviors may include joining a student club, befriending classmates, attending football games, and even abusing alcohol on weekends.

Like statuses, roles are relational (complementary). Playing the role of professor requires teaching, advising students, being present during office hours, responding to email, and grading exams and assignments. Most professors are also expected to serve on committees, do research, publish articles and/or books, and perform services such as giving talks to community groups. Thus, the status of college student or professor involves numerous role requirements that govern who does what, where, when, and how.

Roles can be rigid or flexible. A person who occupies the status of secretary typically plays a role that is defined by rules about when to come to work, how to answer the phone, and when to submit the necessary work and in what format. A boss, on the other hand, usually enjoys considerable flexibility: She or he has more freedom to come in late or leave early, to determine which projects should be completed first, and to decide when to hire or fire employees.

Because roles are based on mutual obligations, they ensure that social relations are fairly orderly. We know what we are supposed to do and what others expect of us. If professors fail to meet their role obligations by missing many classes or coming to classes unprepared, students may respond by studying very little, cutting classes, and submitting negative course evaluations. If students miss classes, don't turn in the required work, or cheat, professors can fail them or assign low grades.

### 5-3a Role Performance

Roles define how we are *expected* to behave in a particular status, but people vary considerably in fulfilling the responsibilities associated with their roles. Many college students succeed, whereas others fail; some professors inspire their students, whereas others put them to sleep. These differences reflect **role performance**, the *actual* behavior of a person who occupies a status. For example, a professor may vary her or his role by demanding more of graduate students than of undergraduates. An instructor may also interact differently with male and female students and with a 19-year-old student than a 40-year-old student who's anxious about returning to college.

### 5-3b Role Set

We occupy many statuses and play many roles associated with each status. A **role set** refers to the different roles attached to a single status. Every role set includes rights and responsibilities associated with people with different statuses and role sets. *Figure 5.2* illustrates six roles of a typical college student. Because a different set of norms governs each of these relationships, the student interacts differently with a classmate than with a reference librarian or a professor. All of these interactions, shaped by explicit or implicit rules, make up a student's role set.

These six roles reflect only one status—that of college student. If you think about other statuses that a college student may occupy (employee, son or daughter, parent, girlfriend or boyfriend, husband or wife), you can see that meeting the expectations of numerous role sets can create considerable role conflict and role strain.

## FIGURE 5.2
## Role Set of a Typical College Student

How does interaction differ with these other role players in a college student's role set?

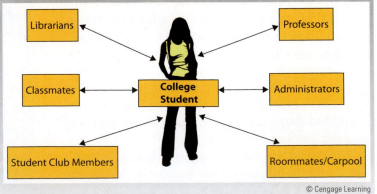

© Cengage Learning

## 5-3c Role Conflict and Role Strain

Playing many roles often leads to **role conflict**, the frustrations and uncertainties a person experiences when confronted with the requirements of two or more statuses. College students who have a job, especially if it's full-time, often experience role conflict because both employers and professors expect them to excel, but it's very difficult to meet these expectations. The role conflict increases if the student is a parent of young children or cares for an aging mother or father who needs help on a daily basis.

Whereas role conflict arises from tensions *between* the roles of two or more statuses, **role strain** is the stress that arises from incompatible demands among roles *within* a single status. Students experience role strain when several exams are scheduled on the same day and they are also involved in time-consuming extracurricular activities. Faculty members experience role strain when the requirements of being a professor—teaching, research, and community service—sap their energy and time. And military chaplains report role strain in preaching about peace while blessing those about to go to war.

Almost all of us experience role strain because many inconsistencies are built into our roles. *Table 5.1* gives examples of some of the factors that create role conflict and role strain.

## 5-3d Coping With Role Conflict and Role Strain

Role conflict and role strain can produce tension, hostility, aggression, and stress-related physical problems such as insomnia, headaches, ulcers, eating disorders, anxiety attacks, chronic fatigue, nausea, weight loss or gain, and drug and alcohol abuse (Weber et al. 1997). Some role conflict and role strain may last only a few weeks, but others are long-lived (e.g., working in a stressful or low-paying job).

To deal with role conflict and role strain, some people *deny that there's a problem*. Employed mothers, especially those who are divorced, often become supermoms—providing home-cooked meals, attending their children's sports activities, and volunteering for a school's fund-raising campaign even though they're exhausted, must do laundry at midnight, and neglect their own health and interests. Supermoms may succeed

over a number of years in managing role conflict and role strain. Eventually, however, they may become angry and resentful, or experience health or emotional problems (Douglas and Michaels 2004).

There are five more effective ways to minimize role conflict and role strain. First, we can reduce role conflict through *compromise* or *negotiation*. To decrease the conflict between work and family roles, for example, many couples draw up schedules that require fathers to do more of the housework and child rearing (see Chapter 12). Second, we can *set priorities*. If extracurricular activities interfere with studying,

**role conflict** the frustrations and uncertainties a person experiences when confronted with the requirements of two or more statuses.

**role strain** the stress that arises from incompatible demands among roles within a single status.

*Glenda M. Powers/Shutterstock.com*

### TABLE 5.1

## Why Do We Experience Role Conflict and Role Strain?

| REASON | EXAMPLE |
|---|---|
| Because many people are overextended, some roles are bound to conflict with others. | Students may study less than they want because employers demand that they work overtime or on weekends. |
| People have little or no training for many roles. | Parents are expected to turn out "perfect" kids even though they receive training for driving a car but none for parenting. |
| Some role expectations are unclear or contradictory. | Some employers pride themselves on having family-friendly policies but expect employees to work 12 hours a day, travel on weekends, and use vacation days to care for a sick child. |
| Highly demanding jobs often create difficulties at home. | Some jobs (such as being in the military, policing, and firefighting) require people to be away from their families for extended periods of time or during crises. |

© Cengage Learning

Like many other people, college students can reduce their role conflict and role strain by setting priorities and compartmentalizing their roles.

Dmitriy Shironosov/Shutterstock.com

which is more important? Succeeding in college always requires making sacrifices, such as attending fewer parties and not seeing friends as often as we'd like.

Third, we can *compartmentalize* our roles. Many college students take courses in the morning, work part-time during the afternoon or evening, and devote part of the weekend to leisure activities. It's not always easy but usually possible to separate our various roles.

Fourth, we can decide *not to take on more roles*. One of the most effective ways to avoid role conflict is to just say "no" for requests to do volunteer work, pressure from family or friends to take on unwanted tasks (such as babysitting), or pleas to become involved in college or community activities.

Finally, we can *exit* a role or status. Withdrawing from community activities and club offices, for example, can decrease role conflict and role strain considerably. There's always pressure to remain in a role, but none of us is indispensable, and there are many people who are eager to replace us. Some role exits are painful and long-lived, however. Divorce, for instance, is not a quick event but a process that is usually spread over many years (sometimes decades), during which two people (and their children) must redefine their expectations and adjust to a different household structure (see Chapter 12).

# 5-4 Explaining Social Interaction

Statuses and roles are two critical components of social structure that shape our everyday relationships, but it is social interaction that provides the basis of these relationships: It affects who you are, how you behave, and what you say. Social interaction seems natural and simple, but why do we interact as we do? And why do people sometimes interpret the same words differently? Three micro-level perspectives—symbolic interactionism, social exchange theory, and feminist theories—provide answers to these and other questions. These perspectives offer distinctive contributions, but they have one characteristic in common: Each explains how people interact in their daily lives (*Table 5.2* summarizes these approaches).

## 5-4a Symbolic Interactionism

For interactionists, the most significant characteristic of all human communication is that people take each other and the context into account (Blumer 1969). When your professors ask, "How are you?" they expect a "Fine, thanks" and probably barely look at you. A doctor, on the other hand, usually looks you in the eye when asking "How are you?" and takes notes as soon as you start to reply. Thus, "How are you?" has different meanings in different social contexts and elicits different responses ("I'm doing okay" vs. "I've been having a lot of headaches during the last few weeks"). In this sense, people construct reality.

### Social Construction of Reality

What people perceive and understand as reality is a creation of the social interaction of individuals and groups. "Human reality is socially constructed reality" because people impose their subjective meanings on interactions to make sense of the world around them (Berger and Luckmann 1966: 172). That is, we produce, interpret, and share the reality of everyday life with others. This social construction of reality typically evolves through direct, face-to-face interaction, but the interaction can also be indirect, as in watching television or participating in online social networks.

Businesspeople, advertisers, politicians, educators, advocacy groups, and even social scientists use words deliberately to shape or change our perceptions of reality. For example, doublespeak is "language that pretends to communicate but really doesn't. [It] makes the bad seem good, the negative appear positive, the unpleasant appear attractive or at least tolerable."

There are several kinds of doublespeak: *Euphemisms* are words or phrases that avoid a harsh, unpleasant, or distasteful reality; *gobbledygook* (or

# DOUBLESPEAK

1. automotive internist
2. internment excavation expert
3. service technician
4. auto dismantler and recycler
5. air support
6. previously owned
7. genuine imitation
8. surveillance
9. equity retreat
10. revenue enhancement

A. repairperson
B. bombing
C. spying
D. gravedigger
E. fake
F. car mechanic
G. junk dealer
H. stock market crash
I. used
J. tax increase

Source: Based on Lutz 1989, and the National Council of Teachers of English Committee on Public Doublespeak 2005.

**Answers: Double Speak**

10. J
9. H
8. C
7. E
6. I
5. B
4. G
3. A
2. D
1. F

© Cory Docken/Spots Illustration/Jupiterimages

> **self-fulfilling prophecy**
> a situation in which if we define something as real and act on it, it can, in fact, become real.

*bureaucratese*) overwhelms the listener with big words and long sentences; and *inflated language* makes everyday things seem impressive and the simple seem complex (Lutz 1989: 1–6). Take the quiz above to see how much doublespeak you understand or use.

## Social Interaction and Self-Fulfilling Prophecies

Our perceptions of reality shape our behavior. In an oft-cited statement, also known as the *Thomas Theorem*, sociologists W. I. Thomas and Dorothy Thomas (1928: 572) observed, "If men define situations as real, they are real in their consequences." That is, our behavior is a result of how we interpret a situation ("My mother-in-law hates me, so I go shopping when she drops in").

Carrying this idea further, sociologist Robert Merton (1948/1966) proposed that our definitions of reality can result in a **self-fulfilling prophecy**: If we define something as real and act on it, it can, in fact, become real. For example, a researcher who interviewed adults ages 21 to 64 found that physical education teachers who publicly humiliated students often turned them off physical fitness for good. One man said, "To this day I feel totally inadequate in team-related activities and have a natural reflex to AVOID THEM AT ALL COSTS" (Strean 2009: 217; capitalization in original). Thus, gym teachers' negative comments made students feel inadequate in sports and

---

## TABLE 5.2

# Sociological Explanations of Social Interaction

| PERSPECTIVE | KEY POINTS |
|---|---|
| Symbolic Interactionist | • People create and define their reality through social interaction.<br>• Our definitions of reality, which vary according to context, can lead to self-fulfilling prophecies. |
| Social Exchange | • Social interaction is based on a balancing of benefits and costs.<br>• Relationships involve trading a variety of resources, such as money, youth, and good looks. |
| Feminist | • The sexes act similarly in many interactions but often differ in communication styles and speech patterns.<br>• Men are more likely to use speech that's assertive (to achieve dominance and goals), while women are more likely to use language that connects with others. |

© Cengage Learning

changed their behavior during adulthood, regardless of their ability.

Our perceptions of reality shape our behavior, but how do people define that reality? Interactionists use two research tools—*ethnomethodology* and *dramaturgical analysis*—to help answer this question.

## Ethnomethodology

A term coined by sociologist Harold Garfinkel (1967), **ethnomethodology** is the study of how people construct and learn to share definitions of reality that make everyday interactions possible. That is, we base our interactions on common assumptions about what makes sense in specific situations (Schutz 1967; Hilbert 1992).

People make sense of their everyday lives in two ways. First, by observing conversations, people discover the general rules that we all use to interact. Second, people can understand interaction rules by breaking them. Over a number of years, Garfinkel instructed his students to purposely violate everyday interaction rules and then to analyze the results. In these exercises, some of his students went to a grocery store and insisted on paying more than was asked for a product. Others were instructed, in the course of an ordinary conversation and without indicating that anything unusual was happening, "to bring their faces up to the subject's until their noses were almost touching" (Garfinkel 1967: 72).

When students violated such everyday interaction rules, they were sanctioned. Grocery clerks became hostile when the students insisted on paying more than the marked price for a product, and people backed off when their noses were almost touching: "Reports were filled with accounts of astonishment, bewilderment, shock, anxiety, embarrassment, and anger" (Garfinkel 1967: 47).

Violating interaction rules, even unspoken ones, can trigger anger, hostility, and frustration. College students become upset if professors ignore teaching norms by using sarcasm and putdowns, coming to class

unprepared or consistently late, misplacing students' homework, constantly reading from the book, or discouraging students' comments and questions (Berkos et al. 2001).

## Dramaturgical Analysis

**Dramaturgical analysis** is a research approach that examines social interaction as if occurring on a stage where people play different roles and act out scenes for the "audiences" with whom they interact. According to sociologist Erving Goffman (1959, 1967), life is similar to a play in which each of us is an actor, and our social interaction is much like theater because we're always on stage and always performing. In our everyday performances, we present different versions of ourselves to people in different settings (audiences).

Because most of us try to present a positive image of ourselves, much social interaction involves *impression management*, a process of suppressing unfavorable traits and stressing favorable ones (see Chapter 4). To control information about ourselves, we often rely on props to convey or reinforce a particular image. For example, physicians, lawyers, and college professors may line their office walls with framed diplomas, medical certificates, or community awards to give the impression that they're competent, respected, and successful.

According to Goffman, the presentation of a performance involves front- and back-stage behaviors. The *front stage* is an area where an actual performance takes place. In front stage areas, such as living rooms or restaurants, the setting is clean and the servers or hosts are typically polite and deferential to guests. The *back stage* is an area concealed from the audience, where people can relax. Bedrooms and restaurant kitchens are examples of back stages. After guests have left, the host and hostess may kick off their shoes and gossip about their company. In restaurants, cooks and servers may criticize the guests, use vulgar language, and yell at each other. Thus, the civility and decorum of the front stage may change to rudeness in the back stage.

Another example of front- and back-stage behavior involves faculty and students. Professors want to create the impression that they're well prepared, knowledgeable, and hardworking. In back-stage areas such as their offices, however, faculty may complain to their colleagues that they're tired of teaching a particular course or grading terrible exams and papers, or they may confess that they don't always prepare for classes as well as they should. Students

Cory Thoman/Shutterstock.com

*The chaotic and crowded back-stage conditions in many restaurant kitchens stand in stark contrast with the efficient and relaxed front-stage behavior of its servers. Can you think of a situation where your front- and back-stage behaviors differ dramatically?*

**social exchange theory**
the perspective whose fundamental premise is that social interaction is based on each person's trying to maximize rewards (or benefits) and minimize punishments (or costs).

often plead with instructors for higher grades: "But I studied very hard" or "I'm an A student in my other courses." In back-stage areas such as dormitories, student lounges, or libraries, however, students may admit to their friends that they barely studied or that they have a low grade-point average.

## 5-4b Social Exchange Theory

The fundamental premise of **social exchange theory** is that social interaction is based on each person's trying to maximize rewards (or benefits) and minimize punishments (or costs). An interaction that elicits a reward, such as approval or a smile, is more likely to be repeated than an interaction that brings a cost, such as disapproval or criticism (Thibaut and Kelley 1959; Homans 1974; Blau 1986). Interactions are most satisfying when there is a balance between giving and taking.

People bring various tangible and intangible resources to a relationship, such as money, status, intelligence, good looks, youth, power, and affection. Any of a person's resources can be traded for more, better, or different resources that another person possesses. Marriages between older men and younger women, for example, often reflect an exchange of the man's power, money, and/or fame for the woman's youth, physical attractiveness, and ability to bear children.

Many of our cost–reward decisions are conscious and deliberate, but others are passive or based on long-term negative interactions. For example, much of the research on domestic violence shows that women stay in abusive relationships because their self-esteem has eroded after years of criticism and ridicule from both their parents and their spouses or partners ("You'll be lucky if anyone marries you," "You're dumb," "You're ugly," and so on). In effect, the abused women (and sometimes men) believe that they have nothing to exchange in a relationship, that they don't have the right to expect benefits (especially when the abuser blames the victim for provoking the anger), or that enduring an abusive relationship is better than being alone (Walker 2000; Bergen et al. 2005).

## 5-4c Feminist Theories

Feminist perspectives offer additional insights on social interaction. Two examples are emotional labor and gender roles.

### Interaction and Emotional Labor

Emotions are important components of social interaction. According to sociologist Arlie Hochschild (1983: 11), we learn *feeling rules* that shape the appropriate emotions for a given role or situation. *Emotional labor*, "the management of feeling to create a publicly observable facial and bodily display," is critical in many occupations that demand continuous interaction and the masking of one's true feelings. Hochschild's examples of such occupations included secretaries, waiters and waitresses, flight attendants, and hotel receptionists. She maintained that because many women occupy low-income jobs with little power that deal with the public, they are more likely than men to perform work that requires emotional labor.

**nonverbal communication** messages that are sent without using words.

There are numerous online and magazine stories about employees who are expected to manage their emotions in dealing with clients, customers, and coworkers, but are the rules for managing emotions the same for everyone? A study that interviewed black professionals found that both women and men believed that they were held to a different standard of "feeling rules" than their white coworkers. Consequently, they kept their emotions in check to avoid being labeled as "too sensitive," "angry," "irritating," or "unpleasant" (Wingfield 2010).

### Interaction and Gender Roles

Women and men are more similar than different in their interactions. A study that recorded conversations of college students in the United States and Mexico found that women and men spoke about the same amount—16,000 words per day. An analysis of studies published since the 1960s concluded that men are generally more talkative than women, but talkativeness depends on the particular situation. During decision-making tasks, men are more talkative than women, but when talking about themselves or interacting with children, women are more talkative than men (Leaper and Ayres 2007; Mehl et al. 2007).

Other studies support the findings that cultural norms and gender role expectations shape our interaction patterns. Generally, women are socialized to be more comfortable talking about their feelings, whereas men are socialized to be dominant and take charge, especially in the workplace. Women often ask questions that probe for a greater understanding of feelings ("Were you glad it happened?"). Women are also much more likely than men to do conversational "maintenance work," such as asking questions that encourage conversation ("What happened at the meeting?") (Lakoff 1990; Robey et al. 1998; Farley et al. 2010).

Compared with female speech, men's speech often reflects *conversational dominance*, speaking more frequently and for longer periods. Men also show dominance by interrupting or ignoring others, reinterpreting the speaker's meaning, or changing the topic (Tannen 1990; Mulac 1998; Toth 2011).

Such interaction differences aren't innate because of one's sex, however. Instead, much depends on women's and men's domestic roles and their position and status in the job hierarchy. For example, men who do much of the parenting have communication styles that are similar to those of mothers. In the workplace, women

Andersen Ross/Jupiterimages

In predominantly white organizations, many black professionals believe that they would be punished or fired for expressing the same feelings and emotions as their white counterparts.

and men who occupy high-level decision-making positions also have similar interaction styles with superiors and subordinates (Cameron 2007).

## 5-5 Nonverbal Communication

**N**onverbal communication refers to messages sent without using words. This silent language, a "language of behavior" that conveys our real feelings, can be more potent than our words (Hall 1959: 15). For example, sobbing sends a much stronger message than saying, "I feel very sad." Some of the most common nonverbal messages are silence, visual cues, touch, and personal space.

### 5-5a Silence

Silence expresses a variety of emotions: agreement, apathy, awe, confusion, disagreement, embarrassment, regret, respect, sadness, thoughtfulness, and fear, to name just a few. In various contexts, and at particular points in a conversation, silence means different things to different people. Sometimes silence saves us from embarrassing ourselves. Think, for example, about the times you fired off an angry email or text message, regretted doing so an hour later, and then dreaded getting a fuming response.

Many of us have experienced the pain of getting "the silent treatment" from friends, family members, coworkers, or lovers. Not talking to people who are important to us builds up anger and hostility. Initially, the "offender" may work hard to make the closemouthed person discuss a problem. Eventually, the target of the silent treatment gets fed up, gives up, or ends a relationship (Rosenberg 1993).

## 5-5b Visual Cues

Visual cues, another form of nonverbal communication, include gestures, facial expressions, and eye contact. Let's consider a few examples of such body language in our daily interactions.

### Gestures in Early Life

How important are gestures even when we're infants? A recent study found an association (not causation) between baby gestures, the children's verbal ability at age 5, and the parents' socioeconomic status (SES). Parents who earned at least $60,000 a year and had a college degree or higher were more likely than those who earned less than $15,000 a year (and were often high school dropouts) to use more gestures, such as pointing, waving, and clapping. The higher SES parents were also more likely than the lower SES parents to associate a baby's gesture with words ("That's right, this is a doll"), which increases the child's vocabulary (Rowe and Goldin-Meadow 2009).

### Gestures in Adulthood

Most of us think we know what certain gestures mean—folding your arms across your chest indicates a closed, defensive attitude; leaning forward often shows interest; shrugging your shoulders signals indifference; narrowing your eyes and setting your jaw shows defiance; and smiling and nodding shows agreement.

Finger-pointing is usually a gesture that directs attention outward, placing blame or responsibility on someone else. The common "talk to the hand" gesture sends a stronger message: "Go away!" or "I'm not listening to you." Few gestures, however, clearly convey a

"This concludes my lecture on non-verbal communication. Any comments or questions?"

Chris Wildt/Cartoonstock

meaning by themselves. Instead, they must be interpreted in context. Because of habit or hearing problems, for instance, a coworker may always lean forward when listening regardless of whether he or she is interested.

The same gesture may have different meanings in different countries:

- Tapping one's elbow several times with the palm of one's hand indicates that someone is sneaky (in Holland), stupid (in Germany and Austria), or mean or stingy (in South America).

- Screwing a forefinger into one cheek signifies a dimple—a traditional sign of feminine beauty—and means "She's beautiful" in Italy and Libya. The same gesture in southern Spain means that a man is effeminate, and in Germany, "You're crazy!"

- In many Middle Eastern countries, people view the shoes and the soles of one's feet as unclean. Thus, stretching out one's legs with the feet pointing at someone or crossing one's legs so that a sole faces another person is considered rude (Morris 1994; Lynch and Hanson 1999; Jandt 2001).

In a remarkable tarmac ritual, and regardless of weather, the ground crews at Japanese airports use many meaningful gestures as a jet pushes back from the gate. First, the crew members line up, snap to attention,

| UNITED STATED | SOUTH AMERICA | JAPAN | FRANCE | GERMANY | OTHER COUNTRIES |
|---|---|---|---|---|---|
| okay | not okay | money | zero | vulgar gesture | better check first |

Hemera Technologies/AbleStock.com/Getty Images

Departing jets get a wave from the ground crew in Tokyo.

Paul Vincent Farrell/endlessness.org

and then, "in perfect unison," they bow. This gesture is directed toward the passengers on the plane ("Thank you for visiting Japan"), toward the plane's flight crew ("We respect your expertise and dedication"), and, ultimately, as recognition of the crew's commitment to service ("We have fulfilled our duties to the best of our abilities"). Then they straighten up, smile broadly, and wave goodbye to show the passengers that they had been welcome to Japan (Gottlieb 2011: 44).

### Facial Expressions

Facial expressions are visual cues of feelings, but they can be deceptive. First, our facial expressions don't always show our true emotions. Parents, for example, tell their children "Don't you roll your eyes at me!" or "Look happy when Aunt Minnie hugs you." Thus, children learn that displaying their real feelings—especially when they're negative—is often unacceptable.

Second, faces can lie about feelings. Parents often know when children are lying because the children avoid eye contact, cry, or blink and swallow frequently. Many adults, in contrast, monitor and control their facial expressions. They can deceive, successfully and over many years, because they have rehearsed the lies in their heads, are smooth talkers with a reputation for being trustworthy, or have gotten away with lying in the past (Ekman 1985; Sullivan 2001).

Third, our facial expressions don't always reflect how we feel because the rules change over time. Consider smiling. Infants learn very quickly that a smile will get a parent's attention and a positive response (the parent smiles back) (Jones and Hong 2001). By kindergarten, most of us have learned to smile (and not smile) only in appropriate situations ("Don't smile during Uncle Fred's funeral, dear."). Finally, facial expressions can be misleading

Ned Frisk/Blend Images/Getty Images

because of cultural variations. American businesspeople have grumbled that Germans are cool and aloof, whereas many German businesspeople have complained that Americans are excessively friendly and hide their true feelings with grins and smiles. The Japanese, who believe that it's rude to display negative feelings in public, smile more than Americans do to disguise embarrassment, anger, and other negative emotions (Jandt 2001).

The only consistent research finding, across cultures, is that women smile more than men. This difference may be due to similar cultural norms that socialize women to hide negative feelings that might make people uncomfortable, and to show deference to men (Szarota 2010).

### Eye Contact

Eye contact serves several social purposes (Eisenberg and Smith 1971; Ekman and Friesen 1984; Siegman and Feldstein 1987). First, we get much information about other people by looking at their eyes. Eyes open wide show surprise, fear, or a flicker of interest. When we are angry, we stare in an unflinching manner. When we are sad or ashamed, our eyes may be cast down.

Second, eye contact is a potent stimulus. Speakers sometimes change their strategy when people in the audience don't look at them. Your instructors, for example, may speak louder, walk around the classroom, or revert to writing on a blackboard when they feel that their students' lack of eye contact indicates boredom or texting. In contrast, people usually evade the eyes of someone whose gaze is disturbingly direct. When two people like each other, they establish eye contact more often and for longer durations than when there is tension in the relationship.

Finally, eye contact often varies across societies. In the United States, making eye contact is a key component of effective body language. Especially during job interviews, we're told, our eyes should signal attentiveness and interest, and "blinking, staring, or looking away whenever you begin speaking makes it hard for you to connect with your interviewer" (Bohannon 2000: 22). In some Asian cultures, like Japan, in contrast, students often avoid making eye contact with their instructors as a sign of respect (Jandt 2001).

## 5-5c Touch

Touch is another important form of nonverbal communication. Touching sends powerful messages about our feelings and attitudes.

### What Touching Communicates

Touching expresses our feelings toward another person. Parents worldwide communicate with their infants

Pictured here are former President Bush and Saudi Crown Prince Abdullah at Bush's Texas ranch in 2002. Many U.S. journalists raised questions about two men holding hands. In response, Arab Americans were quick to point out that Arab society sees the outward display of affection between male friends as an expression of respect and trust. Thus, government officials and military officers often hold hands as they walk together or converse with one another.

Rod Aydelotte-Pool/Getty Images

through touch—stroking, holding, patting, rubbing, and cuddling.

Touching sends many messages, some positive (hugging, embracing, kissing, and holding hands) and some negative (hitting, shoving, pushing, and spanking). Other forms of touching are controlling. One of my students, for example, left a boyfriend because "He said he loved me and trusted me but he gripped my arm tightly every time I talked to another guy and pulled me away." Between intimate partners, a decline in the amount of touching may signal that feelings are cooling off.

### Gendered Touching

Whether touching is viewed as positive or negative depends on the situation and one's gender. In higher education, even when the faculty member is well liked, male and female students perceive touching differently. When female professors touch male students on the arm while talking to them, the students view the gesture as friendly. When a male professor touches a female student on the arm, she may get nervous because she's afraid that the touching may escalate (Lannutti et al. 2001; Fogg 2005). Generally, women are more likely than men to initiate hugs and touches that express support, affection, and comfort. In contrast, men more often use touching women to assert power or show sexual interest (Wood 2011).

### Cross-Cultural Variations in Touching

As with other forms of nonverbal communication, the interpretation of touching varies from culture to culture. In some Middle Eastern countries, people don't offer anything to another with the left hand because it's used to clean oneself after using the toilet. And among many Chinese and other Asian groups, hugging, backslapping, and handshaking aren't as typical as they are in the United States because such touching is seen as too intimate (Lynch and Hanson 1999).

According to one scholar, compared with those in some other societies, people in the United States are "touch-deprived" because they have one of the lowest rates of casual touch in the world:

> If you are talking to a friend in a coffee shop in the United States, you might touch each other once or twice an hour. If you were British and in a London coffee shop, you probably won't touch each other at all. But if you were French and in a Parisian café, you might touch each other a hundred times in an hour! (Jandt 2001: 117)

## 5-5d Personal Space

In the example of airline seats at the beginning of this chapter, you saw that the distance that people establish between themselves is an important aspect of nonverbal communication. Personal space plays a significant role in our everyday nonverbal interactions, reflects power and status, and varies across societies.

### When Is Our Personal Space Violated?

Our living space is public or private. In the public sphere, which is usually formal, we have clearly delineated spaces: "This is your locker," "That's her office," or "You're parking in my spot." We usually decorate our public spaces with businesslike artifacts such as awards or framed photos of an institution's accomplishments.

After a snowstorm, people in densely populated cities who shovel parking spots "mark" their personal spaces with lawn chairs, strollers, and even a table set for two, complete with a bottle of wine. Some of the most aggressive (and locally accepted) retaliation occurs in South Boston, where residents punish violators by slashing their tires or smashing their car windows (Goodnough 2010).

Private spaces send a different message. They convey informality and a relaxed feeling. Private spaces—such as homes, apartments, and dorm rooms—are often mini-museums that reflect people's interests, hobbies, hygiene, and personalities.

People sometimes invade our personal space by standing too close to us in a line, leaning against us, using all of the shared armrest in tight airline seats, or plopping their feet on the chair next to us in a classroom. Not all space intrusions are physical encroachments. Loud cell phone conversations are a good example of an auditory intrusion into our personal space.

In intimate relationships (such as between family members), our personal space is often 2 feet or less because we're at ease with close physical proximity. In contrast, in public situations (such as someone speaking to a large audience or faculty in large classrooms), the personal space is often 12 feet or more because the speaker, who has a higher social status than the audience, has a formal relationship with the listeners and avoids close physical contact (Hall 1966).

## Space and Power

Space usage signifies who has privilege, status, and power (Chapman and Hockey 1999; Falah and Nagel 2005). Wealthy people can afford enormous apartments in the city or houses in the suburbs, whereas the poor are crowded into the most undesirable sections of a city or town or in trailer parks. Executives, including college presidents, usually have huge offices (even entire suites), whereas faculty members often share office space, even though many of their discussions with students are confidential. Generally, the higher the SES, the greater the consumption of space, including large cars, reserved parking spaces, private dining areas, first-class airline seats, and luxurious skyboxes at sports stadiums.

## Cross-Cultural Variations and Space

Cultural values and norms determine how we use space. Americans not only stand in line but also have strict queuing rules: "Hey, the end of the line is back there" and "That guy is trying to cut into the line." In contrast, "along with Italians and Spaniards, the French are among the least queue-conscious in Europe" (Jandt 2001: 109). In these countries and others, people routinely push into the front of a group waiting for taxis, food, and tickets, and nobody objects.

Americans maintain personal space in an elevator by moving to the corners or to the back. In contrast, an Arab male may stand right next to or touching another man even when no one else is in the elevator.

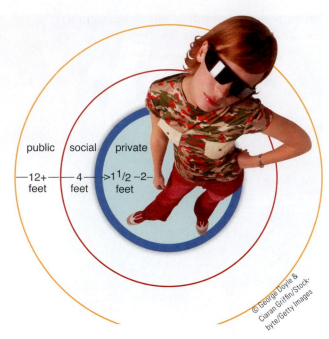

public    social    private

—12+—    —4—    >1½ –2—
feet       feet         feet

© George Doyle & Ciaran Griffin/Stock-byte/Getty Images

Because most Arab men don't share American concepts of personal space in public places and in private conversations, they consider it offensive if the other man steps or leans away. In many Middle Eastern countries, however, there are stringent rules about women and men separating themselves spatially during religious ceremonies and about women avoiding any physical contact with men in public places (Jandt 2001; Office of the Deputy Chief 2006; see also Chapter 3).

# 5-6 Online Interaction

Many of us now interact in *cyberspace*, an online world of computer networks. In cyberspace, social media have mushroomed since 2009.

*Social media* are websites that don't just give information, but allow social interaction. Social media include social networking sites (e.g., Facebook and Twitter), gaming sites (e.g., *Second Life*), video sites (e.g., YouTube), and blogs (websites maintained by people who provide commentary and other material). This final section addresses two questions: Who's online and why? And is online interaction beneficial or harmful?

## 5-6a Who's Online and Why?

The percentage of Americans age 18 and older who are connected to the Internet increased from practically zero in 1994 to 80 percent in 2012 (Pew Internet & American Life Project 2012). More than 83 percent of U.S. adults have a cell phone, but only 35 percent own smartphones that allow them to interact and browse online (Smith 2011). Not having a smartphone may limit a person's immediate online interaction, but the demographic factors discussed in the next section may be even more limiting.

## Variations by Sex, Age, Race, Ethnicity, and Social Class

Almost equal numbers of women (79 percent) and men (81 percent) use the Internet. There has been an enormous growth in Internet use across most age groups, but only 48 percent of Americans aged 65 and older are online compared with 79 percent of those aged 50 to 64, 88 percent of those aged 30 to 49, and 94 percent of those aged 18 to 29. Many older Americans say that they don't need the Internet, whereas others live on fixed incomes and can't afford computers or monthly charges for Web service, or both (Fox 2006; Pew Internet & American Life Project 2012).

As *Table 5.3* shows, Asian Americans are the most wired group in the United States. They're more likely to use the Internet because many parents in this group are professionals who can afford computers and online service, and they encourage their children to use the Internet for education, a major avenue of upward mobility (see Chapters 10 and 13). Latinos with lower educational levels and little English proficiency are the least likely to be connected to the Internet (Livingston et al. 2009).

The biggest digital divide is between social classes. The higher a family's income, the greater the likelihood that its members are Internet users (see *Figure 5.3*). Affluent families can buy home computers, pay for Internet service, and send their children to schools that provide computer-related instruction. Thus, the children from the poorest families are the most likely to lack technological skills.

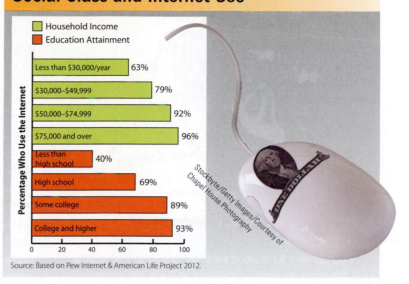

## FIGURE 5.3
## Social Class and Internet Use

Legend:
- Household Income
- Education Attainment

Percentage Who Use the Internet:
- Less than $30,000/year: 63%
- $30,000–$49,999: 79%
- $50,000–$74,999: 92%
- $75,000 and over: 96%
- Less than high school: 40%
- High school: 69%
- Some college: 89%
- College and higher: 93%

Source: Based on Pew Internet & American Life Project 2012.

Stockbyte/Getty Images/Courtesy of Chapel House Photography

## What We Do Online

A *social network* is a website that connects people who share similar personal interests (such as Facebook) or professional interests (such as LinkedIn). According to a recent national survey, however, Americans age 18 and older are more likely to use the Internet to find information on the Web (76 percent) and for email (92 percent) than for getting news, buying a product, or participating in social network sites (Purcell 2011).

## Sociology in Your Life

How have online activities changed since 2002? And what are the variations by sex, race/ethnicity, age, education, and household income? Go to Sociology CourseMate at CengageBrain.com to access this site.

## 5-6b How Beneficial and Harmful Is Online Interaction?

You'll see in Chapter 16 that technological changes bring both benefits and costs. The same is true of online interaction, ranging from family relationships to privacy issues.

## Family Relationships

Some commentators predicted that technology would pull families apart, but 64 percent of adults say that the new communication technology has made their families closer than when they were growing up because they

## TABLE 5.3

## Internet Usage by Race and Ethnicity

| PERCENTAGE OF AMERICANS, AGED 18 AND OLDER, WHO USE THE INTERNET | |
|---|---|
| Asian American | 87 |
| White | 76 |
| Latino | 67 |
| Black | 65 |

Source: Rainie 2011.

"Didn't you get my e-mail?"

Joe Kohl/Cartoonstock.com

can stay in touch with adult children who have moved out (Smith 2011).

Cell phones have increased the frequency of interaction between parents, especially if both are employed, to coordinate schedules and to chat with their children. Also, many parents report spending more time with their children by playing video games with them at home. In the case of young adults who are away at college, parents say that family members interact more often than in the past through email and text messaging (Kennedy et al. 2008).

Others complain that digital technology is cutting into family time. For example, some children complain that they rarely receive their parents' full attention because a parent is often immersed in email, texting, or being online even when pushing a swing, driving, or during dinner (Young 2011). In effect, some parents are paying more attention to their technology than their children.

Social media can have other harmful effects on family members, especially in the case of bad news. When a Baltimore police detective was killed, word spread quickly through police circles and spilled onto Facebook, where the officer's young daughter heard of his death before family members told her in person. Police departments want to be the first to inform relatives about bad news in person and provide immediate grief counseling. In other cases, Facebook friends have posted information about an engagement or a pregnancy before the couple's parents and relatives had been told (Fenton 2010).

### Other Online Social Connections

During the late 1990s, a number of scholars predicted that the Internet would replace close offline relationships, diminish people's involvement in interpersonal and community relationships, and isolate them socially. Contrary to such gloomy forecasts, the Internet has increased interaction, especially among people with similar interests. Millions of Americans use the Internet to plan religious activities, communicate with neighbors, petition politicians, and contact civic organizations. All of these activities have strengthened rather than diluted interpersonal and community ties, and have fostered a more diverse social network of people across social class and ethnic groups (Hampton et al. 2009; Smith 2010).

A national study found an increase since 2008 in the number of U.S. adults who use social networking to discuss important matters, get social support (such as receiving advice), and keep up with or revive dormant relationships (Hampton et al. 2011). Also, some ethnic groups, especially Asian Americans, who are rarely represented on television, have become very successful on YouTube. They've found "millions of eager fans" who follow their comedy sketches and discuss subjects, such as sex and race, that are taboo among older generations (Considine 2011: ST6).

A major disadvantage of cyberspace communication is that it's more impersonal than face-to-face interaction. According to one faculty member, for instance, email has made teaching more isolating because face-to-face interactions have "mostly evaporated": "Students used to drop by my office all the time, and I miss that. . . . After the business part of our conversation was over, we'd chat" (Connolly 2001: B5).

Some also maintain that social networking sites, especially Facebook, are superficial and give people a false sense of friendship and connection to others. You might have 60 "friends" on Facebook, but so what? How many of them really care, for example, about your cat's name or your favorite music, or are interested in viewing photos of your last vacation (Deresiewicz 2009)?

According to some critics, social networking sites may be inflating many people's egos, but they are replacing the time we might take to develop relationships with the few friends that we have (DiSalvo 2010). Others maintain that the social media aren't a substitute for personal contact: "How you spend time with someone is critical since it's over time that you learn about and come to appreciate each other" (Stich 2010: 20).

### Miscommunication

Email is supposed to make communication easier because online interaction lets us think before we speak. During face-to-face and phone discussions, we sometimes regret having been impulsive or rude. Spontaneous email messages can get us into trouble, but we can compose our thoughts and formulate our ideas more clearly in an email than in many face-to-face interactions. In Goffman's terms, we can manage the impressions that we give to others.

On the negative side, even the most neutral exchanges can spark problems. To avoid misunderstandings, email users often use *emoticons* (such as smiley faces or frowning ones) to convey voice inflections and facial expressions.

Still, when we don't see the other person's body language, misinterpretation is common because what we think are witty jokes may be taken as personal insults, and our matter-of-fact comments may be viewed as indifference (Menchik and Tian 2008). For example, a simple, matter-of-fact directive from an instructor, such as "I want you to rewrite this assignment and resubmit it," would usually be perceived as normal by a student in a face-to-face encounter but may be interpreted as abrupt in an email. Also, people sometimes forward private emails and photos to others without the sender's consent, causing embarrassment or anger.

### Cyberbullying and Gossip

In 2010, a Rutgers University freshman committed suicide after his roommate posted a video on the Internet of his sexual encounter with another man. Nationally, 14 percent of adolescents have experienced *cyberbullying*—deliberately using digital media to communicate false, embarrassing, or hostile information about another person. Cyberbullying is less common that traditional bullying, but it has more profound negative outcomes that include depression, anxiety, severe isolation, and, most tragically, suicide (Lenhart et al. 2011; O'Keeffe and Clarke-Pearson 2011; Wang et al. 2011).

In other cases, especially on college campuses, people can post anonymous and vicious racist, sexist, or homophobic comments about students on college network sites. A dean of students has likened the sites to "the worst of junior high," but the websites are protected by free speech (Yan 2009: 98). People who identify themselves can be suspended or expelled if they threaten someone with

physical harm, but being nasty in cyberspace—however offensive or vicious—is legal (Chapman 2010).

### Privacy

The Internet can expose more of our private information to the public than any technology in history. A major cost of online interaction is jeopardizing our privacy because email is neither anonymous nor confidential. Many companies monitor and preserve their employees' email messages and can use them to discipline or fire people. It's also becoming increasingly common for employers to search networking sites such as Facebook before deciding whether to make a job offer to a graduating college student.

People often say things in emails they would never say in person, especially when they're flirting, angry, frustrated, or tired. Emails don't disappear after being deleted but may last indefinitely in cyberspace. Divorce lawyers have successfully retrieved deleted emails to bolster their arguments that a divorcing parent shouldn't receive custody of the children (see Chapter 2).

Many Americans willingly give out personal information on social media sites—photos, phone number, address, age, education and work background, political and religious views, and so on. It takes a private investigator just a few clicks to get a composite picture of someone from public records, email messages, websites, blogs, and networking sites such as Facebook and LinkedIn. Also, computer-savvy individuals can use Web bugs, tags that track users as they move from site to site. The person compiles a profile of what someone likes and dislikes, then sells the information to companies (Lamb 2009).

© Rawdon Wyatt/Alamy

## Even the most neutral electronic exchanges can spark problems.

# SOC

**Social groups**
provide an important part of our
social identity.

AP Photo

# Social Groups, Organizations, and Social Institutions

Every year, about a dozen of my neighbors, all fervent Ravens football fans, get together to buy season tickets and to plan elaborate tailgating parties. Both the adults and children wear Ravens caps and purple jerseys, and most of the drivers attach pennants to their car antennas. These football enthusiasts are a social group.

> **social group** two or more people who interact with one another and who share a common identity and a sense of belonging or "we-ness."

## Key Topics

In this chapter, we'll explore the following topics:

6-1 Social Groups

6-2 Formal Organizations

6-3 Sociological Perspectives on Social Groups and Organizations

6-4 Social Institutions

## 6-1 Social Groups

A **social group** consists of two or more people who interact with one another and who share a common identity and a sense of belonging or "we-ness." Friends, families, work groups, religious congregations, clubs, athletic teams, World War II veterans, and organizations are all examples of social groups.

Some groups are highly organized and stable (political parties); others are fluid and temporary (high school classmates). Interaction, especially face-to-face interaction, is the key ingredient in creating and maintaining groups.

Social groups are essential because they provide an important part of our social identity and help us understand the behavior of other people in our society. Each of us is a member of many social groups, but some are more significant than others because they shape our social and moral beliefs. The most important types of social groups are primary and secondary groups, in-groups and out-groups, and reference groups.

### 6-1a Primary Groups and Secondary Groups

Suppose your car's battery died this morning, your sociology professor gave a pop quiz for which you hadn't studied, your hard drive crashed, and your microwave stopped working. Whom might you call to vent? Your answer reflects the difference between primary and secondary groups.

## What do you think?

It's a good idea for companies to use video surveillance, including in restrooms.

| 1 | 2 | 3 | 4 | 5 | 6 | 7 |

strongly agree      strongly disagree

## Primary Groups

A **primary group** is a relatively small group of people who engage in intimate face-to-face interaction over an extended period. For sociologist Charles Horton Cooley (1909/1983: 24), the most significant primary groups were "the family, the play-group of children, and the neighborhood or community of elders" because they are first and central in shaping a person's social and moral development.

Primary groups are our emotional glue. We call members of our primary group to share good news or to gripe. Primary group members are typically understanding, supportive, and tolerant even when we're in a bad mood or selfish. They have a powerful influence on our social identity because we interact with them on a regular basis over many years, usually throughout our lives. Because primary group members genuinely care about each other, they contribute to one another's personal development, security, and well-being. Our family and close friends, for example, stick with us through good and bad, and we feel comfortable being ourselves with them.

## Secondary Groups

A **secondary group** is a large, usually formal, impersonal, and temporary collection of people who pursue a specific goal or activity. Your sociology class is a good example of a secondary group. You might have a few friends in class, but students typically interact infrequently and relatively formally. When the semester (or quarter) is over and you've accomplished your goal of passing the course, you may not see each other again (especially if you're attending a large college or university). And you certainly wouldn't call your professor if you needed a sympathetic ear at the end of a bad day. Other examples of secondary groups include sports teams, labor unions, and employees of a company.

Unlike primary groups, secondary groups are usually highly structured: There are many rules and regulations, people know (or care) little about each other personally, relationships are formal, and members are

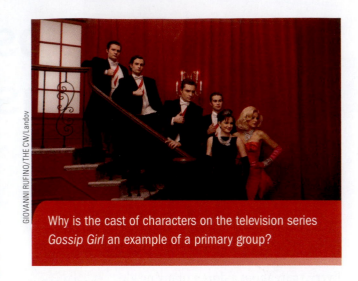

Why is the cast of characters on the television series *Gossip Girl* an example of a primary group?

GIOVANNI RUFINO/THE CW/Landov

© Lew Robertson/Corbis

expected to fulfill particular functions. Whereas primary groups meet our *expressive* (emotional) needs, secondary groups fulfill *instrumental* (task-oriented) needs. Once a task or activity is completed—whether it's earning a grade, turning in a committee report, or building a bridge—secondary groups split up and become members of other secondary groups.

*Table 6.1* summarizes the characteristics of primary and secondary groups. These characteristics are **ideal types**—general traits that describe a social phenomenon rather than every case. Ideal types provide composite pictures of how social phenomena differ rather than specific descriptions of reality. Because primary and secondary groups are ideal types, their characteristics can vary. Thus, members of primary groups may sometimes devote themselves to meeting instrumental needs (such as running a family-owned business), and members of secondary groups (such as military units and athletic teams) can develop lasting ties.

## 6-1b In-Groups and Out-Groups

*All good people agree,*
*And all good people say,*
*All nice people, like Us, are We*
*And everyone else is They.*
*—from* We and They, *by Rudyard Kipling*

Comstock/Jupiter Images

Rudyard Kipling's poem "We and They" captures the essence of in-groups and out-groups. Members of an **in-group** share a sense of identity and "we-ness" that typically excludes and devalues outsiders. **Out-groups** consist of people who are viewed and treated negatively because they are seen as having values, beliefs, and other characteristics different from those of an in-group. For example, "we" vegetarians are healthier than "you" meat eaters, "we" computer nerds are smarter than "you" fraternity and sorority "types," and so on.

Based on ascribed or achieved statuses, almost everyone sees others as members of in-groups and out-groups. From ancient times to the present, people in various parts of the world have made "we" and "they" distinctions based on race or ethnicity, gender, sexual orientation, religion, age, social class, and other social and biological characteristics (Coser 1956; Tajfel 1982; Hinkle and Schopler 1986).

Out-group members are usually aware of being outsiders. For example, overweight people are often viewed by others as lazy, unmotivated, and undisciplined. Many overweight people have internalized such negative attitudes and see themselves as part of an out-group. In a national study of overweight people, nearly 40 percent had negative attitudes about fat people (including themselves), about 33 percent would rather experience a divorce than be obese, 15 percent would give up 10 years or more of life to be thinner, and 4 percent would trade blindness for obesity (Schwartz et al. 2006).

In-groups and out-groups affect our feelings about ourselves and others and our life outcomes. In-groups can be a positive influence, promoting individuals' sense of self-worth and belonging and increasing group solidarity and cohesion. They can also create conflict and provoke inhumane actions, including war. For example, in-group/out-group hostilities have fueled the Palestinian–Israeli battles over the occupation of the West Bank, the eviction of white farmers in Zimbabwe, and ongoing civil wars in some African nations.

> **reference group** a group of people who shape our behavior, values, and attitudes.

## 6-1c Reference Groups

Besides belonging to primary groups, secondary groups, in-groups, and out-groups, we also have reference groups. A **reference group** is a group of people who shape our behavior, values, and attitudes (Merton and Rossi 1950). Reference groups influence who we are, what we do, and who we'd like to be in the future. Unlike primary groups, however, reference groups rarely provide personal support or face-to-face interaction over time.

Reference groups might be people with whom we already associate, such as a college club or a recreational athletic team. They can also be groups that we admire and want to be part of, such as doctors or financial advisors. Each person has many reference groups. Your sociology professor, for example, may be

---

### TABLE 6.1

# Characteristics of Primary and Secondary Groups

|  | CHARACTERISTICS OF A PRIMARY GROUP | CHARACTERISTICS OF A SECONDARY GROUP |
|---|---|---|
| Interaction | • Face to face<br>• The group is usually small | • Face to face or indirect<br>• The group is usually large |
| Communication | • Communication is emotional, personal, and satisfying | • Communication is emotionally neutral and impersonal |
| Relationships | • Intimate, warm, and informal<br>• Usually long-term<br>• Valued for their own sake (expressive) | • Typically remote, cool, and formal<br>• Usually short-term<br>• Goal-oriented (instrumental) |
| Individual Conformity | • Individuals are relatively free to stray from norms and rules | • Individuals are expected to adhere to rules and regulations |
| Membership | • Members are not easily replaced | • Members are easily replaced |
| Examples | • Family, close friends, girlfriends and boyfriends, self-help groups, street gangs | • College classes, political parties, professional associations, religious organizations |

© Cengage Learning

From kindergarten through high school, cliques typically exclude others because of social class, race, religion, ethnicity, sexual orientation, appearance, school activities, or other characteristics. Should schools try to discourage the formation of in-groups? Why or why not?

a member of several professional sociological associations, a golf enthusiast, a parent, and a homeowner. Identification with each of these groups influences her or his everyday attitudes and actions.

Like in-groups and primary groups, reference groups can have a strong impact on our self-identity, self-esteem, and sense of belonging because they shape our current and future attitudes and behavior. We typically add or drop reference groups throughout the life course. If you aspire to move up the occupational ladder, for example, your reference group may change from entry-level employees to managers, vice presidents, and CEOs.

## 6-1d Group Conformity

Most Americans see themselves as rugged individualists who have a mind of their own and don't bow to group pressure (see the discussion of values in Chapter 3). A number of studies have shown, however, that many of us are profoundly influenced by group pressure. Four of the best known of these studies are by Solomon Asch, Stanley Milgram, Philip Zimbardo, and Irving Janis.

### Asch's Research

In a now classic study of group influence, social psychologist Solomon Asch (1952) told subjects that they were taking part in an experiment on visual judgment. After seating six to eight male undergraduates around a table, Asch showed them the line drawn on card 1 and asked them to match

the line to one of three lines on card 2 (see *Figure 6.1*). The correct answer, clearly, is line C.

All but one of the subjects—who usually sat in the last chair—were Asch's confederates, or accomplices. In the first test, all the confederates selected the correct matching line. In the other tests, each of them, one by one, deliberately chose an incorrect line. Thus, the non-confederates faced a situation in which seven other group members had unanimously agreed on a wrong answer. Averaged over all of the trials, 37 percent of the non-confederate subjects ended up agreeing with the group's incorrect answers. When they were asked to judge the length of the lines alone, away from the influence of the group, they made errors only 1 percent of the time.

Asch's research demonstrated the power of groups over individuals. Even when we know that something is wrong, we may go along with the group to avoid ridicule or exclusion. Remember that these experiments were done in a laboratory and with people who didn't know each other. A group's influence on a person's attitudes and behavior can be even stronger when it is a real-life group.

### Milgram's Research

In a well-known laboratory experiment on obedience, psychologist Stanley Milgram (1963, 1965) asked 40 volunteers to administer electric shocks to other study participants. In each experimental trial, one participant was a "teacher" and the other a "learner," one of Milgram's accomplices. The teachers were businessmen, professionals, and blue-collar workers.

The learner was strapped into a realistic-looking chair that supposedly regulated electric currents. The learner was not actually receiving a shock but was told to fake pain and fear. The

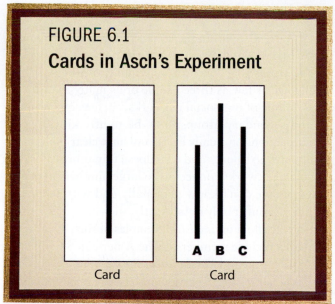

## FIGURE 6.1
### Cards in Asch's Experiment

A B C

Card          Card

Comstock/Jupiter Images

teacher read aloud pairs of words that the learner had to memorize. Whenever the learner gave a wrong answer, the teacher was told to apply an electric shock from a low of 15 volts to a high of 450 volts. When the learners shrieked in pain, the majority of the teachers, although distressed, obeyed the study supervisor and administered the shocks when told to do so.

Milgram's study was controversial. Ordering electric shocks raised numerous ethical questions about the participants' suffering extreme emotional stress (see Chapter 2). However, the results showed that an astonishingly large proportion of the participants obeyed an authority figure's instructions to inflict pain on others.

### Zimbardo's Research

The Stanford Prison Experiment conducted by social psychologist Philip Zimbardo also underscores the influence of groups on behavior (Haney et al. 1973; Zimbardo 1975). Zimbardo recruited volunteers in a local California newspaper for an experiment on prison life. He then selected 24 young men, most of them college students.

On a Sunday morning, nine of the men were "arrested" at their homes as neighbors watched. The men were booked and transported to a mock prison that Zimbardo and his colleagues had constructed in the basement of the psychology building at Stanford University. The "prisoners" were searched, issued an identification number, and outfitted in a dresslike shirt and heavy ankle chains. Those assigned to be "guards" were given uniforms, billy clubs, whistles, and reflective sunglasses. The guards were told that their job was to maintain control of the prisoners but not to use violence.

All the young men quickly assumed the roles of either obedient and docile prisoners or autocratic and controlling guards. The guards became increasingly more cruel and demanding. The prisoners complied with dehumanizing demands (such as eating filthy sausages) to gain the guards' approval and bowed to their authority.

## Sociology in Your Life

What happens when you put "good" people in a dehumanizing situation? Find out by viewing a slide show of the Stanford Prison Experiment. Go to Sociology CourseMate at CengageBrain.com to access this site.

Zimbardo's study was scheduled to run for 2 weeks but was stopped after 6 days because the guards became increasingly aggressive. Among other things, they forced the prisoners to clean out toilet bowls with their bare hands, locked them in a closet, and made them stand at attention for hours. Instead of simply walking out or rebelling, the prisoners became withdrawn and depressed. Zimbardo ended the experiment because of the prisoners' stressed-out reactions.

The experiment raised numerous ethical questions about the harmful treatment of participants. It demonstrated, however, the powerful effect of group conformity: People exercise authority, even to the point of hurting others, or submit to authority if there is group pressure to conform (Zimbardo et al. 2000).

AP Photo, File

Many social scientists have used Milgram's findings to explain the Abu Ghraib prison in Iraq where U.S. soldiers humiliated prisoners because "I was just following orders."

### Janis's Research

Sometimes intelligent people, even those in positions of high responsibility, make disastrous and irrational decisions. Why? Social psychologist Irving Janis (1972: 9) cautioned presidents and other heads of state to be wary of **groupthink**—a tendency of in-group members to conform without critically testing, analyzing, and evaluating ideas, which results in a narrow view of an issue. To preserve friendly relations and to remain loyal to the group, Janis argued, individuals don't raise controversial issues, question weak arguments, or probe "soft-headed thinking." As a result, influential leaders often make decisions, based on their advisors' consensus, that turn out to be political and economic disasters.

A United States Senate study (2004) of intelligence agencies provides an example of groupthink. The report concluded that U.S. leaders' decision to invade Iraq in 2003 was based on a groupthink dynamic that relied on unproven and inaccurate assumptions, inadequate or misleading sources, and a dismissal of conflicting information showing that Iraq had no weapons of mass destruction.

Janis and other researchers have focused on high-level decision making, but groupthink is common in all kinds of groups—student clubs, PTAs, search committees, juries, and community activists, for example (see Hansen 2010). If group members are aware of the negative characteristics of groupthink, they can avoid some of the pitfalls by setting up and following democratic discussion and voting processes, hammering out disagreements, and seeking advice from informed and objective people outside the group.

> Sometimes intelligent people, even those in positions of high responsibility, make disastrous and irrational decisions.

### 6-1e Social Networks

Groups exist within the context of larger social units, such as social networks. A **social network** is a web of social ties that links an individual to others. It may involve as few as three people or as many as millions.

Some of our social networks, such as our primary and secondary groups, may be tightly knit, involve interactions on a daily basis, and have clear boundaries about who belongs and who doesn't. In other cases, our social networks connect us to large numbers of people whom we don't know personally and with whom we interact only rarely or indirectly; such a group's boundaries are fluid or unclear. Examples of such distant networks include members of the American Sociological Association and *Washington Post* readers.

The Internet, in particular, has generated numerous social networks that unite people who have similar interests. Almost 75 percent of all adult Americans are active in some kind of voluntary group or organization. Internet users (80 percent) are more likely to participate in groups than non-Internet users (56 percent) (Rainie et al. 2011). The former are also more likely to say that the Internet has a major impact on their groups' activities (see *Table 6.2*).

## 6-2 Formal Organizations

A **formal organization** is a complex and structured secondary group that has been deliberately created to achieve specific goals in an efficient manner. We depend on a variety of formal organizations to provide goods and services in a stable and predictable way, including city water departments, food producers and grocery chains that stock our favorite items, and garment industries and retailers that produce and sell the clothes we wear.

### 6-2a Characteristics of Formal Organizations

Formal organizations share some common characteristics:

- Social statuses and roles are organized around shared expectations and goals.
- Norms governing social relationships among members specify rights, duties, and sanctions.
- A formal hierarchy includes leaders or individuals who are in charge.

Two of the most widespread and important types of formal organizations in the United States are voluntary associations and bureaucracies.

## TABLE 6.2

# How the Internet Affects Social Networks

Percentage of U.S. adults who said the Internet has a "major impact" on their groups' ability to . . .

| | INTERNET USERS | NON-INTERNET USERS |
|---|---|---|
| Communicate with members | 75 | 44 |
| Draw attention to an issue | 68 | 43 |
| Connect with other groups | 67 | 40 |
| Organize activities | 65 | 41 |
| Impact society at large | 64 | 45 |
| Recruit new members | 55 | 38 |

Source: Based on Rainie et al. 2011: 31.

**voluntary association** a formal organization created by people who share common interests and who are not paid for their participation.

**bureaucracy** a formal organization that is designed to accomplish goals and tasks through the efforts of a large number of people in the most efficient and rational way possible.

## 6-2b Voluntary Associations

A **voluntary association** is a formal organization created by people who share common interests and who are not paid for their participation. Unlike bureaucracies, as you'll see shortly, voluntary associations vary quite a bit in organizational structure. Some small voluntary organizations, such as book clubs, don't have a formal hierarchy, a rigid set of rules or regulations, or even specific goals. Others, like investment clubs, have a constitution, written regulations about the duties and rights of officers and members, minutes, and an agenda that's circulated before each meeting.

Helping others is a strong American value (see Chapter 3). Every year, millions of Americans—many of whom have full-time jobs—are volunteers in a variety of local, regional, and national organizations. These volunteers save government agencies billions of dollars every year by providing needed services (see *Table 6.3*).

Overall, there are four types of voluntary associations: groups that have a religious or altruistic purpose (e.g., Habitat for Humanity), groups that focus on political issues (e.g., the Tea Party), groups that organize around occupations (e.g., American Accounting Association), and groups that have a recreational orientation (e.g., the Beer Can Collectors of America). Millions of Americans also participate in voluntary organizations that include coupon-sharing groups, neighborhood associations, choirs, and cultural groups (see Rainie et al. 2011 for a detailed list of the most popular voluntary associations and some of their demographic traits).

## 6-2c Bureaucracies

When I ask my students to describe a bureaucracy, their answers often include adjectives such as "bungling," "impossible," "buck-passing," "frustrating," "hellish,"

and worse. Such reactions are understandable because, for most of us, dealing with a bureaucracy is usually synonymous with red tape, slow-moving lines, long and tedious forms, and rude employees.

A **bureaucracy** is a formal organization that is designed to accomplish goals and tasks through the efforts of a large number of people in the most efficient and rational way possible. Bureaucracies are not a modern invention but existed thousands of years ago in ancient Egypt, China, and Africa. Some bureaucracies function more smoothly than others, and some formal organizations are more bureaucratic than others. Whether they are relatively small (such as a 500-bed hospital) or huge (such as the Social Security Administration), bureaucracies have some common characteristics.

### Ideal Characteristics of Bureaucracies

Max Weber (1925/1947) identified six key characteristics of the ideal type of bureaucracy. Remember that ideal types describe abstract traits rather than those that fit any specific organization. In Weber's model, the following characteristics describe what an efficient and productive bureaucracy *should* be like:

## TABLE 6.3

# Volunteering in the United States

| | |
|---|---|
| Percentage of adults who volunteer | 26 |
| Total number of volunteers age 16 and older | 63 million |
| Average annual hours per volunteer | 34 |
| Estimated hourly value of volunteer time | $21.79 per hour |
| Total dollar value of volunteer time | $173 billion |

Source: Based on Independent Sector 2012.

- *High degree of division of labor and specialization.* Individuals who work in the bureaucracy perform very specific tasks.

- *Hierarchy of authority.* Workers are arranged in a hierarchy in which each person is supervised by someone in a higher position. The resulting pyramids—often presented in organizational charts—show who has authority over whom and who is responsible to whom. Thus, there is a chain of command, stretching from top to bottom, that coordinates decision making. *Figure 6.2* shows a simplified organizational chart of a small state university of 5,000 students. If this chart included receptionists, food services, plant maintenance, libraries, human resources, student housing, and numerous other offices and staff, the bureaucratic structure would be much more complex. The shaded boxes represent the flow of authority from the top (governor of the state) to the bottom (students in the sociology department).

- *Explicit written rules and regulations.* Detailed written rules and regulations cover almost every possible kind of situation and problem that might arise. They address a variety of issues, including hiring, firing, salary scales, rules for sick pay and absences, and everyday operations. If a person has a question, usually all she or he has to do is look up the answer.

- *Impersonality.* There is no place in a bureaucracy for personal likes or dislikes or tantrums. Instead, employees are expected to follow the rules, to get the work done, and to behave professionally. An impersonal workplace in which all employees are treated equally minimizes conflict and favoritism and increases efficiency.

- *Qualifications-based employment.* People are hired based on objective criteria such as skills, education, experience, and scores on standardized tests. If workers perform well and have the necessary credentials and technical competence, they'll move up the career ladder.

- *Separation of work and ownership.* Neither managers nor employees own the offices they work in,

## FIGURE 6.2
## The Bureaucratic Structure of a Small Public University

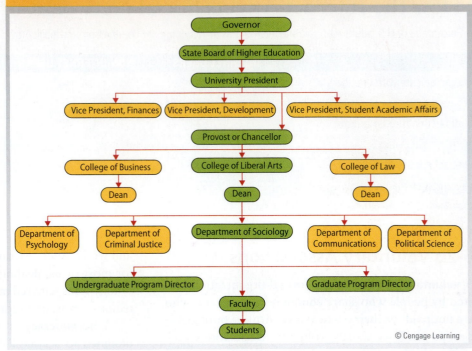

© Cengage Learning

the desks they sit at, the technology they use, or the products that they assemble, invent, or design.

For Weber, all of these characteristics produce an efficient and rational bureaucracy. A "rational matter-of-factness," he maintained, made bureaucracies more productive by "eliminating from official business love, hatred, and all purely personal, irrational, and emotional elements" (Weber 1946: 216). Weber viewed bureaucracies as superior to other forms of organization because they're more efficient and predictable. He worried, however, that bureaucracies could become "iron cages" because people become trapped in them, "their basic humanity denied."

## Shortcomings of Bureaucracies

Weber described the ideal characteristics of a productive and efficient bureaucracy, but what's the reality? As you read through the following list of problems, think about the ones you've experienced while working in a bureaucracy or dealing with one.

- *Weak reward systems* reduce the motivation to do a good job and are thus a major source of inefficiency and lack of innovation (Barton 1980). Besides low wages or salaries, weak reward systems include few or no health benefits, little recognition, unsafe equipment and work environments, and few incentives to be creative.

- *Rigid rules* discourage creativity and can result in **goal displacement**, a preoccupation with rules and regulations rather than achieving the organization's objectives

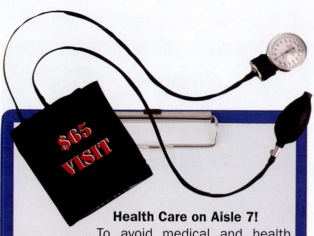

**Health Care on Aisle 7!**

To avoid medical and health insurance bureaucracies, more people are turning to retail clinics (also called "convenient care clinics"). The first one opened in 2000; by 2010, they numbered nearly 1,200 (Porter 2010). The clinics—usually staffed by nurse practitioners—offer low-cost treatment for more than 25 common conditions, such as strep throat and ear infections, and provide health screening tests, immunizations, and physicals. The clinics are located in pharmacies, grocery stores, and "big box" stores such as Target and Walmart. There is no need to make an appointment and minimal (if any) waiting, the patients are in and out in about 15 minutes, and most of the clinics are open evenings and weekends.

(Merton 1968). Instead of questioning whether all the red tape is really necessary, bureaucrats are usually more concerned that people follow established procedures because "we've always done it this way."

- Rigid rules and goal displacement often lead to **alienation**, a feeling of isolation, meaninglessness, and powerlessness. When people are reduced to a "small cog," they feel dehumanized. Alienation—at all levels—may result in high turnover, tardiness, absenteeism, stealing, sabotage, stress, health problems, and in some cases, whistle-blowing (reporting organizational misconduct to legal authorities).

- *Communication problems* are common in bureaucracies. Because communication typically flows down rather than up the hierarchy, employees (and many managers below the highest echelons) rarely know what's going on. Supervisors and their subordinates may be reluctant to discuss problems or offer suggestions for fear of being criticized, demoted, or fired. Those at the top often make ill-informed decisions based on incomplete or inaccurate information, as illustrated by groupthink (Blau and Meyer 1987; Fletcher and Taplin 2002).

- *Parkinson's Law* is the idea that work expands to fill the time available for its completion (Parkinson 1962). This means that even if employees finish an assigned task before the deadline, they'll look busy and act as though they're still working on the task to safeguard their jobs or avoid getting another assignment. Parkinson's Law also explains a bureaucracy's tendency to keep getting larger. Managers, wanting to appear busy, increase their workload by creating rules and forms. They then hire assistants, who in turn require more supervision and paper work. There is also an incentive to spend (even waste) as much money as possible to guarantee an ever-increasing budget and hiring even more people.

## Sociology in Your Life

What are some of an organization's outcomes when people aren't engaged in their jobs? And why are highly educated employees less likely to be engaged in their work than those with a high school diploma or less? Go to Sociology CourseMate at CengageBrain.com to find out.

- A related idea, the *Peter Principle*, proposes that workers are promoted until they reach their level of incompetence (Peter and Hull 1969). In many bureaucracies, employees who perform well are promoted to the next level, usually to administrative positions. Thus, project directors become managers, teachers become principals, and nurses become hospital administrators. Sooner or later, those promoted find themselves in positions for which they lack sufficient knowledge, training, competence, or experience.

"WHAT HAPPENED TO THAT EFFICIENCY REPORT? I HAD IT IN MY HAND NOT TWO MINUTES AGO."

**alienation** a feeling of isolation, meaninglessness, and powerlessness.

**iron law of oligarchy**
the tendency of a bureaucracy to become increasingly dominated by a small group of people.

■ The **iron law of oligarchy** is the tendency of a bureaucracy to become increasingly dominated by a small group of people (Michels 1911/1949). A handful of people can control and rule a bureaucracy because the top officials and leaders monopolize information and resources. As a result, those at the top maintain their power and privilege.

■ The cumulative effect of these and other bureaucratic dysfunctions can result in *dehumanization* because weak reward systems, rigid rules, and other shortcomings limit organizational creativity and freedom. As a result, work becomes more automated and impersonal (see the "The McDonaldization of Society" box).

Once established, bureaucracies are almost impossible to decrease in size or close down. For example, the U.S. Department of Education is a huge bureaucracy that continues to grow even though it has done little to improve public education and even limits college opportunities for many low-income students (see Chapter 13). However, most bureaucracies continue to function, largely because of the internal development of informal groups.

The U.S. workforce has 1.5 engaged employees for every disengaged worker who is emotionally detached from the job because of poor leadership, lack of upward mobility, boredom, and similar problems. Such alienation can lead to poor physical and mental health, low productivity, inferior customer service, and high absenteeism—all of which jeopardize a team's performance and the organization's financial vitality (Harter 2011; Harter and Agrawal 2011).

## 6-2d The Informal Side of Bureaucracy

Informal networks develop at all levels of a bureaucracy, from the bottom to the top. At the top, a CEO or a few members of the board of directors run the organization. They often make decisions informally, and the rest of the board then rubber-stamps the decisions. What about informal networks at the lower ranks?

### Historic Studies of Informal Group Networks

For many years, and well into the 1930s, experts concentrated on organizational efficiency based on the principles of *scientific management* developed by Frederick Winslow Taylor, a mechanical engineer who wanted to improve industrial productivity. Taylor (1911/1967) considered workers—especially those in factories and on assembly lines—mere adjuncts to machines, assumed that people don't like to work, and maintained that employees must be prodded by the promise of financial incentives, close supervision, and clear goals that they can attain with little effort. Taylor believed that management had to enforce cooperation because most workers are incapable of handling even the simplest tasks.

A number of studies conducted by industrial psychologists and sociologists between 1927 and 1932 shook up Taylor's views. The research teams studied employees at the Western Electric Company's Hawthorne plant in Chicago. The *Hawthorne studies*, as they're often called, found that informal groups were critical to the organization's functioning (Roethlisberger and Dickson 1939/1942; Mayo 1945; Landsberger 1958).

Informal social groups can promote an organization's goals if they collaborate, are cohesive, and motivate each other (Barnard 1938). They can also resist an organization's goals and formal rules. In one of the Hawthorne studies, Roethlisberger and Dickson (1939/1942) spent 6 months observing a group of 14 men who wired telephone switchboards in what the company called the "bank-wiring room."

The bank-wiring room work group consisted of nine wiremen, three solderers, and two inspectors. Management offered financial incentives for higher productivity, but the work group pressured people to limit output. The wiremen developed the following norms that controlled the group's behavior:

1. **You should not turn out too much work.** If you do, you are a "rate-buster" and a "speed king."
2. **You should not turn out too little work.** If you do, you are a "chiseler."
3. **You should not tell a supervisor anything that will be detrimental to a coworker.** If you do, you are a "squealer."
4. **You should not "act officious."** If you are an inspector, for example, you should not act like one.

# The McDonaldization of Society

According to sociologist George Ritzer (1996: 1), the organizational principles that underlie McDonald's, the well-known fast-food chain, are beginning to dominate "more and more sectors of American society as well as of the rest of the world." McDonaldization has four components: efficiency, calculability, predictability, and control.

1. *Efficiency* means that consumers have a quick way of getting meals. In a society where people rush from one place to another and where both parents are likely to work or single parents are pressed for time, McDonald's (and similar franchises) offers an efficient way to satisfy hunger and avoid much "fuss and mess."

   Like their customers, McDonald's workers function efficiently: The menu is limited, the registers are automated, and employees perform their tasks rapidly and easily.

2. *Calculability* emphasizes the quantitative aspects of the products sold (portion size and cost) and the time it takes to get the products. Customers often feel that they are getting a lot of food for what appears to be a nominal sum of money and that a trip to a fast-food restaurant will take less time than eating at home. In reality, soft drinks are sold at a 600 percent markup because most of the space in the cup is taken up by ice. It costs at least twice as much and takes twice as long to get to the restaurant, stand in line, and pick up the food than to prepare a similar meal at home.

3. *Predictability* means that products and services will be the same over time and in all locales. "There is great comfort in knowing that McDonald's offers no surprises" (p. 10). The Quarter Pounder is the same regardless where customers order. Outside of the United States, "homesick American tourists in far-off countries can take comfort in the knowledge that they will likely run into those familiar golden arches and the restaurant they have become so accustomed to" (p. 81).

   Workers, like customers, behave in predictable ways: "There are, for example, six steps to window service: greet the customer, take the order, assemble the order, present the order, receive payment, thank the customer and ask for repeat business" (p. 81).

4. *Control* means that technology shapes behavior. McDonald's controls customers subtly by offering limited menus and uncomfortable seats that encourage diners to eat quickly and leave. "Consumers know that they are supposed to line up, move to the counter, order their food, pay, carry the food to an available table, eat, gather up their debris, deposit it in the trash receptacle, and return to their cars. People are moved along in this system not by a conveyor belt, but by the unwritten, but universally known, norms for eating in a fast-food restaurant" (p. 105).

McDonald's controls employees more openly and directly. Workers are trained to do a limited number of things in precisely the same way, managers and inspectors make sure that subordinates toe the line, and McDonald's has steadily replaced human beings with technologies that include soft-drink dispensers that shut off when the glass is full and a machine that rings and lifts the French fries out of the oil when they are crisp. Such technology increases the corporation's control over workers because employees don't have to use their own judgment or need many skills to prepare and serve the food.

Efficiency, calculability, predictability, and control reflect a rational system (remember Weber?) that increases a bureaucracy's efficiency. On the other hand, Ritzer contends, McDonaldization reflects the "irrationality of rationality" because the results can be harmful. For example, the huge farms that now produce "uniform potatoes to create those predictable French fries" of the same size rely on the extensive use of chemicals that then pollute water supplies. And the enormous amount of non-biodegradable trash that McDonald's produces wastes our money because we—and not McDonald's—pay for landfills.

According to Ritzer (2008: 11, 119), McDonald's is such a powerful model that many businesses have cloned its four dimensions of efficiency, calculability, predictability, and control. Examples include "McDentists" and "McDoctors," drive-in clinics designed to deal quickly with minor dental and medical problems; "McChild" care centers like KinderCare; and "McPaper" newspapers, such as *USA Today*.

JOERG KOCH/AFP/Getty Images

© Cengage Learning

If individuals violated any of these norms, the other workers used "binging" (striking a person on the shoulder) to punish them.

Why did the men control productivity rather than take advantage of the management's promise of higher wages? They feared that if they produced at a high level, they would be required to produce at that level regularly. They also worried that high productivity would lead to layoffs because supervisors would conclude that fewer workers could achieve the same output. In addition, the bank-wiring room men experienced high morale and job satisfaction because they felt they had some control over their work. In contradiction to Taylor's scientific management perspective, then, the Hawthorne studies concluded that there is an important relationship between formal and informal organization.

### Modern Work Teams

Since the Hawthorne studies, many managers have recognized that informal groups pervade organizations and affect workers' productivity. As a result, numerous organizations have implemented alternative management strategies.

Today, *self-managing work teams* are the dominant model in most large organizations (Cloke and Goldsmith 2002; Neider and Schriesheim 2005). Contrary to Taylor's notion that workers do and managers think, self-managing work teams, sometimes referred to as *postbureaucratic organizations*, involve groups of 10 to 15 people who take on the duties of their former supervisors. Instead of being told what to do by a boss, self-managing workers gather and interpret information, act on the information, and take collective responsibility for their actions—whether it's designing a new refrigerator or handling food service for a university. Because

they're not held back by unresponsive managers, self-managing teams theoretically become committed to the organization and its success (Wellins et al. 1991).

How effective are self-managing work teams? Well-designed and well-executed team initiatives improve employee morale and the quality of products and services with relatively small costs. The initiatives that fail are marked by low support from management and union leaders, management's interfering with the group's work, or ineffective team leaders who discourage a team's creativity or autonomy (Barker 1993; Lencioni 2002).

## 6-3 Sociological Perspectives on Social Groups and Organizations

How do organizations operate? And how do social groups affect massive bureaucracies? In answering such questions, functionalism, conflict theory, feminist theories, and symbolic interactionism offer different insights. Taken together, these perspectives provide a multifaceted understanding of groups and organizations (*Table 6.4* summarizes these perspectives).

### 6-3a Functionalism: Social Groups and Organizations Benefit Society

Much of what you've read so far comes from functionalist perspectives, which emphasize that social groups and organizations are composed of interrelated, mutually dependent parts. As Weber pointed out, when a bureaucracy operates rationally and efficiently,

Courtesy King Flour Company

Since 1975, the number of employee-owned companies in the United States has grown from 1,600 to more than 11,000, and now represents about 12 percent of the private sector workforce. These firms are profitable because the workers are highly motivated: They have a stake in the business, control the decision making, contribute original ideas and cost-saving solutions, and depend on each other. Pictured here are people learning to bake bread at the King Arthur Flour Company's Baking Education Center in Norwich, Vermont, an employee-owned company that has expanded its product line (Case 2010).

## TABLE 6.4

# Sociological Perspectives on Groups and Organizations

| THEORETICAL PERSPECTIVE | LEVEL OF ANALYSIS | MAIN POINTS | KEY QUESTIONS |
|---|---|---|---|
| **Functionalist** | Macro | Organizations are made up of interrelated parts and rules and regulations that produce cooperation in meeting a common goal. | • Why are some organizations more effective than others? <br> • How do dysfunctions prevent organizations from being rational and effective? |
| **Conflict** | Macro | Organizations promote inequality that benefits elites, not workers. | • Who controls an organization's resources and decision making? <br> • How do those with power protect their interests and privileges? |
| **Feminist** | Macro and micro | Organizations tend not to recognize or reward talented women and regularly exclude them from decision-making processes. | • Why do many women hit a glass ceiling? <br> • How do gender stereotypes affect women in groups and organizations? |
| **Symbolic Interactionist** | Micro | People aren't puppets but can determine what goes on in a group or organization. | • Why do people ignore or change an organization's rules? <br> • How do members of social groups influence workplace behavior? |

© Cengage Learning

workers and bosses cooperate to turn out a final product, whether it's an automobile or a can of soup.

A primary factor in job satisfaction is effective leadership. Even in our sprawling government bureaucracies, and across all organizational levels, the happiest workers are those whose bosses reward motivation and commitment, encourage integrity, manage people fairly, and promote the employees' professional development and creativity (Partnership for Public Service 2011).

Such leaders aren't always successful themselves, however. For example, a recent survey of military leadership found that 97 percent of Army officers and sergeants have seen "exceptional" leaders, but 83 percent had directly observed "toxic" leaders. These were commanders who put their own needs first, micromanaged subordinates, behaved in a mean-spirited manner, or made poor decisions. Toxic leaders "may create a self-perpetuating cycle with harmful and long-lasting effects on morale, productivity, and retention of quality personnel." Nonetheless, about half the soldiers expected that the toxic commanders would be promoted, presumably because they are more likely than constructive leaders to accomplish a group's goals (Center for Army Leadership 2011: 1).

Functionalists acknowledge that organizations (and especially bureaucracies) can be dysfunctional. Workers may be alienated because of weak reward systems, favoritism, and incompetent supervisors. Employers waste more than $5,000 a year per worker because employees who aren't challenged spend more than 2 hours a day pretending to work: They shop online, do email, and even surf networking sites hoping to find leads for better jobs (Rothlin and Werder 2008).

Many organizations are trying to remedy such dysfunctions. Despite many people's complaints, employers have a legal right to use surveillance technology to monitor workers' use of company computers and cell phones to increase productivity (Barnes 2010). Since 2008, business spending on employees has grown only 2 percent compared with 26 percent on equipment and software (Rampell 2011). Many people complain that doing so increases unemployment rates. From a functionalist perspective, however, investing more in equipment than workers is a rational bureaucratic response to the rising costs of health care benefits and competition from other countries (see Chapter 11).

## Critical Evaluation

Functionalism is useful in understanding how groups and organizations fulfill essential functions such as motivating people to get work done and achieving a common goal. For critics, especially conflict theorists, functionalists exaggerate cooperation and tend to gloss over dysfunctions such as worker dissatisfaction and alienation.

A related issue is whether informal social networks improve worker morale and control as much as functionalists claim. After all, a tedious and monotonous job is tedious and monotonous regardless of the degree of informal coworker interaction.

From a conflict perspective, product placement on TV shows and in movies manipulates viewers. Political satirist Stephen Colbert often ridicules this strategy by telling his audience that his blatant product placement ads are supposed to influence their attitudes and behavior.

## 6-3b Conflict Theory: Some Benefit More Than Others

Conflict theorists contend that organizations are based on vast differences in power and control. In addition to sex and ethnicity, which we'll examine shortly, social class is important in determining individuals' places in organizations. In many companies, those at the higher levels are more comfortable hiring and promoting others similar to themselves. Phrases such as "fitting the mold" and "having common interests" are often code words for belonging to an in-group.

Inequality in income, status, and other rewards means that owners and managers can easily exploit workers. Those at the top dictate to those at the middle and the bottom. Because organizations serve elites, conflict theorists argue, they are usually undemocratic and ignore workers' needs and interests. Also, writes a management professor, "I've examined scores of empirical studies since the early 1980s and have not found convincing evidence that performance reviews are fair, accurate or consistent across managers, or that they improve organization effectiveness." Instead, subjective evaluations "are intimidating tools that make employees too scared to speak their minds" (Culbert 2011: A25). Conflict theorists would agree with such conclusions because, they maintain, a corporation's main objective is "to pursue, relentlessly and without exception, its own self-interest, regardless of the often harmful consequences it might cause to others" (Bakan 2005: 1–2).

According to some analysts, there is considerable incompetence, waste, and corruption in organizations. In 2009 alone, for example, Social Security made $6.5 billion in overpayments to people not entitled to receive them. In 2010, throughout the federal government, improper payments totaled $125 billion, but no one seems to know whether the overpayments occurred because of "simple errors," outright fraud, or both (Ohlemacher 2011).

A survey of federal workers and contractors concluded that many government agencies are dysfunctional. Besides wasting billions of taxpayer dollars, the agencies provide generous subsidies and tax concessions to the oil, coal, and agriculture industries as well as religious groups and foreign countries; don't fire incompetent workers; and have numerous (and unnecessary) levels of executives and managers rather than people working in the field. An employee in the Environmental Protection Agency complained, for instance, "The layers of management are insane. . . . It takes 13 steps and five layers to get a signature from our office director" (Rein 2011: B4). Because of these and similar issues, many Americans are disillusioned with a number of organizations (see *Table 6.5*).

Large corporations wield enormous power. When revenue for the major U.S. airlines dropped in 2008, they started charging fees for baggage, a tiny pillow or thin blanket, and selecting seats more than 24 hours before a flight. The profits and CEO salaries of major U.S. airlines have soared since 2008, but some airlines are now considering a general fee for lavatory use and to bring an infant on board (Goodale 2010; Matlin 2010; Carey 2011). The average American feels powerless about the rising costs because the Transportation Department claims that "it has no authority to regulate the prices that airlines charge for transportation services" (Maynard 2009: 3).

### Critical Evaluation

Among its contributions, conflict theory has underscored the inequality within groups and organizations that saps workers' motivation and limits their economic success. Critics, however, fault conflict theory for assuming that greater equality always leads to a more successful and productive organization. As you saw earlier, even self-managing work teams can fail because of ineffective team leaders or management interference with teams.

In addition, do conflict theorists focus too much on organizational deficiencies? There are, after all, many organizations that are efficient and profitable as well as corporations that don't exploit their workers (Kaplan 2010).

## 6-3c Feminist Theories: Men Benefit More Than Women

Feminist scholars agree with functionalists that organizations can be effective in attaining common goals. They also agree with conflict theorists that those with power protect their own interests. However, feminist analyses emphasize that across all social classes, women (and especially minority women) consistently fare worse than men, especially in not having leadership roles.

Women have enjoyed increased success in organizations since the 1980s, but rarely at the highest levels. For example, women hold more than half of all management and professional positions in the United States, but make up only 2 percent of *Fortune* 500 CEOs and 15 percent of the board directors. At *Fortune 100* companies, only 18 percent of board directors are women (Alliance for Board Diversity 2011; Lipman 2011).

Large numbers of women still hit a **glass ceiling**—attitudes or organizational biases in the workplace that prevent them from advancing to leadership positions. Organizational barriers reflect, in large part, stereotypes about gender roles. For example, senior-level U.S. executives—both women and men—typically describe women as better at stereotypically feminine "caretaking," such as supporting and encouraging others, and men as excelling at stereotypically masculine "taking charge," such as influencing superiors and problem solving. Because many promotions are based on taking charge rather than caretaking skills, women are often overlooked in decisions on promotions and salary increases (Sabattini et al. 2007; Foust-Cummings et al. 2008).

### Critical Evaluation

Feminist theories have expanded conflict explanations of groups and organizations by showing that many talented women are still treated like outsiders. One weakness of feminist theories, however, is that much of the emphasis is still on white and black women in professional and managerial positions, even though Latinas and Asian American women comprise a large segment of these workers and those in the general labor force.

A second limitation is that even when feminist scholars say that both sexes suffer from organizational stereotypes, they offer little data on how such stereotypes affect men. In addition, some critics fault feminist scholars for spending too much time focusing on women's

ongoing struggles rather than their progress (see Helgesen 2008).

### 6-3d Symbolic Interactionism: People Define and Shape Their Situations

Whereas functionalist, conflict, and (some) feminist theorists examine groups and organizations on a macro level, symbolic interactionists focus on interpersonal relationships. Interactionists emphasize that an individual's perception and definition of a situation shape group dynamics and, consequently, organizations. Group leaders or members can create or reinforce conformity (as shown in the Asch and Milgram studies). Informal groups can also determine what goes on in an organization by refusing to obey the rules and implementing their own (as in the Hawthorne studies). Thus, according to symbolic interactionists, individuals make choices, change rules, and mold their own identities instead of being manipulated as passive members of a group (Kivisto and Pittman 2001).

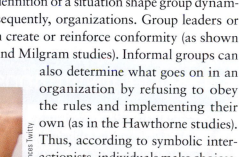
© iStockphoto.com/Frances Twitty

> **glass ceiling** attitudes or organizational biases in the workplace that prevent women from advancing to leadership positions.

---

**TABLE 6.5**

## Many Americans Have Little Confidence in Organizations

Percentage who said, in 2011, that they have "a great deal" or "quite a lot" of confidence in . . .

| | |
|---|---|
| The church or organized religion | 48 |
| The medical system | 39 |
| The U.S. Supreme Court | 37 |
| The presidency | 35 |
| The public schools | 34 |
| The criminal justice system, newspapers, and television news | 28 (each) |
| Banks | 23 |
| Organized labor | 21 |
| Big business | 19 |
| Health maintenance organizations (HMOs) | 19 |
| Congress | 12 |

Note: A majority of Americans expressed high esteem only for the military (78 percent), small business (64 percent), and the police (56 percent).

Source: Based on Jones 2011.

Symbolic interactionists also note that people's outcomes depend on how coworkers and bosses interpret the same behavior. For example, whether women are supervisors or clerical workers, if they lose their temper, they are overwhelmingly seen as too emotional, incompetent, out of control, weak, and worth less pay. Their angry male counterparts, in contrast, are often viewed as authoritative and in control (Brescoll and Uhlmann 2008).

### Critical Evaluation

Symbolic interactionism has made important contributions in describing how members of groups and organizations interpret the world around them and, as a result, affect what goes on. Social networking sites, especially Facebook, have empowered employees to discuss everyday problems despite management's control (Kirkpatrick 2010).

Despite these contributions, interactionist perspectives have several weaknesses. Because symbolic interactionism is a micro-level theory, it ignores macro-level factors that exploit workers and consumers. About 54 percent of young people ages 8 to 21 buy products—whether they need them or not—because of media ads, a macro-level factor (Martin 2006). In addition, and in contrast to interactionists' claims, most people have little control in shaping or changing their situations. Instead, formal organizations (such as the U.S. government and large companies) often invade the privacy of citizens, workers, or consumers by collecting information and monitoring people's behavior.

We've looked at the general characteristics of social groups and formal organizations, both of which are part of larger structures that sociologists call *social institutions*. This final section introduces you to this important sociological concept and lays the foundation for discussions of social institutions in later chapters.

# 6-4 Social Institutions

**A** **social institution** (or simply, *institution*) is an organized and established social system that meets one or more of a society's basic needs. Some social institutions are almost universal because they ensure a society's survival and practically all members of a society participate. According to functionalists, there are five major social institutions worldwide:

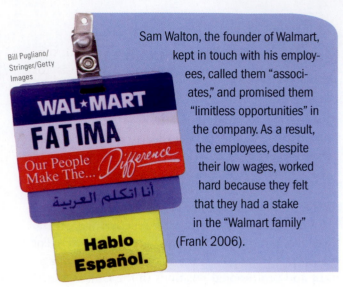

Bill Pugliano/Stringer/Getty Images

Sam Walton, the founder of Walmart, kept in touch with his employees, called them "associates," and promised them "limitless opportunities" in the company. As a result, the employees, despite their low wages, worked hard because they felt that they had a stake in the "Walmart family" (Frank 2006).

- The *family* replaces the members of a society through procreation, socializes children, and legitimizes sexual activity between adults.
- The *economy* organizes a society's development, production, distribution, and consumption of goods and services.
- *Political institutions* maintain law and order, pass legislation, and form military groups for internal and external defense.
- *Education* helps to socialize children, transmits knowledge, and provides information and training for jobs and other work-related activities.
- *Religion* encompasses beliefs and practices that offer a sense of meaning and purpose in life.

Besides these core universal institutions (which we'll examine in Chapters 11–13), others that have emerged include law, science, medicine and health care, sports, criminal justice, and the military.

## 6-4a Why Social Institutions Are Important

Social institutions are abstractions, but they have an organized purpose, weave together norms and values, and, consequently, guide behavior. No two families are exactly alike, but the family institution reflects broadly shared cultural agreements about what a family should be and do. In the same vein, no two grocery stores are exactly alike, but the economy as an institution creates and maintains a variety of rules that usually make food shopping predictable.

## 6-4b How Social Institutions Are Interconnected

Social institutions are linked to one another. Through taxes, the economy provides income for families that

On average, an 8 ounce soft drink contains at least 3 tablespoons of sugar. Over the course of a year, drinking one soda a day can make you 10 pounds fatter.

**5 major social institutions:**
- **family**
- **economy**
- **political system**
- **education**
- **religion**

need financial support and funds for political and education institutions. Education, in turn, trains children (and adults) in the general and specific skills required for productive participation in the economy. The family instills many of the values of hard work and success that educational institutions reinforce. Families and schools could not survive without the goods and services provided by the economy, and the economy needs workers who have been socialized by the family and trained by schools to enter the labor force. And in many communities, criminal justice, religious, and legal institutions have formed coalitions to decrease substance abuse and other social problems.

Understanding institutions can tell us a lot about how a society functions and how we're connected to one another. Consider, for example, the linkages among six social institutions—the economy, the political system, medicine, education, family, and the media—in addressing Americans' weight. Stated simply, overweight and obesity refer to ranges of weight that are greater than what is generally considered normal for one's height, as measured by the body mass index (BMI), an indicator of body fat (see Ogden and Carroll 2010).

About 32 percent of adult Americans say that they are somewhat overweight, and only 4 percent report being very overweight (i.e., obese) (Mendes 2010). In stark contrast, medical researchers find that about 7 in 10 American adults are either overweight (34 percent), obese (34 percent), or extremely obese (6 percent), and that the obesity rate has more than doubled since the late 1970s (Pan et al. 2009; Ogden and Carroll 2010; see also Chapter 14).

Children and adolescents who are overweight or obese are likely to be overweight or obese in adulthood. Media organizations (such as television networks, magazines, and newspapers) routinely broadcast such results and their negative health outcomes, including heart disease, strokes, asthma, diabetes, and early death. In 2010, First Lady Michelle Obama announced a national initiative to combat childhood obesity that included medical professionals, business and government leaders, grassroots activists, celebrity public service announcements, and cartoon characters (Givhan

2010). Some schools had already insisted that companies provide lunches that are lower in salt and fat and offer more whole grains and more fresh fruit snacks and drinks (Jackson 2010). To curb obesity, Baldwin Park, California, the home of the first-ever drive-through restaurant (In-N-Out Burger), even banned construction of any new drive-in establishments (Goodale 2010).

Have such efforts to curb obesity been successful? No. Among other reasons, some political groups complained that the First Lady shouldn't tell them what to eat, and Congress hasn't passed laws to rid school vending machines of sugary snacks and beverages. Almost all politicians rely on the food industry to help finance their reelection campaigns and to hire them as lobbyists after they leave public office (see Chapter 11). The food industry had promised to regulate itself and offer children healthier snacks and fast-food options, but this hasn't happened yet (Simpson and Baertlein 2012; see also Paulson 2012 on some progress by the Obama administration to implement healthier school lunches).

As you've seen in this chapter, social groups, formal organizations, and social institutions shape our behavior. Nonetheless, we're not robots who succumb to bureaucracies. Instead, and despite numerous constraints, we can make better decisions if we understand how groups, organizations, and social institutions affect our everyday lives.

## ⬛ Study Tools

Ready to study? Go to **Sociology CourseMate** at **www.cengagebrain.com** to complete practice quizzes, review flashcards, watch videos, and more.

SOC

All of us violate
some of society's rules.

Blair Seitz/Alamy

# Deviance, Crime, and the Criminal Justice System

*[handwritten: diff. { deviance / crime } connection]*

Flight attendant Jack Slater got into a dispute with a passenger who ignored instructions on an overhead baggage compartment. When the passenger's baggage struck Slater on the head, he got on the intercom, announced "I'm done! I quit!" After grabbing a beer from the beverage cart, Slater deployed the exit chute, slid onto the tarmac, and drove home. Flight attendants around the country applauded Slater, but he was arrested. In exchange for no jail time, he agreed to attend a yearlong mental health program and to pay a $10,000 fine to JetBlue to help replace the chute (Hesse 2010; Kaplan 2010).

> **deviance** traits or behavior that violates expected rules or norms.

At one time or another, all of us violate some of society's rules. We'll examine the characteristics of deviance and crime, discuss why people violate laws, and then look at institutional attempts to control and change rule-breaking behavior. First, however, take the quiz on page 118 to see how much you know about U.S. deviance and crime.

## Key Topics

In this chapter, we'll explore the following topics:

## 7-1 What Is Deviance?

**H**ave you ever driven above the speed limit? Cheated on an exam? Engaged in underage drinking? All are examples of deviance—traits or behavior that violate expected rules or norms. The word *deviance* has a derogatory connotation for the general public, but sociologists don't make such value judgments. Instead, they are interested in understanding and explaining deviance.

### 7-1a Some Key Characteristics of Deviance

Deviance is universal because it exists in every society. However, the key characteristics of deviance can vary quite a bit over time, from situation to situation, from group to group, and from culture to culture (Sumner 1906; Schur 1968).

## What do you think?

There would be less crime if the punishments were more severe.

| 1 | 2 | 3 | 4 | 5 | 6 | 7 |
|---|---|---|---|---|---|---|

strongly agree      strongly disagree

■ *Deviance can be a trait, a belief, or a behavior.* People usually do something to be considered deviant. We can also be treated as outsiders simply because of our appearance, skin color, religious beliefs, or sexual orientation (Becker 1963; see also Chapters 9, 10, and 13).

■ *Deviance is accompanied by social stigmas.* A **stigma** is a negative label that devalues a person and changes her or his self-concept and social identity. Stigmatized individuals may react in many ways: They may alter their appearance (as through cosmetic surgery), associate with others like themselves who accept them (as in gangs), hide information about some aspect of their deviance (as an ex-convict who does not reveal that status), or divert attention from a stigma by excelling in some area (as music or sports) (Goffman 1963).

■ *Deviance varies across and within societies.* What is appropriate or tolerated in one society may be deviant in another. For example, in 2007, only 1.6 percent of all births in South Korea were to unmarried women, compared with nearly 41 percent in the United States. As a 33-year-old single mother in South Korea explained, "Once you become an unwed mom, you're branded as immoral. . . . People treat you as if you had committed a crime. You fall to the bottom rung of society" (Sang-Hun 2009: 6). In the United States, in contrast, there's little stigma in having a child out of wedlock (see Chapters 9 and 12).

## Sociology in Your Life

Is sexting a deviant act? Check out these recent data to find out just how pervasive sexting is by going to CourseMate at CengageBrain.com.

■ *Deviance varies across situations.* What is seen as normal in one context may be stigmatized in another. Almost 40 percent of Americans ages 18 to 40 have at least one tattoo, but many employers—especially business firms and federal agencies—disapprove of visible "body art" (Hendrix 2009).

■ *Deviance is formal or informal. Formal deviance* is behavior that violates laws. A major example is crime, a topic we'll examine shortly. In contrast, *informal deviance* is behavior that disregards accepted social norms, such as picking one's nose or teeth or scratching one's private parts in public, belching loudly, and not dressing appropriately (e.g., wearing jeans to a wedding reception or job interview). Even when people are no longer formally deviant, they may describe themselves as informally deviant ("I'm a recovering alcoholic").

■ *Perceptions of deviance can change over time.* Many behaviors that were acceptable in the past are now seen as deviant. Only during the 1980s and 1990s did U.S. laws define date rape, marital rape, stalking, and child abuse as crimes. Smoking—widely accepted in the past—has been prohibited in many public places. In 2011, for the first time, a majority of Americans (59 percent) supported a smoking ban in *all* public places, a considerable increase from 31 percent in 2003 (Newport 2011). Also, increasing numbers of employers don't hire people who fail urine tests for nicotine usage (Koch 2012).

On the other hand, most Americans now shrug off behaviors that were stigmatized in the past. Cohabitation and out-of-wedlock births, seen as sinful and immoral only a few decades ago, are now widespread and accepted by most Americans. And Americans' support for legalizing marijuana increased from only 12 percent in 1970 to 50 percent in 2011 (Newport 2011).

### HOW MUCH DO YOU KNOW ABOUT U.S. DEVIANCE AND CRIME?

**True or False?**

1. Most crime victims are women.

2. Crime rates have decreased during the last decade.

3. People are more likely to be arrested for a drug violation than for driving while drunk.

4. Serious crimes such as murder and assault are more common than less serious crimes such as illegal gambling and prostitution.

5. People in affluent neighborhoods are more likely to experience burglary than those in poor neighborhoods.

6. Death penalties deter crime.

© Cengage Learning

The answers for #2 and #3 are true; the others are false. You'll see why as you read this chapter.

© iStockphoto/nicholas belton

More than 1,000 women in Ghana live in exile in witch camps. Most are poor widows or older women who are blamed for outbreaks of disease in their villages and the illness or death of relatives or neighbors (LaFraniere 2007). In the Democratic Republic of Congo, thousands of children whose parents have died in local wars or because of AIDS are living on the streets because their relatives have accused them of witchcraft that caused the parents' deaths. Such charges are usually because of the relatives' living in poverty and not being able to feed the children (Shapiro 2009).

Markus Matzel/Peter Arnold Inc.

## 7-1b Who Decides What's Deviant?

Because deviance is culturally relative and the standards change over time, who decides what's right or wrong? Those who have authority or power. During our early years, parents and teachers specify acceptable and unacceptable behavior. As we get older, laws also define what is deviant (e.g., being forbidden to drive until age 16 or to purchase or consume alcohol until age 21).

## 7-2 What Is Crime?

**W**hat comes to mind when you hear the word *crime*? Most of us typically imagine a murder or a violent physical attack, but such offenses constitute a minority of crimes. **Crime** is a violation of societal norms and rules for which punishment is specified by law. Many sociologists are **criminologists**, researchers who use scientific methods to study the nature, extent, cause, and control of criminal behavior. Measuring crime may seem straightforward, but the task isn't as simple as it appears.

## 7-2a Measuring Crime

There are several primary sources of crime statistics, but two of the most important are the FBI's Uniform Crime Report (UCR) and victimization surveys. Each has its strengths and weaknesses.

### Uniform Crime Report

The best known and most widely cited source of official crime statistics is the UCR, which includes crimes reported to the police and arrests made each year. Because the UCR has been published since 1930, the statistics are useful in examining trends over time. However, the UCR doesn't include federal offenses such as corporate crime, kidnapping, and Internet crimes, nor any of the roughly 60 percent of all crimes that aren't reported to the police (Hart and Rennison 2003).

Most offenders aren't caught. There are no witnesses, or the police don't investigate because they

*Are any of these people deviant? Why or why not?*

© BananaStock/Jupiterimages/ © Seth Kushner/Workbook Stock/Jupiterimages/ © nicholas belton/iStockphoto /© Image copyright Dmitry Naumov, 2009. Used under license from Shutterstock.com

must devote most of their limited resources to violent crimes (murder, rape, robbery, and aggravated assault) rather than property crimes (larceny-theft, burglary, and motor vehicle theft). In 2010, only 47 percent of those arrested for violent crimes and 18 percent of those arrested for property crimes were turned over to the courts for prosecution (Federal Bureau of Investigation 2011).

In effect, then, the UCR data don't provide accurate measures of the prevalence (extent) of crime. Because of these limitations, many criminologists also use victimization surveys to measure the extent of crime.

### Victimization Surveys

A **victimization survey** involves interviewing people about their experiences as crime victims. Every year, the U.S. Department of Justice sponsors the most widely used survey, the National Crime Victimization Survey (NCVS). Because the response rates are at least 90 percent, the NCVS offers a more accurate picture of many offenses than does the UCR. The NCVS also includes both reported and unreported crime, and it's not affected by police discretion in deciding whether to arrest an offender.

Still, many crimes are underreported. Some people don't want to admit having been victims—especially when the perpetrator is a family member or friend. Others may be too embarrassed to tell the interviewer that they were victimized while drunk or engaged in drug sales (Hagan 2008).

## 7-2b How Much Crime Is There?

No one knows the extent of U.S. crime. Even the best sources (such as the UCR and NCVS) are only estimates, but some offenses are more common than others.

### Incidence of Serious Crimes

Of the almost 1.4 million crimes reported to the police in 2010, 88 percent were property crimes. The media are most likely to cover violent crimes, which inspire the greatest fear, but Americans are much more likely to be victimized by theft or burglary than to be murdered, raped, robbed, or assaulted with a deadly weapon (see *Table 7.1*). In fact, the average person is 433 times more likely to experience a

theft than to be murdered (Federal Bureau of Investigation 2009, 2011; Glassner 2010).

### Victimless Crimes

Offenses that are the least likely to be reported—such as illicit drug use, prostitution, drunkenness, and illegal gambling—are called *public order crimes*, or *victimless crimes*. **Victimless crimes** are acts that violate laws but involve individuals who don't consider themselves victims. For example, prostitutes argue that they are simply providing services to people who want them, and substance abusers claim that they aren't hurting anyone but themselves.

Some contend that this term is misleading because victimless crimes often lead to property and violent crimes, as when addicts commit burglary, rob people at gunpoint, and engage in identity theft to get money for drugs. In a national survey, 70 percent of the respondents said that a family member's drug abuse had a negative effect on the emotional or mental health of at least one other family member, and 39 percent experienced financial problems because they went into debt to pay for an uninsured addict's treatment (Saad 2006).

## 7-2c Victims and Offenders

The media often overstate the amount of crime in our society. For example, a prominent *New York Times* journalist recently described American culture as "soaked in blood" and "insanely violent" because of the widespread availability of guns (Herbert 2009:

### TABLE 7.1

## Serious Crime in the United States, by Volume and Rate, 2010

| CRIMES | VOLUME (NUMBER OF CRIMES) | RATE (PER 100,000 INHABITANTS) |
|---|---|---|
| **Violent Crime** | **1.3 million** | **404** |
| Murder | 14,748 | 5 |
| Forcible rape | 84,767 | 28 |
| Robbery | 367,832 | 119 |
| Aggravated assault | 778,901 | 252 |
| **Property Crime** | **10.3 million** | **3,514** |
| Burglary | 2.1 million | 730 |
| Larceny-theft | 6.9 million | 2,362 |
| Motor vehicle theft | 1.2 million | 422 |

Source: Based on Federal Bureau of Investigation 2011, Table 1.

19). It's true that we have high violent crime rates compared with many other societies, but as you just saw, the incidence of property crimes is considerably higher than the incidence of violent crimes.

Between 2001 and 2010, U.S. violent crimes decreased by 20 percent and property crimes by 13 percent (Federal Bureau of Investigation 2011). In fact, the odds of being murdered or robbed are now less than half of what they were in the early 1990s, when violent crimes peaked (Oppel 2011).

Despite the drop in crime rates since 2000, 66 percent of Americans say that crime is increasing, and 60 percent believe that the U.S. crime problem is "extremely serious" or "very serious," up from 42 percent in 2005 (Jones 2010). The discrepancy between perception and reality may be due to a number of factors, such as media sensationalism and viewing nonfictional television crime stories or crime dramas such as the *Law and Order* and *CSI* series (Kort-Butler and Hartshorn 2011). Regardless of the reasons, crime affects some groups much more than others. Let's begin by looking at crime victims.

### Who Are the Victims?

Crime victimization isn't random. Instead, there are trends by sex, race/ethnicity, age, and social class.

1. **Sex.** Men are more likely than women to be crime victims, but the type of victimization differs. Males experience higher rates of homicide, robbery, and aggravated assault, whereas females are more likely to be victims of rape, sexual assault, and intimate partner violence (Truman 2011; White House Council on Women and Girls 2011).

2. **Race/Ethnicity.** About half of all homicide victims are black men, a majority of them between the ages of 17 and 29. The victimization rates for violent crimes are higher among African Americans (26 per 1,000 persons age 12 and older) than among Latinos (16) or whites (18) (Harrell 2007; Rand 2009).

3. **Age.** Most crime victims are younger than 25. They are 20 times more likely than people 65 and older to experience nonfatal violent crime (such as robbery). However, crimes against older people are especially traumatic because the victims are usually physically unable to fight back, and it takes them longer to recover from a crime both physically and emotionally (Klaus 2005; Madigan 2009).

4. **Social Class.** In general, lower income households experience higher victimization rates than their higher income counterparts. For example, as *Figure 7.1* shows, the poorest households are considerably more likely than those earning $75,000 or more per year to be victims of property crime.

### Who Are the Offenders?

Most offenders are never caught, but arrest rates show patterns by age, sex, race/ethnicity, and social class.

1. **Age.** Generally, the older Americans get, the less likely they are to commit a crime. In 2010, for example, 28 percent of all of those arrested were ages 15 to 21, and 26 percent were ages 22 to 29. In contrast, only 4 percent of those arrested were 55 or older, and many of their crimes involved drug abuse violations and drunk driving. Among those arrested for murder, 43 percent were between the ages of 17 and 24 compared with only 8 percent of those age 50 and older (Federal Bureau of Investigation 2011).

2. **Sex.** Of all those arrested in 2010, 75 percent were men. Men make up 82 percent of persons arrested for violent crime and 65 percent of those arrested for property crime. In other crimes, 95 percent of carjackers, 90 percent of those who rob, and 92 percent of those arrested for weapons violations and sexual offenses are men (Rennison 2003; Klaus 2006; Federal Bureau of Investigation 2011).

3. **Race/Ethnicity.** Of those arrested in 2010, 69 percent were white, 28 percent were African American, and the remainder was American Indian or Asian American. (The FBI doesn't provide data on Latinos because it collects information on race but

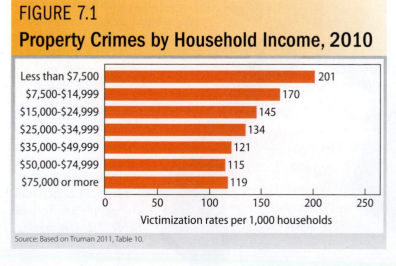

## FIGURE 7.1
## Property Crimes by Household Income, 2010

| Household Income | Victimization rates per 1,000 households |
|---|---|
| Less than $7,500 | 201 |
| $7,500–$14,999 | 170 |
| $15,000–$24,999 | 145 |
| $25,000–$34,999 | 134 |
| $35,000–$49,999 | 121 |
| $50,000–$74,999 | 115 |
| $75,000 or more | 119 |

Source: Based on Truman 2011, Table 10.

not ethnicity.) Proportionately, African Americans make up about 13 percent of the general population but account for 29 percent of violent crimes (Federal Bureau of Investigation 2011).

**4. Social Class.** Prison statistics consistently show that those who are incarcerated have low education levels. For example, 68 percent of state prisoners don't have a high school diploma, whereas only 11 percent have attended college or have a college degree (Harlow 2003).

Crime rates are higher in low-income areas, but does this mean that poor people are more deviant? Because police devote more resources to poor neighborhoods, those at the lower end of the socioeconomic ladder are more likely to be caught, arrested, prosecuted, and incarcerated. Middle-class criminals are more elusive. For example, the theft of *intellectual property* is a primarily middle-class crime that includes offenses such as software piracy, bootlegging musical recordings and movies, selling company trade secrets, and copyright violations (Motivans 2004). As you'll see shortly, many middle- and upper-class people commit serious crimes for which they're never prosecuted.

Most of us don't commit crimes but conform to laws and cultural expectations. We usually conform, however, not because we're innately good, but because of social control.

# 7-3 Controlling Deviance and Crime

**S**ocial control refers to the techniques and strategies that regulate people's behavior in society. The purpose of social control is to eliminate, or at least reduce, deviance and crime. Ensuring people's conformity to group or societal expectations involves both informal and formal social control, as well as positive and negative sanctions.

## 7-3a Informal and Formal Social Control

Most conformity is due to the internalization of norms during the powerful process of socialization. Because most of us genuinely care about the opinions of family members, friends, and teachers, we try to live up to their expectations (see Chapter 4). In this sense, we conform because of *informal social controls* that we learn and internalize during childhood.

In addition to informal mechanisms, *formal social control* regulates social behavior. Unlike informal social control, formal social control exists outside of the individual. For example, a college dean might threaten a student with suspension or expulsion, or a business might install security cameras.

## 7-3b Positive and Negative Sanctions

Most of us conform because of **sanctions**—rewards or punishments for obeying or violating a norm. *Positive sanctions* are rewards for desirable behavior. They include a variety of facial expressions (such as smiling), body language (such as hugging), comments (such as "Congratulations!"), and other forms of recognition (such as good grades, trophies, and promotions). Positive sanctions are very effective because they increase our self-confidence, self-esteem, and motivation, especially when the rewards are deserved.

*Negative sanctions* are punishments that convey disapproval for violating a norm. Negative sanctions range from mild and informal expressions (such as frowns and gossip) to more severe and formal reactions (such as fines, arrests, and incarceration). Some American Indian reservations have revived the traditional penalty of banishment to deal with gangs and drugs. The troublemakers can be ordered off the reservation and stripped of their tribal membership (Snell 2007). The ultimate formal negative sanction in most societies, including the United States, is execution.

The next sections examine four important sociological perspectives that help us understand why people are deviant. *Table 7.2* summarizes these theories.

Comstock/Jupiter Images

Men are much more likely than women to be victims of violent crime.

## TABLE 7.2

# Sociological Explanations of Deviance and Crime

| THEORETICAL PERSPECTIVE | LEVEL OF ANALYSIS | KEY POINTS |
|---|---|---|
| Functionalist | Macro | • Anomie increases the likelihood of deviance.<br>• Crime occurs when people experience blocked opportunities to achieve the culturally approved goal of economic success. |
| Conflict | Macro | • Laws protect the interests of the few (primarily those in the upper classes) rather than the rights of the many.<br>• Law enforcement is rarely directed at the illegal activities of the powerful. |
| Feminist | Macro and micro | • Crimes committed by women reflect their general oppression due to social, economic, and political inequality.<br>• Many women are criminal offenders or victims because of culturally organized beliefs and practices that are sexist and patriarchal. |
| Symbolic Interactionist | Micro | • People learn deviant and criminal behavior from others, such as parents and friends, who are important in their everyday lives.<br>• If people are labeled or stigmatized as deviant, they are likely to develop deviant self-concepts and engage in criminal behavior. |

© Cengage Learning

# 7-4 Functionalist Perspectives on Deviance and Crime

For functionalists, deviance and crime are normal parts of the social structure. Functionalists don't endorse undesirable behavior, but they view deviance and crime as both functional and dysfunctional.

## 7-4a Deviance and Crime Can Be Dysfunctional and Functional

*Dysfunctions* are the undesirable consequences of behavior, but what is dysfunctional for one group or individual may be functional for another. For example, gangs are functional because they provide members with a sense of belonging, identity, and protection. Gangs are also dysfunctional because they commit violent and property crimes (Egley and O'Donnell 2009).

Crime and deviance are *dysfunctional* because they:

■ *Create tension and insecurity.* Any violation of norms—a babysitter who cancels at the last minute or the theft of your laptop—makes life unpredictable and increases anxiety.

■ *Erode trust in personal and formal relationships.* Crimes such as date

Robert Nickelsberg/Getty Images

rape and stalking make many women suspicious of men. In 2009, identity theft was a top-ranked crime among Americans: 66 percent worried about identity theft compared with having a car broken into or stolen (47 percent) or being murdered (19 percent) (Saad 2009; see also Langton 2011). Such concerns are justified: Many victims have had problems obtaining banking services or credit cards because financial institutions don't trust them (Baum 2006).

■ *Damage confidence in institutions.* Since the scandals involving Enron and other corporations in 2001, and the 2008 crash in the stock market when taxpayers had to pay for the financial industry's corporate fraud and mismanagement, millions of people, even those who didn't lose money, worry that their retirement funds may disappear in the future (see Chapter 12).

■ *Are costly.* Besides personal costs to victims (including fear, emotional trauma, and physical injury), deviance is expensive. All of us pay higher prices for consumer goods and services (such as auto and property insurance), as well as taxes for prosecuting criminals and for building and maintaining prisons. In 2007, the most recent year for which data are available, illicit drug use alone cost the U.S. economy more than $193 billion. Some of the costs were due to criminal justice expenses, hospitalization, treatment, and lower workforce productivity (National Drug Intelligence Center 2011).

**anomie** the condition in which people are unsure of how to behave because of absent, conflicting, or confusing social norms.

**strain theory** the idea that people may engage in deviant behavior when they experience a conflict between goals and the means available to obtain the goals.

Deviance and crime can also be *functional* because they provide a number of societal benefits (Durkheim 1893/1964; Erikson 1966; Sagarin 1975). For example, deviance and crime

- *Affirm cultural norms and values.* Negative reactions, such as expelling a college student who's caught cheating or incarcerating a convenience store robber, assert a society's norms and values about being honest and law-abiding.

- *Provide temporary safety valves.* Some deviance is accepted under certain conditions, enabling people to "blow off steam." Typically, a community and police will tolerate noise, underage drinking, loud music, and obnoxious behavior during college students' spring break because it's a short-lived nuisance.

- *Create social unity.* Some deviant behavior, especially when there is a common enemy, unites a group, community, or society. After the U.S. Navy Seals killed Osama bin Laden in 2011, most Americans—regardless of age, political views, social class, or race and ethnicity—applauded his death.

- *Improve the economy.* Deviance and crime can benefit a community financially. Some towns welcome new prisons because they generate jobs and stimulate the economy (Semuels 2010). Unlike a manufacturing plant that might close down or move to another country, prisons are rarely dismantled.

- *Trigger social change.* After several studies showed that an estimated 450,000 people were killed or injured in 2009 because of "distracted driving" collisions, a number of states have passed antitexting laws (Hanes 2010; Insurance Institute for Highway Safety 2010; Ceasar 2012).

## 7-4b Anomie and Social Strain

Functionalists offer many theories of deviance and crime, but two of the most influential are anomie and strain theories. Both analyze why so many people commit crimes and engage in deviant behavior even though they share many of the same goals and values as those who conform to social norms.

### Durkheim's Concept of Anomie

Émile Durkheim (1893/1964, 1897/1951) introduced the term **anomie** to describe the condition in which people are unsure of how to behave because of absent,

John Lund/Jupiter Images

conflicting, or confusing social norms. During periods of rapid social change, such as industrialization in Durkheim's time, societal rules may break down. As many young people moved to the city to look for jobs in the nineteenth century, norms about proper behavior that existed in the countryside crumbled. Even today, as you'll see in Chapter 15, many urban newcomers experience anonymity and miss the neighborliness that was common at home.

### Merton's Concept of Social Strain

Robert Merton (1938) elaborated on the concept of anomie to explain how social structure helps to create deviance. According to Merton, Americans are socialized to believe that anyone can realize the American dream of accumulating wealth and being successful economically. To achieve the *cultural goal* of economic success, society emphasizes legitimate and *institutionalized means* such as education, hard work, saving, starting at the bottom and working one's way up, and making sacrifices instead of seeking pleasure and quick gratification.

In many countries, including the United States, however, people don't always have access to institutionalized means for financial success because of poverty, low wages, or long-term unemployment. How do people respond? Merton's **strain theory** posits that people may engage in deviant behavior when they experience a conflict between goals and the means available to obtain the goals. Not all people turn to deviance to resolve social strain (see *Table 7.3*). Most of us *conform* by working harder and longer to become successful. The fact that you're reading this textbook shows that, using Merton's language, your mode of adaptation to strain is conformity—one of achieving success through an institutionalized means such as getting a college education.

## TABLE 7.3
## Merton's Strain Theory of Deviance

| MODE OF ADAPTATION | ACCEPT CULTURAL GOALS? | ACCEPT INSTITUTIONALIZED MEANS TO ACHIEVE GOALS? | EXAMPLES |
|---|---|---|---|
| Conformity | Yes | Yes | College graduate |
| Innovation | Yes | No | Thief |
| Ritualism | No | Yes | Low-level bureaucrat |
| Retreatism | No | No | Alcoholic |
| Rebellion | No—seeks to replace goals | No—seeks to replace means for achieving goals | Antigovernment militia member |

Source: Based on Merton 1938.

Merton's other four modes of adaptation reflect deviance. *Innovation* occurs when people endorse the cultural goal of economic success but turn to illegitimate means, especially crime, to achieve the goal. For many people living in inner-city ghettos, education, hard work, and deferred gratification are often unachievable. For others, crime is a quick way to become richer. For innovators, "it's not how you play the game but whether you win or lose."

In *ritualism,* people don't expect to be rich but get the necessary education and experience to obtain or keep a job. In many bureaucratic positions, employees don't make decisions and can't move up the ladder, but their jobs are relatively secure. Ritualists aren't criminals, but they're deviant because they've given up on becoming financially successful. Instead, they do what they're told and "go along to get along."

In *retreatism*, people reject both the goals and the means for success. Merton used examples such as vagrants, alcoholics, and drug addicts, some of whom give up because they believe it's impossible to succeed.

In *rebellion*, people feel so alienated that they want to change the social structure entirely by substituting new goals and means for the current ones. Contemporary examples are terrorists and U.S. militia groups that oppose the federal government.

### 7-4c Critical Evaluation

A major contribution of functionalism is showing how social structure, not just individual attitudes and behavior, produces or reinforces deviance and crime. Functionalism also helps us understand current and emerging forms of deviance. For instance, computer-related crimes have increased for a number of structural reasons such as Americans' greater use of the Internet and offenders' overcoming security technology.

Functionalist theories also have weaknesses: (1) the concepts of *anomie* and *strain theory* are limited because they overlook the fact that not everyone in the United States embraces financial success as a major goal in life; (2) the theories don't explain why women's crime rates are much lower than men's, especially since women generally have fewer legitimate opportunities for financial success than men; (3) crime rates have declined since 2000 despite chronic poverty and unemployment; and (4) functionalism doesn't explain why people commit some crimes (such as setting fires just for kicks or murdering an intimate partner) that have nothing to do with being successful (Anderson and Dyson 2002; Williams and McShane 2004).

The most consistent criticism is that functionalism typically focuses on lower-class deviance and crime. Conflict theorists have filled this gap by examining middle- and upper-class crime.

## 7-5 Conflict Perspectives on Deviance and Crime

Functionalists ask, "Why do some people commit crimes and others don't?" Most conflict theorists focus on who makes the laws and ask, "Why are some acts defined as criminal while others are not?" (Akers 1997).

## 7-5a Capitalism, Power, Social Inequality, and Crime

For conflict theorists, the most powerful groups in society control the law, which defines what's deviant and who will be punished. The higher a person's social class, the greater the likelihood that her or his offense will be ignored or treated lightly (Chambliss and Seidman 1982; Reiman and Leighton 2010).

### White-Collar Crime

**White-collar crime** refers to illegal activities committed by high-status individuals in the course of their occupation (Sutherland 1949; Cressey 1953). A number of sociologists have shown that there is a strong association between capitalism and white-collar crime. That is, people with economic and political power define as criminal any behavior that threatens their own interests (Vold 1958; Turk 1969, 1976; Quinney 1980; Kraska 2004). As a result, newspapers and TV news stations (which are owned by the affluent) routinely focus on "street crimes" rather than corporate crimes.

### Corporate Crimes

**Corporate crimes** (also known as *organizational crimes*) are illegal acts committed by executives to benefit themselves and their companies. Corporate crimes include a vast array of illicit activities such as conspiracies to stifle free market competition, price-fixing, tax evasion, and false advertising. The target of the crime can be the general public, the environment, or even a company's own workers. Often, corporate offenders commit multiple crimes such as stock fraud, insider trading, and perjury.

Corporate crimes are fairly common. As early as 2002, for example, Toyota Motor Corporation received but ignored numerous complaints about sudden acceleration that resulted in deaths. In 2010, Toyota finally admitted that the sticky gas pedals could cause drivers to lose control of their vehicles but delayed notifying the car owners and the U.S. Transportation Department (Vartabedian and Bensinger 2010).

AP Photo/Kathy Willens

In 2009, Bernard Madoff—a highly respected stock-broker and investment advisor—pleaded guilty to defrauding thousands of investors of almost $170 billion between 1991 and 2008. Madoff was sentenced to 150 years in a federal prison, but his clients haven't recovered their money (Tresniowski et al. 2011).

### Cybercrime

**Cybercrime** (also called *computer crime*) refers to illegal activities that are conducted online. These high-tech crimes include defrauding consumers with bogus financial investments, embezzling, being paid to recommend stock on chat rooms, and stealing business data. Other offenses include sabotaging computer systems, hacking (gaining unauthorized access to computers), and stealing confidential information.

In 2010 alone, the Internet Crime Complaint Center (2011) received almost 304,000 complaints. The most common offenses were undelivered merchandise or payments, identity theft, and scams in which someone posed as the FBI to get information about addresses, Social Security numbers, and other personal and financial data.

### Organized Crime

**Organized crime** refers to activities of individuals and groups that supply illegal goods and services for profit. Organized crime includes drug distribution, loansharking (lending money at illegal rates), prostitution, illegal gambling, pornography, theft rings, the hijacking of cargo, and the laundering of money obtained from illegal activities through legitimate enterprises.

The *Godfather* films and popular television shows such as *The Sopranos* romanticized organized crime and presented the perpetrators as primarily Italian men. Italian-run organized crime exists, but Latino, Asian American, and African American men now run most organized crime. Some Russian and other Eastern European groups have been operating on U.S. soil since the 1970s.

## 7-5b Law Enforcement, Power, and Crime

Why do so many people commit white-collar crimes? Because they can, according to conflict theorists. First, most white-collar crimes are *not criminalized* (Turk 1969; Lilly et al. 1995). Federal prosecutors rarely bring charges against many corporate criminals, because lying isn't a crime and it's difficult to prove that an executive intentionally violated a specific law (Quick 2010; Nocera 2011).

Second, there are *few penalties*. When prosecutors bring civil charges, the fines often represent less than half of 1 percent of annual revenue. In 2009, for example, top officials at Goldman Sachs agreed to pay a $550 million fine to settle accusations of securities fraud that made almost $52 billion in profits. The fine represented four days of the bank's revenues (Summers 2010; see also Prins 2011). If, using a comparable percentage, you made $100,000 a year illegally and had to pay a $1,000 fine, would there be a financial incentive for honesty?

Third, white-collar crimes thrive because of *privilege*. The common cultural background shared by lawmakers and many white-collar defendants leads to greater leniency for these offenders than for street criminals. Also, the federal government awards billions in contracts to companies that repeatedly violate air and water pollution laws, cause deaths and injuries to their workers because of unsafe equipment, and even defraud the government (Silverstein 2002; see also Chapter 8).

Finally, law enforcement has been inadequate because of *limited resources*. After 9/11, nearly one-third of the FBI's best white-collar-crime specialists were reassigned to focus on national security. As a result, prosecutions of financial institutions for fraud dropped 48 percent from 2000 to 2007, and insurance fraud cases plummeted 75 percent (Lichtblau et al. 2008). To save time and money, the Justice Department and the Securities and Exchange Commission (SEC) often let companies investigate themselves. Because many of the internal investigators are former Justice and SEC employees, it's easy for them to omit evidence that might lead to criminal prosecution (Hilzenrath 2011).

## 7-5c Critical Evaluation

Conflict theories have explained the linkages between capitalism, power, and social class that may lead to criminal laws that benefit those at the top. Critics, however, point to several weaknesses. First, some contend that conflict theory exaggerates the importance of capitalism in explaining white-collar crime. Other capitalist societies, such as Japan, have much lower crime rates because of greater social solidarity and control of deviant behavior through shaming and restrictions on individual

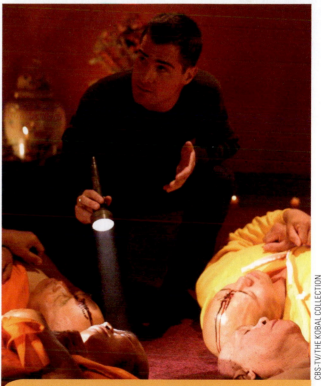

*CSI: Crime Scene Investigation* premiered in 2000. Some of the main characters have changed over the years, but *CSI* is the most watched drama series in the world, airing in 70 countries from Argentina to Vietnam (Gorman 2011). Why do you think this show is so popular?

In 2009, 36-year-old Ryan LeVin, a super-rich heir to a family costume jewelry fortune, killed two men in Florida when he lost control of his Porsche while hurtling down a street at more than 100 miles an hour, then fled the scene. In 2011, he pleaded guilty to two counts of vehicular homicide and faced 30 years in prison. Because LeVin's attorney convinced the widows to accept a financial settlement (the amount is unknown), LeVin was sentenced to 2 years of house arrest, which he'll serve at one of his parents' two luxury seaside condos. LeVin never apologized to the victims' families, and his lawyer asked the court to return his Porsche (Shepherd 2011a; Holland 2011).

Joe Cavaretta/Sun Sentinel/MCT via Getty Images

men who saw women as not worthy of much analytical attention or assumed that explanations of male behavior were equally applicable to females (Flavin 2001; Simpson and Gibbs 2006; Belknap 2007). To remedy such omissions, feminist scholars have concentrated on girls and women as victims and offenders.

## 7-6a Women as Victims

One of the most publicized events of 2009 was the discovery of Jaycee Dugard, a California girl who was kidnapped in 1991 while waiting for a school bus. The kidnapper was a convicted sex offender who repeatedly raped Jaycee for 18 years and fathered her two daughters, then ages 11 and 15.

The Dugard case is a recent example of the physical and sexual victimization of women and children. As you saw earlier, many of the serious crimes (such as murder and robbery) are committed by men against men, but women and girls are almost always the victims of sexual assault, rape, intimate partner violence, stalking, sexual exploitation, female infanticide, and other crimes that degrade women and deny them basic human rights. Even in neighborhoods where the household income is $75,000 or higher, 36 percent of women, compared with only 13 percent of men, fear walking alone at night near their homes (Saad 2010).

Feminist scholars offer several explanations for women's victimization. One is **patriarchy**, a hierarchical system in a society in which cultural, political, and economic structures are controlled by men. In patriarchal societies, including the United States, because women have less access to power, they are at a disadvantage in creating and implementing laws that are sympathetic to female victims (Price and Sokoloff 2004; DeKeseredy 2011). For example, of the 193 countries in the world, only 104 have made rape a crime, and the existing laws are rarely enforced (United Nations Development Fund for Women 2007). Such inequity diminishes women's control over their lives and increases their invisibility as victims.

Another reason for female victimization is the effect of culture on gender roles. Many girls and women have been socialized to be victims of male violence because

**patriarchy** a hierarchical system in a society in which cultural, political, and economic structures are controlled by men.

freedom (Leonardsen 2004). In effect, some critics say, capitalism is not a major reason for crime.

A second criticism is that conflict theory deemphasizes the crimes committed by the poor. The national cost of robbery, rape, and other street crimes is more than $4 billion a year in legal expenses, victim injuries, wage losses, and crime prevention activities (Mokhiber 2007). Thus, some critics maintain, low-income people are just as deviant as the rich, and their crimes are costly to society. A related criticism is that conflict theories tend to ignore the fact that some affluent people, including corporate executives, don't get away with their crimes (see, for example, Leopold 2011 and Tucker 2011).

Third, some critics say that conflict theory ignores the ways in which crime is functional for society as a whole. As you saw earlier, deviance provides jobs and affirms law-abiding cultural norms and values. Finally, many contend, the most influential conflict theories focus almost entirely on men as both victims and offenders (Moyer 2001; Belknap 2007). Feminist theories have filled this gap.

## 7-6 Feminist Perspectives on Deviance and Crime

For much of its history, sociology focused almost entirely on male offenders. As in most other academic disciplines, nearly all sociologists were

of societal images of women as weaker, less intelligent, and less valued than men: "Girls are rewarded for passivity and feminine behavior, whereas boys are rewarded for aggressiveness and masculine behavior" (Belknap 2007: 243). In effect, then, both sexes internalize the belief that male victimization of females is normal.

But, you might be thinking, there are many women who reject sex stereotypes and could, therefore, escape or avoid victimization. Feminist theorists point to several problems about such beliefs. One is that low-income women often fear that leaving an abuser could result in greater economic hardship for themselves and their children, and could even get themselves killed.

Second, because patriarchal societies often don't punish many males who commit violence against women, or give them light sentences, many female victims feel trapped. Recently, for example, a Maryland judge sentenced a 20-year-old man to 3 years in prison for killing his girlfriend's puppy and 3 years in prison for beating the girlfriend over a 2-year period (Siegel 2009).

Siede Preis/Photodisc/ Getty Images

## 7-6b Women as Offenders

Men are much more likely than women to commit crimes, especially more serious crimes, but girls' and women's arrest rates have increased for some offenses. Between 2001 and 2010, for example, female arrests rose 19 percent for burglary, 29 percent for robbery, 32 percent for larceny-theft, and 36 percent for drunken driving (Federal Bureau of Investigation 2011).

Some analysts posit that girls and women are becoming more deviant because of a breakdown of family, religion, and community; less adequate schooling, resulting in high dropout rates; greater assertiveness; and the pervasive violence in much of today's entertainment. Others maintain that the higher arrest rates are at least partly a by-product of policy changes, such as more aggressive policing and the greater likelihood that parents and school officials will call the police to deal with girls' unruly behavior (Prothrow-Stith and Spivak 2005; Steffensmeier et al. 2005; Chesney-Lind and Jones 2010).

Some of the explanations of female offenders parallel those of female victims. According to some feminist criminologists, women who commit crimes experience mistreatment that begins in early childhood: Girls are more likely than boys to be victims of family-related sexual abuse, the assaults begin at a younger age, and the abuse lasts longer. These factors can lead to suicide as well as to delinquent or criminal offenses such as running away from home, truancy, drug abuse, and prostitution (Snyder and Sickmund 2006; Chesney-Lind and Irwin 2008; Urbina 2009).

Other feminists emphasize patriarchy and women's limited economic opportunities. Because of their marginalization in the economy, some women resort to criminal activities—especially shoplifting, petty theft, and prostitution—to survive financially or to support a family. Women's offenses are highest in cities, where women's economic oppression and poverty are greatest (Heimer et al. 2006; Flower 2010).

## 7-6c Critical Evaluation

Feminist sociologists have been at the forefront of studying female offenders and victims: "The bottom line is that gender shapes human behavior in all arenas,

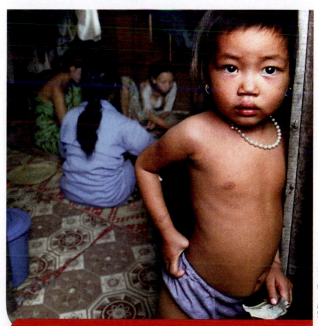

AP Photo/David Longstreath

Holding money exchanged for sex, the child of a prostituted woman stands in the doorway of a brothel in Phnom Penh, Cambodia. Children, especially girls, who are raised in brothels, are highly vulnerable to sexual exploitation. Many of the customers are Western men.

**differential association theory** a perspective that asserts that people learn deviance through interaction, especially with significant others.

**labeling theory** a perspective that holds that society's reaction to behavior is a major factor in defining oneself or others as deviant.

and crime and victimization are no exceptions" (Heimer and Kruttschnitt 2006: 1). If it had not been for feminist scholars, there would probably still be little awareness of crimes such as date rape, stalking, intimate partner violence, and the international sex trafficking of women and girls. Women's advocates have also been instrumental in the FBI's revising its 85-year-old definition of rape. The new definition, which will go into effect in 2012, includes men as victims; oral, anal, and vaginal penetration; and penetration by any body part or object (Kilar and Fenton 2012).

Some critics contend that because concepts such as patriarchy are difficult to measure, feminist research has yet to show specifically how patriarchy affects female criminality. Another criticism is that most feminist analyses emphasize direct male violence against women (such as partner abuse and rape) and street crime but say little about women's white-collar crimes (Belknap 2001; Moyer 2001; Friedrichs 2004).

breaking into liquor stores for their dad or stealing tractor-trailer trucks, hundreds of them. The girls turned to petty crimes to support their drug addictions" (Butterfield 2002: 1).

People become deviant, according to sociologist Edwin Sutherland, if they have more contact with significant others who violate laws than with those who are law-abiding. In effect, people learn techniques for committing criminal behavior, along with the values, motives, rationalizations, and attitudes that reinforce such behavior (Sutherland and Cressey 1970). Thus, in Rooster's clan, almost everyone wound up in prison because they grew up in an environment that taught deviance rather than conformity.

Sutherland emphasized that differential association doesn't occur overnight. Instead, people are most likely to engage in crime if they are exposed to deviant values (1) early in life, (2) frequently, (3) over a long period of time, and (4) from significant others and reference groups (such as parents, siblings, close friends, and important business associates).

Considerable research supports differential association theory. Almost 47 percent of state prisoners have a parent or other close relative who has also been incarcerated. Even before age 13, children who associate with peers who smoke, use drugs, or commit crimes are more likely to do so themselves because they learn these behaviors from their friends (Conway and McCord 2005; Adamson 2010; McGloin and Piquero 2010). Thus, according to differential association theory, we are products of our socialization.

Steve Cole/Photodisc/Getty Images

# 7-7 Symbolic Interaction Perspectives on Deviance and Crime

For symbolic interactionists, deviance is socially constructed because it's in the eye of the beholder. They offer many theories to explain deviance, but two of the best known are differential association theory and labeling theory.

## 7-7a Differential Association Theory

**Differential association theory** is a perspective that asserts that people learn deviance through interaction, especially with significant others such as family members and friends. Dale "Rooster" Bogle's family in Oregon is an example of this theory. Even though Rooster, the father, served time in prison for theft, he taught his sons and daughters to survive by stealing: "By the time the boys were 10 years old they were

## 7-7b Labeling Theories

Have you ever been accused of something wrong that you didn't do? What about getting credit for something that you and others know you didn't deserve? In either case, did people start treating you differently? The reactions of others are the crux of **labeling theory**, which holds that society's reaction to behavior is a major factor in defining oneself or others as deviant.

A good example of labeling is the American Psychiatric Association's *Diagnostic and Statistical Manual of Mental Disorders (DSM),* which defines what behaviors are normal or abnormal. Some new disorders in the 2013 edition may include "hypersexuality," especially for men, and "binge eating," especially for women.

Lisa F. Young/Shutterstock.com

The *DSM* has far-reaching effects, according to one of the past editors: "Anything you put in that book . . . has huge implications not only for psychiatry but for pharmaceutical marketing, research, for the legal system, for who's considered to be normal or not, for who's considered disabled. *And it has huge implications for stigma because the more disorders you put in, the more people get labels*" (Carey 2010: 1, emphasis added; see also Chapter 14).

Some of the earliest sociological studies found that teenagers who misbehaved were tagged as delinquents. Such tagging changed the child's self-concept and resulted in more deviance and criminal behavior (Tannenbaum 1938). During the 1950s and 1960s, two influential sociologists—Howard Becker and Edwin Lemert—extended labeling theory.

### Becker: Deviance Is in the Eyes of the Beholder

According to Howard Becker (1963: 9), being a deviant or a criminal depends on how others react: "The deviant is one to whom that label has successfully been applied; *deviant behavior is behavior that people so label*" (emphasis in original). That is, it's not an act that determines deviance, but whether and how others react.

Some people are never caught or prosecuted for crimes they commit, and thus aren't labeled deviant. In other cases, people who are not breaking the law (such as cheating on taxes) may be falsely accused and stigmatized. In effect, then, deviance is in the eye of the beholder because societal reaction, rather than an act, labels people as law-abiding or deviant. Moreover, labeling can lead to secondary deviance.

### Lemert: Primary and Secondary Deviance

Edwin Lemert (1951, 1967) expanded on the effects of labeling by differentiating between primary and secondary deviance. **Primary deviance** is the initial act of breaking a rule. Primary deviance can range from relatively minor offenses, such as not attending a family member's funeral, to serious offenses, such as rape and murder.

Even if people aren't guilty of primary deviance, labeling can result in **secondary deviance**, rule-breaking behavior that people adopt in response to the reactions of others. A teenager who is caught trying marijuana may be labeled a "druggie." If the individual is rejected by others, he or she may accept the label, associate with drug users, and become involved in a drug-using subculture. According to Lemert, a single deviant act will rarely result in secondary deviance. The more times a person is labeled, however, the higher the probability that she or he will accept the label and engage in deviant behavior.

There's considerable evidence that labeling affects people's lives. For example, nearly half of all Americans who experience severe health problems such as

> **primary deviance** the initial act of breaking a rule.
>
> **secondary deviance** rule-breaking behavior that people adopt in response to the reactions of others.

© Colin young-wolff/Alamy

Almost 18 million low-income Americans live in trailer parks because they can't afford other housing. Most realize that they're stigmatized as "trailer trash," but they try to maintain their dignity despite the negative views. For example, they describe themselves as homeowners, differentiate themselves from those who commit crimes, and take care of their property (Kusenbach 2009).

depression never seek treatment because they fear being stigmatized as mentally ill (U.S. Department of Health and Human Services 1999). Most recently, those who have been jobless for a year or longer report facing stigmas in job searches: Employers assume that they're lazy, they don't really want to work, or there's something wrong with them. Some companies have explicitly barred the long-term unemployed from certain job openings, telling them outright in job ads that they need not apply (Hunsinger 2011).

## 7-7c Critical Evaluation

Despite symbolic interaction's contributions in explaining crime and deviance, critics point to several weaknesses. Differential association theory doesn't explain impulsive crimes of rage (such as domestic murder), those committed by people who have grown up in law-abiding families, or why crime rates are much higher in some inner-city racial and ethnic communities than others. This theory also ignores the possibility that deviant values and behaviors can be unlearned, as when young children model teachers or other adults whom they respect (Anderson and Dyson 2002; Williams and McShane 2004; Graif and Sampson 2009).

Critics have faulted labeling theory on other points: The theory exaggerates the importance of judgments in altering a person's self-concept, doesn't explain why crime rates are higher in the South than in other parts of the United States or at particular times of the year (such as before holidays), and doesn't tell us *why* people commit crimes. Some critics also contend that social reactions are the *result* rather than the *cause* of deviant behavior, and that social control agents (such as police) are most likely to label people who have committed serious crimes or who have a long criminal record (Schur-

Andrey Eremin/Shutterstock.com

man-Kauflin 2000; Benson 2002). Conflict theorists, in particular, criticize symbolic interactionists for ignoring structural factors—such as poverty and low-paid jobs—that create or reinforce deviance and crime (Currie 1985).

# 7-8 The Criminal Justice System and Social Control

Social institutions such as the family, education, and religion try to maintain *social control* over behavior; the criminal justice system has the *legal* power to control crime and punish offenders. The **criminal justice system** refers to government agencies—including the police, courts, and prisons—that are charged with enforcing laws, passing judgment on offenders, and changing criminal behavior. The criminal justice system relies on three major strategies to control crime: prevention and intervention, punishment, and rehabilitation. Punishment, as you'll see, is the least effective approach.

## 7-8a Prevention and Intervention

Crime prevention is much less expensive than punishment. For example, many crimes involve substance abuse, but the average annual cost per person for treatment of alcohol or drug abuse for outpatients is $1,443, compared with $24,000 for incarceration (National Institute on Drug Abuse 2009).

Some crimes have plummeted because of technology. Car theft, for example, has decreased 40 percent since 2003 because most new automobiles are impossible to hot-wire: They come equipped with immobilizer systems that require an authorized key to start (Klein and White 2011). Technology, however, hasn't stopped other crimes such as burglary because many offenders have become more sophisticated in avoiding home security systems.

Because most criminal behavior first occurs in adolescence (or earlier), criminal justice agencies often focus many of their prevention and intervention efforts on juveniles and their families. The most common agents of such efforts are social service agencies, community outreach programs, and the police.

### Social Service Agencies and Community Outreach Programs

Numerous organizations—composed of law enforcement professionals, social workers, and nonprofit groups—try to prevent crime. Because of high prison costs, the U.S. Department of Justice, among other federal agencies, funds numerous initiatives that try to decrease youth violence and delinquency, drug dealing, rape, robbery, prostitution, and domestic violence (Office of Juvenile Justice and Delinquency Prevention 2010; Office of Justice Programs 2011).

How well do prevention and intervention programs work? A national study of adolescents ages 12 to 17 found that the grades of those who participated in a 5-month drug treatment program improved by about 30 percent, often to a level of B or better (Spiess 2003).

What about "scared straight" programs that are still very popular in this country as a get-tough response to juvenile crime? These programs include tours of prisons and face-to-face encounters with inmates, including murderers, who try to keep high-risk adolescents from becoming lifelong criminals. An evaluation of nine scared-straight programs concluded that they don't deter the teenage participants from deviant behavior (Petrosino et al. 2003; Robinson and Slowikowski 2011). It's not clear, however, whether the negative effects are due to the adolescent's long history of offenses, not having a positive role model who helps them change their behavior, or other reasons.

## Police

The primary role of the police is to enforce society's laws, but can the police prevent crimes? Police can head off some crimes by cruising high-risk areas in patrol cars or by having more officers on foot patrols. Concentrating on *hot spots,* areas of high criminal activity, reduces crime only temporarily, however, because criminals simply move to other parts of the city (U.S. Department of Justice 1996; Braga 2003). Also, police can't prevent most crimes because they have no control over macro-level factors that lead to criminal behavior, such as poverty, unemployment, low educational opportunities, and neighborhood deterioration.

## 7-8b Punishment

Those who endorse a **crime control model** believe that crime rates increase when offenders don't fear apprehension or punishment. This perspective emphasizes protecting society and supports a tough approach toward criminals in sentencing, imprisonment, and capital punishment.

### Sentencing

After someone has been found guilty of a criminal offense or has pleaded guilty, a judge (and sometimes a jury) imposes a *sentence,* or penalty. A sentence can be a fine, *probation* (supervision instead of serving time in jail), incarceration, or the death penalty. Those who receive *parole* are released from prison before the end of their full sentence on condition that they check in regularly with an officer and obey the law.

About 31 percent of those sentenced get probation; the others serve a term in a local jail or a state or federal prison. Many people question the fairness of sentencing because those convicted of similar crimes can receive widely different sentences depending, among other factors, on race/ethnicity, social

$17.25 million
Murder

$448,532
Rape

$335,733
Robbery

$145,379
Aggravated Assault

$41,288
Burglary

These are just the economic costs of the five major U.S. crimes, including victim costs, criminal justice costs, and the lost productivity of both victims and offenders (DeLisi et al. 2010).

Do you think that court-ordered public shaming strategies, such as this one, deter crime and rehabilitate the offender?

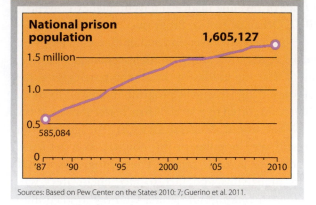

## FIGURE 7.2
## The Number of U.S. Prisoners Has Nearly Tripled Since 1987

National prison population — 1,605,127

585,084

'87 '90 '95 2000 '05 2010

Sources: Based on Pew Center on the States 2010: 7; Guerino et al. 2011.

class, variations in state sentencing laws (including how many times a person has been arrested for a crime), and how juries and judges evaluate the seriousness of a crime (Durose and Langan 2004; Whelan 2007). For example, after a jury found the defendant guilty, a New Orleans judge recently handed down a life sentence to a man who, for the fourth time, was arrested for possessing marijuana (Shepherd 2011b).

As state and local budgets shrink, even states with reputations for being tough on crime are embracing more lenient punitive policies. Some states have begun to shorten the average number of years on probation and parole, which decreases supervision costs. Other states are reducing the number of people sent to prison because it costs an average of $79 a day to keep an inmate in prison but about $3.50 a day to monitor the same person on probation (Richburg 2009; Pew Center on the States 2009). Despite greater leniency, the number of Americans on probation and parole surged from 1.6 million in 1982 to almost 5 million in 2010 (Glaze and Bonczar 2011).

It's tempting to assume that the more people behind bars, the less crime there will be.

### Incarceration

Since 2008, and for the first time in history, 1 in every 100 American adults is in prison (see *Figure 7.2*). The United States has less than 5 percent of the world's population, but almost 25 percent of the planet's prisoners. Besides the sheer number of inmates, the United States is also the global leader in inmates per capita, ahead of nations such as Russia and England (Pew Center on the States 2008).

Of all U.S. state and federal prisoners, 97 percent are men (Guerino et al. 2011). However, imprisonment rates vary by race/ethnicity and sex. As *Table 7.4* shows, incarceration rates are highest for African Americans, both women and men.

In 2010, for the first time in nearly 40 years, the number of state prisoners decreased by almost 1 percent (Guerino et al. 2011). Why the drop? Crime rates have recently declined, but prison counts dropped in 26 states primarily because of economic reasons. Driven by budget crises, many states have pursued a variety of strategies such as releasing low-level offenders before their sentences were complete, reducing the severity of sentences, and accelerating the release of inmates who completed educational, vocational, and substance abuse treatment programs. Releasing prisoners early has been controversial, but state officials maintain that their budgets can't afford the high costs of incarceration (Haq 2010; Pew Center on the States 2011).

### Capital Punishment

About 61 percent of Americans support *capital punishment* (the death penalty), a percentage that has been fairly consistent since 1936, and 40 percent say

that the death penalty isn't imposed often enough (Newport 2011). By 2008, 137 nations had abolished the death penalty or suspended executions. Of all the executions worldwide in 2007, 88 percent took place in China, Iran, Saudi Arabia, Pakistan, and the United States (Campbell 2008).

In the United States, 34 states have the death penalty, but the number of executions declined from 98 in 1999 to 43 in 2011 (Death Penalty Information Center 2012). There are several reasons for fewer executions. First, the nation's police chiefs believe that the death penalty doesn't deter crime. Second, many opponents of the death penalty contend that minorities receive death penalty sentences more often than whites. Third, some argue that the death penalty is a waste of money because inmates can spend 15 to 20 years appealing the sentence. In California, for example, it costs taxpayers $137 million per year for inmates who appeal a death sentence, compared with $12 million per year for those serving life sentences (Death Penalty

Information Center 2009, 2012).

It's tempting to assume that the more people behind bars, the less crime there will be. State spending on corrections surged from $20 billion in 1982 to $74 billion in 2007 (Bureau of Justice Statistics 2011). However, *recidivism* (being arrested for committing another offense after being released) has barely changed since 1999: Almost half of released prisoners are back behind bars within 3 years (Pew Center on the States 2011).

## 7-8c Rehabilitation

**Rehabilitation**, a third approach to controlling deviance, maintains that appropriate treatment can change offenders into productive, law-abiding citizens. Advocates argue that public assistance, educational opportunities, job training, and intervention programs can reduce recidivism because most inmates come out of prison with no job skills, money, or other resources to reintegrate into society. According to a criminology professor who served 9 years in federal prisons for distributing marijuana, to lower recidivism rates, prisoners should be released with Social Security cards, current drivers' licenses, sufficient money to cover rent and food for 3 months, and professional services (employment assistance, personal and family counseling, drug and alcohol treatment programs, and medical benefits) (Ross and Richards 2002).

Are rehabilitation programs successful? Only if they provide employment after release. Other effective rehabilitation efforts have offered training for a trade while in prison, earning a high school diploma or college degree while in prison or after release, and services that address several needs (such as housing, employment, and medical services) rather than just one (such as counseling for drug abuse) (Lowenkamp and Latessa 2005; Cumberworth 2010; Flower 2010).

> **rehabilitation** a social control approach that holds that appropriate treatment can change offenders into productive, law-abiding citizens.

---

### TABLE 7.4
## U.S. Imprisonment Rates by Race/Ethnicity and Sex, 2010

| IMPRISONMENT RATE PER 100,000 U.S. RESIDENTS | | |
|---|---|---|
| RACE/ETHNICITY | MALE | FEMALE |
| Total | 938 | 67 |
| White | 456 | 47 |
| African American | 3,059 | 133 |
| Latino | 1,252 | 77 |

Note: "Total" includes American Indians, Alaska Natives, Asians, Native Hawaiians, other Pacific Islanders, and persons identifying as two or more races.

Source: Based on Guerino et al. 2011, Table 14.

---

### Study Tools

Ready to study? Go to **Sociology CourseMate** at **www.cengagebrain.com** to complete practice quizzes, review flashcards, watch videos, and more.

# SOC

**Having resources** can mean the difference between life and death.

© Roger Hutchings/Alamy

# Social Stratification: United States and Global

# 8

## Key Topics

In this chapter, we'll explore the following topics:

| | |
|---|---|
| **8-1** | What Is Social Stratification? |
| **8-2** | Dimensions of Stratification |
| **8-3** | Social Class in America |
| **8-4** | Poverty in America |
| **8-5** | Social Mobility |
| **8-6** | Global Inequality |
| **8-7** | Sociological Explanations: Why There Are Haves and Have-Nots |

In the United States, Chrissandra, 50, used to make $100,000 a year as a nursing home executive. She lost her job and now she and her daughter live on $11,000 a year. In contrast, a casino billionaire purchased a $68 million Boeing jet for family vacations (Bertoni 2010; Haygood 2010). In India, where 37 percent of the population is poor, five of the richest people live in a 27-story tower that has three helipads on the roof, nine elevators, a grand ballroom, and hundreds of servants and staff (Nanda 2010; Yardley 2010).

As these examples show, many people around the world are struggling to survive while others enjoy astonishing wealth. This chapter examines social stratification. It considers why people move up or down the social class ladder, looks at global inequality, and discusses some of the sociological theories on why there are haves and have-nots. First, however, take the brief quiz on the next page to see how much you know about U.S. economic inequality.

## 8-1 What Is Social Stratification?

**S**ocial stratification is the hierarchical ranking of people in a society who have different access to valued resources, such as property, prestige, power, and status. All societies are stratified, but some more than others.

An **open stratification system** is based on individual achievement and allows movement up or down. In a **closed stratification system**, movement from one social position to another is limited by ascribed statuses such as sex, skin color, and family background. Closed stratification systems are considerably more fixed than open ones, but no stratification system is completely open or closed.

**social stratification** the hierarchical ranking of people in a society who have different access to valued resources, such as property, prestige, power, and status.

**open stratification system** a system that is based on individual achievement and allows movement up or down.

**closed stratification system** a system in which movement from one social position to another is limited by ascribed statuses such as sex, skin color, and family background.

## What do you think?

Americans who are poor just aren't working hard enough.

| 1 | 2 | 3 | 4 | 5 | 6 | 7 |
|---|---|---|---|---|---|---|

strongly agree        strongly disagree

On the bottom rung were the Dalits (formerly referred to as "untouchables"), who were very poor and performed the most menial and unpleasant jobs, such as collecting waste and cleaning streets. Fearing being "polluted," persons of higher castes would not interact with the Dalits (Mendelsohn and Vicziany 1998).

India outlawed the caste system in 1949, but social distinctions are deeply entrenched, and most people socialize with and marry within their own castes (Banerjee et al. 2009). Castes are so common in some parts of India, the national census now collects data on them (Haub 2011). Dalits are becoming a potent political force, but caste discrimination, including at many universities, is pervasive (Neelakantan 2011).

## 8-1a Closed Stratification Systems

> **social class** a category of people who have a similar standing or rank in a society based on wealth, education, power, prestige, and other valued resources.

Two closed stratification systems exist today: slavery and castes. In *slavery*, an extreme form of inequality, people own others as property and have almost total control over their lives. In *chattel slavery*, people are bought and sold as commodities, sometimes multiple times. Chattel slaves are often abducted from their homes, inherited, or given as gifts to pay a debt. The United Nations banned all forms of slavery worldwide in 1948, but it persists in many countries in the Middle East, Africa, the Balkans, and Asia (U.S. Department of State 2006; Lampman 2007; Dixon 2009).

*Castes*, a second type of closed stratification system, are social categories based on heredity. Because social status is ascribed at birth, caste members are severely restricted in their choice of occupation, residence, and social relationships. A good example is India, where a caste system has existed for more than 3,000 years. At the top were the Brahmins (priests, scholars, and educated class), followed by Kshatriyas (kings and warriors), Vaishyas (merchants and farmers), and Sudras (peasants, laborers, and craftspeople).

## 8-1b Open Stratification Systems

In open stratification systems, social classes are relatively fluid because they are based on achieved rather than ascribed statuses. A **social class** is a category of people who have a similar standing or rank in a society based on wealth, education, power, prestige, and other valued resources. Theoretically, people in open stratification systems can move from one class to another. As you'll see throughout this chapter, however, the more resources someone has at birth, the greater her or his chance of moving into a higher social class. In effect, then, open stratification systems are not as open as many people believe.

# 8-2 Dimensions of Stratification

The Occupy Wall Street (OWS) protest began in mid-2011 in New York City's Wall Street financial district. The protestors' slogan, "We are the 99%," referred to the growing U.S. income inequality between the richest 1 percent and the rest of the population.

Income is a critical stratification factor, but it's not the only one. Instead, sociologists use a multidimensional approach that includes wealth, prestige, and power.

## 8-2a Wealth

**Wealth** is the abundance of economic assets and material possessions that a person or family owns, including property and income. *Property* comes in many forms, such as buildings and land; personal possessions such as furniture, jewelry, and works of art; and stocks, bonds, and retirement savings. *Income* is money a person receives regularly, usually in the form of wages or a salary, rents, interest on savings accounts, dividends on stock, royalties, or the proceeds from a business.

Income and wealth differ in several important ways:

- Wealth is *cumulative*. It increases over time, especially through investment, whereas income is usually spent on everyday expenses.

- Because wealth is accumulated over time, much of it can be *passed on to the next generation*. With an estimated $61 billion income in 2011, Bill Gates, cofounder of Microsoft Corporation, is one of the wealthiest people in the world. If each of Gates's three children inherits only 1 percent of his fortune, each will get at least $610 million!

- *Wealth ensures economic security and future prosperity* because affluent people can provide their children with an education, start a business, and live luxuriously in retirement.

U.S. wealth and income inequality is staggering. In 2011, the country's 413 billionaires had a collective wealth of $1.5 trillion, an average of $3.7 billion per person ("Slowing Giant . . ." 2011). As *Figure 8.1* shows, the top 1 percent of U.S. households owns 36 percent of all wealth, compared with only 13 percent for the bottom 80 percent. Income

**wealth** abundance of economic assets and material possessions that a person or family owns, including property and income.

India's Dalits, such as those pictured here, still perform unpleasant tasks such as burning corpses and removing rubbish and human waste.

## FIGURE 8.1
## U.S. Wealth and Income Inequality

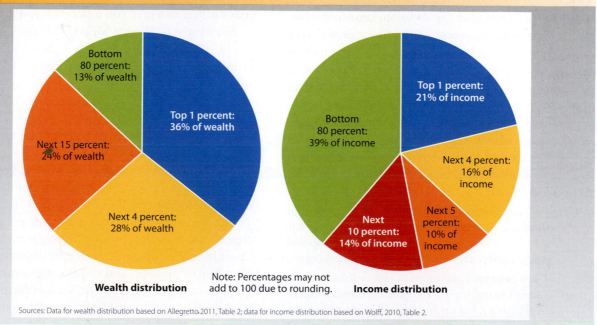

**Wealth distribution**

Bottom 80 percent: 13% of wealth
Top 1 percent: 36% of wealth
Next 15 percent: 24% of wealth
Next 4 percent: 28% of wealth

Note: Percentages may not add to 100 due to rounding.

**Income distribution**

Top 1 percent: 21% of income
Bottom 80 percent: 39% of income
Next 4 percent: 16% of income
Next 5 percent: 10% of income
Next 10 percent: 14% of income

Sources: Data for wealth distribution based on Allegretto, 2011, Table 2; data for income distribution based on Wolff, 2010, Table 2.

**prestige** respect, recognition, or regard attached to social positions.

**power** the ability of individuals or groups to achieve goals, control events, and maintain influence over others despite opposition.

inequality is also huge. In 1973, the richest 1 percent of Americans received 9 percent of all income (Atkinson et al. 2011). They now receive 21 percent of all income, compared with only 39 percent for the bottom 80 percent.

Despite such stark disparities, 52 percent of Americans believe that our society is *not* divided into haves and have-nots ("No Consensus About . . ." 2011). And only 45 percent say that wealth and income inequality is a problem "that needs to be fixed," down from 52 percent in 1998 (Morin 2012).

There has also been a huge disparity in wealth gains over the last generation. Between 1983 and 2009, the richest 5 percent of households obtained almost 82 percent of all the nation's gains in wealth. The bottom 60 percent had less wealth in 2009 than in 1983 (Mishel 2011; see also Kochhar et al. 2011).

## Sociology in Your Life

How much inequality is there in all facets of U.S. society? Go to Sociology CourseMate at CengageBrain.com to get more of the facts.

## 8-2b Prestige

A second dimension of social stratification is **prestige**—respect, recognition, or regard attached to social positions. Prestige is based on many criteria, including wealth, family background, power, occupation, and accomplishments. Regarding accomplishments, for example, every college convocation acknowledges students who graduate *cum laude, magna cum laude,* and *summa cum laude.*

We typically evaluate people according to the kind of work they do. *Table 8.1* presents a sample of prestige scores for U.S. occupations. Because no occupation is worthless or perfect, the scores range from 20 to 86 rather than from 0 to 100. If you examine *Table 8.1,* you'll notice several characteristics of the most prestigious occupations:

- They *require more formal education* (such as college or postgraduate degrees) and/or extensive training. Physicians (86), for example, must fulfill internship and residency requirements after receiving a medical degree.
- They are primarily *nonmanual rather than manual and require more abstract thought.* All jobs have some physical activity, but an architect (73) must use more imagination in designing a building than a carpenter (39), who usually performs very specific tasks.
- They *pay more,* even though there are some exceptions. A realtor (49) or a truck driver (30) may earn more than a registered nurse (66), but registered nurses are likely to earn more over a lifetime because they have steady employment, good health benefits, retirement programs, and more opportunities to find jobs during layoffs or career changes.
- They *are seen as more socially important.* An elementary school teacher (64) may earn less than the school's janitor (22), but the teacher's job is more prestigious because teachers contribute more to a society's well-being.
- They *involve greater self-expression, autonomy, and freedom from supervision.* A dentist (72) has considerably more freedom in performing her or his job than a dental hygienist (52). In effect, then, higher prestige occupations provide more privileges.

## 8-2c Power

A third important dimension of social stratification is **power**—the ability of individuals or groups to achieve goals, control events, and maintain influence over others despite opposition. In every society, power is based on social class, but there are other sources of power. One is *custom* or *tradition,* as when the chief of a tribe has total authority based on family lineage. Another source of power is *charisma.* Leaders such as Mahatma Gandhi and Martin Luther King Jr. inspired millions of people to demand change, peace, and social justice. Power is also tied to *particular occupations:* Your professor has the power to give you an A or an F, and a police officer can give you a ticket or just a warning.

Sociologist C. Wright Mills (1956) coined the term *power elite* to describe a small, tightly knit group of white men—especially corporate executives, national political leaders, and high-ranking military officers—who make all the important decisions in U.S. society. Many contemporary sociologists agree that a socially cohesive and very wealthy group of white men continues to be a "ruling class" that has the power to dominate much of the American economy and government (see, for example, Zweigenhaft and Domhoff 1998; we'll examine power elites in Chapter 11).

A person's ranking may be about equal on the dimensions of wealth, prestige, and power. A Supreme Court judge, for instance, is usually affluent, enjoys a great deal of prestige, and wields considerable power. In many cases, however, there is *status inconsistency,* meaning that a person ranks differently on the various stratification dimensions. Consider funeral directors. Their prestige is relatively low, but most have higher incomes than college professors, who are among the most educated people in U.S. society and have relatively high prestige (see *Table 8.1*).

We've looked at stratification systems and their dimensions, but how, specifically, do people in different social classes behave? And how does social class affect our behavior?

# 8-3 Social Class in America

**B**ecause social class, like social stratification, is multidimensional, sociologists typically use **socioeconomic status (SES)**—an overall ranking of a person's position in the class hierarchy based on income, education, and occupation—to measure social class. Some sociologists ask people to identify the social classes in their communities (*reputational approach*), some ask people to place themselves in one of a number of classes (*subjective approach*), but most use SES indicators (*objective approach*).

Because there are different ways of measuring social class, sociologists don't always agree on the number of social classes in the United States. However, there is consensus on four general social classes: upper, middle, working, and lower. Except for the working class,

many sociologists divide these classes further into more specific strata.

Sociologists Dennis Gilbert and Joseph Kahl (1993) have proposed a teardrop model of the American class structure, based primarily on income and occupation (see *Figure 8.2*). They caution that social class is a complex concept and that specifying the dividing lines between classes is as much art as science, but the model provides an overview of the U.S. social class structure. Besides income, education, and occupation, social classes also differ in values, power, prestige, social networks, and *lifestyles* (tastes, preferences, and ways of living).

## 8-3a The Upper Class

You saw earlier that very rich Americans control a vastly disproportionate amount of the total U.S. wealth and income. This group comprises two classes: the upper-upper class and the lower-upper class.

### The Upper-Upper Class

Upper-upper-class people rarely appear on the lists of wealthiest individuals issued by *Forbes* or other sources. Because they value their privacy, some upper-upper-class members refuse to be listed even in *The*

> **socioeconomic status (SES)** an overall ranking of a person's position in the class hierarchy based on income, education, and occupation.

| TABLE 8.1 |
|---|

## Prestige Scores for Selected U.S. Occupations

Do you agree with these rankings? Are there occupations that you think should rank higher or lower? If so, why?

| HIGHER PRESTIGE JOBS | MEDIUM PRESTIGE JOBS | LOWER PRESTIGE JOBS |
|---|---|---|
| 86 Physician | 59 Police officer | 39 Carpenter |
| 85 Supreme Court judge | 58 Actor | 36 Child-care worker |
| 78 Lawyer | 55 Radio/TV announcer | 36 Hairdresser |
| 74 College professor; computer systems analyst | 54 Librarian | 35 Assembly-line worker |
| 73 Architect | 53 Firefighter | 33 Cashier in supermarket |
| 72 Dentist | 52 Dental hygienist | 31 Auto body repairperson |
| 69 Member of the clergy | 52 Social worker | 30 Truck driver |
| 66 Registered nurse | 51 Electrician | 28 Garbage collector; waiter/waitress |
| 66 High school teacher | 49 Funeral director; realtor | 25 Bartender |
| 65 Accountant | 48 Manager of a supermarket | 23 Cleaner, private home |
| 64 Elementary school teacher | 47 Mail carrier | 22 Janitor |
| 62 Veterinarian | 46 Secretary | |

Sources: Based on Nakao and Treas 1992; J. Davis et al. 2005.

## FIGURE 8.2
## The American Class Structure

| TYPICAL OCCUPATIONS | | TYPICAL ANNUAL HOUSEHOLD INCOME |
|---|---|---|
| **UPPER CLASS 3%** | **Upper Class 3%** | |
| *Upper-Upper Class 1%* | | *Upper-Upper* Hundreds of millions to billions of $ |
| Heirs of "old money" | | |
| *Lower-Upper Class 2%* | | *Lower-Upper* Millions of $ |
| Executives | | |
| High-level managers | | |
| Owners of corporations | | |
| **MIDDLE CLASS 40%** | | *Upper-Middle* $76,000+ |
| *Upper-Middle Class 14%* | | |
| Upper-level managers | | |
| Professionals | | |
| Medium-sized business owners | | |
| *Lower-Middle Class 26%* | **Middle Class 40%** | *Lower-Middle* $46,000–$75,999 |
| Lower-level managers | | |
| Semiprofessionals | | |
| Craftspeople | | |
| Foremen | | |
| Nonretail workers | | |
| **WORKING CLASS 30%** | **Working Class 30%** | *Working* $19,000–$45,999 |
| Semiskilled workers | | |
| Clerical | | |
| Retail sales | **Lower Class 27%** | |
| **LOWER CLASS 27%** | | |
| *Working Poor 13%* | **Working Poor** | *Working Poor* $9,000–$18,999 |
| Low-paid manual | | |
| Retail and sales workers | | |
| *Underclass 14%* | **Underclass** | *Underclass* Under $9,000 |
| Unemployed, part-time low wage earners | | |
| Public assistance recipients | | |

Sources: Based on O'Hare 2002; Wellner 2003; Gilbert 2008, p. 13.

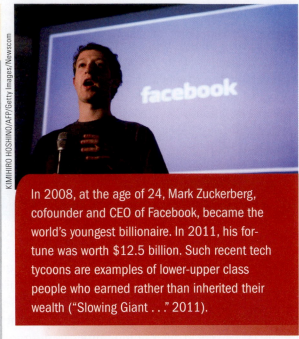

In 2008, at the age of 24, Mark Zuckerberg, cofounder and CEO of Facebook, became the world's youngest billionaire. In 2011, his fortune was worth $12.5 billion. Such recent tech tycoons are examples of lower-upper class people who earned rather than inherited their wealth ("Slowing Giant . . ." 2011).

KIMIHIRO HOSHINO/AFP/Getty Images/Newscom

*Social Register*, an inventory of America's social elite that has been published since 1887.

An inherited fortune brings power (McNamee and Miller 1998). Upper-upper-class white males, in particular, shape the economic and political climate through a variety of mechanisms: dominating the upper levels of business and finance, holding top political positions in the federal government, underwriting thousands of think tanks and research institutes that formulate national policies, and shaping public opinion through the mass media (Zweigenhaft and Domhoff 2006; see also Hacker and Pierson 2010).

### The Lower-Upper Class

The lower-upper class—which is much more diverse than the upper-upper class—is the *nouveau riche*, those with "new money." Some, like the Kennedys, amassed fortunes several generations ago, but many of the new rich—such as Oprah Winfrey, Bill Gates, and Mark Zuckerberg—worked for their income rather than inherited it. Besides business entrepreneurs, the lower-upper class also includes high-level managers of huge corporations, self-made millionaires, and some

highly paid athletes and actors, but their lifestyles vary considerably.

Some lower-upper class members live modestly, but many flaunt their new wealth. Sociologist Thorstein Veblen (1899/1953) coined the term *conspicuous consumption*, a lavish spending on goods and services to display one's social status and enhance one's prestige.

Because they lack the "right" ancestry and have usually made their money by working for it, lower-uppers are not accepted into "old money" circles that have strong feelings of in-group solidarity. Still, lower-upper class members engage in lifestyles and rituals that try to parallel those of the upper-upper class, such as having personal chefs, taking exotic vacations (often in private planes), and joining country clubs (Sherwood 2010).

### 8-3b The Middle Class

Most Americans describe themselves as middle class, including 41 percent of those with incomes below $20,000 and 33 percent of those with incomes above $150,000 (Pew Research Center 2008). There are several strata in the middle class, but sociologists often distinguish between the *upper-middle class* and the *lower-middle class*.

### The Upper-Middle Class

Upper-middle-class members, although rich, live on earned income rather than accumulated or inherited wealth. Their salaries are high enough to provide economic stability and sizable savings, but usually because both spouses have high educational levels and careers.

The occupations of this group—mainly professional and managerial—usually require a Ph.D. or advanced degree in business, law, or medicine. People in this class include corporate executives and managers (but not those at the top), high government officials, owners of large businesses, physicians, and successful lawyers and stockbrokers. Most of these occupations have considerable on-the-job autonomy and freedom from supervision, but these people are three times more likely than those in the general population to work 50 or more hours per week (Dewan and Gebeloff 2012).

## Sociology in Your Life

How much annual income do you need to live comfortably? Compare your answer to those of other Americans by going to Sociology CourseMate at CengageBrain.com.

### The Lower-Middle Class

The lower-middle class, more diverse than the upper-middle class, is composed of people in nonmanual and semiprofessional occupations. Nonmanual jobs include office staff, low-level managers, owners of small businesses, medical and dental technicians, secretaries, police officers, and sales workers (such as insurance and real estate agents). Examples of semiprofessional

Among America's working poor are hotel housekeepers. On average, those employed even at expensive hotels earn less than $19,000 a year working full-time.

Status symbols—cars, clothes, vacations, and so on—are not limited to the super-rich. Many upper-middle-class members, in particular, use status symbols to show that they've made it by buying "almost rich" cars (such as Jaguar X-type sedans that start at $30,000), upscale kitchen appliances, designer handbags, and expensive jewelry.

occupations are nursing, social work, and teaching. Almost all of these jobs require training beyond high school and many, especially the semiprofessional occupations, require a college degree. Most families in the lower-middle class rely on two incomes to maintain a comfortable standard of living.

Unlike upper-middle-class jobs, those in the lower-middle class have less autonomy and freedom from supervision, and there is little chance for advancement. People are more likely to follow orders than to give them. Except for some retirement funds, most have only modest savings to cover emergencies. Many buy used or inexpensive late-model cars, eat out at middle-income restaurants, and take occasional vacations, but they rarely have the income to buy luxury products without going deeply into debt.

## 8-3c The Working Class

The working class consists of skilled and semiskilled laborers, factory employees, and other blue-collar workers in manual occupations. They are construction workers, assembly-line workers, truck drivers, auto mechanics, repair personnel, bartenders, and skilled craft workers such as carpenters and electricians. Most of the jobs are blue collar, but some—such as clerks and retail sales workers in the service sector—are white collar.

People who fill working-class jobs often have a high school diploma but no college education, and many of the positions provide little or no opportunity for advancement. Most see their work as boring and routine—a source of income to survive on rather than a means of attaining personal fulfillment. Most of the semiskilled jobs require little training, are mechanized, and are closely supervised.

Working-class people who purchase homes, including mobile homes, may experience foreclosure because of delinquent payments. Many use credit cards to pay off bills each month but then can barely pay the monthly minimum. Debts become overwhelming when borrowers who live from paycheck to paycheck suffer setbacks such as divorce, illness, or job loss.

## 8-3d The Lower Class

People in the lower class are at the bottom of the economic ladder because they have little education and few occupational skills, work in minimum wage jobs, or are often unemployed. Most of the lower class is poor, but sociologists often distinguish between the *working poor* and the *underclass*.

### The Working Poor

About 26 percent of all employees are the **working poor**—people who work at least 27 weeks a year but receive such low wages that they live in or near poverty. Almost half of American families headed by a married couple with at least one full-time, full-year worker are among the working poor (Holzer 2007; "A Profile of the Working Poor . . ." 2009).

Up to half of the people who visit food pantries and soup kitchens in some states are the working poor (Ryan 2008). Many attribute their situation to bad luck or fate and feel powerless over their economic insecurity.

### The Underclass

The **underclass**, which occupies the bottom rung of the economic ladder, consists of people who are persistently poor, residentially segregated, and relatively isolated from the rest of the population. Most rarely work, are chronically unemployed, or drift in and out of jobs. Social scientists often use the term *underclass* to describe inner-city minorities, but it applies to people of any race or ethnicity who are destitute and have little chance of moving out of abject poverty. They may work erratically or at part-time jobs, but their lack of skills, low educational levels, and in many cases, disabilities make it difficult for them to find regular, full-time jobs (Beeghley 2000; Gilbert 2008).

For some sociologists, being in the underclass or one of "the ghetto poor" is the result of joblessness rather than poverty, because the underclass grew when most middle-class people—both white and black—moved to the suburbs, the number of factory jobs declined, and the poor became more isolated (Wilson 1996). Others argue that the underclass is locked in a "culture of poverty" that has more to do with personal inadequacies (such as laziness) than structural factors (such as unemployment) (see Murray 2012). Most social scientists agree, however, that the underclass experiences a wide range of social problems—crime, welfare, family dissolution, drug abuse, poor health, and domestic violence.

## 8-3e How Social Class Affects Us

Our social class, more than any other single variable, affects just about all aspects of our lives, because having resources can mean the difference between life and death. Max Weber referred to the consequences of social stratification as **life chances**—the extent to which people have positive experiences and can secure the good things in life (such as food, housing, education, and good health) because they have economic resources. Regarding health, for example,

- Poor children are less healthy than those at higher SES levels and are more likely to be unhealthy in adulthood. Unhealthy adults earn less, spend less time in the labor force, and must often retire earlier (Currie 2011; Ferrie and Rolf 2011; Redd et al. 2011).

- Living in a lower-SES neighborhood increases biological "wear and tear" due to chronic stress and accelerates the onset of diseases (Bird et al. 2010).

- Americans who earn $90,000 or more a year are almost three times as likely as those who earn less than $24,000 a year to say that they have good emotional and physical health and access to medical care (Mendes 2010).

The United States is the wealthiest nation in the world, but the gulf between the rich and the poor is at its widest since the 1920s (Bruder 2011).

## 8-4 Poverty in America

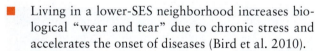

In 2010, the number of poor Americans (46.2 million) was the largest since the government began collecting poverty data in 1959. Nationally, more than

15 percent of the population lives in poverty, compared with 11 percent in 1975 (DeNavas-Walt et al. 2011). Between the ages of 20 and 75, nearly 60 percent of Americans will experience at least one year in poverty (Rank 2011). But what is poverty? Who are the poor? And why does poverty exist?

## 8-4a What Is Poverty?

There are two ways to define poverty: absolute and relative. **Absolute poverty** is not having enough money to afford the most basic necessities of life, such as food, clothing, and shelter ("what I need"). **Relative poverty** is not having enough money to maintain an average standard of living ("what I want").

### The Poverty Line

The **poverty line** is the minimal level of income that the federal government considers necessary for basic subsistence. To determine the poverty line, the Department of Agriculture (DOA) estimates the annual cost of food that meets minimum nutritional guidelines and then multiplies this figure by three to cover the minimum cost of clothing, housing, health care, and other necessities. Anyone whose income is below this line is considered officially poor and is eligible for government assistance (such as food stamps).

The poverty line, which in 2010 was $22,113 for a family of four (two adults and two children), is adjusted every year to reflect cost-of-living increases. If a family makes a dollar more than the poverty line figure, it is not officially categorized as poor. Also, many people earn considerably less than the poverty line. In 2010, 44 percent of poor people—those in "extreme poverty"—earned less than half of the poverty threshold. That's about $11,000 for a family of four (DeNavas-Walt et al. 2011).

### Is the Poverty Line Realistic?

Some people believe that the official poverty line is too high. They argue, for example, that many people aren't as poor as they seem because poverty levels don't include the value of noncash benefits such as food stamps, medical services (such as Medicare and Medicaid), public housing subsidies, and unreported income (Eberstadt 2009).

Others claim that the poverty line is too low because it doesn't include child care and job transportation costs or the higher cost of housing and other living expenses in some states (such as Massachusetts and California), regions (the Northeast compared with the South), and

metropolitan areas. Some also argue that poverty estimates are too low because they don't include other measures of economic hardship, such as the number of people who aren't defined as poor but rely on food banks and soup kitchens to survive (Kneebone et al. 2011).

## 8-4b Who Are The Poor?

Poverty isn't random. Both historically and currently, the poor share some common characteristics that include age, sex, family structure, and race and ethnicity.

### Age

Children under age 18 make up only 24 percent of the U.S. population but 36 percent (more than 16 million) of all people living in poverty. Among older Americans, people 65 and older make up 13 percent of the total population and 9 percent (3.5 million) of the poor (DeNavas-Walt et al. 2011). The poverty rate of older Americans is at an all-time low, lower than any other age group, because government programs for the elderly, especially Medicare and Medicaid, have generally kept up with the rate of inflation. In contrast, many programs for children living in poverty have been reduced or eliminated since 1980 (Rank 2011).

> **absolute poverty** not having enough money to afford the most basic necessities of life.
>
> **relative poverty** not having enough money to maintain an average standard of living.
>
> **poverty line** the minimal level of income that the federal government considers necessary for basic subsistence.

Justin Sullivan/Getty Images

As unemployment and poverty rates have increased, "tent cities," such as this one in Sacramento, California, have sprung up across the country.

## Gender and Family Structure

Of all people 18 years and older who are poor, 58 percent are women, and women's poverty rates are twice as high as those of men (15 percent and 8 percent, respectively) (National Women's Law Center 2011).

Researcher Diana Pearce (1978) coined the term **feminization of poverty** to describe the higher likelihood that female heads of households will be poor. Because of increases in divorce and particularly unmarried childbearing, single-mother families are four to five times more likely to be poor than married-couple families, and more likely to be extremely poor over many years (National Women's Law Center 2011).

## Race and Ethnicity

In absolute numbers, there are more poor whites (20 million) than poor Latinos (13 million), blacks (11 million), or Asians and Pacific Islanders (1.7 million). Proportionately, however, 27 percent of blacks and Latinos are poor, compared with 10 percent of whites and 12 percent of Asians (DeNavas-Walt et al. 2011).

## 8-4c Why Are People Poor?

Social scientists have a number of theories about poverty (see Cellini et al. 2008 for a nontechnical summary). Sociologists offer two general explanations: One blames the poor; the other emphasizes societal factors.

### Blaming the Poor: Individual Failings

About 69 percent of Americans believe that it's possible to start out poor and get rich through hard work because this is a land of opportunity (Jones 2007).

In a recent national survey, 43 percent of Americans said that wealthy people became rich because of their hard work, ambition, and education (Morin 2012).

Such views reflect those of proponents of an influential *culture of poverty* perspective, which contends that the poor don't succeed because they're deficient: They share certain values, beliefs, and attitudes about life that differ from those of people who aren't poor; are more permissive in raising their children; and are more likely to seek immediate gratification instead of planning for the future (Lewis 1966; Banfield 1974). The assertion that these values, beliefs, and attitudes are transmitted from generation to generation implies that the poor create their own problems through a self-perpetuating cycle of poverty ("like father, like son").

### Blaming Society: Structural Characteristics

In contrast to blaming the poor, many sociologists argue that macro-level structural factors create and sustain poverty. In a classic article on poverty, sociologist Herbert Gans (1971) maintained that inequality is functional because the poor

- Ensure that society's dirty yet necessary work gets done (such as dishwashing and cleaning bedpans in hospitals).

### FIVE WAYS THE POOR PAY MORE

**FOOD.** Not having a car to get to a supermarket chain or discount store means getting groceries at the corner store where staples are more expensive and produce is limited and often less fresh.

**DOING BUSINESS.** Most neighborhood stores charge more for products because real estate and insurance are more expensive. Poor people also pay more for financial services. Banks are scarce and "check-cashing stores" charge up to 10 percent of a check's value to cash it.

**CREDIT.** Because poor people don't have credit, they must rely on "cash advance" stores if they can prove that they get a regular paycheck. If poor people qualify for such a loan, they pay an effective annual percentage rate of about 825 percent.

**EVERYDAY HASSLES.** The poor are hassled almost every day by bill collectors because of overdue payments, deal with Laundromat trips, carry groceries long distances, and contend with the threat of harm in high-crime areas.

**WAITING.** Poor people spend much of their time waiting—in food pantry lines, at blood banks to sell their blood, at bus stops to get to work, and for apartments in safer neighborhoods (Jeffery 2006; Brown 2009; Schwartz 2011).

© Cengage Learning

© Frank van den Berghz/iStockphoto

- Subsidize the middle and upper classes by working for low wages.

- Buy goods and services that would otherwise be rejected (such as day-old bread, used cars, and the services of incompetent professionals).

- Absorb the costs of societal change and community growth (as when low-income people are pushed out of their homes by urban renewal and construction of expressways, parks, and stadiums).

Thus, according to Gans, poverty persists in the United States (and elsewhere) because many people benefit from the consequences.

Which perspective is more accurate, blaming the poor or blaming society? Some people are poor because they're lazy and would rather get a handout than a job. However, many researchers show that most people are poor because of economic conditions such as low wages, job loss, a lack of affordable housing, physical or mental disabilities, or an inability to afford health insurance, which in turn can result in acute health problems that interfere with employment (Acs and Nichols 2010; Tavernise 2011). Perhaps the best evidence for the structural argument is that more working-class people are slipping into poverty because of our floundering economy, which was due largely to financial mismanagement of some of the largest financial corporations, not the flaws of poor people (Allegretto 2011; see also Chapter 11).

## 8-5 Social Mobility

**S**ocial mobility is a person's ability to move up or down the social class hierarchy. The movement is due to a number of factors, but let's begin by looking at the types of social mobility that sociologists examine.

### 8-5a Types of Social Mobility

There are different types of social mobility: horizontal and vertical mobility, and intergenerational and intragenerational mobility.

**Horizontal mobility** means moving from one position to another at the same class level, or making a lateral move. Ashley, a nurse, might move from the pediatrics to the obstetrics department, but her salary won't change much because the position is similar in responsibilities and qualifications. Sociologists are generally not very interested in horizontal mobility because it involves little change in one's social class.

**Vertical mobility** refers to moving up or down the class hierarchy. If Ashley wants a higher salary, she may decide to undergo more training and become a physician's assistant (PA). Working under the supervision of a doctor, a PA examines, diagnoses, and treats patients. If Ashley is laid off and can't find another job, she might experience downward mobility. Vertical mobility can be intragenerational and intergenerational.

**Intragenerational mobility** refers to moving up or down the class hierarchy over one's lifetime. If Ashley begins as a nurse's assistant, becomes a registered nurse,

**social mobility** a person's ability to move up or down the social class hierarchy.

**horizontal mobility** moving from one position to another at the same class level.

**vertical mobility** moving up or down the class hierarchy.

**intragenerational mobility** moving up or down the class hierarchy over one's lifetime.

Bruce Ayres/Getty Images

One of the most devastating consequences of poverty is homelessness. An estimated 3.5 million people (about 1 percent of Americans) are likely to experience homelessness in a given year. Almost one in four of America's homeless are families with children. Because there aren't enough shelters, some live in their cars (National Coalition for the Homeless 2009; United States Conference of Mayors 2010).

**intergenerational mobility** moving up or down the class hierarchy relative to the position of one's parents.

and then a PA, she experiences intragenerational mobility. **Intergenerational mobility** is moving up or down the class hierarchy relative to the position of one's parents. It is a change in social class that occurs across two or more generations. If Ashley's parents were blue-collar workers, her upward movement to the middle-class is an example of intergenerational mobility.

Intragenerational and intergenerational mobility can be downward or upward. When Ashley moves up from being a registered nurse to a PA and is in a higher social class than her blue-collar parents, she is an example of both intragenerational and intergenerational mobility. If, however, she is paralyzed after a car crash and winds up on public assistance, Ashley will experience both downward intragenerational and intergenerational mobility.

## 8-5b Recent Trends in Social Mobility

From the 1940s to the 1970s, U.S. upward mobility increased. Since the 1980s, however, Americans have been less likely to rise to a higher social class (Sawhill 2008; Kopczuk et al. 2009; Winship 2011). For example, only 7 percent of children born to parents in the bottom fifth of the income distribution make it to the top fifth by adulthood (Haskins 2008b). When social mobility occurs, most moves are short (Nichols and Favreault 2009). A child from a working-class family, for example, is more likely to move up to the lower-middle class than to jump to the lower-upper class. The same is true of downward mobility: Someone from the middle class is more likely to slide into the working class than to drop to the underclass.

There is less upward mobility in the United States, the United Kingdom, and Italy than in many other industrialized countries such as Australia, Canada, Denmark, Finland, Norway, and Spain, as well as some developing countries such as Ecuador and Peru. For example, it takes an average of six generations for American children to move up to the next fifth of the income distribution, compared with three generations in Canada, Finland, Norway, and Denmark (Jäntti et al. 2006; Isaacs 2008d).

## 8-5c What Affects Social Mobility?

Vertical social mobility doesn't always reflect people's talents, intelligence, or hard work. Instead, much social mobility depends on structural, demographic, and individual factors.

### Structural Factors

Macro-level variables, over which individuals have little or no control, affect social mobility in many ways. First, *changes in the economy* spur upward or downward mobility. During an economic boom, the number of jobs increases, and many people have an opportunity to move up. During recessions, such as the Great Recession between late 2007 and mid-2009, long-term unemployment promoted downward mobility (Rose and Winship 2009).

Second, *government policies and programs* affect social mobility. Unlike the United States, the European countries with the highest upward mobility rates encourage greater equality. For example, they fund universal health care, which reduces the chance of people falling into poverty because of medical emergencies or poor health; provide affordable housing; and have invested in technical schools where young people can learn a high-paying trade (Foroohar 2011; Deparle 2012).

Third, *immigration* stimulates social mobility. Because many immigrants take low-paying jobs, groups that are already here advance to higher positions. Such upward mobility continues as poor immigrants take the least desirable jobs, allowing others to move into higher-status occupations (Haskins 2008c).

AP Photo/Kirsty Wigglesworth

J. K. Rowling worked as a teacher. After a divorce, she lived on public assistance, writing *Harry Potter and the Philosopher's Stone* during her daughter's naps. The book was published in 1997; by 2007, Rowling, a billionaire and one of the world's richest women, was considerably wealthier than the Queen of England.

## Demographic Factors

Demographic factors, which are usually interrelated, also affect social mobility. Three of the most important are education, sex, and race and ethnicity.

*Education.* Especially when the economy is slumping, people with college and graduate degrees fare better than those with a high school education or less. Those who don't graduate from high school often face long and frequent bouts of unemployment, must get by with temporary employment, and may move down the socioeconomic ladder (Haskins 2008a).

*Sex.* Women's massive entry into the labor force since the 1980s has increased family income and many single women's upward mobility. The median income for men has decreased since the 1970s, but men's mobility is less affected than women's by a divorce, out-of-wedlock children, or widowhood. For both sexes, but especially women, whom one marries also affects upward or downward mobility (Isaacs 2008c).

*Race and ethnicity.* African Americans and Latinos, especially those from low-income backgrounds, usually experience little upward mobility. Black and Latino middle classes have grown since the 1970s, but both groups still lag significantly behind whites in median family income (Isaacs 2008a, 2008b; DeLeire and Lopoo 2010).

## Individual Factors

*Family background.* The best way to be upwardly mobile is to choose the right parents. For example, only 7 of the 44 U.S. presidents came from the lower-middle class or below. Abraham Lincoln, although born in a log cabin, had a father who was one of the wealthiest people in the community. And about a third of the students with low grades at Ivy League universities wouldn't be there if their parents weren't celebrities, well-known politicians, or others who donated at least $25 million to the school (Golden 2006).

*Socialization.* Our upbringing influences the interests and activities that determine what French sociologist Pierre Bourdieu (1984) called *habitus*—the habits of speech and lifestyle that determine where one will feel comfortable and knowledgeable. Upper-class parents emphasize flexibility, autonomy, and creativity because they expect their children to step into positions that require such characteristics. Poor and working-class parents stress obedience, honesty, and appearance—the marks of respectability that many employers expect.

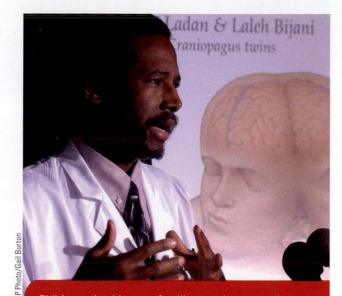

AP Photo/Gail Burton

Children raised in poor families can be upwardly mobile. Dr. Benjamin S. Carson is a world-renowned physician who is currently Director of Pediatric Neurosurgery at the prestigious Johns Hopkins Hospital in Baltimore, Maryland. Benjamin was 8 years old when his mother, who married when she was only 13, divorced his father. Mrs. Carson sometimes worked three jobs at a time to provide for Benjamin and his older brother, Curtis. When the boys' grades fell, Mrs. Carson limited their watching television and wouldn't let them play until they had finished their homework. She required her sons to read two library books a week and to give her written reports, even though she could barely understand the reports because she had left school after the third grade. In 2008, President Bush awarded Dr. Carson a Medal of Freedom, the nation's highest civilian award, for his "groundbreaking contributions to medicine and his inspiring efforts to help America's youth fulfill their potential" (Nitkin 2008: 7B).

*Connections and chance.* Many people get jobs by word of mouth rather than by searching newspaper and online ads. In a candid autobiographical sketch, sociologist S. M. Miller (2001: 1, 3) described his rise in academia, where connections and networking, rather than merit, "made the difference." He concluded that many employment decisions are based not on a person's individual abilities but on the "inequitable distribution of access to people who can help you get a good job."

## TABLE 8.2

# Global Economic Inequality, 2011

| | PERCENTAGE OF WORLD POPULATION | TOTAL GNI* | PERCENTAGE OF WORLD GNI | GNI PER CAPITAL | SOME COUNTRIES IN THIS CATEGORY |
|---|---|---|---|---|---|
| **World** | **7 billion (total population)** | **$62 billion** | ——- | **$9,097** | ——— |
| Low income | 12% (796 million) | $417 million | 2 | $510 | Primarily in Central and East Africa, parts of Asia (e.g., Pakistan, India, and China), and Indonesia (e.g., Java, Sumatra, and Borneo) |
| Middle income | 71% (5 billion) | $18 billion | 29 | $3,754 | Russia, Baltic states, Mexico, Latin America, some countries in northern and southern Africa (Algeria, Libya, and Angola), and many countries in the Middle East (e.g., Turkey, Iran, Iraq, and Saudi Arabia) |
| High income | 17% (1.1 billion) | $43 billion | 69 | $38,658 | Primarily industrialized nations (e.g., United States, Canada, Great Britain, Japan, Australia, New Zealand, and Israel) |
| **United States** | **5% (312 million)** | **$15 billion** | **24** | **$47,170** | |

*GNI is the Gross National Income.
Source: Based on material in World Bank 2011.

# 8-6 Global Inequality

All societies are stratified, but inequality is greater in some countries than in others. As a result, many people live in very different worlds.

## 8-6a Living Worlds Apart

The richest 2 percent of the world's adults own more than half of all global wealth; the bottom half owns barely 1 percent of global wealth (Davies et al. 2008). The super-rich (those with assets of $30 million or more) represent less than 1 percent of the world's millionaires, but hold 36 percent of all global wealth (Capgemini and Merrill Lynch Global Management 2011). Such differences in global wealth are astounding, but what about variations across nations?

The World Bank (2008) describes global stratification in terms of high-income, middle-income, and low-income countries. There are many poor people in high-income countries and some very affluent people in low-income countries, but *Table 8.2* provides an overview of the enormous economic disparities across countries.

*High-income countries*, which have a developed industrial economy, have only 17 percent of the world's population but 69 percent of the world's income. Since 2006, the wealth gap has widened in the United Kingdom, China, and India, but it's the most pronounced in the United States. Although we're a very rich country, we have the fourth highest per capita income inequality rate in the world, and are behind developing countries such as Chile, Mexico, and Turkey (OECD 2011; see also *CIA World Factbook* 2010).

*Middle-income countries*, which comprise 71 percent of the world's population, have a developing industrial economy and a considerably lower GNI per person than high-income countries. Most people in middle-income countries have less access to education, health services, food, and amenities such as electricity, automobiles, and telephones than people in high-income countries, but their standard of living is higher than in low-income countries.

*Low-income countries* are the least industrialized and largely agricultural. They have an average GNI of only $510 per person. Many people are impoverished, with a low standard of living and little access to health services, education, and clean water.

Worldwide, 48 percent of people live on less than $2 a day, and almost 20 percent live on less than $1 a day. Many of these people live in South Asia and sub-Saharan Africa. The children who survive starvation experience stunted physical growth, permanent mental retardation, lower intelligence levels, and a reduced capacity to learn by age 2. Such problems create a vicious cycle: Childhood hunger can lead to permanently dulled minds, an inability to pursue a livelihood, and a greater likelihood of having children with a diminished mental capacity (United Nations Development Programme 2010; Chandy and Gertz 2011; World Bank 2011).

El Greco/Shutterstock.com

their workers depend on external markets for jobs. High-income countries can extract raw materials (such as diamonds and oil) with little cost. They can also set the prices for the agricultural products that low-income countries export regardless of market prices, forcing many small farmers to abandon their fields because they can't pay for labor, fertilizer, and other costs (Carl 2002).

**Davis–Moore thesis** the functionalist view that social stratification benefits a society.

None of these perspectives explains inequality across *all* societies. Instead, global inequality is due to a combination of factors, including a country's values and customs and exploitation by high-income countries.

## 8-6b Why Is Inequality Universal?

Many theories try to explain why inequality is universal, but three of the most influential have been modernization theory, dependency theory, and world-system theory.

*Modernization theory* claims that low-income countries are poor because their leaders don't have the attitudes and values that lead to experimentation and the use of modern technology. Instead, policy makers adhere to traditional customs that isolate them and prevent them from competing in a global economy. In effect, modernization theory blames poor nations for their poverty and other problems. After the key foundations of modernity and capitalism are in place, this perspective maintains, low-income countries will prosper.

*Dependency theory* contends that the main reason why low-income countries are poor is because they are pawns that high-income countries exploit and dominate. Rich nations wield an enormous amount of power by exporting jobs overseas, manipulating foreign aid, draining less powerful countries of their resources, penetrating other countries with multinational corporations, and coercing national governments to comply with their interests (e.g., by not passing environmental laws). In effect, according to dependency theorists, high-income countries benefit because the poor provide cheap labor and aren't powerful enough to protest even though they work in hazardous conditions and earn less than $1 a day.

More recently, *world-system theory*, similar to dependency theory, argues that "the economic realities of the world system help rich countries stay rich while poor countries stay poor" (Bradshaw and Wallace 1996: 44). That is, those countries (such as the United States) that dominate the world economy control the economies of low-income countries because

# 8-7 Sociological Explanations: Why There Are Haves and Have-Nots

Why are societies stratified? *Table 8.3* summarizes the key points of functionalist, conflict, feminist, and symbolic interaction theories. Let's begin by looking at a long-standing debate between functionalists and conflict theorists on why there are haves and have-nots.

## 8-7a Functionalist Perspectives: Stratification Benefits Society

Functionalists see stratification as both necessary and inevitable because social class provides each individual a place in the social world and motivates people to contribute to society. Without a system of unequal rewards, functionalists argue, many important jobs wouldn't be performed.

### The Davis–Moore Thesis

Sociologists Kingsley Davis and Wilbert Moore (1945) developed one of the most influential functionalist perspectives on social stratification that persists today. The **Davis–Moore thesis**, as it's commonly called, asserts that social stratification benefits society. The key arguments of the Davis–Moore thesis can be summarized as follows:

1. **Every society must fill a wide variety of positions and ensure that people accomplish important tasks.** Societies need teachers, doctors, farmers, trash collectors, engineers, secretaries, plumbers, police officers, and so on.

During the past few years, we've heard about the booming economies of India and China, but not everyone has benefited. A tent city in India houses many of the workers who earn about $1.30 a day building new office towers for the affluent nearby (left). In China, a man in Shanghai begs as wealthier residents pass by (right).

4. **Society must offer greater rewards to motivate the most qualified people to fill the most important positions.** People won't undergo many years of education or training unless they are rewarded by money, power, status, and/or prestige. If doctors and nurses earned the same salaries, there wouldn't be much incentive for people to spend so many years earning a medical degree.

According to the Davis–Moore thesis and other functionalist perspectives, then, stratification and inequality are necessary to motivate people to work hard and to succeed. In open class systems, functionalists claim, inequality is based on **meritocracy**—a belief that individuals are rewarded for what they do and how well, rather than on the basis of their ascribed status. In closed class systems, ascribed statuses perpetuate inequality but also maintain order and stability because people perform the same jobs as their parents and ancestors.

## Critical Evaluation

Melvin Tumin (1953) was the first sociologist to challenge the Davis–Moore thesis. First, he argued, societies don't always reward the positions that are the most important for the members' survival. If the

**meritocracy** a belief that individuals are rewarded for what they do and how well, rather than on the basis of their ascribed status.

2. **Some positions are more important than others for a society's survival.** Doctors, for example, provide more critical services to ensure a society's continuation than do lawyers, engineers, or bankers.

3. **The most qualified people must fill the most important positions.** Some jobs require more skill, training, or intelligence than others because they are more demanding, and it's often difficult to replace the workers. Pilots, for example, must have more years of training and aren't replaced as easily as flight attendants.

## TABLE 8.3

# Sociological Explanations of Social Stratification

| PERSPECTIVE | LEVEL OF ANALYSIS | KEY POINTS |
|---|---|---|
| **Functionalist** | Macro | • Fills social positions that are necessary for a society's survival<br>• Motivates people to succeed and ensures that the most qualified people will fill the most important positions |
| **Conflict** | Macro | • Encourages workers' exploitation and promotes the interests of the rich and powerful<br>• Ignores a wealth of talent among the poor |
| **Feminist** | Macro and micro | • Constructs numerous barriers in patriarchal societies that limit women's achieving wealth, status, and prestige<br>• Requires most women, not men, to juggle domestic and employment responsibilities that impede upward mobility |
| **Symbolic Interactionist** | Micro | • Shapes stratification through socialization, everyday interaction, and group membership<br>• Reflects social class identification through symbols, especially products that signify social status |

Between mid-2010 and mid-2011, singer Lady Gaga earned $260 million. Do her earnings reflect her contribution to society, especially compared with teachers, physicians, dentists, trash collectors, computer scientists, and others?

ChinellatoPhoto/Shutterstock.com

highest-paid professional athletes, actors, and pop musicians went on strike, many of us would probably barely notice. If, on the other hand, those who earn only a fraction of all income—such as garbage collectors, teachers, doctors, truck drivers, and mail carriers—refused to work, society would grind to a halt. Thus, according to Tumin, there's little association between earnings and the jobs that keep a society going.

Second, Tumin claimed, Davis and Moore overlook the many ways that stratification limits the discovery of talent. Where wealth is differentially distributed, for instance, access to education, especially higher education, depends on the wealth of one's parents. As a result, large segments of the population are likely to be deprived of the chance to even discover what their talents are, and society loses (see also Chapter 13).

Third, Tumin criticized Davis and Moore for ignoring the critical role of inheritance. In upper social classes, sons and daughters don't have to work because their inherited wealth guarantees a lifetime income and perpetuates privileges over generations.

Functionalist theories are limited for other reasons. For example, they don't explain why (1) social mobility is more limited in the United States than in other industrialized (and even some developing) countries, as you saw earlier; (2) so many college graduates can find only low-paying jobs (Rank 2011); and (3) racial/ethnic income gaps persist (e.g., during the Great Recession, the median net worth of households decreased by 16 percent for whites but 53 percent for blacks and 66 percent for Latinos) (Kochhar et al. 2011).

Functionalist theories also ignore the possibility that inequality undermines people's trust in political and economic institutions and erodes national solidarity (see Wilkinson and Pickett 2009).

## 8-7b Conflict Perspectives: Stratification Harms Society

Like Tumin, conflict theorists maintain that social stratification is dysfunctional because it hurts individuals and societies. Karl Marx's (1934) analysis of social class and inequality has had a profound influence on modern sociology, especially conflict theory.

### Capitalism Benefits the Rich

Marx was aware that a diversity of classes can exist at any one time, but he predicted that capitalist societies would ultimately be reduced to two social classes: the capitalist class, or bourgeoisie; and the working class, or proletariat. The **bourgeoisie**, those who own and control capital and the means of production (such as factories, land, banks, and other sources of income), are able to amass wealth and power. The **proletariat**, workers who sell their labor for wages, earn barely enough to survive.

For conflict theorists, the economic struggles of the U.S. middle and working classes since the late-1970s were not primarily the result of globalization and technological changes but rather of a long series of government policies that overwhelmingly favored the very rich (Hacker and Pierson 2010). This policy of **corporate welfare** consists of an array of subsidies, tax breaks, and assistance that the government has created for businesses. Here are a few examples:

- American taxpayers will have to pay $12.2 trillion for the federal government's bailing out of mismanaged financial institutions ("Adding Up . . ." 2011).

- U.S. corporate income tax rates decreased from 53 percent in 1952 to 11 percent in 2010, even though companies benefit enormously from the taxpayer-supported highways they use, law enforcement and judicial systems that protect their intellectual and physical property, public education that provides a workforce, and military personnel who protect corporate assets abroad (Anderson et al. 2011).

- Some corporations pay no taxes at all. In 2010, General Electric—the nation's largest corporation—reported worldwide profits of $14.2 billion, received

**bourgeoisie** those who own and control capital and the means of production.

**proletariat** workers who sell their labor for wages.

**corporate welfare** an array of subsidies, tax breaks, and assistance that the government has created for businesses.

a federal tax benefit of $3.2 billion, paid its top CEO almost $12 million, and spent almost $42 million to lobby Congress for tax breaks (Anderson et al. 2011; Kocieniewski 2011; Lublin 2011). In contrast, the tax burden of an employee ranges from 33 percent to 41 percent (Buffet 2011).

Recently, billionaire investor Warren Buffet (2011: B21) criticized a "billionaire-friendly" Congress for coddling the super-rich: "While the poor and middle class fight for us in Afghanistan, and while most Americans struggle to make ends meet, we mega-rich continue to get our extraordinary tax breaks. . . . It's nice to have friends in high places."

Some rich executives make money even after they die. Whether CEOs die while working or after retirement, "golden coffin" arrangements provide their already wealthy families with up to $300 million in death benefits. In contrast, even when unionized companies pay the family if a worker dies on the job, the payouts are usually barely enough to cover the costs of the funeral (Smith 2011).

## Critical Evaluation

Some scholars have criticized Marxian and later conflict theories for several reasons. First, Marx predicted that as the numbers of oppressed workers increased, the proletariat would overthrow the bourgeoisie. Even though the concentration of corporate wealth has increased during the past 100 years and many people have become poorer, there have been some protests but no revolutions in capitalist countries, and many Americans go into debt to purchase luxury items such as TVs, jewelry, and pet supplies (Clifford and Martin 2011). Thus, according to critics, conflict theorists are exaggerating the effects of economic inequality.

Second, some people question whether the wealthy always act primarily out of economic self-interest. For example, billionaires such as Bill Gates, Warren Buffet, Jon Hunstman Sr. (the founder and CEO of a chemical company), and George Soros (a businessman) have donated billions of dollars to improve education, health care, and human rights in the United States and other countries (Whelan 2011). Third, functionalists in particular contend that conflict theorists underrate the capacity of individuals for upwardly mobile. They maintain that if people really want to succeed, they can do so by working hard and making sacrifices.

Finally, some critics point out that conflict theorists ignore the fact that 46 percent of American households pay no federal income taxes (and these tax breaks have nearly doubled since 1975) because they receive public

assistance, get deductions for raising children under age 18 and/or paying for their education, don't report income, or have very low wages (Johnson et al. 2011; Montgomery 2011).

## 8-7c Feminist Perspectives: Women Are Almost Always at the Bottom

For feminist scholars, functionalist and conflict theories are limited because they typically focus on men in describing and analyzing social stratification and social class. As a result, women are largely invisible.

### Patriarchy Benefits Primarily Men, Not Women

The majority of U.S. and world adults who are poor are women. In many countries, girls and women continue to suffer discrimination in the economy, politics, and access to medical services. Especially in low- and middle-income countries, almost 4 million girls and women are "missing" each year because of high death rates. About 20 percent are aborted because of a preference for sons, and more than a third die from complications of childbirth and pregnancy. Besides excessive deaths, most women are blocked from entering higher-paying occupations (such as finance and business), earn less than men in all jobs, have less access to land and credit, and exercise little decision-making power in households and politics (World Bank 2012).

Kar/Shutterstock.com

The Great Recession hit men especially hard. Since 2009, however, men have outpaced women in finding employment, including in historically female-dominated occupations such as retail trade, education, and health services. Minority women have been the least likely to benefit from the slow economic recovery. Since 2009, the unemployment rate has decreased for white women but increased for Latina, black, Asian, and foreign-born women (Kochhar 2011; Sanburn 2011; "Second Anniversary . . ." 2011).

Feminist theorists contend that, in a patriarchal system, men shape the stratification system because they control a disproportionate share of wealth, prestige, and power. The feminization of poverty often results in women's downward mobility. In addition, women often have to overcome economic inequities as well as juggle domestic and workplace responsibilities (see Chapters 11 and 12). The gender gaps in wealth, income, and household burdens make it harder for women to access resources and opportunities to build a strong future for themselves and their children.

## Critical Evaluation

Some critics point out that feminist theorists often focus only on poor women in showing how patriarchy affects social stratification and social class (see Kendall 2002). Another criticism is that many feminist scholars don't explain why so many women succeed despite patriarchal barriers. Third, feminist theories don't account for some dramatic cross-cultural variations. In a study of 165 countries, for example, Canada, Australia, the Netherlands, and the United States ranked in the top 10 regarding women's treatment under the law, workplace participation, and access to education and health care, but well behind developing countries such as Cuba, Rwanda, and Burundi in women's representation in political offices (Ellison 2011).

## 8-7d Symbolic Interaction Perspectives: People Create and Shape Stratification

Symbolic interactionists focus on how people reproduce social classes. They address micro-level issues such as how people learn their social positions in everyday life and how such learning affects their attitudes, behavior, and lifestyles.

### People's Beliefs and Actions Affect Their Social Class

People in upper, middle, working, and lower classes interact and socialize their children differently because of family background, education level, and income. Children as young as 4 years old are aware of and enact their social class. Among preschoolers, upper-middle-class children speak, interrupt, ask for help, and argue more often than working-class children. Such behavior receives more adult attention and gives upper-middle-class children more opportunities to develop their language skills (Streib 2011).

Social contexts also affect mobility. The occupational worlds into which children are born have consequences for the aspirations they develop, the skills they value and to which they have access, and the networks and resources on which they can draw. Doctors' children, for example, are more likely to become physicians themselves because they are exposed to medical discussions at home, are encouraged to pursue medicine as an occupation, and are embedded in social networks that provide them with information about how to become physicians (Jonsson et al. 2009).

Social class is internalized so deeply that most Americans believe they live in a far more equal country

© Danita Delimont/Alamy

According to symbolic interactionists, people in lower social classes show deference to those in higher social classes. Doing so confirms the inequality of the relationship and reinforces each person's position in the social hierarchy.

than in fact they do. A recent national survey found that Americans thought that the richest 20 percent control about 59 percent of the wealth, while the real number is closer to 84 percent (Norton and Ariely 2011). Not understanding our nation's extreme wealth disparity helps explain why people aren't demanding greater economic equality (DeGraw 2011).

## Critical Evaluation

Symbolic interaction theories are important in understanding the everyday processes that underlie social stratification, but there are several weaknesses. First, the theories don't explain why—despite the same family background, resources, and socialization—some siblings are considerably more upwardly mobile than their brothers and sisters.

Second, conflict theorists, especially, fault symbolic interactionists for ignoring structural factors—such as the economy, corporate welfare, and inherited wealth—that create and reinforce inequality. As you'll see in Chapter 13, for example, many talented and intelligent teenagers drop out of high school because of limited family income and other resources.

## Study Tools

Ready to study? Go to **Sociology CourseMate** at **www.cengagebrain.com** to complete practice quizzes, review flashcards, watch videos, and more.

Gender, gender roles,
and sexuality
are social creations.

Chip Wass

# Gender and Sexuality

For the average American male, compared with his female counterpart, three pairs of shoes are usually enough, he can shower and be ready in 10 minutes, his underwear is $10 for a three pack, and he rarely encounters a line in public restrooms. Do such descriptions stereotype women and men? Or contain a kernel of truth? This chapter examines how gender, gender roles, and sexuality shape our lives. First, however, take the short quiz on the next page to see how much you know about gender and sexuality.

**sex** the biological characteristics with which we are born.

**gender** learned attitudes and behaviors that characterize women and men.

## Key Topics

In this chapter, we'll explore the following topics:

**9-1** How Women and Men Are Similar and Different

**9-2** Contemporary Gender Stratification and Inequality

**9-3** Sexuality

**9-4** Some Current Controversies About Sexuality

**9-5** Gender and Sexuality Across Cultures

**9-6** Sociological Explanations of Gender and Sexuality

## 9-1 How Women and Men Are Similar and Different

Many people use *sex* and *gender* interchangeably, but they have distinct meanings. The terms are related, but sex is a biological designation whereas gender and gender roles are social creations (Caplan and Caplan 2009; Fine 2010; Muehlenhard and Peterson 2011).

### 9-1a Sex and Gender

**Sex** refers to the biological characteristics with which we are born—chromosomes, anatomy, hormones, and other physical and physiological attributes. These attributes *influence* our behavior (such as shaving beards and wearing bras), but do *not determine* how we think or feel. Whether we see ourselves and others as feminine or masculine depends on gender, a considerably more complex concept than sex.

**Gender** refers to learned attitudes and behaviors that characterize women and men. Gender is based on social and cultural expectations rather than on physical traits. Thus, most people are *born* either male or female, but we *learn* to be women or men because we internalize behavior patterns expected of each sex. In many societies, for example, women are expected to look young, thin, and attractive, and men are expected to amass as much wealth as possible.

## What do you think?

Today, women have more education and employment advantages than men.

| 1 | 2 | 3 | 4 | 5 | 6 | 7 |
|---|---|---|---|---|---|---|
| strongly agree | | | | | | strongly disagree |

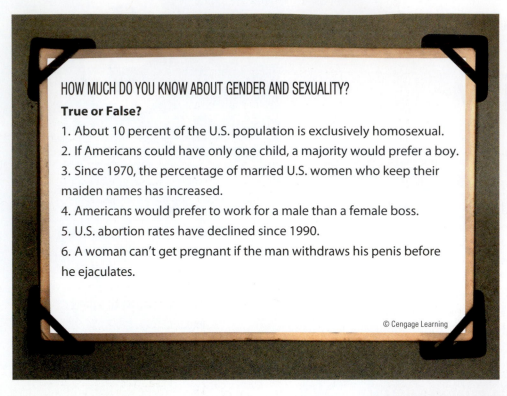

on a number of characteristics, including cognitive abilities, verbal and nonverbal communication, leadership traits, and self-esteem (Hyde 2005; see also Caplan and Caplan 2009).

Americans are more likely now than in the past to pursue jobs and other activities based on their ability and interests rather than their sex. For the most part, however, our society still has fairly rigid gender roles and widespread **gender stereotypes**—expectations about how people will look, act, think, and feel because of their sex. For example, many stay-at-home dads report feeling stigmatized because their wives are the breadwinners

**gender identity** a perception of oneself as either masculine or feminine.

**gender roles** the characteristics, attitudes, feelings, and behaviors that society expects of females and males.

**gender stereotypes** expectations about how people will look, act, think, and feel based on their sex.

# 9-1b Gender Identity and Gender Roles

People develop a **gender identity**, a perception of themselves as either masculine or feminine, early in life. Many Mexican baby girls but not boys have pierced ears, for example, and hairstyles and clothing for American toddlers differ by sex. Gender identity, which typically corresponds to a person's biological sex, usually remains relatively fixed throughout life.

About 90 percent of Americans believe that women are more emotional than men. In fact, both sexes *experience* emotions such as anger, happiness, and sadness just as deeply. What differs is how women and men *express* their emotions because men are more likely to suppress their feelings, whereas women tend to show their emotions more openly (Newport 2001; Simon and Nath 2004).

Such differences are largely the result of **gender roles**—the characteristics, attitudes, feelings, and behaviors that society expects of females and males. A review of the research done between 1990 and 2004 on the differences between the sexes concluded that women and men are much more alike than different

UPI/Kevin Dietsch/Newscom

Speaker of the House John Boehner cried on election night in 2010 when he thanked his supporters. A month later, he broke down twice during a *60 Minutes* interview when he spoke about his rise from humble origins and his concern about schoolchildren. Some said the tears showed what Boehner felt passionately about, but many mocked him. A cohost of *The View,* a popular television show targeted at women, dubbed him "Weeper of the House," and late night comedians did tearful impressions (Jensen 2010). Do such reactions illustrate sexism?

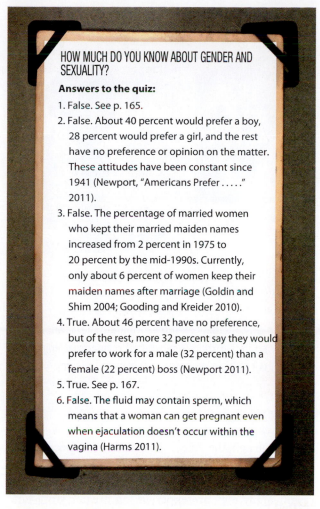

### HOW MUCH DO YOU KNOW ABOUT GENDER AND SEXUALITY?

**Answers to the quiz:**

1. False. See p. 165.
2. False. About 40 percent would prefer a boy, 28 percent would prefer a girl, and the rest have no preference or opinion on the matter. These attitudes have been constant since 1941 (Newport, "Americans Prefer . . . . ." 2011).
3. False. The percentage of married women who kept their married maiden names increased from 2 percent in 1975 to 20 percent by the mid-1990s. Currently, only about 6 percent of women keep their maiden names after marriage (Goldin and Shim 2004; Gooding and Kreider 2010).
4. True. About 46 percent have no preference, but of the rest, more 32 percent say they would prefer to work for a male (32 percent) than a female (22 percent) boss (Newport 2011).
5. True. See p. 167.
6. False. The fluid may contain sperm, which means that a woman can get pregnant even when ejaculation doesn't occur within the vagina (Harms 2011).

(Mundy 2012). However, 43 percent of U.S. men, compared with 33 percent of women, say that it's OK for men to be stay-at-home dads (Databeast 2012). Do such data suggest that men are less sexist than women?

# 9-2 Contemporary Gender Stratification and Inequality

**S**exism is an attitude or behavior that discriminates against one sex, usually women, based on the assumed superiority of the other sex. Much sexism is *subtle* because it undervalues females. For example, only about 15 percent of Wikipedia contributors, members of Congress, authors of op-ed articles in influential national newspapers, and even Rock and Roll Hall of Fame inductees are women (Carmon 2011; Cohen 2011). Because most people have internalized beliefs that women are inferior to men, we often assume that such inequality is normal.

Considerable sexism is still *blatant* because it's visible, intentional, and easily documented. According to numerous female bloggers, for example, cybersexism (including stalking, death threats, and hate speech)

is prevalent, and women whose user names indicate their sex are 25 times more likely than men to experience online harassment (Valenti 2007).

Men also experience sexism. For example, one of my students wrote the following during an online discussion of gender roles:

*Some parents live their dreams through their sons by forcing them to be in sports. I disagree with this but want my [9-year-old] to be "all boy." He's the worst player on the basketball team at school and wanted to take dance lessons, including ballet. I assured him that this was not going to happen. I'm going to enroll him in soccer and see if he does better.*

Is this mother suppressing her son's natural dancing talent? We'll never know because she, like many parents, expects her son to fulfill gender roles that meet with society's approval. Thus, our social roles are *gendered* in that males and females are often treated differently because of their sex (Howard and Hollander 1997).

According to the mainstream media, women are dominating society as never before, men are in a "state of crisis," and some even predict the "end of men" (see Rosin 2010). These writers worry that women are outpacing men in education, becoming the family breadwinners, and doing less housework and child care than men (see Mukhopadhyay 2011; Schwyzer 2011).

Others argue that there is still widespread **gender stratification**—people's unequal access to wealth, power,

**sexism** an attitude or behavior that discriminates against one sex, usually females, based on the assumed superiority of the other sex.

**gender stratification** people's unequal access to wealth, power, status, prestige, and other valued resources because of their sex.

At an early age, sex-appropriate activities prepare girls and boys for future adult roles. As a result, there are few male ballet dancers and female auto mechanics.

© Corbis Bridge/Alamy

status, prestige, opportunity, and other valued resources because of their sex. We'll examine some of this stratification in greater depth in later chapters. For now, let's look briefly at gender inequality in the family, education, the workplace, and politics, remembering that these institutions intersect and affect each other.

## 9-2a Gender and Family Life

Americans have more choices today, but is family life less gendered than in the past? Probably more so than many of us think, particularly in family structure, sharing domestic tasks, and experiencing family-work conflicts.

### Family Structure

Compared with the 1950s, people are marrying about 5 years later on average (age 27 for women and 29 for men). College graduates marry even later (30 for women and 31 for men). Postponing marriage increases the likelihood that many women will give birth for the first time in their 30s instead of their 20s, and that they will have fewer children or none (White House Council on Women and Girls 2011). Though family sizes are smaller, 70 percent of married mothers are in the labor force, compared with only 40 percent in 1970 (U.S. Census Bureau 2012). This means that there is less time for child care and housework.

### Child Care and Housework

Employed women, especially mothers, often complain that they have to do *everything*—work outside the home, raise children, and do all the housework. Are they right? In 2010, among full-time workers who were parents of children under age 18, on average fathers spent almost 17 hours a week compared to mothers' 26 hours on household activities (such as housework, food preparation and cleanup, lawn and garden care, and household management). Fathers spent almost 11 hours a week on child care and domestic chores compared with mothers' 17 hours. Since 2003, both mothers and fathers have been spending less time on child care and domestic tasks, and there is greater sharing of the household labor. Nonetheless, women spend from 33 percent to 40 percent more time than men do on child care and household tasks (Bureau of Labor Statistics 2005, 2011).

### Family-Work Conflicts

Since 2003, men's participation in household activities and child rearing has stayed about the same or increased slightly, but many are now experiencing

The Butler Café is very popular in Japan. It's a place for young women to unwind from the stresses of the outside world, including pressure to conform and marry. The only men present are butlers, who respond in less than 6 seconds if a customer rings a golden bell located on each table.

more pressure to "do it all to have it all." Unlike even a generation ago, the "ideal" man today is expected to be not only a successful breadwinner but also an involved and nurturing husband/partner, father, and son. As a result, many men are experiencing greater work-family conflict than in the past. In 2008, 49 percent of employed fathers reported some or a lot of work-family conflict, up from 34 percent in 1977. The conflict is even greater for fathers in two-income households. Their work-family conflict increased from 35 percent in 1977 to 60 percent in 2008, compared with 41 percent and 47 percent, respectively, for mothers (Aumann et al. 2011).

## 9-2b Gender and Education

U.S. girls and women have made substantial educational progress in the last few decades. Despite the headway, there are gender differences at all educational levels.

### Primary and Secondary Education

Many teachers and schools send gendered messages to children that follow them from preschool to college. When children enter kindergarten, they perform similarly on both reading and mathematics tests. By the third grade, however, boys, on average, score higher than girls in math and science assessment tests and lower than girls in reading tests. These gaps increase throughout high school (Dee 2006).

Boys ages 6 to 14 are more than twice as likely as girls to have a developmental disability and three times as likely to be diagnosed with mental retardation (Sparks 2011). At school, twice as many females as males are victims of electronic bullying that includes social networking and text messaging. However, 10 percent of male high school students, compared with 5 percent of female high school students, have reported being threatened or injured with a weapon on school property (White House Council on Women and Girls 2011).

Such problems may partly explain why boys are more likely than girls to disengage from school. In public K–12 schools, boys are three times as likely as girls to be expelled; they are less likely to be enrolled in high school after age 16 and less likely to finish high school (Mortenson 2011).

### Higher Education

Since 1981, there have been more females than males enrolled in college, and women—across all racial/ethnic groups—now earn a larger percentage of associate, bachelor's, and master's degrees than men (Kim 2011; see also Chapter 13). There are large gender differences by major, however. Since 2000, women have received only 24 percent of undergraduate STEM (science, technology, engineering, and math) degrees. Women in STEM jobs earn 16 percent less than men in STEM jobs, but 33 percent more than women in other jobs (Beede et al. 2011; see also Kaminski and Geisler 2012). Thus, over a lifetime, the income differences are considerable, particularly between women in STEM and other occupations.

| TABLE 9.1 |
|---|
| **As Rank Increases, the Number of Female Faculty Members Decreases** |

| RANK | PERCENTAGE OF FEMALE FACULTY MEMBERS |
|---|---|
| Professor | 28 |
| Associate Professor | 41 |
| Assistant Professor | 48 |
| Instructor | 55 |

Note: Of the almost 729,000 full-time faculty members in 2009, 43 percent were female.

Source: Based on Snyder and Dillow 2011, Table 260.

Even when women earn Ph.D.'s in male-dominated fields, they are less likely to be hired than men. For example, women have received more than 45 percent of all doctorates in biology since 1990, but only 14 percent of full-time biology professors are women (Handelsman et al. 2005). Once hired, female faculty members are less likely than men to be promoted. Since 1982, *one-third of all recipients of Ph.D. degree recipients have been women*, but men still dominate the rank of full professor (see *Table 9.1*).

## 9-2c Gender and the Workplace

There has been progress toward greater workplace equality, but we still have a long way to go. In the United States (as around the world), many jobs are segregated by sex, there are ongoing gender pay gaps, and numerous women experience sexual harassment and pregnancy discrimination.

### Gender-Segregated Work

A number of U.S. occupations are filled almost entirely by either women or men. For example, between 92 and 98 percent of all registered nurses, child care workers, secretaries, dental hygienists, and preschool and kindergarten teachers are women. At least 98 percent of all pilots, mechanics, plumbers, and loading machine operators are men. Women have made some progress in the higher paying professional occupations that require at least a college degree, but 78 percent of physicians and surgeons, 75 percent of architects,

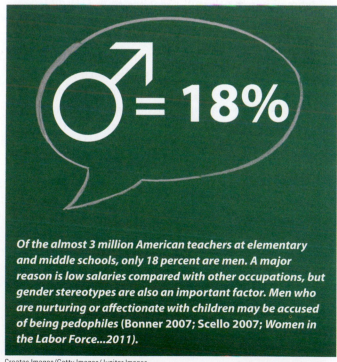

Of the almost 3 million American teachers at elementary and middle schools, only 18 percent are men. A major reason is low salaries compared with other occupations, but gender stereotypes are also an important factor. Men who are nurturing or affectionate with children may be accused of being pedophiles (Bonner 2007; Scello 2007; *Women in the Labor Force...2011*).

Creatas Images/Getty Images/Jupiter Images

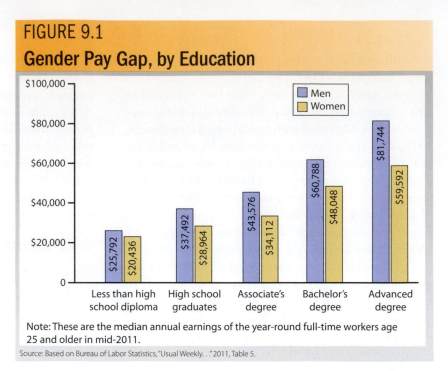

## FIGURE 9.1
## Gender Pay Gap, by Education

Note: These are the median annual earnings of the year-round full-time workers age 25 and older in mid-2011.

Source: Based on Bureau of Labor Statistics, "Usual Weekly. . ." 2011, Table 5.

80 percent of dentists, and 93 percent of chemical engineers are men (*Women in the Labor Force . . .* 2011).

### Gender Pay Gaps

In 1979, the first year for which comparable earnings data are available, women who worked full-time year-round earned 62 percent of what men earned. In 2010, women who worked full-time year-round had a median annual income of $34,788, compared with $42,848 for men ("Highlights . . ." 2011). This means that women earned 81 cents for every dollar men earned, and it took 32 years to decrease the disparity by only 19 cents. In effect, in 2010, *the average woman had to work 13 extra weeks to have the same earnings as a man.*

This income difference between women and men is the **gender pay gap** (also called the *wage gap, pay gap,* and *gender wage gap*). Over a lifetime, the average woman who works full-time year-round for 47 years is deprived of a significant amount of money because of the gender pay gap: $700,000 for high school graduates, $1.2 million for college graduates, and more than $2 million for women with a professional degree (such as business, medicine, or law). Lower wages and salaries reduce women's savings, purchasing power, and quality of life, and bring them lower Social Security benefits after retirement (Murphy and Graff 2005; Soguel 2009).

Not only do women earn less than men at all educational levels, the gender pay gap increases as the level of educational attainment increases (see *Figure 9.1*). Female high school graduates earn 79 percent of what their male counterparts earn, but women's earnings drop to 73 percent of men's for those with advanced degrees. (Note in *Figure 9.1* that, as a group, women with advanced degrees earn less than men with a college degree.)

Why is there a gender pay gap? A common explanation is that women choose fields with lower earnings (such as health care and teaching in elementary and middle schools), whereas men tend to choose higher paying fields (such as the STEM occupations). Second, some scholars maintain that women—especially those in professional, managerial, and executive positions—are getting stuck under glass ceilings. Even in federal agencies, for example, men with much less experience and lower educational levels are routinely promoted over more accomplished women (Stone 2007; Waldref 2008). Third, mothers are more likely than fathers (or women with no children) to work part-time, take leaves, or take a break from the workforce to raise children—factors that reduce wages and salaries (Dey and Hill 2007).

The wage gap can be partially explained by differences in education, experience, and time in the labor force, but about 41 percent of the gap is the result of sex discrimination in hiring, promotion, and pay, and occupational segregation. That is, a wage gap remains even when women and men have the same education, occupation, number of years in a job, seniority, marital status, number of children, and are similar on numerous other factors (Boraas and Rodgers 2003; Blau and Kahn 2006).

### Sexual Harassment

**Sexual harassment** is any unwanted sexual advance, request for sexual favors, or other conduct of a sexual nature that makes a person uncomfortable and interferes with her or his work. It includes *verbal behavior* (such as pressure for dates or demands for sexual favors in return for hiring, promotion, or tenure, as well as the threat of rape), *nonverbal behavior* (such as indecent gestures or the display of posters, photos, or drawings of a sexual nature), and *physical contact* (such as pinching, touching, or rape). Between 1997 and 2011, the

## TABLE 9.2

# U.S. Women in Elective Offices, 2012

| POLITICAL OFFICE | TOTAL NUMBER OF OFFICE HOLDERS | PERCENTAGE WHO ARE WOMEN |
|---|---|---|
| U.S. Congress | 535 | 17 |
| Senate | 100 | 17 |
| House of Representatives | 435 | 17 |
| Governor | 50 | 12 |
| State Legislator | 7,382 | 24 |
| Attorney General | 50 | 14 |
| Secretary of State | 50 | 24 |
| State Treasurer | 50 | 16 |
| State Comptroller | 50 | 8 |
| Mayor (100 largest cities) | 100 | 12 |

Source: Based on material at the Center for American Women and Politics 2012.

## 9-2d Gender and Politics

In 1872, Victoria Chaflin Woodhull of the Equal Rights Party was the first female presidential candidate. Since then, 36 women have sought the nation's highest office, but none has broken the glass ceiling of a male-dominated U.S. presidency.

Unlike a number of other countries (including Great Britain, Germany, India, Israel, Pakistan, Argentina, Chile, and Philippines), the United States has never had a woman serving as president or even vice president. In the U.S. Congress, 83 percent of the members are men. In several other important elective offices (such as governor, mayor, or state legislator), only a handful of the decision makers are women (see *Table 9.2*), and this number hasn't changed much since the early 1990s (Center for American Women and Politics 2011).

Equal Employment Opportunity Commission (EEOC) received almost 206,000 formal sexual harassment complaints, 84 percent of them from female employees (U.S. Equal Employment Opportunity Commission 2011b). Lawyers say the statistics would be much higher, but many companies now require new employees to agree to arbitrate complaints, including sexual harassment, as a condition of being hired (Green 2011).

Nearly two-thirds of Americans say that sexual harassment is a serious problem in this country. About 25 percent of women have been sexually harassed at work. Only 40 percent of them reported the incident, however, because they believed that doing so wouldn't do any good or they feared being fired or demoted (Clement 2011).

### Pregnancy Discrimination

The federal Pregnancy Discrimination Act of 1978 makes it illegal for employers with more than 15 workers to fire, demote, or penalize a pregnant employee. Some state laws extend this protection to companies with as few as four employees. In 2011 alone, nearly 5,800 women filed complaints that they had been fired, demoted, or had some of their responsibilities taken away when employers learned that they were pregnant (U.S. Equal Employment Opportunity Commission 2011a). The EEOC complaints about both sexual harassment and pregnancy discrimination represent only the tip of the iceberg because only a fraction of women ever take action. Many aren't aware of their rights, and others fear losing their jobs or don't have the resources to pursue lengthy lawsuits.

*"Are you hiring me because I'm cheap, I'm qualified, or I'm cheap and qualified?"*

© Liza Donnelly

## Sociology in Your Life

How does your state compare with others in the proportion of women in the state legislature? To find out, go to Sociology CourseMate and CengageBrain .com (once you are on the Rutgers site, click on "State-by-State Fact Sheets" on the left menu). How would you explain the difference between your state and others?

**sexual identity** our awareness of ourselves as male or female and the ways that we express our sexual values, attitudes, feelings, and beliefs.

**sexual orientation** a preference for sexual partners of the same sex, of the opposite sex, of both sexes, or neither sex.

**homosexuals** those who are sexually attracted to people of the same sex.

**heterosexuals** those who are sexually attracted to people of the opposite sex.

**bisexuals** those who are sexually attracted to members of both sexes.

**asexuals** those who lack any interest in or desire for sex.

**transgender people** those who are transsexuals, intersexuals, or transvestites.

Women's voting rates in the United States have been higher than men's since 1984 (U.S. Census Bureau 2012). Why, in contrast, are there so few women in political office? One reason may be that women, socialized to be nurturers and volunteers, see themselves as supporters rather than active doers. As a result, they may spend many hours organizing support for a candidate rather than running for political office themselves.

Second, women are less likely than men to receive encouragement to run for office (such as from a political party or its leaders). As a result, even successful women are twice as likely as men to rate themselves as not qualified to run for office. In contrast, men are two-thirds more likely than women with similar credentials to consider themselves qualified or very qualified to run for office (Lawless and Fox 2005).

Third, there is a lingering sexism, among both men and women, that "from the pulpit to the presidency," men are better leaders (Tucker 2007: A9).The pervasive sexism in media coverage of political candidates was especially evident during the 2008 presidential campaign. Compared with men, women received considerably less media coverage of their political campaigns, or the coverage was more disparaging (Zurbriggen and Sherman 2010).

## 9-3  Sexuality

In the movie *Annie Hall*, a therapist asks two lovers how often they have sex. The man rolls his eyes, and complains, "Hardly ever, maybe three times a week!" The woman exclaims, "Constantly, three times a week!" Sexuality is considerably more complex than just having sex, however, because it's a product of our sexual identity, sexual orientation, and sexual scripts.

### 9-3a Sexual Identity

Our **sexual identity** is our awareness of ourselves as male or female and the ways that we express our sexual values, attitudes, feelings, and beliefs. It involves placing ourselves in a category created by society (such as female and heterosexual) and learning, both consciously and unconsciously, how to act in that category.

Most people's sexual identity corresponds with their biological sex, gender identity (seeing oneself as feminine or masculine), sexual attraction, and sexual behavior, but not always. Among Americans ages 15 to 44, for example, 99 percent report being sexually attracted only to the opposite sex, but 14 percent identify themselves as homosexual or bisexual, and 13 percent of women and 5 percent of men have had same-sex experiences (Chandra et al. 2011). Thus, sexual identity, sexual attraction, and sexual behavior can differ.

### 9-3b Sexual Orientation

Our sexual identity incorporates a **sexual orientation**—a preference for sexual partners of the same sex, of the opposite sex, of both sexes, or neither sex:

- **Homosexuals** (from the Greek root *homo*, meaning "same") are sexually attracted to people of the same sex. Male homosexuals prefer to be called *gay*, female homosexuals are called *lesbians*, and both gay men and lesbians are often referred to as *gays*. *Coming out* is a person's public announcement of a gay or lesbian sexual orientation.

- **Heterosexuals**, often called *straight*, are attracted to partners of the opposite sex.

- **Bisexuals**, sometimes called *bis*, are attracted to members of both sexes.

- **Asexuals** lack any interest in or desire for sex.

Heterosexuality is the predominant sexual orientation worldwide, but homosexuality exists in all known cultures.

### Transgender Sexual Orientations

Our cultural expectations dictate that we are female or male, but a number of people are "living on the boundaries of both sexes" (Lorber and Moore 2007: 141). **Transgender people** encompass several groups:

lev radin/Shutterstock.com

Isis King, born Darrell Walls, was the first transgender contestant to appear as a finalist on the reality show *America's Next Top Model*.

- *Transsexuals* are people born with one biological sex but who choose to live their life as another sex—either by consistently cross-dressing or by having their sex surgically altered.
- *Intersexuals* are people whose medical classification at birth is not clearly either male or female (this term has replaced *hermaphrodites*).
- *Transvestites* are people who cross-dress at times but don't necessarily consider themselves a member of the other sex.

Transgender people include gays, heterosexuals, bisexuals, and men and women who don't identify themselves with any specific sexual orientation.

### Prevalence of Homosexuality

How many Americans are gay men, lesbians, and bisexuals? No one knows for sure, largely because it's difficult to define and measure sexual orientation. For example, are people homosexual if they have ever engaged in same-sex behavior? What about heterosexuals who are attracted to people of the same sex?

Researchers measure the prevalence of homosexuality by simply asking people to identify their sexual orientation. Americans estimate that 25 percent of the population is gay or lesbian (Morales 2011). National studies vary somewhat on the prevalence of homosexuality, but about 6 percent of Americans identify themselves as homosexual, almost 8 percent describe themselves as bisexual, and 18 percent have had same-sex contact (Chandra et al. 2011).

## 9-3c Sexual Scripts

We like to think that our sexual behavior is spontaneous, but all of us have internalized sexual scripts. A **sexual script** specifies the formal or informal norms for acceptable or unacceptable sexual activity, which individuals are eligible sexual partners, and the boundaries of sexual behavior. Social scripts can change over time and across groups, but are highly gendered in two ways—women's increasing hypersexualization and a persistent sexual double standard.

### Sexual Scripts and the "Sexy Babes" Trend

Sexualized social messages are reaching ever younger audiences, teaching or reinforcing the idea that girls and women should be valued for how they look rather than their personalities, abilities, and other nonphysical traits. For example, padded bras for 7- and 8-year-olds (that's right, 7- and 8-year-olds!) are sold by retailers such as Abercrombie & Fitch and Amazon.com; 43 percent of girls ages 6 to 9 use lipstick or lip gloss; and 80 percent of girls ages 13 to 18 list shopping as their favorite hobby (Hanes 2011; see also Levin and Kilbourne 2009, and Orenstein 2011).

Many girls are obsessed about their looks from an early age, for a variety of reasons (including their mothers' role modeling), but media images play a large role in the hypersexualization of girls. Girls and boys see cheerleaders (with increasingly sexualized routines) on TV far more than they see female basketball players or other athletes (Hanes 2011). And, in a study of the 100 top-grossing films in 2008, the researchers found that 13- to 20-year-old females were far more likely than their male counterparts to be depicted in revealing clothes (40 percent vs. 7 percent), partially naked (30 percent vs. 10 percent), and physically attractive (29 percent vs. 11 percent) (Smith and Choueiti 2011).

What about adult roles? In television shows, women are represented in far more diverse roles—such as doctors, lawyers, and criminal investigators—but they're often sexy ("hot"). Also, star female athletes regularly pose naked or semi-naked for men's magazines (Hanes 2011).

Who benefits from girls' and women's hypersexualization? Marketers who convince girls (and their parents) that being popular and "sexy" requires the right clothes, makeup, hair style, and accessories, create a young generation of shoppers and consumers who will increase business profits more than ever before (Lamb and Brown 2007; Levin and Kilbourne 2009).

**sexual script** specifies the formal or informal norms for acceptable or unacceptable sexual activity, which individuals are eligible sexual partners, and the boundaries of sexual behavior.

CB2/ZOB/WENN.com/Newscom

Galia Slayen built a life-sized Barbie to show what she would look like if she were a real woman. (Slayen used a toy for the head because she wasn't able to create a proportional head) ("Life Size Barbie . . ." 2011).

## Sexual Scripts and the Double Standard

Among adolescents, the higher the number of sexual partners, the greater a boy's popularity. In contrast, girls who have more than eight partners are far less popular than their less experienced female peers (Kreager and Staff 2009).

*Hooking up*—which can mean anything from kissing to sexual intercourse—is now more common than dating at many high schools and colleges (England et al. 2007). A variation of hooking up is "friends with benefits" (FWB) that involves regular sexual contact. Hooking up and FWB have advantages for both sexes: Both are cheaper than dating, require no emotional or time commitment, and make people feel sexy and desirable (Bogle 2008).

The main disadvantage is that a sexual double standard is common in both hooking up and FWB. Among college students, for example, men are more likely than women to initiate sex and to be expected to do so, to experience an orgasm because there's less concern than in an ongoing relationship about women's sexual satisfaction, both sexes describe the women as "sluts" whereas the men "score," and women often get a bad reputation as being "easy" (England and Thomas 2009; Armstrong et al. 2010).

## 9-3d Heterosexism and Homophobia

In 2011, a record number of Americans, 58 percent, said that society should accept homosexuality; 69 percent of those younger than age 30 felt this way (Kohut et al. 2011). Numerous municipal jurisdictions, corporations, and smaller companies now extend more health care and other benefits to gay and lesbian employees and their partners than to unmarried heterosexuals who live together. In 2008, 125 of the *Fortune 500* companies included "gender identity" in their nondiscrimination policies, compared with almost none in 2002. And among the largest corporations, 58 percent provided transgender benefits such as paying for regular hormone treatment, psychological counseling, and gender reassignment surgery (Belkin 2008). Why, however, do millions of Americans condemn homosexuality? Two concepts—*heterosexism* and *homophobia*—offer some answers.

**Heterosexism** is a belief that heterosexuality is superior to and more natural than homosexuality or bisexuality. Like sexism, heterosexism pervades societal practices, laws, and institutions. A national survey of

U.S. adolescents in grades 7 through 12 found that gay and lesbian teens were 40 percent more likely than their straight peers to be punished by schools, police, and the courts for similar misbehavior such as truancy, drinking, shoplifting, burglary, selling drugs, and physical violence (Himmelstein and Brückner 2011). Other examples of heterosexism include the widespread hostility toward same-sex marriages (which we'll discuss shortly) and sex education curricula in many middle schools and high schools that rarely address homosexuality.

Because heterosexism is deeply engrained in our culture, even individuals who aren't prejudiced against homosexuals often use slurs such as "faggot" (for males) and "dyke" (for females) to dismiss men and women who seem to be not masculine or feminine enough, and exchange jokes about homosexuals (Pascoe 2007). As a result, many gays and lesbians still hide, deny, or try to suppress their sexual orientation.

A manifestation of heterosexism is **homophobia**, the fear and hatred of homosexuality. Homophobia is less overt today than in the past but is still widespread. It often expresses itself in *gay bashing*—threats, assaults, or acts of violence directed at homosexuals. In 2010, of the almost 7,690 hate crimes reported to law enforcement agencies, 19 percent were against homosexuals, but much gay bashing isn't reported (Federal Bureau of Investigation 2011).

## 9-4 Some Current Controversies About Sexuality

Most Americans see sex as a private act, but many also believe that that there should be government control of some sexual expressions and decisions. Three of the most controversial issues today are abortion, same-sex marriage, and pornography.

## 9-4a Abortion

**Abortion** is the expulsion of an embryo or fetus from the uterus. It can occur naturally—in *spontaneous abortion* (miscarriage)—or can be induced medically. After the United States outlawed abortion in the nineteenth century, the procedure remained illegal in most states until 1973, when the U.S. Supreme Court legalized abortion as a nationwide constitutional right (*Roe v. Wade*).

### Trends

Nearly half (49 percent) of all U.S. pregnancies are unintended. Many are unwanted, whereas others occur sooner than planned. Over a lifetime about 43 percent of the unintended pregnancies and 22 percent of all pregnancies end in abortion; 30 percent of women have an abortion by age 45 (Finer and Zolna 2011; Jones and Kavanaugh 2011).

The number of abortions declined from 1.6 million in 1990 (the all-time high) to 1.2 million in 2008, when 2 percent of all women had an abortion. The *abortion rate*, or the number of abortions per 1,000 women ages 15 to 44, was 29 in 1981; it fell steadily and reached an all-time low of 19.6 in 2008 (Jones and Kavanaugh 2011).

Abortion is most common among women who are 29 and younger, white, and have at least some college education (see *Figure 9.2*). Abortion cuts across all religious groups, but 25 percent of the women are Roman Catholic, 26 percent are born-again/evangelical Christians, and 17 percent have no religious affiliation. Also, 43 percent live below the poverty level (Jones and Kavanaugh 2011).

### Why Is Abortion Controversial?

Since its legalization, abortion has been one of the most persistently contentious issues in U.S. politics and culture. More Americans describe themselves as "pro-life" (49 percent) than "pro-choice" (45 percent), with 6 percent expressing no opinion. Both anti and pro-abortion groups agree on some issues, such as requiring a patient's informed consent, but 27 percent want abortion legal in all cases and 22 percent want it illegal in all cases (Saad 2011).

> **abortion** the expulsion of an embryo or fetus from the uterus.

Antiabortion groups believe that the embryo or fetus is not just a mass of cells but a human being from the time of conception and, therefore, has a right to life. In contrast, many abortion rights advocates point out that, at the moment of conception, the organism lacks a brain and other specifically and uniquely human attributes, such as consciousness and reasoning. Abortion rights proponents also believe that a pregnant woman has a right to decide what will happen to her body (Almond 2007).

Antiabortion groups maintain that abortion is immoral and endangers a woman's physical and emotional health. Whether abortion is immoral is a religious and philosophical question. Safety, however, can be measured on two levels: physical and emotional. On the physical level, legal abortions in the first trimester (up to 12 weeks of pregnancy) are safer than driving a car, using oral contraceptives, undergoing sterilization, or continuing a pregnancy. They pose virtually no long-term risk in future pregnancies of miscarriage, birth defects, or preterm or low-birth-weight babies. There is also no evidence, despite the claims of abortion opponents, that having an abortion increases the risk of breast cancer or causes infertility (Boonstra et al. 2006; Guttmacher Institute 2011).

What about emotional health? Antiabortion activists argue that abortion leads to postabortion stress disorders, depression, and even suicide. Several national studies conducted since the 1980s have consistently concluded that abortion poses no hazard to an adolescent or adult woman's mental health. Abortion doesn't increase emotional problems such as depression or low self-esteem and doesn't lead to drug or alcohol abuse (Major et al. 2008; Warren et al. 2010; Munk-Olsen et al. 2011).

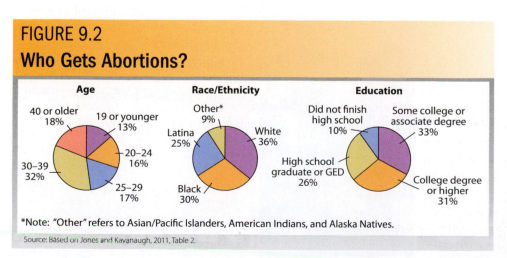

## FIGURE 9.2
## Who Gets Abortions?

*Note: "Other" refers to Asian/Pacific Islanders, American Indians, and Alaska Natives.

Source: Based on Jones and Kavanaugh, 2011, Table 2.

About 85 percent of Americans want to keep abortion legal, but it's virtually impossible for some women to get legal abortions in at least 10 states. Since 1973, states have passed 372 laws that restrict access to abortion; 19 states enacted 80 restrictions in the first half of 2011 alone. One of the restrictions requires women to view an ultrasound image of the fetus at least 24 hours before the abortion (Marcotte 2011; "States Enact . . ." 2011).

## 9-4b Same-Sex Marriage

*Same-sex marriage* (also called *gay marriage*) is a legally recognized marriage between two people of the same sex. Although still controversial, same-sex marriage is becoming more acceptable in the United States and some other countries.

### Trends

Some states have legalized *civil unions* (sometimes called *domestic partnerships* or *registered partnerships*) that give gay couples the same legal rights as married couples. Some of these rights include joint ownership of homes and other property, a share of the partner's medical or life insurance, a right to inheritance, and survivors' benefits if a partner dies. However, voters in 31 states have passed laws limiting marriage to a man and a woman, and 29 states have amended their constitutions to explicitly prohibit gay marriage (Drogin 2009; Farrell 2010).

Same-sex marriage is legal in Connecticut, Iowa, Massachusetts, New Hampshire, New York, Vermont, Washington State, the District of Columbia, and, most recently, Maryland, but Maryland voters may repeal the law during the November 2012 election. Worldwide, since 2001, 10 countries have legalized same-sex marriage (Argentina, Belgium, Canada, Iceland, the Netherlands, Norway, Portugal, South Africa, Spain, and Sweden).

### Why Is Same-Sex Marriage Controversial?

Half of Americans support same-sex marriage (up from 37 percent in 2006). Opposition is greatest among those who regularly attend religious services, live in the South, are 55 and older, and hold conservative views on family issues (Newport 2011, 2012).

Rob Melnychuk/Jupiterimages

Those who favor same-sex marriage argue that people should have the same legal rights regardless of sexual orientation. They also believe that marriage may increase the stability of same-sex couples and lead to better physical and mental health for gays and lesbians. Those who oppose same-sex marriage contend that such unions are immoral, weaken our traditional notions of marriage, and are contrary to religious beliefs. *Table 9.3* provides a summary of some of the major pro and con arguments in this debate.

## 9-4c Pornography

**Pornography** is the graphic depiction of images that may lead to sexual arousal. The images include photographs, videos (including those on the Internet), and other visual materials. The *pornography industry*, a broader concept, includes massage parlors, prostitution rings, stripping, live sex shows and lap dances, street prostitution, escort services, peep shows, phone sex, international and domestic trafficking of children (especially girls and women), mail-order bride services, and prostitution tourism (Whisnant and Stark 2004).

### Trends

Worldwide, there are more than 4 million pornographic websites, an estimated 420 million pages, and 68 million search requests for pornography each day. The pornography industry brings in more revenue than Microsoft, Google, Amazon, eBay, Yahoo, Apple, and Netflix combined (Dines 2010).

An estimated 25 percent of Americans look at Internet pornography every day; more than 90 percent of the consumers are men, and some spend at least 6 hours a day viewing these sites (Paul 2005). Among U.S. college students, 20 percent of men view online pornographic materials every day or nearly every day, compared with only 3 percent of women (Carroll et al. 2008). The average American boy first views pornography at age 11, which means that pornography is the major source of sex education (or miseducation) for many boys (Dines 2010).

### Why Is Pornography Controversial?

Many proponents argue that pornography augments the U.S. economy, enhances some people's sexual lives, and provides a safe outlet for men's sexual fantasies about oppressing women rather than resorting to violence. Thus,

## TABLE 9.3

# Why Are People For or Against Same-Sex Marriages?

What do *you* think? What other reasons can you add for each side of the debate?

| SAME-SEX MARRIAGE SHOULD BE LEGAL BECAUSE . . . | SAME-SEX MARRIAGE SHOULD NOT BE LEGAL BECAUSE . . . |
|---|---|
| • Attitudes and laws change. Until 1967, for example, interracial marriages were prohibited. | • Interracial marriages are between women and men; gay marriages violate many people's notions about male-female unions. |
| • Gay marriages would strengthen families that already exist. Children would be better off with parents who are legally married. | • Gay households are not the best place to raise children. Children raised in gay households might imitate their parents' homosexual behavior. |
| • It would be easier for same-sex couples to adopt children, including those with emotional and physical disabilities. | • All adopted children—with and without disabilities—are better off with parents who can provide heterosexual gender role models. |
| • Every person should be able to marry someone that she or he loves. | • People can love each other without getting married. |
| • Gay marriages are good for the economy because they boost businesses such as restaurants, bakeries, hotels, airlines, and florists. | • What's good for the economy isn't necessarily good for society, especially its moral values and religious beliefs. |

Sources: King and Bartlett 2006; Semuels 2008; Sullivan 2011; Whitehead 2011.

pornography meets many people's, especially men's, needs. Pornography can also be lucrative for women who strip at expensive casinos and who act in popular adult films, providing a living for women with low educational levels and few market skills. Child pornography is illegal, but adult pornography is legal under the First Amendment guarantee of free speech (Paul 2005; Urbina 2007).

Those who oppose pornography, on the other hand, point to mounting scientific research which shows that the burgeoning demand for pornography endangers children and women and warps personal relationships. Several studies have found that many pedophiles were regular users of child pornography sites before they sexually victimized children (Bourke and Hernandez 2009; Olver 2010).

A recent study compared men who buy sex (especially prostitution and pornography) and those who don't. The researchers found that the sex buyers were considerably more likely than their counterparts to be sexually violent, to have received their sex education from pornography, to believe that women secretly want to be raped, to see women as sexual objects rather than people, and to enjoy having power over women (Farley et al. 2011).

## 9-5 Gender and Sexuality Across Cultures

In 2011, three women (one from Yemen and two from Liberia) shared the prestigious Nobel Peace Prize. Of the 101 individuals who have been awarded this honor since 1901, however, only 15 have been women. The prize is a good example of women's progress on some fronts,

Jamie McCarthy/WireImage/
© iStockphoto.com/Gordan Poropat

Jenna Jamison left home at 16. At 19, she gave up posing nude for magazines and stripping in clubs, and began to act in pornographic films. At 31, she published a memoir, *How to Make Love Like a Porn Star*, which was on the *New York Times* best-seller list for 6 weeks. Her porn site, adult films, sex phone messages, and sex toys have made her a billionaire (Miller 2005).

but there is still considerable inequality between women and men regarding both gender roles and sexuality.

## Sociology in Your Life

Worldwide, how many women and men believe that wife beating is acceptable? How many girls are married before age 18? And how many women are legislators? You can find these answers at Sociology CourseMate at CengageBrain.com.

### 9-5a Gender Inequality

Of 134 countries (representing more than 90 percent of the world's population), Iceland, Norway, Finland, Sweden, and New Zealand have the greatest gender equality in economic participation and opportunity, educational attainment, health and survival, and political empowerment. The United States ranks only 19th, and behind some poor countries such as Lesotho, Philippines, and Sri Lanka. The biggest gender gaps are in the Middle East, North Africa, and sub-Saharan Africa. In all countries and regions, the areas of greatest inequality between women and men are in economic participation and political leadership (Hausmann et al. 2011).

### Economic Participation

Across 131 countries, 33 percent of men, compared with 18 percent of women, have "good jobs"—those that are full-time, provide health insurance, and ensure labor rights. Of those in the workforce, 23 percent of women, compared with 16 percent of men, are *underemployed*—working part-time when they want a full-time job and working below their level of education, skill, and experience. Women's underemployment rates are highest in sub-Saharan Africa and Latin America, and lowest in East Asia and the countries that broke away from the Soviet Union (Marlar 2011).

Globally, female-owned firms account for 40 percent of all privately held businesses and generate $1.9 trillion in annual sales ("The X Factor" 2011). However, only 3 percent of the CEOs of *Fortune 500* companies and 16 percent of all U.S. corporate board members are women—about the same as in 1998. In contrast, Finland, Sweden, and Norway have laws requiring 40 percent of all board members to be women (Kowitt and Arora 2011; McGregor 2011).

### Political Leadership

Worldwide, 19 percent of those holding seats in national legislatures are women. Rwanda has 56 percent,

TABLE 9.4

## Countries With Highest Percentage of Women in National Legislatures

| | |
|---|---|
| Rwanda | 56% |
| Sweden | 46% |
| South Africa | 45% |
| Cuba, Iceland | 43% |
| Netherlands | 42% |
| Finland, Norway | 40% |

Source: Based on Clifton and Frost 2011, 10–13.

followed by seven countries in which women comprise 40 to 46 percent of those in high-level political positions (see *Table 9.4*). In contrast, only 17 percent of all U.S. legislators are women (Clifton and Frost 2011).

At the other end of the continuum, Saudi Arabia, with a score of 0, is the lowest ranking country in the world on women's political empowerment (Hausmann

Christian Heeb/laif/Redux

A few days after King Abdullah of Saudi Arabia announced that women would be able to vote in 2015, a woman was arrested for defying the country's ban on female driving. After protests from the international community, the king overturned the sentence of 10 lashes. Saudi Arabia is the only country in the world that bans women—including foreign women—from driving (Winter 2011). This country has some of the most educated women in the world, including those with STEM college degrees. Because paid employment is an all-male preserve, Saudi women comprise only 7 percent of the labor force, although half of all women are looking for work (Charles 2011; Rashad 2012).

Warren Goldswain/Shutterstock.com

et al. 2011). In 2011, King Abdullah announced that, beginning in 2015, women will be able to run as candidates in municipal elections, serve on the king's advisory board, and, with a male family member's approval, "will even have the right to vote" ("Saudi Arabia and Its Women" 2011; "Saudi King Announces . . ." 2011).

## 9-5b Sexual Violence and Oppression

Globally, women have fewer rights and opportunities than men. There's been more acceptance of homosexuality in some countries, but heterosexism prevails.

### Women Are Often Victimized

An estimated 100 million girls worldwide are "missing" because of sex-selective abortions in India, neglect and abandonment of baby girls in South Korea, and female infanticide (killing of newborns) in China (Mack 2011). In many countries, women experience sexual violence and male dominance because of custom, laws, or minimal government protection. For example,

- In Afghanistan, about 80 percent of girls—some as young as 8 years old—are forced into marriages (Zirulnick 2011; Schnall 2012). About 87 percent of Afghan women report experiencing physical, psychological, or sexual abuse.

- In Pakistan, 90 percent of women experience domestic violence in their lifetimes (Human Rights Now 2011).

- In the Democratic Republic of Congo, almost 1,160 women are raped every day by soldiers, strangers, and intimate partners (Peterman et al. 2011).

### Homosexuality Around the World

You saw earlier that many Americans are more accepting of homosexuality than in the past. In many non-Western countries, however, homosexuals, especially men, may be stoned, imprisoned, or killed. In Africa, four nations impose a death penalty on gay men, and 20 nations may imprison gay men and lesbians from a month to a lifelong sentence (Baldauf 2010).

Despite such sanctions, some attitudes are changing. For example, Australian passports now designate "M" for male, "F" for female, and "X" for transgender ("Australian Passports . . ." 2011). In India, the 2011 national census for the first time offered three gender options: male, female, or a "third sex" that includes people who are gay, lesbian, or transgender (Cohn 2011). And in Thailand, which has the world's biggest transsexual population, an airline has recently recruited "third sex" flight attendants (Mutzabaugh 2011).

## 9-6 Sociological Explanations of Gender and Sexuality

Gender and sexuality affect all people's lives, but why is there so much variation over time and across cultural groups? The four sociological perspectives answer this and other questions about gender and sexuality somewhat differently (*Table 9.5* summarizes these perspectives).

## 9-6a Functionalism

Functionalists view women and men as having distinct roles that ensure a family's and society's survival. These roles help society operate smoothly, and they have an impact on the types of work that women and men do.

### Division of Gender Roles and Human Capital

Some of the most influential functionalist theories, developed during the 1950s, proposed that gender roles

## TABLE 9.5

## Sociological Explanations of Gender Inequality and Sexuality

| THEORETICAL PERSPECTIVE | LEVEL OF ANALYSIS | KEY POINTS |
|---|---|---|
| **Functionalist** | Macro | • Gender roles are complementary, equally important for a society's survival, and affect human capital.<br>• Agreed-on sexual norms contribute to a society's order and stability. |
| **Conflict** | Macro | • Gender roles give men power to control women's lives instead of allowing the sexes to be complementary and equally important.<br>• Most societies regulate women's, but not men's, sexual behavior. |
| **Feminist** | Macro and micro | • Women's inequality reflects their historical and current domination by men, especially in the workplace.<br>• Many men use violence—including sexual harassment, rape, and global sex trafficking—to control women's sexuality. |
| **Symbolic Interactionist** | Micro | • Gender inequality is a social construction that emerges through day-to-day interactions and reflects people's gender role expectations.<br>• The social construction of sexuality varies across cultures because of societal norms and values. |

© Cengage Learning

differ because women and men have distinct roles and responsibilities. A man (typically a husband and father) plays an *instrumental role* of economic provider; he is competitive and works hard, even if he is overwhelmed by multiple roles such as the responsible breadwinner, the devoted husband, and the dutiful son. A woman (typically a wife and mother) plays an *expressive role*; she provides the emotional support and nurturance that sustain the family unit and support the father/husband. For functionalists, the instrumental and expressive roles are complementary. The duties are specialized, but both roles are equally important in meeting a family's needs and ensuring a society's survival (Parsons and Bales 1955; Gaylin 1992; Betcher and Pollack 1993).

Such traditional gender roles help to explain occupational sex segregation. People differ in the amount of human capital that they bring to the labor market. *Human capital* is the array of competencies— such as education, job training, and experience—that have economic value and increase productivity.

From a functionalist perspective, what individuals earn is the result of the choices they make and,

consequently, the human capital that they accumulate to meet labor market demands. Women diminish their human capital because they choose lower paying occupations (such as social work rather than computer science), as well as postpone or leave the workforce for childbearing and child care. When they return to work, women have lower earnings than men because, even in higher paying occupations, their human capital (knowledge, skills, experience) has deteriorated or become obsolete (Kemp 1994).

### Why Is Sexuality Important?

For functionalists, sexuality is critical for reproduction, but people should limit sex to marriage and forming families. Functionalists view sex outside of marriage as dysfunctional because most unmarried fathers don't support their children. These offspring often experience instability and poverty, as well as a variety of emotional, behavioral, and academic problems (Seltzer 2004; Avellar and Smock 2005).

You might dismiss the functionalist view of limiting sex to marriage as outdated.

Paul Hakimata Photography/Shutterstock.com

Worldwide, however, sex outside of marriage is prohibited and *arranged marriages*—in which parents or relatives choose their children's future mates—are the norm. Most children agree to arranged marriages out of respect for their parents' wishes and social custom and because such matches solidify relationships with other families and ensure that the woman's sexual behavior will be confined to her husband, avoiding any doubt about the offspring's parentage (see Benokraitis 2011).

George Marx/Getty Images

### Critical Evaluation

Instrumental and expressive roles are useful in understanding some of the differences between traditional and nontraditional gender roles. In traditional roles, each person knows what is expected of him or her. Men and women don't have to argue over who does what. If the house is clean, she's a "good wife"; if the bills are paid, he's a "good husband."

One of the criticisms of the functionalist perspective is that even during the 1950s, white middle-class male sociologists ignored almost a third of the labor force that was composed of working-class, immigrant, and minority women who played *both* instrumental and expressive roles. A related criticism is that functionalists tend to overlook the fact that most people don't have the choice of playing strictly instrumental or expressive roles. Only 19 percent of Americans say that women should return to their traditional roles in society (Parker 2009), probably because most families rely on two incomes for economic survival. A major limitation of the human capital model is its assumption that women have lower earnings than men because they "choose" lower paying occupations. As you saw earlier, there is a gender pay gap across all occupations, even those dominated by women.

Functionalism frowns on sexual relationships outside of marriage. For example, many functionalists see cohabitation (living together outside of marriage) as deviant because it doesn't always result in marriage (Popenoe and Whitehead 2006). As you'll see in Chapter 12, however, marriage doesn't guarantee long or happy relationships.

## 9-6b Conflict Theory

Conflict theorists see gender inequality as built into the social structure, both within and outside the home.

In many developing nations, women perform the vast majority of the agricultural work but have virtually no property rights because they aren't allowed, by law, to own land (Seager 2009). In industrialized countries, men control most of a society's resources. Like functionalists, conflict theorists see sexuality as a key component of a society's organization, but they view sexuality as reflecting and perpetuating gender inequality.

### Capitalism and Gender Inequality

Unlike functionalists, conflict theorists maintain that capitalism, not complementary roles, explain gender roles and men's social and economic advantages. A full-time stay-at-home American mother would earn at least $135,000 a year if paid for her work as housekeeper, cook, van driver, psychologist, day care provider, and so on (Wulfhorst 2006). In effect, traditional gender roles

© Julio Etchart/Alamy

चोवीस तास फक्त स्त्रियांसाठी सारे समय केवल महिलाओं के लिये
LADIES ONLY
FOR ALL THE TWENTY FOUR HRS

Since 1995, the number of employed women in India has surged. One of their most difficult tasks is getting to work on crowded trains because men routinely pinch and grope them, or shout insults because the women aren't full-time housewives (Ridge 2009: 4). To decrease the sexual harassment, the government has introduced women-only commuter trains—known as Ladies Specials—in four of India's largest cities. The trains represent a tiny fraction of the nation's commuter trains, but vandalism and harassment persist (Yardley 2009: A1).

are profitable for business. Companies can require their male employees to work long hours or make numerous overnight business trips and don't have to worry about workers' demanding child care services.

Most conflict theorists agree that all men are not equally privileged, and that women in upper classes have more economic power, status, and prestige than men in lower classes. However, within social classes, men typically enjoy more power and control than women.

### Is Gender Inequality Linked to Sexual Inequality?

According to many conflict theorists, gender inequality gives men economic, political, and/or interpersonal power to control or dominate women's sexual lives. Most of the victims of domestic violence and rape, in the United States and around the world, are women and girls. In workplace sexual harassment cases, the offender is typically a male supervisor. In prostitution and sex trafficking, 80 percent of the victims worldwide are women and girls who live in poverty (Farr 2005).

In some societies, especially in the Middle East and some African countries, men dictate how women should dress and whether they can travel, work, receive health care, attend school, or start a business; dismiss women's charges of sexual assaults (including gang rapes); punish real or imagined instances of women's, but not men's, marital infidelity; and blame girls for child rape because they're "seducing" older men (Neelakantan 2006). Many women have internalized such gender inequality. For example, between 31 to 40 percent of the women in Uganda believe that wife beating is acceptable if the wife argues with her husband or refuses to have sex with him (Clifton and Frost 2011).

### Critical Evaluation

Conflict theory is useful in showing how social structures reinforce men's domination of women. Some, however, have criticized conflict theorists for focusing on male-female competition rather than cooperation. At home, many women (43 percent) have the final say in many decisions, and 80 percent of both sexes report being happy with this situation (Morin and Cohn 2008).

Outside the home, women aren't always as submissive as some conflict theorists claim. Like men, women often barter with their employers to get what they want

("I'll work late all this week if I can have Friday off"). A related criticism is that conflict theory emphasizes the differences between women and men rather than their common goals, similar attitudes, and focuses on power plays for economic and political dominance instead of negotiations.

Another criticism is that conflict theory usually ignores how women exploit other women. In the United States, for instance, women comprise 19 percent of those involved in the sex trafficking of women and children (Banks and Kyckelhahn 2011). In other cases, women provide men with sexual favors in exchange for resources for themselves or their children (Barrows 1989; Rhoads 2004).

## 9-6c Feminist Theories

Feminist scholars agree with conflict theorists that gender stratification benefits men and capitalism. Feminist theorists emphasize, however, that women's subordination also includes their daily vulnerability to violence: "If they are not getting harassed on the street, living in an abusive relationship, recovering from a rape, or in therapy to deal with the sexual abuse they suffered as children, [women] are ordering their daily lives around the *threat* of men's violence" (Katz 2006: 5). Feminist scholars, more than any other group of theorists, are especially concerned about men's controlling women's sexual lives.

### Living in a Gendered World

All feminists (female and male) agree on three general points: (1) men and women should be valued equally; (2) women should have more control over their lives; and (3) political, economic, family, and other institutions can reduce gender inequality. Let's look briefly at three prominent feminist explanations of gender roles.

*Liberal feminism* maintains that gender equality can be achieved through equal rights and opportunities. This approach emphasizes that gendered socialization creates inequality by teaching children culturally accepted masculine and feminine attitudes and behavior that limit their options in the family, education, and the workplace (Jagger and Rothenberg 1984).

*Radical feminism* contends that patriarchy is the major reason for men's dominance over women, including sexual exploitation. The control is deliberate, brings men many benefits, and is supported by oppression in medicine, religion, science, law, and other institutions. For radical feminists, all women are potential victims of violence and all men are capable of violence. In this sense, patriarchy is more critical than social class in giving men power over women's lives (Bart and Moran 1993; Lorber 2005).

*Multicultural feminism* (sometimes referred to as *racial ethnic feminism* or *multiracial feminism*) asserts that gender, race, and social class intersect to form a hierarchical stratification system that shapes women's and men's attitudes, experiences, and behavior. Privileged women have less status than privileged men, but upper-class white men *and* women subordinate lower-class women *and* minority men (Lorber 2005; Andersen and Collins 2010).

Understanding the interconnections between gender, race, ethnicity, and social class provides a comprehensive picture of people's experiences. For example, working-class women (across all racial/ethnic groups) are generally less likely than their middle-class counterparts to be in the labor force for a combination of reasons: They usually grew up in families with gender ideologies that emphasized traditional breadwinner/housework roles, with lower expectations about working continuously, and with fewer resources to obtain a higher education that prepares people for higher paying occupations (England 2010; Damaske 2011).

## Is Sexuality Linked to Social Control?

Feminist theorists often point to sexuality as the root of inequality between women and men, both interpersonally and within institutions. Men assert their power through rape, intimate partner violence, sexual harassment, and the exploitation of women through prostitution and pornography. About 140 million girls—usually between the ages of 4 and 12—in Africa and some countries in Asia and the Middle East have experienced female genital mutilation/cutting (FGM/C), which involves removing all or part of the girl's external genitalia, such as the clitoris, to control her sexual desire and preserve her chastity before marriage. Sometimes the girls hemorrhage and die. In other cases, they become infertile because of infections or experience lifelong health problems (Feldman-Jacobs and Clifton 2010; Benokraitis 2011).

Feminist theories have gone much further than conflict theory in documenting men's control of women's sexuality across cultures and over time. Examples include forcing women to marry against their will; *honor killings* (murders committed by male family

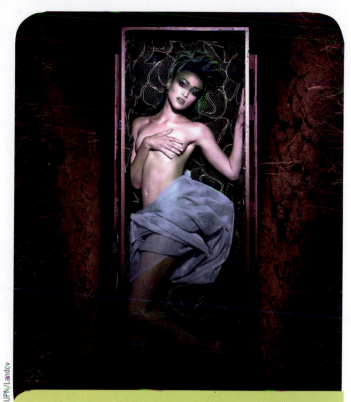

UPN/Landov

For feminist scholars, some popular TV shows have sunk to an all-time low in their portrayal of women. In *America's Next Top Model*, for example, the contestants pretended to be dead by drowning, shooting, electrocution, and even decapitation. In a similarly objectionable episode, models were shown in coffins and depicting the seven deadly sins (pictured here is envy).

members if they believe that a female has brought shame on the family by having sex outside of marriage or refusing an arranged marriage); assaulting women sexually; and punishing women for seeking a divorce or committing adultery (Amnesty International 2006; Magnier 2009). Feminist scholars maintain that as long as men control women's bodies and behavior, women's sexual well-being will be inferior to men's (hooks 2000).

## Critical Evaluation

Feminist perspectives provide insightful analyses of gender inequality, but have limitations. *Liberal feminism* typically emphasizes women's, but not men's, oppression in capitalistic societies. *Radical feminism* has been criticized for being too narrow; for example, men are not universally violent or sexually exploitative.

Lisa S./Shutterstock.com

## 9-6d Symbolic Interactionism

Whereas functionalist, conflict, and some feminist theories are macro level, symbolic interactionists focus on the everyday processes that produce and reinforce gender roles. We "do" gender, sometimes consciously and sometimes unconsciously, by adjusting our behavior and our perceptions depending on the sex of the person with whom we're interacting (West and Zimmerman 1987, 2009). Our expression of our sexuality, similarly, is not inborn but reflects socialization, and what families and other societal groups deem as appropriate and inappropriate behavior (Hubbard 1990).

### Gender Is a Social Construction

For symbolic interactionists, gender is a social creation, and gender inequality is shaped by learning gender roles through daily social interaction. For example, negative stereotypes about girls' abilities in math and science can significantly lower girls' test performance and lower their aspirations in STEM careers. When teachers tell girls and boys that both are equally capable in math and science, "the difference in performance essentially disappears" (Hill et al. 2010: 2). Thus, social interaction is important in cultivating or stifling abilities and interests.

In effect, believing in gender differences can actually *produce* differences. Women who begin college intending to become engineers are more likely than men to change their major and choose another career because they lack confidence, not competence. As one of the most sex-segregated professions, engineering carries ingrained cultural stereotypes that men are more naturally suited to the field. As a result, women's professional self-confidence falters and they change majors (Cech et al. 2011; see also Chapter 13).

### How Do We Construct Sexuality?

For symbolic interactionists, sexuality is also socially constructed. As with gender roles, we *learn* to be sexual and to express our sexuality differently over time and across cultures because the people around us guide and limit our sexual behavior. For example, a study of mothers of 3- to 5-year-old children found that mothers who are affiliated with conservative religions try to prevent homosexuality in their children, generally by telling them that being gay is wrong and violates God's laws (Martin 2009).

*Multicultural feminism* is very inclusive, but this strength can also be a weakness. When it comes to economic issues such as the gender pay gap, for example, which variable is the most important—race, ethnicity, or social class? Because gender inequality has many causes, it's not clear which factors should be given priority in implementing social change.

Some critics also contend that feminist theorists focus too much on men's sexual domination and power and minimize the importance of sexuality that reflects love and affection. And, like conflict theorists, feminist scholars are sometimes accused of glossing over women's exploitation of others. In Iraq, for example, sex traffickers are often women. They target the youngest girls because virgins bring the highest prices (Naili 2011).

One might also question many feminist theorists' contention that men control women's sexuality. For example, a recent study found that powerful women in business were just as likely as their male counterparts to commit adultery because they had the opportunity and self-confidence to do so (Lammers et al. 2011).

© 2009 Los Angeles Times/Luis Sinco

In some parts of Indonesia, the police—who enforce religious rules about women's and men's segregation—go after women not properly covered by head scarves and couples engaging in public displays of affection. Pictured here, a policeman lectures a young man for sitting too close to a woman while on an outing at the beach (Glionna, 2009).

In many Middle East countries, men have premarital sex but don't marry women who aren't virgins (Fleishman and Hassan 2009). In the United States, as you saw earlier, men who have casual sex are "studs" whereas women are "sluts." Thus, sexual double standards are socially constructed.

Because sexuality is socially constructed, attitudes and behavior can change. Many Americans have become more comfortable with categories such as bisexual, lesbian, and gay. In China, a university that planned to publicly shame students who engaged in "uncivilized behavior" such as hugging or kissing in public withdrew the policy after widespread protests that such surveillance violated the students' privacy (Meng 2011).

*Sergii Figurnyi/Shutterstock.com*

### Critical Evaluation

Symbolic interaction theory is important in explaining how gender and gender roles shape our everyday lives. Despite their contributions, interactionists have been criticized for ignoring the social structures that create and maintain gender inequality. U.S. military policies dictate whether women can participate in ground combat, an important criterion in promotions. Female soldiers in the Iraq and Afghanistan wars "have done nearly as much in battle as their male counterparts: patrolled streets with machine guns, served as gunners on vehicles, disposed of explosives, and driven trucks down bomb-ridden roads" (Alvarez 2009: A1). So far, however, and despite a military commission's recommendation to do so, the Pentagon has not eliminated its combat exclusion policies for women (Daniel 2011).

Symbolic interactionists have enhanced our understanding of sexuality by showing how behavior is socially constructed and, consequently, why there are cross-cultural variations in sexual attitudes and practices. However, the theories don't explain why siblings, even identical twins—who are socialized similarly—have different sexual orientations. A related weakness is that symbolic interactionists don't explain why, historically and currently, women around the world are considerably more likely than men to be subjected to sexual control and exploitation. Such analyses require macro-level analyses that examine family, religious, political, and economic institutions.

### Study Tools

Ready to study? Go to **Sociology CourseMate** at **www.cengagebrain.com** to complete practice quizzes, review flashcards, watch videos, and more.

SOC

America's multicultural umbrella includes at least 150 distinct ethnic or racial groups.

John Lund/BlendImages/Jupiter Images

# Race and Ethnicity

Two researchers randomly assigned either a stereotypically African American name (Lakisha Washington or Jamal Jones) or a name that sounded typically white (Emily Walsh or Greg Baker) to both high-quality and low-quality fictitious résumés. They then sent almost 5,000 of the résumés in response to advertisements for a variety of jobs in Chicago and Boston. The results were startling.

- Applicants with white-sounding names received 50 percent more calls for interviews. In fact, a white-sounding name yielded as many more calls as an additional 8 years of work experience for an applicant with a black-sounding name.
- Applicants with black-sounding names received fewer calls across all occupations (from managers to clerical workers) and in all industries (communications, manufacturing, finance, and social services).
- Employers who posted "Equal Employment Opportunity Employer" in their ad discriminated as much as other employers (Bertrand and Mullainathan 2003).

As this example shows, some of us enjoy more opportunities than others simply because of the color of our skin. The situation has improved during the last 50 years or so, but not as much as most people think. This chapter examines the impact of race and ethnicity on our lives, why racial-ethnic inequality is still widespread, and the growth of interracial and interethnic relationships. First, however, take the brief quiz on the next page to see how much you know about U.S. race and ethnicity.

## 10-1  U.S. Racial and Ethnic Diversity

The United States is the most multicultural country in the world, a magnet that draws people from hundreds of nations and is home to millions of Americans who are bilingual or multilingual. Of the almost 313 million U.S. population, 13 percent are foreign born, but this population is expected to grow to 23 percent by 2050. As a result, America's multicultural umbrella includes at least 150 distinct ethnic or racial groups (Passel and Cohn 2008; U.S. Census Bureau 2012).

By 2025, only 58 percent of the U.S. population is projected to be white—down from 86 percent in 1950 (see *Figure 10.1*). By 2050, whites may make up only 47 percent of the total population because Latino and Asian populations are expected to triple in size (Passel et al. 2012). The multiracial population is also increasing. The number of Americans who identify themselves as being two or more races is projected to more than triple—from almost 6 million in 2010 to more than 16 million in 2050 (2 percent and almost 4 percent of the total population, respectively) (U.S. Census Bureau News 2008; Passel et al. 2011).

## Key Topics

In this chapter, we'll explore the following topics:

## What do you think?

The United States is a melting pot.

| 1 | 2 | 3 | 4 | 5 | 6 | 7 |
|---|---|---|---|---|---|---|

strongly agree          strongly disagree

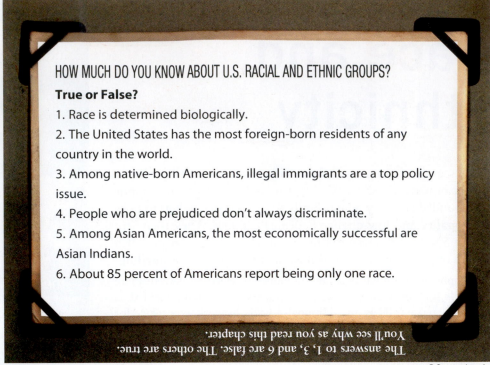

HOW MUCH DO YOU KNOW ABOUT U.S. RACIAL AND ETHNIC GROUPS?

**True or False?**

1. Race is determined biologically.

2. The United States has the most foreign-born residents of any country in the world.

3. Among native-born Americans, illegal immigrants are a top policy issue.

4. People who are prejudiced don't always discriminate.

5. Among Asian Americans, the most economically successful are Asian Indians.

6. About 85 percent of Americans report being only one race.

The answers to 1, 3, and 6 are false. The others are true. You'll see why as you read this chapter.

© Cengage Learning

# 10-2 The Social Significance of Race and Ethnicity

All of us identify with some groups and not others in terms of sex, age, social class, and other factors. Two of the most common and influential sources of self-identification, as well as labeling by others, are race and ethnicity.

## 10-2a Race

**Race** refers to people who share physical characteristics, such as skin color and facial features, that are passed on through reproduction. Contrary to the popular belief that race is determined biologically, it is a *social construction*, a societal invention that labels people based on physical appearance. Possibly only 6 of the human body's estimated 35,000 genes determine the color of a person's skin. Because all human beings carry 99.9 percent of the same genetic material (DNA), the "racial" genes that makes us look different are miniscule compared with the genes that make us similar (Graves 2001; Pittz 2005).

If our DNA is practically identical, why are we so obsessed with

race? People react to the physical characteristics of others, and those reactions have consequences. Skin color, hair texture, and eye shape, for example, are easily observed and mark groups for unequal treatment. As long as we sort people into racial categories and act on the basis of these characteristics, our life experiences will differ in access to jobs and other resources, how we treat people, and how they treat us (Duster 2005).

## 10-2b Ethnicity

An **ethnic group** (from the Greek word *ethnos*, meaning "nation") is a group of people who identify with a common national origin or cultural heritage that includes language, geographic roots, food, customs,

## FIGURE 10.1

## Racial and Ethnic Composition of the U.S. Population, 1950–2025

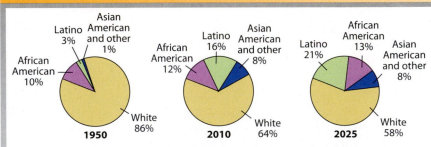

Note: "Asian American and other" includes American/Indian/Alaskan Native, Native Hawaiians and and Pacific Islanders, some other race, and those who identify themselves with two or more races.

Sources: U.S. Census and Population Division, U.S. Census Bureau, 2008; Passel et al. 2011.

© Cengage Learning

There's much variation in skin color across and within groups. People of African descent have at least 35 different shades of skin tone (Taylor 2003). So, can you determine someone's race simply by looking at her or him?

© Getty Images/Comstock/Jupiter Images

increase in interracial relationships and marriages, as you'll see later in this chapter, our definitions and measurements of race and ethnicity are becoming increasingly complex.

traditions, and/or religion. Ethnic groups in the United States include Puerto Ricans, Chinese, Serbs, Arabs, Swedes, Hungarians, Jews, and many others. Like race, ethnicity can be a basis for unequal treatment, as you'll see shortly.

### 10-2c Racial-Ethnic Group

People who have distinctive physical and cultural characteristics are a **racial-ethnic group**. Some people use the terms *racial* and *ethnic* interchangeably, but remember that *race* refers to physical characteristics with which we are born, whereas *ethnicity* describes cultural characteristics that we learn. The term *racial-ethnic* includes both physical and cultural traits.

Describing racial-ethnic groups has become more complex because the U.S. government allows people to identify themselves in terms of both race and ethnicity. In the 2000 and 2010 U.S. Census, for example, some Latinos could check off "Black" for race and "Cuban" for ethnic origin. Such choices generate dozens of racial-ethnic categories.

Most people maintain that they are color-blind, but we learn at an early age that some physical attributes are more valued than others. A notable example is light-colored skin, which confers *white privilege*, the advantages that white people enjoy simply because they happen to be in a particular category. Most white people don't feel privileged because they aren't wealthy. Nonetheless, they enjoy everyday benefits that they take for granted, such as not being followed by store detectives when they shop and being able to easily buy products that represent their own race—such as greeting cards, dolls, toys, and children's magazines (McIntosh 1995; Johnson 2008; Rothenberg 2008).

Race, ethnicity, and racial-ethnic group aren't fixed in how we identify ourselves or others. Because of the

## 10-3 Our Changing Immigration Mosaic

The United States has the most foreign-born residents of any country in the world, 1 in 8 people. More notable than the growth of the foreign-born population is the change in the countries of origin. In 1960, 75 percent of the foreign born were from Europe. By 2010, more than 80 percent were from Asia (mainly China and the Philippines) and Latin America (mainly Mexico) (Gibson and Jung 2006; Grieco and Trevelyan 2010).

### 10-3a Unauthorized Immigrants

The United States admits more than 1 million immigrants every year—more than any other nation. A major change has been the rise of unauthorized (also called *undocumented* and *illegal*) immigrants—from 180,000 in the early 1980s to 11.2 million in 2010. Unauthorized immigrants make up 30 percent of all foreign-born U.S. residents, 4 percent of the nation's population, and 5 percent of its labor force (Passel and Cohn 2011). About 62 percent of undocumented immigrants are from Mexico, 17 percent are from Central and Latin America, 7 percent are from Asia, and 14 percent are from other countries, including Canada and Europe (Hoefer et al. 2011).

Among the roughly 700 million adults worldwide who would like to relocate permanently to another country if they could, the United States is the top destination. Of the 14 million Mexicans who would like to resettle somewhere else, 44 percent would like to move permanently to the United States (Clifton 2010; Esipova et al. 2010). Most immigrants—legal and illegal—come to the United States for economic opportunities; less than 10 percent arrive as refugees who are fleeing political, religious, or other oppression in their own countries (Martin and Midgely 2010).

## 10-3b Reactions to Unauthorized Immigrants

In 2008, 39 percent of Americans favored cutbacks in legal immigration (Jones 2008). How do they feel about illegal immigration? Recently, 46 percent of Americans said that unauthorized immigration should be a policy priority, but placed it far behind the economy (87 percent), jobs (84 percent), terrorism (73 percent), Social Security and education (66 percent each), and six other issues ("Economy Dominates . . ." 2011).

Because federal measures to deal with illegal immigration have been languishing in Congress, some states passed their own laws. In 2010, Arizona passed the broadest and strictest anti-illegal immigration law in recent U.S. history. In 2011, five states passed laws that sometimes went even further by barring illegal immigrants from receiving state or local public benefits and getting a driver's license (Fausset 2011; Hing 2011). In 2012, the Supreme Court, saying that the federal government has the sole power to enforce illegal immigration laws, struck down Arizona's law, but allowed police to check the immigration status of people lawfully stopped for other reasons.

About 61 percent of Americans approve of laws that would require police to verify the legal status of someone suspected of being in the country illegally, but 57 percent oppose changing the U.S. Constitution to bar children born to illegal immigrants from becoming citizens automatically ("Public Favors . . ." 2011). Public opinion polls show widespread dissatisfaction with our "broken" immigration system, but there is no consensus on how to fix the problem.

Of those who want stricter enforcement of current laws, 40 percent say that their biggest concern is that illegal immigration places a burden on government services. Far fewer worry that illegal immigration contributes to crime (9 percent) or hurts America's customs and way of life (6 percent) ("Public Favors Tougher . . ." 2011).

Those who endorse policies that provide a "path to citizenship" for unauthorized immigrants maintain that undocumented immigrants provide numerous economic benefits for their host countries. They clean homes and offices, toil as nannies and busboys, serve as nurses' aides, and pick fruit—all at low wages and in jobs that most American-born workers don't want. Illegal immigrants are also more likely than U.S.-born people to work in dangerous jobs (such as mining, logging, and construction) that have high fatality rates due to accidents. Shortly after Arizona cracked down on illegal immigrants, the state lost revenue of at least $141 million in the tourist industry alone because native-born workers in this sector depended on illegal immigrants to work in the lowest paying jobs (Orrenius and Zavodny 2009; Zuehlke 2009; Wolgin and Kelley 2011).

Many scholars argue that, in the long run, easing illegal immigrants' path to citizenship bring more benefits than costs. For example, undocumented immigrants constitute an important labor force for an aging (and primarily white) U.S. population that will require many workers to support Social Security and Medicare payments for older Americans. Naturalized citizens earn higher wages, which decreases the likelihood of family poverty; pay billions in state and federal taxes; and contribute to U.S. culture (Mather 2009; Pastor et al. 2010; Shierholz 2010).

Guest workers, who are permitted to work in the United States on a temporary basis because of labor shortages, especially in agriculture, often live in little more than shacks (left). After millions of undocumented immigrants fled to other states from Arizona, Alabama, and Georgia because of their harsh laws, many employers complained that their produce was rotting in the fields because native-born Americans were unwilling to work for low wages (middle). Other employers couldn't find enough laborers—in construction and landscaping—to clean up and rebuild communities in the aftermath of devastating tornados (right), and Alabama alone lost at least $130 million in taxes that undocumented immigrants would have paid ("The Impact of . . ." 2011; Shepherd 2011).

© iStockphoto.com/nicholas belton; AP Images: Melanie Stetson Freeman/The Christian Science Monitor/Getty Images; AP Photo/Butch Dill

# 10-4 Dominant and Minority Groups

The president of the Boston City Council tried, unsuccessfully, to strike the word *minority* from official city documents because, he argued, the term is insulting and inaccurate in a city whose population is 51 percent people of color (Wiltenburg 2002). For sociologists, *minority group* and *dominant group* are descriptive terms that have little to do with a group's size.

## 10-4a What Is a Dominant Group?

A **dominant group** is any physically or culturally distinctive group that has the most economic and political power, the greatest privileges, and the highest social status in a society. As a result, it can treat other groups as subordinate. For example, in most societies, men are a dominant group because they have more status, resources, and power than women (see Chapter 9).

Dominant groups aren't necessarily the largest in number. From the seventeenth century until 1994, about 10 percent of the population in South Africa was white and had almost complete control of the black population. Because of **apartheid**, a formal system of racial segregation, the black inhabitants couldn't vote, lost their property, and had minimal access to education and politics. Apartheid ended in 1994, but most black South Africans are still a minority because whites "hold the best jobs, live in the most expensive homes, and control the bulk of the country's capital" (Murphy 2004: A4).

## 10-4b What Is a Minority?

Sociologists describe Latinos, African Americans, Asian Americans, Middle Eastern Americans, and American Indians as minorities. A **minority** is a group of people who may be subject to differential and unequal treatment because of their physical, cultural, or other characteristics, such as sex, sexual orientation, religion, ethnicity, or skin color. Minorities may be larger than a dominant group in number, but they have less power, privilege, and social status. For example, American minorities have fewer choices than dominant group members in finding homes and apartments because they are less likely to get help—from either a real estate agent or a bank—with the intricacies of mortgage financing that most people need (Massey 2007).

## 10-4c Patterns of Dominant-Minority Group Relations

To understand some of the complexity of dominant-minority group relations, think of a continuum. At one end of the continuum is genocide; at the other end is pluralism (see *Figure 10.2*).

### Genocide

**Genocide** is the systematic effort to kill all members of a particular ethnic, religious, political, racial, or national group. By 1710, for example, the colonists in

**dominant group** any physically or culturally distinctive group that has the most economic and political power, the greatest privileges, and the highest social status.

**apartheid** a formal system of racial segregation.

**minority** a group of people who may be subject to differential and unequal treatment because of their physical, cultural, or other characteristics, such as sex, sexual orientation, religion, ethnicity, or skin color.

**genocide** the systematic effort to kill all members of a particular ethnic, religious, political, racial, or national group.

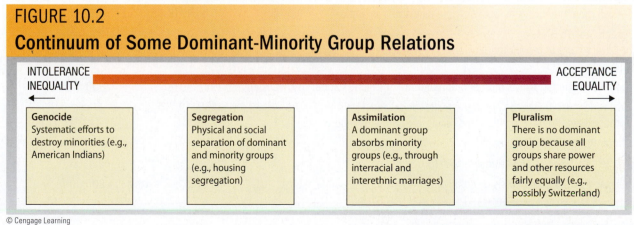

## FIGURE 10.2
## Continuum of Some Dominant-Minority Group Relations

INTOLERANCE INEQUALITY → ACCEPTANCE EQUALITY

**Genocide**
Systematic efforts to destroy minorities (e.g., American Indians)

**Segregation**
Physical and social separation of dominant and minority groups (e.g., housing segregation)

**Assimilation**
A dominant group absorbs minority groups (e.g., through interracial and interethnic marriages)

**Pluralism**
There is no dominant group because all groups share power and other resources fairly equally (e.g., possibly Switzerland)

© Cengage Learning

**segregation** the physical and social separation of dominant and minority groups.

**assimilation** the process of conforming to the culture of the dominant group, adopting its language and values, and intermarrying with that group.

America had killed thousands of Indians in skirmishes, poisoned others, and promoted scalp bounties. In 1851, the governor of California officially called for the extermination of all Indians in the state (de las Casas 1992; Churchill 1997). As *Table 10.1* shows, well over 74 million people have been victims of genocide during the twentieth century alone. The total number of people killed in these countries equals about 24 percent of the current U.S. population.

## Segregation

**Segregation** is the physical and social separation of dominant and minority groups. In 1954, the Supreme Court ruling in *Brown v. Board of Education* declared *de jure*, or legal, segregation unconstitutional. It was followed by the passage of a variety of federal laws that prohibited racial segregation in public schools, as well as discrimination in employment, voting, and housing.

*De facto*, or informal, segregation has replaced *de jure* segregation in the United States and many other countries. Some *de facto* segregation may be voluntary, as when members of racial or ethnic groups prefer to live among their own group. In most cases, however, *de facto* segregation is due to discrimination, as when realtors steer minorities away from white neighborhoods. Such residential segregation deprives minorities of access to quality schools, retail stores, leisure activities, and burgeoning suburban job opportunities (Farley and Squires 2005; Rugh and Massey 2010).

## Assimilation

Many minority group members blend into U.S. society through **assimilation**, the process of conforming to the culture of the dominant group, adopting its language and values, and intermarrying with that group. There are some indications that newcomers may be assimilating more slowly than in the past. For example, the federal government offered online help in filling out the 2010 Census in five languages—English, Spanish, Chinese, Korean, and Russian; the number of people who don't speak English at home has more than doubled since 1980; and 25 percent of Americans report not speaking English well or at all (Shin and Kominski 2010).

Others maintain that immigrants who arrived after 1995 are assimilating more rapidly than their predecessors, but that the degree of assimilation varies by country of origin, educational level, and legal status. Mexicans, for example, are considerably less likely to assimilate than immigrants from the Philippines, Vietnam, or South Korea because they are more likely to have entered the country illegally, which cuts them off from getting a good job, most public assistance programs, and eventual citizenship. Still, although most Mexican immigrants speak little English in the first generation, English dominates the second generation, and Spanish fades in the third generation (Vigdor 2008; Telles 2010).

## TABLE 10.1

# Twentieth-Century Genocide

| EVENT | YEARS | ESTIMATED NUMBER OF DEATHS |
|---|---|---|
| Turkish government's massacre of Armenians living in Turkey | 1915–1918 | 1.5 million |
| Joseph Stalin's (Russian dictator) massacre of nearly 25 percent of the population of Ukraine and of many other Eastern European countries (including Latvia, Lithuania, and Estonia) | 1932–1953 | 50+ million |
| Japanese Imperial Army's murder of inhabitants of China's capital city, Nanking | 1937–1938 | 300,000 |
| Nazi Holocaust in Germany (about 6 million Jews and 11 million Soviet civilians, homosexuals, Romani, political and religious opponents, and others) | 1938–1945 | 17 million |
| Khmer Rouge leader Pol Pot's slaughter of Cambodians | 1975–1979 | 2+ million |
| Murders of Muslims by Serb majority in former Yugoslavia | 1992–1995 | 200,000+ |
| Massacres of minority Tutsis by dominant Hutu group in Rwanda, Africa | 1994 (nine months) | 800,000 |

Sources: Niewyk and Nicosia 2000; United Human Rights Council 2004.

### Pluralism

In **pluralism**, sometimes called *multiculturalism*, minority groups retain their culture but have equal social standing in a society. One historian describes the United States as a pluralistic society because it is multicultural, multicolored, and multilingual. Thus, an American might have diverse neighbors such as a German architect and his Iranian wife, a Palestinian contractor, a Korean scientist, and a car salesman from Madagascar (Karnow 2004).

The United States is pluralistic because most racial and ethnic communities ("Little Italy," "Greek Town," "Little Korea," "Spanish Harlem") live peacefully side by side, have numerous ethnic newspapers and radio stations, and have the same constitutional rights (such as freedom of speech). On the other hand, people of various skin colors and cultures don't always experience the same social standing, and there is considerable racial-ethnic friction in the United States.

## 10-5 Some Sources of Racial-Ethnic Friction

In 2012, a Montana federal judge circulated a racist email that compared President Obama's mother to an animal because she gave birth to a biracial son. And soon after Whitney Houston, a highly successful black singer, died, a number of journalists, talk radio hosts, and Fox News reporters immediately attributed her death to a drug overdose because "most drug abusers are black," even though whites are more likely than blacks to be drug addicts (Adams 2012; DeVega 2012; McCarthy 2012; Uwimana 2012). A few of these people, including the judge, later apologized, but such public comments, especially by highly visible people, show that racism is still common in U.S. culture.

### 10-5a Racism

**Racism** is a set of beliefs that one's own racial group is naturally superior to other groups. It is a way of thinking about racial and ethnic differences that justifies and preserves the social, economic, and political interests of dominant groups (Essed and Goldberg 2002).

Whereas blacks see racism as continuing, many whites view it as a problem that has been pretty much "solved." After all, we have an African American president and a Latina Supreme Court justice, many law firms advertise summer positions that are limited to minority candidates, and minorities now own a large proportion of

small businesses. One outcome of such changes is that many whites now believe that antiwhite bias is a bigger social problem than antiblack bias (Bernstein 2011; Norton and Sommers 2011). You'll see that such beliefs aren't supported by data, but they fuel prejudice and discrimination.

### 10-5b Prejudice

**Prejudice** is an attitude, positive or negative, toward people because of their group membership. We often prejudge those who are different from us in race, ethnicity, or religion ("Asians are really hard workers" or "White people

> **pluralism** minority groups retain their culture but have equal social standing in a society.
>
> **racism** a set of beliefs that one's own racial group is naturally superior to other groups.
>
> **prejudice** an attitude, positive or negative, toward people because of their group membership.

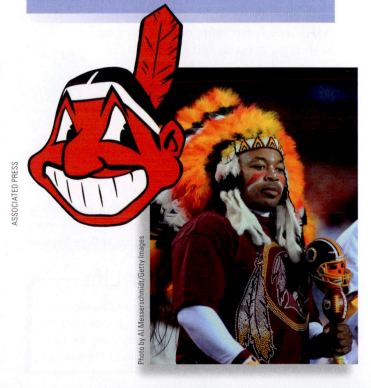

A number of colleges and professional teams have replaced their logos, nicknames, and mascots because many American Indians have denounced them as demeaning stereotypes. Some exceptions are the Cleveland Indians in baseball (left) and the Washington Redskins in football (right). Why do many people see the Indian images and team names as offensive? Why do you think that the Minnesota Vikings and University of Notre Dame's Fighting Irish are acceptable?

ASSOCIATED PRESS

Photo by Al Messerschmidt/Getty Images

CHAPTER 10: RACE AND ETHNICITY **185**

**stereotype** an oversimpli-fied or exaggerated generaliza-tion about a category of people.

**ethnocentrism** the belief that one's own culture, society, or group is inherently superior to others.

**scapegoats** individuals or groups whom people blame for their own problems or shortcomings.

**discrimination** any act that treats people unequally or unfairly because of their group membership.

**individual discrimina-tion** harmful action on a one-to-one basis by a member of a dominant group against a member of a minority group.

**institutional discrimi-nation** unequal treatment and opportunities that members of minority groups experience as a result of the everyday operations of a society's laws, rules, policies, practices, and customs.

can't be trusted"). Prejudice isn't one-sided because *any-one* can be prejudiced, use stereotypes, and engage in ethnocentric behavior.

A **stereotype** is an oversimplified or exagger-ated generalization about a category of people (see, for example, www.stuffwhite peoplelike.com). Stereo-types can be positive ("All African Americans are athletic") or negative ("All African Americans are lazy"). Whether positive or negative, stereotypes distort reality. Some blacks are great athletes and some are lazy, just like people in other groups. Once estab-lished, however, stereo-types are difficult to dis-card because people often dismiss any evidence to the contrary as an exception ("For an Asian, Jeremy Lin is a terrific basketball player").

**Ethnocentrism** is the belief that one's own culture, society, or group is inher-ently superior to others. If we are ethno-centric, we reject those outside our group as strange, deviant, and inferior: "*Our* customs, *our* laws, *our* food, *our* tra-ditions, *our* music, *our* religion, *our* beliefs and values, and so forth, are somehow better than those of other societies" (Smedley 2007: 32).

Stereotypes and ethnocentrism can result in a displacement of anger and aggression on **scapegoats**, individuals or groups whom people blame for their own problems or shortcomings ("They didn't hire me because the company wants blacks" or "I didn't get into that college because

Asians Americans are at the top of the list"). Minorities are easy scapegoating targets because they typically dif-fer in physical appearance and are usually too powerless to strike back (Allport 1954; Feagin and Feagin 2008).

At one time or another, almost all newcomers to the United States have been scapegoats. Especially in times of economic hardship, the most recent immi-grants often become scapegoats ("Latinos are replac-ing Americans in construction jobs"). Stereotypes, ethnocentrism, and scapegoating are attitudes, but they often lead to discrimination.

## 10-5c Discrimination

**Discrimination** is any act that treats people unequally or unfairly because of their group membership. Dis-crimination encompasses all sorts of actions, rang-ing from social slights (such as not inviting minority coworkers to lunch) to rejection of job applications and hate crimes. Discrimination can be subtle (such as not sitting next to someone) or blatant (such as racial slurs), and it occurs at individual and institutional levels.

**Individual discrimination** is harmful action on a one-to-one basis by a member of a dominant group against a member of a minority group. In a national survey, for example, more than half of blacks said that they face everyday discrimination when eating in restaurants, shopping, renting an apartment, buying a house, or applying for a job (Kohut et al. 2007).

In **institutional discrimination** (also called *insti-tutionalized discrimination*, *systemic discrimina-tion*, and *structural discrimination*), minority group members experience unequal treat-ment and opportunities as a result of the everyday operations of a society's laws, rules, policies, practices, and customs. Institutional discrimination is wide-spread. In health services, for instance, minorities tend to receive lower quality care than whites, even when they have private health insurance and are treated by the same doctors (*National Healthcare Disparities Report* 2003; Sequist et al. 2008).

maska/Shutterstock.com

## 10-5d Relationship Between Prejudice and Discrimination

About 65 years ago, sociologist Robert Merton (1949) described the relationship between prejudice and dis-crimination in a way that is still useful today. Merton's model includes four types of people and their possible response patterns (see *Table 10.2*).

## Sociology in Your Life

Do you have unconscious stereotypes and biases? Find out by taking the Implicit Association Test by going to Sociology CourseMate at CengageBrain.com, but note the site's comments about the validity of the results.

*"You look like this sketch of someone who's thinking about committing a crime."*

© David Sipress/The New Yorker Collection from cartoonbank.com

*Unprejudiced nondiscriminators* are "all-weather liberals": They aren't prejudiced and don't discriminate. They believe in the American creed of freedom and equality for all and cherish egalitarian values. They may not do much, however, individually or collectively, to change discrimination. In contrast, but equally consistent in attitude and action, are *prejudiced discriminators*, or "active bigots." They are prejudiced, and they discriminate. They are willing to break laws to express their beliefs and protect their vested interests.

*Unprejudiced discriminators* are "fair-weather liberals": They aren't prejudiced, but they discriminate because it's expedient or in their own self-interest to do so. If, for example, an insurance company charges higher automobile insurance premiums to people with low occupational and educational levels (often minorities), agents will implement these policies even though they themselves aren't prejudiced.

## TABLE 10.2

# Relationship Between Prejudice and Discrimination

| | | DOES THE PERSON DISCRIMINATE? | |
|---|---|---|---|
| | | Yes | No |
| **IS THE PERSON PREJUDICED?** | Yes | Prejudiced discriminator (e.g., a prejudiced person who attacks minority group members verbally or physically) | Prejudiced nondiscriminator (e.g., a prejudiced person who goes along with equal employment opportunity policies) |
| | No | Unprejudiced discriminator (e.g., an unprejudiced person who joins a club that excludes minorities) | Unprejudiced nondiscriminator (e.g., an unprejudiced employer who hires minorities) |

Source: Based on Merton 1949.

*Prejudiced nondiscriminators* are "timid bigots": They're prejudiced, but don't discriminate. Despite their negative attitudes, they hire minorities and are civil in everyday interactions because they believe they must conform to antidiscrimination laws or situational norms. If, for example, most of their neighbors or coworkers don't discriminate, prejudiced nondiscriminators will go along with them.

## 10-6 Major Racial and Ethnic Groups in the United States

Americans absorb many aspects of immigrants' cultures: "Our everyday lexicon is sprinkled with Spanish words. We are now just as likely to grab a burrito as a burger. Hip-hop is tinged with South Asian rhythms. And Chinese New Year and Cinco de Mayo are taking their places alongside St. Patrick's Day as widely celebrated American ethnic holidays" (Jiménez 2007: M1). Of the major U.S. racial-ethnic groups, some encounter more barriers than others, but all have numerous strengths that enhance U.S. society.

### 10-6a European Americans: A Declining Majority

During the seventeenth century, English immigrants settled the first colonies in Massachusetts and Virginia. Other white Anglo-Saxon Protestants (WASPs), who included people from Wales and Scotland, quickly followed. Most of these groups spoke English. Some of the immigrants were affluent, but many were poor or had criminal backgrounds.

#### Diversity

About 58 percent of the U.S. population has a European background. The largest groups have ancestors from Germany, Ireland, England, Italy, Poland, France, and the Scandinavian countries (see *Table 10.3*).

#### Constraints and Strengths

WASPs generally looked down on later waves of immigrants from southern and eastern Europe. They viewed the newcomers as inferior, dirty, lazy, and uncivilized because they differed in language, religion, and customs. New England, which was 90 percent Protestant, was particularly

## TABLE 10.3

## Americans of European Descent, 2009

Of the many European ancestries that Americans report, the following comprise at least 1.5 percent of the U.S. population.

| ANCESTRY | NUMBER (IN MILLIONS) | PERCENTAGE OF TOTAL U.S. POPULATION |
|----------|----------------------|-------------------------------------|
| German | 50.7 | 16.5 |
| Irish | 36.9 | 12.0 |
| English | 27.7 | 9.0 |
| Italian | 18.1 | 5.8 |
| Polish | 10.1 | 3.3 |
| French | 9.4 | 3.0 |
| Scottish | 5.8 | 1.9 |
| Dutch | 5.0 | 1.6 |
| Norwegian | 4.6 | 1.5 |

Note: This survey was based on a total estimated population of almost 307.1 million Americans.

Source: Based on U.S. Census Bureau 2012, Table 52.

hostile to Irish Catholics, characterizing them as irresponsible and shiftless (Feagin and Feagin 2008).

All the later waves of European immigrants faced varying degrees of hardship in adjusting to the new land because the first English settlers had a great deal of power in shaping economic and educational institutions. In response to prejudice and discrimination, many of the immigrants founded churches, schools, and recreational activities that maintained their language and traditions (Myers 2007).

Despite stereotypes, prejudice, and discrimination, European immigrants began to prosper within a few generations. They surmounted numerous obstacles and became influential in all sectors of American life. Overall, they now fare much better financially than most other groups in the United States. This doesn't mean that all are rich. In fact, in absolute numbers, poor whites outnumber those of other racial-ethnic groups (see Chapter 8).

## 10-6b Latinos: A Growing Minority

About 1 in 3 Americans is a member of a racial or ethnic minority group, but Latinos are the fastest growing group and constitute almost 17 percent of the nation's population. The growth of the Latino population this century is due mainly to births in the United States, not recent immigration, because, on average, Latinas have three births each, one more than white, black, and Asian or Pacific Islander women (Saenz 2010; Taylor and Lopez 2011; U.S. Census Bureau Population Division 2012).

### Diversity

More than half of the growth in the total U.S. population between 2000 and 2011 was due to the increase in the Latino population. Worldwide, only Mexico (112 million) has a larger Latino population than the United States (52 million) (Ennis et al. 2011; U.S. Census Bureau Population Division 2012). And more persons of Puerto Rican origin now live in the 50 states and the District of Columbia than in Puerto Rico (Lopez and Velasco 2011b).

Some Latinos trace their roots to the Spanish and Mexican settlers who established homes and founded cities in the Southwest before the arrival of the first English settlers on the East Coast. Others are recent immigrants or children of the immigrants who arrived in large numbers at the beginning of the twentieth century. Of the Latinos living in the United States, most are from Mexico (see *Figure 10. 3*), but Spanish-speaking people from many different countries vary widely in their customs, cuisines, and cultural practices.

### Constraints and Strengths

Most Latinos are taking their place in mainstream America, but many still encounter obstacles. The median

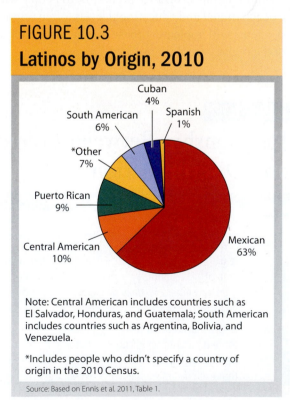

## FIGURE 10.3

## Latinos by Origin, 2010

Note: Central American includes countries such as El Salvador, Honduras, and Guatemala; South American includes countries such as Argentina, Bolivia, and Venezuela.

*Includes people who didn't specify a country of origin in the 2010 Census.

Source: Based on Ennis et al. 2011, Table 1.

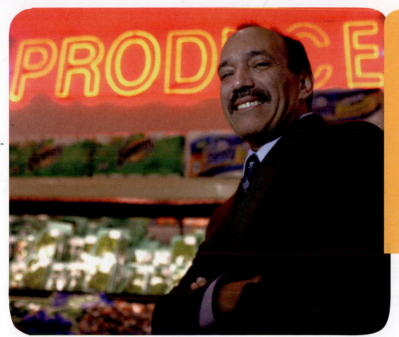

Jeffrey Macmillian for U.S. News & World Report

Dominican-born Alfredo Rodriguez is one of numerous successful Latino businessmen who are rebuilding neglected inner-city neighborhoods. In 1985, Rodriguez bought his first grocery store in Queens, New York, with the $25 a week his mother had been setting aside for him for a decade. In 2002, he purchased a supermarket in Newark, New Jersey, to meet the needs of local Latino shoppers. Five years later, his Xtra Supermarket had annual sales of $9 million (Rayasam 2007).

household income of Latinos is 59 percent of that of Asian Americans, but higher than that of American Indians and African Americans (see *Figure 10.4*). Almost 39 percent of Latino families earn $50,000 a year or more, up considerably from only 7 percent in 1972, but almost 25 percent of Mexican Americans and Puerto Ricans and 17 percent of Cuban Americans live below the poverty line (U.S. Census Bureau 2012). For the first time in U.S. history, more Latino children are living in poverty—6.1 million in 2010—than children of any other racial or ethnic group, including whites (Lopez and Velasco 2011a).

## FIGURE 10.4
## U.S. Median Household Income by Race and Ethnicity, 2010

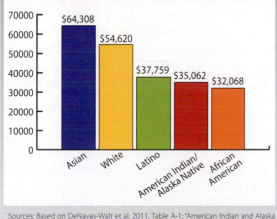

Sources: Based on DeNavas-Walt et al. 2011, Table A-1; "American Indian and Alaska Native..." 2011.

As with other groups, the socioeconomic status of Latinos reflects a number of interrelated factors, especially education level, English language proficiency, recency of immigration, and occupation. Many Latinos who were professionals in their native land find only low-paying jobs. Often they don't have time to both work and learn English well enough to pass accreditation exams to practice as doctors, lawyers, and accountants. In Miami, for example, many Cuban physicians have learned English and obtained licenses to practice medicine. But, says one Cuban American doctor, "I know neurosurgeons who are working in warehouses or factories or as gas attendants" (Ojito 2009: D1).

Despite their generally lower economic and educational attainment (see Chapters 11 and 13), many Latinos are successful. They own almost 1.6 million businesses, or 15 percent of all U.S. businesses. Between 1995 and 2005, many foreign-born Latino immigrants earned better hourly wages than their predecessors because they tended to be older, better educated, and more likely to be employed in construction than in agriculture (*Hispanic-Owned Firms . . .* 2006; Kochhar 2007).

Among their many contributions, Latinos have breathed life into a number of small Midwestern towns that "seemed to be staggering toward the grave" because of declining numbers of white residents. The newcomers have reopened shuttered storefronts, bought long-vacant houses, and saved some elementary and middle schools from closing because of low native-born enrollments (Sulzberger 2011: A1).

## 10-6c African Americans: A Major Source of Diversity

The 44 million African Americans are the second largest minority group in the United States, making

Michael Ochs Archives/Getty Images; AP Images

Few people know that Madame C. J. Walker (1867–1919) was a manufacturer of hair care products for African American women and one of the first American women to become a millionaire. Or that Dr. Charles R. Drew (1904–1950) was a renowned surgeon, teacher, and researcher. He was responsible for founding two of the world's largest blood banks, saving untold lives during and since World War II.

up almost 14 percent of the population (U.S. Census Bureau 2012). This percentage includes those of more than one race.

## Diversity

Most African Americans share a common characteristic: They are members of the only group ever brought to the United States involuntarily and legally enslaved. The term *African American* encompasses tremendous diversity, including native-born Americans with black, white, American Indian, and/or Latino ancestors, as well as recent immigrants from Africa and elsewhere. Among foreign-born blacks, nearly two-thirds are from the Caribbean or a Central or South American country, nearly one-third were born in Africa, and the remainder has come from Europe and other regions (Kent 2007).

## Constraints and Strengths

The effects of 350 years of slavery and legal segregation are still evident. African Americans make up 14 percent of the U.S. population but 27 percent of those living in poverty (U.S. Census Bureau 2012). The median family income of African Americans is the lowest of all racial-ethnic groups (see *Figure 10.4* on p. 189). Especially in the country's inner cities, many young black men are high school dropouts, jobless, or incarcerated (Mincy 2006; U.S. Census Bureau 2012).

The gap in educational achievement between whites and blacks has narrowed during the last four decades. For example, the proportion of African Americans who earned at least a high school diploma increased from 26 percent in 1964 to 82 percent in 2010. During the same period, those with at least a college degree increased from 4 percent to 18 percent (U.S. Census Bureau 2012). Some economists estimate that at least 30 percent of the black-white wage gap is due to differential treatment in hiring, firing, and pay rather than differences in formal schooling or specific job skills (Fryer et al. 2011).

Despite such discrimination and centuries of oppression, many African Americans are successful. A third of all black households have annual incomes of $50,000 or more. Blacks own almost 6 percent of U.S. businesses. Between 2002 and 2007, the number of black-owned businesses increased by 61 percent to almost 2 million, more than triple the overall rate of 18 percent, and these businesses generated almost $137 billion in sales in 2007 (*Black-Owned Firms . . .* 2006; "Census Bureau Reports . . ." 2010; U.S. Census Bureau 2012).

## 10-6d Asian Americans: A Model Minority?

The more than 18 million Asian Americans comprise almost 6 percent of the U.S. population. As a group, Asian immigrants are more likely than their counterparts from non-Asian countries to report speaking English "well" or "very well" and to become U.S. citizens (Batalova 2011; Humes et al. 2011; U.S. Census Bureau Population Division 2012).

## Diversity

Asian Americans encompass a broad swath of cultural groups. They come from at least 26 countries in East and Southeast Asia (including China, Taiwan, Korea, Japan, Vietnam, Cambodia, and the Philippines) and South Asia (especially India, Pakistan, and Sri Lanka), and there are at least 19 Asian languages spoken in the United States ("Asian/Pacific American . . ." 2011).

These diverse origins mean that there are tremendous differences in languages and dialects (and even alphabets), religions, cuisines, and customs. Asian Americans also include Native Hawaiian and other Pacific Islanders from Guam and Samoa. Chinese form the largest Asian American group, followed by Filipinos and Asian Indians (see *Figure 10.5*). Combined, these three groups account for 55 percent of all Asian Americans.

Charlie Edward/Shutterstock.com

## Constraints and Strengths

The most successful Asian Americans are those who speak English relatively well *and* have high educational levels. Asian Americans have the highest median income of all U.S. racial-ethnic groups (see *Figure 10.4* on p. 189). Asian Indians—two-thirds of whom have advanced degrees—have the highest annual median household income (almost $74,000) of Asian American subgroups, compared with less than $40,000 for Cambodians, many of whom are less educated, experience language barriers, and have few marketable skills (U.S. Census Bureau 2010).

Overall, Asian Americans have higher educational levels than any other U.S. racial-ethnic group. Half have at least a college degree compared with 28 percent of the general population. They are also more likely than other racial-ethnic groups to be concentrated in highly skilled occupations such as information technology, science, engineering, and medicine (Batalova 2011; "Asian/Pacific American . . ." 2012).

Because of their educational and economic success, Asian Americans are often hailed as a "model minority." Such labels are misleading, however, because there is considerable variation across subgroups. For example, among Asian Americans age 25 and older, more than 63 percent of Hmong, Laotians, and Cambodians and 50 percent of Native Hawaiians and Vietnamese haven't attended college, compared with only about 23 percent of Filipinos and Asian Indians (Teranishi 2011; see also Chapter 13).

Asian Indians own about 43 percent of the 47,000 hotels and motels in the United States. In many cases, the owners bought run-down lodgings and converted them to upscale Sheraton and Hilton hotels (Yu 2007). Asian Americans have also been more successful than other minority groups in penetrating corporate suites. In 1995, for example, all of the *Fortune 500* CEOs were white. By 2008, four were black men, five were Latino men, and seven were Asian Americans, two of them women (DiversityInc 2008).

## 10-6e American Indians: A Growing Nation

American Indians used to be called the "vanishing Americans," but they have "staged a surprising comeback" due to higher birth rates, a longer life expectancy, and better health services (Snipp 1996: 4). The 6.3 million American Indians and Alaska Natives make up almost 2 percent of the U.S. population and are expected to increase to about 9 million by 2050 ("American Indian and Alaska Native . . ." 2011).

### Diversity

Like Asian Americans, American Indians and Alaska Natives (AIANs) are a heterogeneous group. Of the 565 federally recognized tribes, 8 have more than 100,000 members. The Cherokee, with almost 820,000 members, is the largest, followed by the Navajo and Choctaw

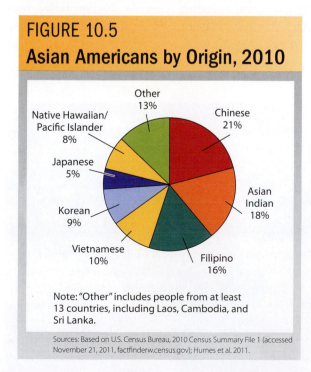

## FIGURE 10.5
## Asian Americans by Origin, 2010

- Other 13%
- Native Hawaiian/Pacific Islander 8%
- Japanese 5%
- Korean 9%
- Vietnamese 10%
- Filipino 16%
- Chinese 21%
- Asian Indian 18%

Note: "Other" includes people from at least 13 countries, including Laos, Cambodia, and Sri Lanka.

Sources: Based on U.S. Census Bureau, 2010 Census Summary File 1 (accessed November 21, 2011, factfinderw.census.gov); Humes et al. 2011.

(Norris et al. 2012). Tribes speak 150 native languages (although many are quickly vanishing) and vary widely in their religious beliefs and cultural practices. Thus, a Comanche-Kiowa educator cautions, "Lumping all Indians together is a mistake. Tribes . . . are sovereign nations and are as different from another tribe as Italians are from Swedes" (Pewewardy 1998: 71; Ashburn 2007).

### Constraints and Strengths

AIANs are a unique minority group because they're not immigrants and have been in what is now the United States longer than any other group. Nevertheless, they have experienced centuries of subjugation, exploitation, and political exclusion (such as not having the right to vote until 1924) and have endured a legacy of broken treaties, stolen lands, and tribal extinction (Wilkinson 2006).

Many AIANs are better off today than they were a decade ago, but long-term institutional discrimination has been difficult to shake. For example, 23 percent live below the poverty line compared with 11 percent of the general population. The median household income of AIANs is slightly higher than that of African American households, but lower than that of other racial-ethnic groups (see *Figure 10.4* on p. 189). This group's educational levels have increased, but only 13 percent of AIANs have a bachelor's degree or higher, compared with 28 percent of the general population (U.S. Census Bureau 2012).

Despite numerous obstacles, AIANs have made considerable economic progress by insisting on self-determination and the rights of tribes to run their own affairs. American Indian tribes now own 37 percent of the U.S. gambling industry. Few tribes benefit, however, because many casinos are in remote areas that don't attract tourists. In other cases, powerful tribes have cast out some clans to increase their own share of the profits (Dadigan 2011; Dao 2011). Outside of gaming, the number of AIAN-owned businesses grew from fewer than 5 in 1969 to nearly 237,000 in 2007, most of them in construction and repair, maintenance, and personal and laundry services (Taylor and Kalt 2005; Kestin and Franceschina 2007; "American Indian and Alaska Native . . ." 2011).

## 10-6f Middle Eastern Americans: An Emerging Group

The Middle East is "one of the most diverse and complex combinations of geographic, historical, religious, linguistic, and even racial places on Earth" (Sharifzadeh 1997: 442). The Middle East encompasses about 30 countries, including Armenia, Turkey, Israel, Iran, Afghanistan, Pakistan, and 22 Arab nations (such as Algeria, Iraq, Kuwait, Saudi Arabia, and the United Arab Emirates).

Mario Tama/Getty Images

Mohegan Sun in southern Connecticut is one of the largest casinos in the United States. It has spent some of its profits on college scholarships, a $15 million senior center, and health insurance for tribal members. In contrast, some of the poorest tribes, such as the Navajo and Hopi, who have rejected gaming for religious reasons, have many members who live in poverty without kitchen facilities (such as stoves or refrigerators) or indoor plumbing.

AP Images

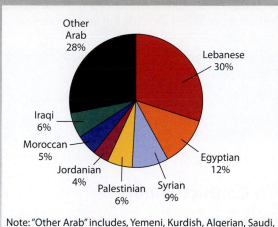

Other Arab 28%

Lebanese 30%

Iraqi 6%

Moroccan 5%

Jordanian 4%

Palestinian 6%

Syrian 9%

Egyptian 12%

Note: "Other Arab" includes, Yemeni, Kurdish, Algerian, Saudi, Tunisian, Kuwaiti, Libyan, Berber, United Arab Emirati, Omani, Qatari, Bahraini, Alhuceman, Bedouin, Rio de Oro, and others from the Middle East and North Africa.

Source: Based on U.S. Census Bureau 2012, Table 52.

## Diversity

Those from the Middle East comprise a heterogeneous population that is a "multicultural, multiracial, and multiethnic mosaic" (Abudabbeh 1996: 333). Most are Muslims, but many are Christians or Jews. Arabic is the most common language, but people from the Middle East also speak Turkish, Farsi, Kurdish, and other languages. People from the Middle East encompass a multitude of ethnic and linguistic groups with very different customs and cultural practices.

Most Middle Eastern Americans are of Arab ancestry. About 1.2 million Americans, less than 0.5 percent of the total population, report that their ancestry is solely or partly Arab. Almost half were born in the United States, but nearly half arrived during the 1990s (de la Cruz and Brittingham 2003; Brittingham and de la Cruz 2005). Those who identify themselves as Arab Americans come from many different countries (see *Figure 10.6*).

## Constraints and Strengths

Middle Eastern Americans tend to be better educated and wealthier than other Americans. Arab Americans are nearly twice as likely as the average U.S. resident to have a college degree: 41 percent compared with 24 percent. As in other groups, there are wide variations. For example, Lebanese have higher median family incomes (almost $61,000) than Moroccans ($41,000). Because of their generally higher educational levels, the

proportion of Arab Americans working in management jobs is higher than the proportion of all Americans: 42 percent versus 34 percent. Not all Middle Eastern Americans are successful, of course. Lebanese and Syrians have the lowest poverty rates at 11 percent, compared with more than 26 percent of those from Iraq and many other Middle East countries (Brittingham and de la Cruz 2005).

Generally, Middle Eastern Americans are well integrated into American life. Besides their high educational levels, three of four speak only English at home or speak it very well, more than half are homeowners, 75 percent of the men are in the labor force, and more than half are U.S. citizens (Brittingham and de la Cruz 2005).

# 10-7 Sociological Explanations of Racial-Ethnic Inequality

As with other topics, the four major sociological theories help us understand racial-ethnic relations. (*Table 10.4* summarizes the key points of each perspective.) Each has its strengths and weaknesses.

## 10-7a Functionalism

Those who criticize immigrants for not becoming Americanized quickly enough reflect a functionalist view of racial-ethnic relations. That is, if a society is to work harmoniously, newcomers must assimilate by adopting the dominant group's values, goals, and especially, language. Otherwise, a society will experience discord and conflict rather than social solidarity.

## TABLE 10.4

# Sociological Explanations of Racial-Ethnic Inequality

| THEORETICAL PERSPECTIVE | LEVEL OF ANALYSIS | KEY POINTS |
|---|---|---|
| **Functionalist** | Macro | Prejudice and discrimination can be dysfunctional, but they provide benefits for dominant groups and stabilize society. |
| **Conflict** | Macro | Powerful groups maintain their advantages and perpetuate racial-ethnic inequality primarily through economic exploitation. |
| **Feminist** | Macro and micro | Minority women suffer from the combined effects of racism and sexism. |
| **Symbolic Interactionist** | Micro | Hostile attitudes toward minorities, which are learned, can be reduced through cooperative interracial and interethnic contacts. |

© Cengage Learning

### Stability and Cohesion

From a functionalist perspective, racial-ethnic inequality sustains a pool of cheap labor for jobs that require little or no training. Many farmers rely on immigrants to work in fields, orchards, and vineyards at low wages. Otherwise, many crops would rot in the fields, leading to farmers' economic ruin and increasing the cost of food (Bustillo 2006). Thus, dominant group members benefit from inequality by avoiding undesirable jobs, and minorities with low educational or skill levels find employment.

Racial-ethnic inequality also maintains or increases many dominant group members' current status, power, and profits. Keeping minorities from acquiring higher paying jobs, as well as education and housing, makes more of these scarce resources available to the dominant group. In inner-city neighborhoods, as you saw in Chapter 8, fringe bankers (check advance stores and pawnshops) profit by offering financial services to low-income residents who can't get bank loans or credit cards.

### Critical Evaluation

Functionalism helps us understand how inequality benefits some groups over others and why, as a result, it persists. Functionalists acknowledge that inequality is dysfunctional; for example, racism prevents a society from recognizing or rewarding talented people who could make important contributions. This isn't functionalists' major focus, however. By emphasizing that racial-ethnic inequality helps to maintain a society's stability, functionalists seem to accept discrimination as inevitable (Chasin 2004). Another weakness is that functionalists don't explain why many members of some minority groups—especially blacks, Latinos, and American Indians—who have similar educational levels to those of whites are still at the bottom of the economic ladder.

## 10-7b Conflict Theory

Conflict theorists see ongoing strife between dominant and minority groups. Dominant groups try to protect their power and privilege, whereas subordinate groups struggle to gain a larger share of societal resources. Once a system of racial oppression is in place (as through segregation), racial hierarchies are supported and perpetuated through economic inequality, which reinforces current social stratification (see Chapter 8).

### Economic and Social Class Inequality

For most conflict theorists, economic inequality generates racial-ethnic inequality. According to a classic explanation, there's a "split labor market." Jobs in the *primary labor market*, held primarily by white workers, provide better wages, health and pension benefits, and some measure of job security. In contrast, workers in the *secondary labor market* (such as fast-food employees) are largely minorities and easily replaced. Their wages are low, there are few fringe benefits, and working conditions are generally poor (Doeringer and Piore 1971; Bonacich 1972).

Such economic stratification pits minorities against each other and low-income whites. Because these groups compete with each other instead of uniting against exploitation, capitalists don't have to worry about increasing wages or providing safer work environments. Some conflict theorists maintain that race is a more important factor than social class in perpetuating inequality. They note, for example, that even middle-class African Americans (and their counterparts in other racial-ethnic groups) experience discrimination on a daily basis that reminds them of their subordinate position in U.S. society (Feagin and Sikes 1994).

### Critical Evaluation

Conflict theories help explain why discrimination occurs and persists, as when members of privileged groups benefit by subordinating minorities economically. However, conflict theorists often assume that racial inequality is conscious, deliberate, widespread, and inescapable. In contrast, large majorities of both African Americans and Latinos say that immigration and income, not race or ethnicity, are the primary sources of social conflict in the United States (Morin 2009). Such data suggest a substantial consensus among some minorities that racial inequality is not due entirely to racism.

There is ample evidence that racial inequities persist in employment, but the discrimination isn't always as conscious as some conflict theorists maintain. For example, a study of 700 retail stores found that store managers tended to hire members of their own racial or ethnic group. The researchers attributed the outcomes to factors such as segregated neighborhoods and hiring networks (i.e., institutional discrimination), rather than to prejudice and deliberate discrimination (Giuliano et al. 2009).

## 10-7c Feminist Theories

Walk through almost any hotel, large discount store, nursing home, or fast-food restaurant in the United States. You'll notice two things: Most of the low-paid employees are women, and predominantly minority women. For feminist scholars, such segregation of minority women is due to gendered racism.

### Gendered Racism

**Gendered racism** refers to the overlapping and cumulative effects of inequality due to racism *and* sexism. Many white women encounter discrimination on a daily basis (see Chapter 9). Minority women, however, are also members of a racial-ethnic group, bringing them a double dose of inequality. If social class is included, some minority women experience *triple oppression*. Many affluent women, in particular, have no qualms about exploiting recent immigrants, especially Latinas, who perform demanding housework at very low wages (Segura 1994; Hondagneu-Sotelo 2001).

Gendered racism also occurs *within* racial-ethnic groups. According to a black male sociologist, for example, scholars rarely discuss black male privilege, which is characterized by having advantages over black women. Examples include being promoted more often and getting higher pay than their black female counterparts who are equally skilled and educated (National Public Radio 2010).

### Critical Evaluation

Feminist perspectives have sharpened our understanding of the effects of gendered racism and minority women's

subordinate status in U.S. society (Newman 2005). Like conflict theorists, however, feminist scholars often assume that gendered inequality is deliberate. In some cases, as you saw earlier, gender discrimination can be unintentional.

**gendered racism** the overlapping and cumulative effects of inequality due to racism *and* sexism.

Because all of us have internalized institutional discrimination, minority group members may also be guilty of reinforcing gendered racism in schools, workplaces, and other situations. For example, African American girls as young as 15 report having to cope with black men's sexual harassment, stereotypes that black girls and women are sexually promiscuous, and many black men's preference for women with lighter complexions, facial features and hair that are closer to European standards of beauty (Friedman 2011; Thomas et al. 2011). A related weakness is that feminist explanations seldom explore oppression of minorities by other minorities (such as affluent African American women and Latinas who exploit domestic and farm workers).

## 10-7d Symbolic Interactionism

According to symbolic interactionists, we learn attitudes, norms, and values throughout the life course. Because, as you saw earlier, race and ethnicity are

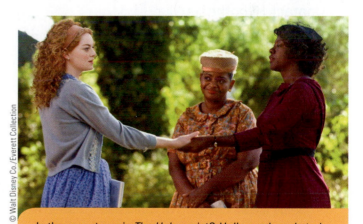

Is the recent movie *The Help* racist? Hollywood marketed the film as a progressive story of triumph over racial injustice as a white woman forges a bond with black maids and exposes the racism they are faced with during the early 1960s. The Association of Black Women Historians, however, denounced the movie as distorting, ignoring, and trivializing the experiences of black domestic workers and black life. What do you think?

© Walt Disney Co./Everett Collection

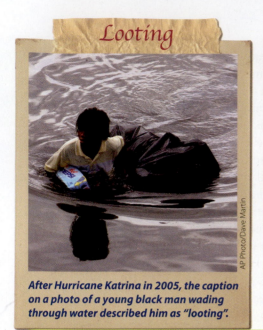

**Looting**

*After Hurricane Katrina in 2005, the caption on a photo of a young black man wading through water described him as "looting".*

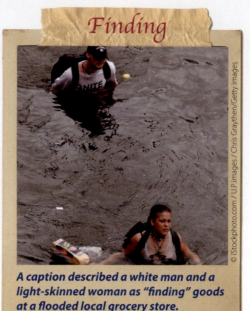

**Finding**

*A caption described a white man and a light-skinned woman as "finding" goods at a flooded local grocery store.*

For symbolic interactionists, images shape our perceptions of racial and ethnic groups.

against that group. Such contacts are most effective when dominant and minority group members have approximately the same status (such as coworkers or bosses), when they share common goals (such as working on a project), when they cooperate rather than compete, and if an authority figure supports intergroup interaction (such as when an employer requires white supervisors to mentor minority workers) (Allport 1954; Kalev et al. 2006).

### Critical Evaluation

Symbolic interactionism is valuable in helping us understand how race and ethnicity shape our everyday lives. If we're aware of the negative effects of labeling and selective perception, for example, we can change our attitudes and behavior. It's not clear, however, why labeling, selective perception, and racial bias are more common among some people than others, especially when they're similar on a number of variables such as social class, gender, age, religion, race, and ethnicity (Dovidio 2009).

Another weakness is that symbolic interactionism tells us little about the social structures that create and maintain racial-ethnic inequality. For instance, people who aren't prejudiced can foster discrimination by simply going along with the inequitable policies that have been institutionalized in education, the workplace, and other settings (i.e., the unprejudiced discriminators described earlier).

**contact hypothesis** the idea that the more people get to know members of a minority group personally, the less likely they are to be prejudiced against that group.

constructed socially, labeling, selective perception, and social contact can have powerful effects on everyday intergroup relations.

### Labeling, Selective Perception, and the Contact Hypothesis

We learn attitudes toward dominant and minority groups through labeling and selective perception, both of which can increase prejudice and discrimination. For example, a comprehensive study of major U.S. news magazines (such as *Time* and *U.S. News & World Report*) concluded that labeling immigrants as a "problem" and a "menace" ignored "the broader array of Latino roles and contributions to American communities" (Gavrilos 2006: 4; see also "Media Coverage of Hispanics" 2009).

Some media critics have lambasted black sitcoms (such as *House of Payne* and *Let's Stay Together*) as stereotypical and artificial portrayals of African American life (Weisbuch et al. 2009; Abrams 2011; Crockett 2011). Both positive and negative images influence many viewers' perceptions of race and reinforce racial bias.

People can decrease labeling and selective perception. For example, the **contact hypothesis** posits that the more people get to know members of a minority group personally, the less likely they are to be prejudiced

## 10-8 Interracial and Interethnic Relationships

President Obama, the son of a black father from Kenya and a white mother from Kansas, identified himself as "black" on the 2010 census questionnaire (Roberts and Baker 2010). Although there was no category specifically for mixed race or biracial, he could have checked off both black and white. The president is just one of the growing number of Americans who are biracial or multiracial.

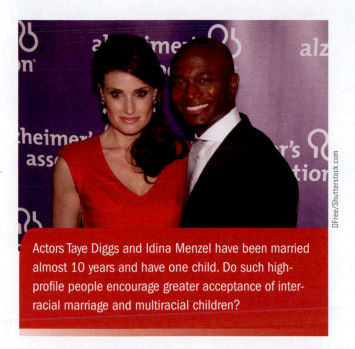

Actors Taye Diggs and Idina Menzel have been married almost 10 years and have one child. Do such high-profile people encourage greater acceptance of interracial marriage and multiracial children?

## 10-8a Growing Multiracial Diversity

The 2000 U.S. Census for the first time allowed people to mark more than one race. In 2010, 97 percent of Americans reported being only one race, but almost 3 percent (9 million people) self-identified as being of two or more races, up from 2.4 percent in 2000. Every state saw its multiple-race population jump by at least 8 percent. Overall, 3 percent of whites, 7 percent of African Americans, 8 percent of Latinos, 15 percent of Asians, 44 percent of American Indians/Alaska Natives, and 56 percent of Native Hawaiians/Pacific Islanders reported two or more races (Humes et al. 2011).

## 10-8b Interracial Dating and Marriage

Laws against **miscegenation**, marriage or sexual relations between a man and a woman of different races, existed in America as early as 1661. It wasn't until 1967, in the U.S. Supreme Court's *Loving v. Virginia* decision, that antimiscegenation laws were overturned nationally. Recently, however, a small Baptist congregation in Kentucky voted to ban interracial couples from becoming members or participating in worship activities (Waldron 2011).

In 1958, only 4 percent of Americans approved of black-white marriages, compared with 86 percent in 2008 (Jones 2011). An overwhelming majority (85 percent) of young adults born in 1981 and later say

that they'd be fine with a family member's marriage to someone of a different racial or ethnic group (Keeter et al. 2010).

Attitudes about interracial marriages are changing, but what about behavior? Racial-ethnic intermarriages have increased slowly—from only 0.7 percent of all marriages in 1970 to 10 percent in 2010. In 2010, also, 18 percent of heterosexual unmarried couples were of different races and 21 percent of same-sex couples were interracial (Fields and Casper 2001; Lofquist et al. 2012; Wang 2012).

There's some evidence that intermarriage rates may increase in the future. In 2010, for example, a record 15 percent (1 in 7) of new marriages were between spouses of different races or ethnicities, but the lowest intermarriage rates continue to be between blacks and whites (Wang 2012).

The increase in intermarriage reflects many interrelated factors—both micro and macro level—that include everyday contact and changing attitudes. For example, we tend to date and marry people we see on a regular basis. The higher the educational level, the greater the potential for intermarriage, because educated minority group members often attend integrated colleges, and their workplaces and neighborhoods are more integrated than in the past (Qian 2005; Passel et al. 2010).

People often marry outside of their racial-ethnic groups because of a shortage of potential spouses within their own group. Because the Arab American population is so small, for example, 80 percent of U.S.-born Arabs have non-Arab spouses. In contrast, intermarriage rates for Latinos and Asian Americans have decreased since 1990—for both women and men—because the influx of new immigrants has provided a larger pool of eligible mates (Kulczycki and Lobo 2002; Qian and Lichter 2011).

**miscegenation** marriage or sexual relations between a man and a woman of different races.

## Study Tools

Ready to study? Go to **Sociology CourseMate** at **www.cengagebrain.com** to complete practice quizzes, review flashcards, watch videos, and more.

# SOC

The economy and politics are closely linked social institutions that affect all of us.

BRIAN NGUYEN/Reuters/Landov

# The Economy and Politics

In 2012, almost half of all Americans said that they were worse off financially than in the past and that they couldn't get ahead through hard work (Saad 2012). Almost 83 percent said they have less trust in politics than they did 10 or 15 years ago, and only 8 percent described the economy as "good" or "excellent," compared with 38 percent in 2004 ("Frustration with Congress . . ." 2011; Ford 2012).

Such national data illustrate the close linkage between a person's life; the **economy**, a social institution that determines how a society produces, distributes, and consumes goods and services; and **politics**, a social institution through which individuals and groups acquire and exercise power and authority and make decisions. Societies worldwide differ in the kinds of economic and political systems they develop because of many factors such as globalization and a revolution in communications technology. Let's begin by looking at global economic systems.

> **economy** a social institution that determines how a society produces, distributes, and consumes goods and services.
>
> **politics** a social institution through which individuals and groups acquire and exercise power and authority and make decisions.
>
> **capitalism** an economic system in which wealth is in private hands and is invested and reinvested to produce profits.

## Key Topics

In this chapter, we'll explore the following topics:

- **11-1** Global Economic Systems
- **11-2** Corporations and the Economy
- **11-3** Work in U.S. Society Today
- **11-4** Sociological Explanations of Work and the Economy
- **11-5** Global Political Systems
- **11-6** Politics, Power, and Authority
- **11-7** Politics and Power in U.S. Society
- **11-8** Sociological Perspectives on Politics and Power

## 11-1 Global Economic Systems

The two major economic systems around the world are capitalism and socialism. In actual practice, economies are usually some mixture of both.

### 11-1a Capitalism

**Capitalism** is an economic system in which wealth is in private hands and is invested and reinvested to produce profits. Ideally, capitalism has four essential characteristics (Smith 1937; Heilbroner and Thurow 1998):

- *Private ownership of property.* Property—such as real estate, banks, and utilities—belongs to individuals or organizations rather than the state or the community.

- *Competition.* Capitalists compete in producing goods and services that offer consumers the greatest value in price and quality.

- *Profit.* Selling something for more than it costs to produce generates profits and an accumulation of wealth for individuals and companies.

- *Investment.* By investing profits, capitalists can increase their own wealth. Workers also can accumulate savings and make investments that supplement their income.

## What do you think?

U.S. politics decreases people's economic well-being.

$$\boxed{1 \quad 2 \quad 3 \quad 4 \quad 5 \quad 6 \quad 7}$$

strongly agree        strongly disagree

**monopoly** domination of a particular market or industry by one person or company.

**oligopoly** domination of a market by a few large producers or suppliers.

**socialism** an economic system based on the principle of the public ownership of the production of goods and services.

**communism** a political and economic system in which all members of a society are equal.

In reality, capitalism doesn't function ideally because of abuses, greed, and worker exploitation. It usually results in monopolies and oligopolies rather than a free market that encourages competition. Federal laws prohibit the establishment of a **monopoly**, the domination of a particular market or industry by one person or company. With little or no competition, a company can raise prices and reduce production.

An **oligopoly**, which is legal, is the domination of a market by a few large producers or suppliers. For example, six companies—including CBS, Time Warner, and the Walt Disney Corporation—now own 90 percent of the U.S. media market. And until fairly recently, a few large American companies dominated the automobile and steel industries. New entrepreneurs have little chance of breaking into an industry dominated by an oligopoly. Consumers have fewer choices in buying products or services (such as fuel and its delivery) because a small number of firms control most of the market.

## 11-1b Socialism

**Socialism** is an economic system based on the principle of the public ownership of the production of goods and services. Ideally, socialism is the opposite of capitalism and has the following characteristics:

- *Collective ownership of property.* The community, rather than the individual, owns property. The government owns utilities, factories, land, and equipment, but distributes them equally among all members of a society.

- *Cooperation.* Working together and providing social services to all people are more important than competition.

- *No profit motive.* Goods and services are distributed equally, and private profits that are fueled by greed and exploitation of workers are forbidden.

- *Collective goals rather than individual investments.* The state is responsible for all economic planning and programs. People are discouraged from accumulating individual profits and investments and are expected to work for the greater good.

There have been many socialist governments during the last 150 years, but none has reflected pure socialism. Competition and individual profits are officially forbidden, but government officials, top athletes, and high-ranking party members enjoy more freedom, larger apartments, higher incomes, and greater access to education and other resources than the rank and file.

## 11-1c Mixed Economies

After World War II, many countries nationalized much private industry. Great Britain, Germany, Sweden, Belgium, the Netherlands, and some countries in Latin America, Asia, and Africa established socialist programs that included national health care and government-owned enterprises, such as education and child care programs. All of these countries, however, incorporated capitalist features such as competition, private ownership of property, and free market trade.

Karl Marx predicted that **communism**, a political and economic system in which all members of a society are equal, would become dominant around the world because exploited workers would revolt against the yoke of capitalism. His prediction hasn't come true.

Walt Disney Corporation is part of an oligopoly. Among other holdings, it owns 4 publishing companies, 19 major television and cable stations, at least 60 radio stations, numerous magazines, and 6 theme parks and resorts (in the United States, Tokyo, Paris, and Hong Kong). All of these outlets promote and sell Disney products, increasing the corporation's profits even further.

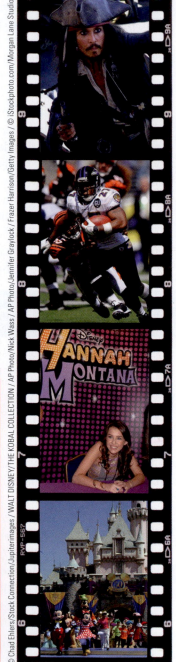

© Chad Ehlers/Stock Connection/Jupiterimages / WALT DISNEY/THE KOBAL COLLECTION / AP Photo/Nick Wass / AP Photo/Jennifer Graylock / Frazer Harrison/Getty Images / © iStockphoto.com/Morgan Lane Studios

China is a good example of a country that espoused communism and practiced socialism but is now endorsing many aspects of capitalism. Many state-owned enterprises are becoming private, and political leaders approve of foreign-owned factories that produce electronic goods, clothing, toys, and other products for export. As communist-run industries have dissolved, workers no longer have the security they enjoyed in the past. The income gap between the rich and the poor has increased dramatically, and there has been a surge in corruption, land seizures, arbitrary taxes, and social unrest (Landsberg 2007; Lee 2007).

Large companies drive both capitalism and mixed economies. Corporations, in particular, wield enormous influence both at home and abroad.

## 11-2 Corporations and the Economy

**W**hen Bob Thompson sold his construction company for $442 million, he distributed almost a third of the profits to his 550 workers, as well as some retirees and widows. About 80 became instant millionaires ("Boss Sells His Company. . ." 1999). Unlike Thompson, most corporate heads have luxurious lifestyles, cut their employees' pensions, retire as billionaires, and pass on their massive wealth to their heirs (see Chapter 8). Corporate and political power are interwoven, but let's begin by looking at some of the characteristics of corporations.

### 11-2a Corporations

A **corporation** is a social entity that has legal rights, privileges, and liabilities apart from those of its members. Until the 1890s, there were only a few U.S. corporations—in textiles, railroads, and the oil and steel industries. Today, there are almost 6 million, most created for profit, but a mere 0.5 percent of corporations bring in 90 percent of all corporate income (U.S. Census Bureau 2012).

One legal scholar describes corporations as governing our lives: "They determine what we eat, what we watch, what we wear, where we work, and what we do" (Bakan 2004: 5). Whereas many industrialized nations have closed corporate tax loopholes, the United States has expanded them. As a result, many corporations exercise more power than do governments, and have amassed enormous wealth. A number of U.S. corporations—including General Electric, which in 2010 reported worldwide profits of $14.2 billion—pay no taxes; 29 companies alone have more cash than the U.S. Treasury Department; and the profits of the *Fortune 500* corporations soared by 81 percent in 2011 (Diamond 2011; Durden 2011; Jilani 2011; Kocieniewski 2011).

### 11-2b Conglomerates

A **conglomerate** is a giant corporation that owns a collection of companies in different industries. Conglomerates emerged during the 1960s and grow by acquiring companies through mergers. Mergers might increase the value of shareholders' stock, but typically make chief executives "truly, titanically, stupefyingly rich" (Morgenson 2004: C1). An example of a conglomerate is Kraft Foods, which owns companies that produce snacks, beverages, pet foods, a variety of groceries, and convenience foods, and has ties with other corporations such as Starbucks (see www.kraft.com/brands).

### 11-2c Interlocking Directorates

Conglomerates aren't the only locus of corporate power and wealth. In an **interlocking directorate**, the same people serve on the boards of directors of several

> **corporation** a social entity that has legal rights, privileges, and liabilities apart from those of its members.
>
> **conglomerate** a giant corporation that owns a collection of companies in different industries.
>
> **interlocking directorate** a situation in which the same people serve on the boards of directors of several companies or corporations.

REUTERS/Nir Elias/Landov

China's economic boom has not benefited all of its citizens. Many urban centers are thriving—offering those in upper and middle classes high-rise apartments, stores, and restaurants. In contrast, millions of Chinese, including many older people, like the one pictured here, survive by scouring trash bins for plastic bottles to recycle.

In a 5-4 decision, in 2010 the U.S. Supreme Court ruled (*Citizens United v. Federal Election Commission*) that corporations and unions could make unlimited contributions to support or denounce individual candidates in elections. According to the decision, corporations should have the same rights to free speech as humans. Do you agree with the ruling and Mitt Romney's statement that "Corporations are people, too"?

**transnational corporation** (sometimes called a *multinational corporation* or an *international corporation*) a large company that is based in one country but operates across international boundaries.

**transnational conglomerate** a corporation that owns a collection of different companies in various industries in a number of countries.

**work** a physical or mental activity that produces or provides either goods or services.

companies or corporations. Some interlocking directorates are especially powerful because they include past U.S. presidents and members of Congress. For example, Sam Nunn was a partner at a prestigious law firm in Atlanta, Georgia, that represented numerous corporations. He was a Democrat senator from Georgia for almost 25 years, chaired several influential Senate committees, and then became a board member at ChevronTexaco Corporation, Coca-Cola, and Dell Inc., among others. Because they sit on each other's boards, the members of interlocking directorates can set the prices for food, gasoline, automobiles, and even movie tickets (Krugman 2002; Draffan 2003).

## 11-2d Transnational Corporations and Conglomerates

Interlocking directorates have become more influential than ever because of the proliferation of transnational corporations. A **transnational corporation** (sometimes called a *multinational corporation* or an *international corporation*) is a large company that is based in one country but operates across international boundaries. By moving production plants abroad, large U.S. corporations can avoid trade tariffs, bypass environmental regulations, and pay low wages. The political leaders of many countries welcome transnational corporations to stimulate their economies, create jobs, and enrich their personal bank accounts (Caston 1998).

The most powerful are **transnational conglomerates**, corporations that own a collection of different companies in various industries in a number of countries. For example, General Electric, which owns hundreds of companies in the United States, has subsidiaries in at least 27 countries and on every continent.

A handful of U.S. transnational corporations dominate world trade. Of the 25 most profitable companies worldwide, 17 are based in the United States. In effect, then, a very small group of corporations and interlocking directorates holds considerable global economic and political power (McGregor and Hamm 2008).

# 11-3 Work in U.S. Society Today

Poll after poll shows that financial hardship is a top issue for many Americans. Many worry about the economy in general, losing their job, and paying off debt, and 40 percent have used college or retirement savings to pay bills (Morales 2011; *Time/Money* Poll 2011; Newport 2012). In early 2012, 25 percent of homeowners were still "underwater," owing more on their mortgages than their homes were worth, due to the housing market collapse that began in 2008 (Bennett 2011/2012). Economic problems vary by social class, but food pantries and homeless shelters report that they've been swamped with unprecedented numbers of people, many of them middle-income earners, who have depleted their savings and can't keep up with mounting utility, food, and gas prices (Reich 2011).

There are many reasons for such economic hardship in America. Much of the hardship has to do with changes in **work**, a physical or mental activity that produces or provides either goods or services.

## 11-3a Deindustrialization and Globalization

At age 19, Jane Knudsen went to work in a textile mill in North Carolina in 1973. When the mill shut down in 2008, Knudsen found other jobs but was laid off because of a recession. The nation has lost almost 1 million textile and apparel jobs since 2000; the Bureau of Labor Statistics expects the number of these jobs to decline another 48 percent to only 259,000 by 2018 (Wiseman 2010).

AP Photo/M. Spencer Green

James Tucker, a machinist, has lost three full-time jobs in the past 25 years. After the last layoff, unable to find a job for 7 months, he takes part in mock interview sessions in Rockford, Illinois. Tucker hopes that the local workforce development office can send him to school so he can learn to program the milling machines he used to run.

## Deindustrialization

Jane Knudsen, like many others, is a casualty of **deindustrialization**, a process of social and economic change because of the reduction of industrial activity, especially manufacturing. Since 2000, 32 percent of U.S. manufacturing jobs have disappeared (Hargrove 2011). One reason for this decline is that, beginning in the early 1960s, companies have been spending more on machines than on people. According to a manager at a large company that makes plastic products, machines don't require interviews and drug tests, training, and expensive health care benefits and safety programs (Rampell 2011). Globalization has also accelerated deindustrialization.

## Globalization

One of the most significant economic changes of the late twentieth century was **globalization**, the growth and spread of investment, trade, production, communication, and new technology around the world.

Globalization also affects political systems, culture, the environment, and many other aspects of life. One example of globalization is a motor vehicle that is assembled in the United States with practically all of its parts manufactured and produced in Germany, Japan, South Korea, or developing countries.

Proponents argue that everyone benefits from globalization because it increases economic freedom and democracy, creates millions of jobs, and brings affordable goods and services (such as cell phones) to millions of households around the world. Critics contend that globalization displaces workers in Western economies, exploits poor people in developing countries, spreads pollution, and destroys indigenous cultures, natural habitats, and animal species (Bivens 2008; see also Chapter 15). They also maintain that, by and large, globalization benefits only the world's most powerful companies' profits by expanding their worldwide base of consumers (Hunter and Yates 2002; Schaeffer 2003). Among other consequences, deindustrialization and globalization have weakened labor unions and spurred offshoring.

## 11-3b Labor Unions

The changing U.S. economy has affected *labor unions*, organized groups that seek to improve wages, benefits, and working conditions. Union membership has decreased since the 1970s, and today includes only 7 percent of private sector workers and 36 percent of the total workforce (Klein 2011; U.S. Census Bureau 2012).

In the 1950s, 75 percent of Americans approved of labor unions, compared with only 52 percent today (Jones 2011). Five states prohibit unionization, and in 2011, at least 12 states significantly restricted collective bargaining rights (Allegretto et al. 2011; Ripley 2011).

Do we still need unions? Opponents argue that unions have too much influence, that its members are overpaid and drain state resources because of the pension benefits and high salaries of public sector employees such as teachers, nurses, sanitation workers, and police (Greeley 2011; Jones 2011; McKinnon 2011; Schlesinger 2011). Others contend that unions have limited employers' flexibility in hiring and firing decisions and that ever-increasing labor and health care costs have forced employers to move their operations overseas to remain competitive (Rosenfeld 2010; Greeley 2011;

**deindustrialization** a process of social and economic change because of the reduction of industrial activity, especially manufacturing.

**globalization** the growth and spread of investment, trade, production, communication, and new technology around the world.

"Labor Unions Seen . . ." 2011).

**offshoring** sending work or jobs to another country to cut a company's costs at home.

Proponents argue that unions are critical for many workers and their families. Some maintain that states have experienced budget deficits because of the housing crisis and a recession that was due to Wall Street speculation and not overpaid public sector union members, who are simply trying to make a modest living. Others point out that union members have made numerous concessions such as decreasing their wages and benefits. Most important, historically, unions have benefited almost all workers by insisting on paid holidays and vacations, greater workplace safety, overtime pay, and by challenging corporations that are becoming more dominant and powerful in both the economy and politics (Allegretto et al. 2011; Klein 2011; Schlesinger 2011; Welch 2011).

## 11-3c Offshoring

**Offshoring** refers to sending work or jobs to another country to cut a company's costs at home. Sometimes called *international outsourcing* or *offshore outsourcing*, the transfer of manufacturing jobs overseas has been going on since at least the 1970s. Labor unions can do little more than watch.

Many jobs were offshored to Canada, Hungary, the Philippines, Poland, Russia, Egypt, Venezuela, Vietnam, and South Africa, but most have gone to China and India. Between 2001 and 2008, 2.4 million American jobs, most of them in manufacturing, were lost to China: "The impact has affected essentially all production workers . . . roughly 70 percent of the private-sector workforce, or about 100 million workers" (Scott 2010: 1).

Initially, most of the offshored jobs were blue-collar manufacturing jobs. Between 2000 and 2010, however, U.S. firms have offshored 28 percent of high-level, well-paid information technology (IT) jobs, including those in accounting, computer science, and engineering (National Science Board 2012). Large wage differences (see *Figure 11.1*) make it very attractive for companies to reduce costs by replacing U.S. employees with lower cost overseas workers (Hira 2008).

Proponents maintain that offshoring has few negative effects on American workers and that consumers benefit by buying products and services at low prices (Liu and Trefler 2008). Critics contend that there's something wrong with a government that provides

### FIGURE 11.1

## Average Annual Earnings of an Accountant in the United States and India

$63,000

$5,000

© Cengage Learning

corporate tax incentives to move U.S. industries and jobs to other countries. Corporations profit, but slash workforces and reduce wages at home (Meyerson 2011).

## 11-3d How Americans' Work Has Changed

Across the country, many Americans are struggling to survive. They have adopted a variety of techniques, including taking low-paying jobs and working part-time. If these strategies fail, they find themselves among the unemployed.

### Low-Wage Jobs

One observer has described the United States as "a nation of hamburger flippers" because of the growth of low-wage jobs (Levine 1994: 1E). These are jobs that are safe from offshoring because they must be done on-site, such as child care and hotel and restaurant work.

When a recent Gallup poll asked in an open-ended question to name the nation's most important problem, 42 percent of Americans said low wages, unemployment, and a lack of money—up from 24 percent in 2008. These responses far exceeded other concerns such as health care (4 percent) and education (3 percent) (Jacobe and Jones 2009; Jones 2011).

The federal minimum wage, which rose from $6.55 to $7.25 an hour in 2009, increased the wages of less than 4 percent of the workforce. When adjusted for inflation, the new federal minimum wage is about 18 percent lower than the minimum wage between 1961 and 1981, and lower than the current minimum wage required by 18 states and the District of Columbia (U.S. Department of Labor, Wage and Hour Division 2012).

A person who works full-time at the minimum wage earns $14,500 a year, far below the 2010 federal poverty level of $22,113 for a family of two (De-Navas Walt et al. 2011). If the nation's largest low-wage employer, Walmart, paid its 1.4 million U.S. workers $12 per hour and passed every single penny of the costs to consumers, the average Walmart customer would pay just 46 cents more per shopping trip, or about $12 a year (Jacobs et al. 2011).

## Sociology in Your Life

How does your state or jurisdiction compare with others regarding minimum wage rates? How might you explain the variations? Go to the link at CengageBrain.com to find out.

During the Great Recession (June 2007 to December 2009), the U.S. economy lost 21 percent of jobs in lower-wage occupations (those paying $8 to $13 an hour), 60 percent in mid-wage occupations ($14 to $20 an hour), and 19 percent in higher-wage occupations ($21 to $53 an hour). Since the recession, 73 percent of the 1.7 million jobs added have been in lower-wage occupations (such as stocking clerk and food preparer), and the earnings declined by 2.3 percent (Bernhardt 2011). In effect, then, low-paid jobs are replacing many of those in the middle, and even low-wage workers are earning less than before the recession.

### Part-Time Work

Of the almost 28 million part-timers (those who work less than 35 hours a week), 31 percent are involuntary because people can't find suitable full-time employment or employers have reduced their hours because of a slowdown in business (Bureau of Labor Statistics News Release 2012). Part-time work, traditionally parceled out to lower-level hourly employees such as cashiers and administrative assistants, has spread to white-collar and professional sectors and now includes marketing directors, engineers, and financial officers (Davidson 2011).

Employers save on costs, such as health insurance, vacations, and unemployment insurance, by hiring more (or all) part-time workers, but admit that the downside includes higher turnover and employees who are often less committed to turning out a better product or providing a better service. For their part, involuntary part-timers say that the irregular hours disrupt family life. Although 20 percent of part-timers work at least two jobs to get by, many teeter near poverty (Davidson 2011).

### Unemployment

The U.S. unemployment rate surged from less than 5 percent in 2007 to 10 percent in early 2010, and dropped to 8.1 percent in mid-2012 (Bureau of Labor Statistics 2010; Bureau of Labor Statistics News Release 2012). The economy has fewer jobs than it did in 2001 (Shierholz 2009, 2011; Goodman 2010).

Layoffs are common in the workplace, but a recent development is widespread **downsizing**, a euphemism for firing large numbers of employees at once. When a factory closes or a company sends jobs offshore, thousands of people lose their jobs. Unemployment causes hardship, but it hits some groups and sectors harder than others. Recent job losses have been greatest among blacks and Latinos, state and local government employees, people in construction and manufacturing industries, and those with only a high school education (Shierholz 2012b).

Today's unemployment rate of about 8 percent is misleading because it doesn't include discouraged and underemployed workers. When official statistics include these numbers, the unemployment rate in late 2011 was closer to 19 percent (Thornton 2011).

### Underemployed Workers

The **underemployed** are people who have part-time jobs but want full-time work or whose jobs are below their experience, skill, and education level. Some researchers estimate that about 20 percent of Americans are underemployed, and they expect the numbers to increase (Mendes and Marlar 2011).

**downsizing** a euphemism for firing large numbers of employees at once.

**underemployed** people who have part-time jobs but want full-time work or whose jobs are below their experience and education level.

*"The job comes with a thirty-day guarantee."*

In good times and in bad, African American men have the highest unemployment rates. And compared with their white counterparts, black workers are unemployed longer after their unemployment insurance benefits end. Pictured here are laid-off workers examining listings at a job fair.

Even in the current weak economy, many employers say that a large number of jobs are going unfilled because companies can't find skilled workers (Mullaney 2012). However, national data show that this "skills mismatch" explanation doesn't jibe with the facts. For example, *"there are no jobs for more than three out of four unemployed workers, no matter what the job seekers do . . . and in every major industry"* (Shierholz 2012a: 1, emphasis in original).

Among the underemployed, almost 33 percent of those who graduated from college between 2006 and 2010 said that their postcollege job was below their education and skill level (Godofsky et al. 2011). Such self-reports may be accurate because the underemployment rate for workers with a bachelor's degree or higher grew from 4 percent when the recession began in 2007 to almost 9 percent in mid-2011 (Green and Bivens 2011). Moreover, 4 in 10 workers say that their company is understaffed (Newport 2010).

It seems, then, that many companies are trying to get by with fewer employees than in the past, even when business picks up, because workers fear being laid off if they complain that they are overworked and underpaid. Such worries affect job satisfaction and stress.

## 11-3e Job Satisfaction and Stress

A whopping 71 percent of Americans report being "not engaged" or "actively disengaged" in their work, meaning they are emotionally disconnected from their workplaces and are less likely to be productive (Blacksmith and Harter 2011). Those who say they are very satisfied typically work in occupations that involve teaching, caring for and protecting others, and creativity. The least satisfying jobs are usually in low-skilled manual and service occupations, especially jobs involving retail customer service and food/beverage preparation.

In most European countries, the typical work week is 35 hours—comparable with the definition of part-time work in the United States. Americans work more hours (1,804 per year) and more weeks per year (48 weeks) than people in any other industrialized country except Japan, and they have the highest productivity rate in the world (Begala 2011).

The United States is the only industrialized nation in the world that doesn't legally guarantee workers a paid vacation. In contrast, the typical worker, especially in Western Europe, has at least 6 weeks of paid vacation, regardless of job seniority or the number of years worked (Ray and Schmitt 2007). U.S. workers have the least vacation days in the industrialized world—14 days, on average—and a third don't take all the time they've earned (Expedia.com 2008).

About 60 percent of Americans shrink their vacations to a few days or long weekends because they have too much to do on the job, fear being laid off, or worry about not being promoted when their company keeps pushing employees to put in more hours (Alesina et al. 2005; Egan 2006; Joyce 2006). A few corporations encourage their employees to take all of their vacation days to avoid job burnout, reduce stress, and recharge, but employers save about $65 billion a year when workers don't take the paid vacations to which they're entitled (Expedia.com 2008).

## 11-3f Women and Minorities in the Workplace

One of the most dramatic changes in the United States during the twentieth century was the increase of women in the labor force (see *Table 11.1*). Many factors have contributed to the surge in women's employment, especially since the 1970s, including the growth in the number of college-educated women (and, consequently, more job opportunities), an increase in the number of working single mothers, and the higher costs of homeownership, requiring two incomes.

Largely because of higher educational attainment, 29 percent of women in two-income marriages bring home the bigger paycheck, up from only 4 percent in 1970 (Fry and Cohn 2010; *Women in the Labor Force*

## TABLE 11.1

## Women and Men in the Labor Force, 1890–2011

| YEAR | PERCENTAGE OF MEN AND WOMEN IN THE LABOR FORCE | | WOMEN AS A PERCENTAGE OF ALL WORKERS |
|------|------|------|------|
| | MEN | WOMEN | |
| 1890 | 84 | 18 | 17 |
| 1900 | 86 | 20 | 18 |
| 1920 | 85 | 23 | 20 |
| 1940 | 83 | 28 | 25 |
| 1960 | 84 | 38 | 33 |
| 1980 | 78 | 52 | 42 |
| 1990 | 76 | 58 | 45 |
| 2011 | 71 | 58 | 47 |

Source: Based on *Women in the Labor Force. . . 2011*; U.S. Bureau of Labor Statistics, "Employment Status" 2012; U.S. Census Bureau 2012.

. . . 2011). As a group, however, women have lower earnings than men in both the highest and lowest paying occupations, and the wage gaps are greater in high-income jobs (see *Figure 11.2*).

As you saw in Chapter 9, a number of factors help explain the gender pay gap, such as women taking leaves to have and raise children. Still, a pay gap remains even when women and men have the same education, number of years in a job, seniority, marital status, number of children, and are similar on numerous other factors (Glynn and Powers 2012).

There are earnings disparities across racial-ethnic groups, but the differences are especially striking by sex. As you examine *Figure 11.3*, note two general characteristics. First, earnings increase—across all racial-ethnic groups and for both sexes—as people go up the occupational ladder. But across all occupations, men have higher earnings than women of the same racial-ethnic group. At the bottom of the occupational ladder are African American women and Latinas, with the latter faring worse than any of the other groups. Thus, both sex *and* race-ethnicity affect earnings.

## FIGURE 11.2

## Women Earn Less Than Men Whether They're CEOs or Cooks

These are five of the highest and lowest paid occupations of full-time, year-round workers in the United States. How might you explain why the earnings differ by sex, especially in the highest paid jobs?

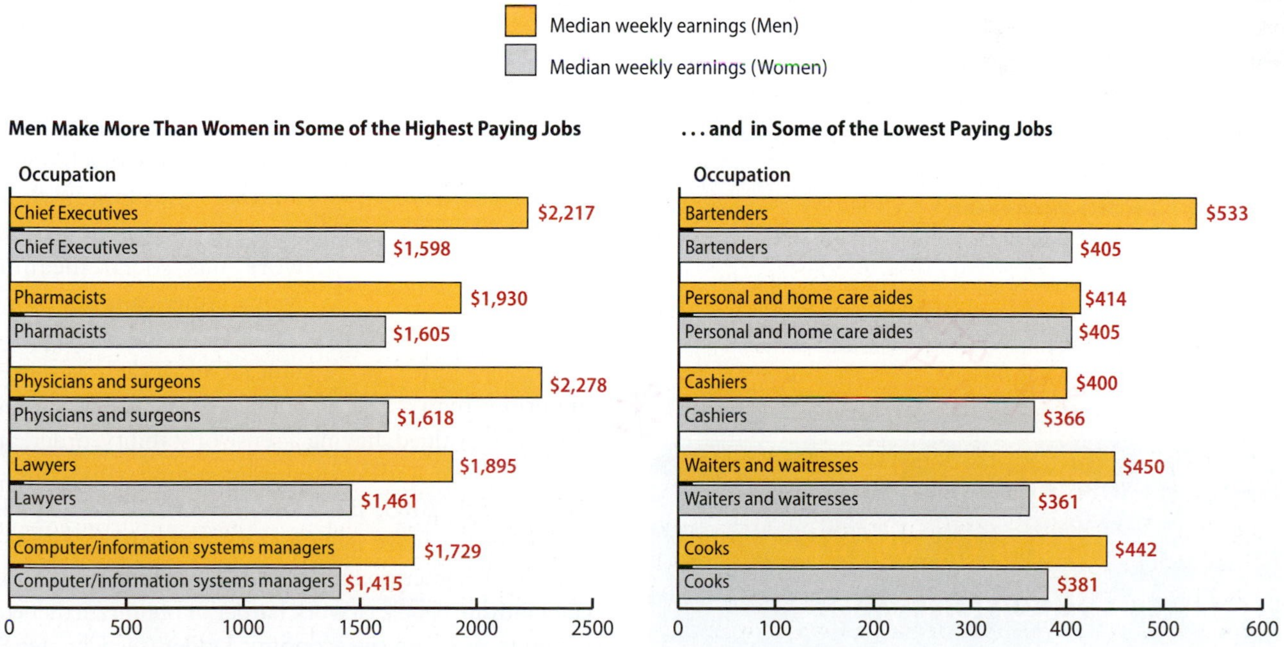

■ Median weekly earnings (Men)
■ Median weekly earnings (Women)

**Men Make More Than Women in Some of the Highest Paying Jobs**

Occupation

| Chief Executives | $2,217 |
| Chief Executives | $1,598 |
| Pharmacists | $1,930 |
| Pharmacists | $1,605 |
| Physicians and surgeons | $2,278 |
| Physicians and surgeons | $1,618 |
| Lawyers | $1,895 |
| Lawyers | $1,461 |
| Computer/information systems managers | $1,729 |
| Computer/information systems managers | $1,415 |

0   500   1000   1500   2000   2500

**. . . and in Some of the Lowest Paying Jobs**

Occupation

| Bartenders | $533 |
| Bartenders | $405 |
| Personal and home care aides | $414 |
| Personal and home care aides | $405 |
| Cashiers | $400 |
| Cashiers | $366 |
| Waiters and waitresses | $450 |
| Waiters and waitresses | $361 |
| Cooks | $442 |
| Cooks | $381 |

0   100   200   300   400   500   600

Note: Some of the differences between men's and women's median weekly earnings may seem small, but multiply each figure by 52 weeks. Thus, in annual earnings, male physicians and surgeons average more than $90,000 a year compared with only about $55,000 for female physicians and surgeons.

Source: Based on U.S. Bureau of Labor Statistics 2008, Table 39.

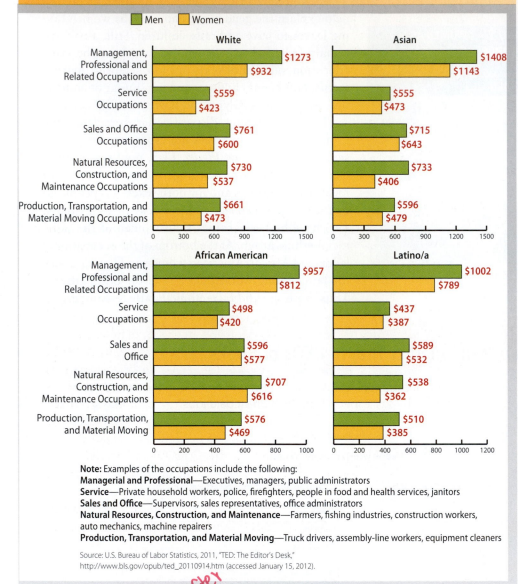

## FIGURE 11.3
## Median Weekly Earnings of Full-Time Workers by Occupation, Sex, and Ethnicity, 2010

■ Men  ■ Women

**White**

| Occupation | Men | Women |
|---|---|---|
| Management, Professional and Related Occupations | $1273 | $932 |
| Service Occupations | $559 | $423 |
| Sales and Office Occupations | $761 | $600 |
| Natural Resources, Construction, and Maintenance Occupations | $730 | $537 |
| Production, Transportation, and Material Moving Occupations | $661 | $473 |

**Asian**

| Occupation | Men | Women |
|---|---|---|
| Management, Professional and Related Occupations | $1408 | $1143 |
| Service Occupations | $555 | $473 |
| Sales and Office Occupations | $715 | $643 |
| Natural Resources, Construction, and Maintenance Occupations | $733 | $406 |
| Production, Transportation, and Material Moving Occupations | $596 | $479 |

**African American**

| Occupation | Men | Women |
|---|---|---|
| Management, Professional and Related Occupations | $957 | $812 |
| Service Occupations | $498 | $420 |
| Sales and Office | $596 | $577 |
| Natural Resources, Construction, and Maintenance Occupations | $707 | $616 |
| Production, Transportation, and Material Moving | $576 | $469 |

**Latino/a**

| Occupation | Men | Women |
|---|---|---|
| Management, Professional and Related Occupations | $1002 | $789 |
| Service Occupations | $437 | $387 |
| Sales and Office | $589 | $532 |
| Natural Resources, Construction, and Maintenance Occupations | $538 | $362 |
| Production, Transportation, and Material Moving | $510 | $385 |

**Note:** Examples of the occupations include the following:
**Managerial and Professional**—Executives, managers, public administrators
**Service**—Private household workers, police, firefighters, people in food and health services, janitors
**Sales and Office**—Supervisors, sales representatives, office administrators
**Natural Resources, Construction, and Maintenance**—Farmers, fishing industries, construction workers, auto mechanics, machine repairers
**Production, Transportation, and Material Moving**—Truck drivers, assembly-line workers, equipment cleaners

Source: U.S. Bureau of Labor Statistics, 2011, "TED: The Editor's Desk,"
http://www.bls.gov/opub/ted_20110914.htm (accessed January 15, 2012).

*stop after 11-4*

# 11-4 Sociological Explanations of Work and the Economy

Sociological theories offer differing perspectives on work and the economy, each lending insight into understanding their impact on society. Table 11.2 summarizes each perspective's key points.

## 11-4a Functionalist Theories: The Economy Provides Many Societal Benefits

Functionalists typically emphasize the positive aspects of the economy rather than its constraints. They also see capitalism as bringing prosperity to society as a whole.

### Key Characteristics

For functionalists, work is important because it enhances a society's economy and defines many of its members' roles. Work is also functional because it bonds people. As jobs become more specialized, a small group of workers is responsible for getting the work done, and coworkers can get to know one another (Parsons 1954, 1960; Merton 1968). Such networks increase workplace solidarity and enhance a sense of belonging to a group where people listen to each other's ideas, interact, and "share a vision for the work [they] do together" (Gardner 2008: 16).

Besides providing income, work has social meaning. Many Americans see money as only the fourth most important factor in their jobs—less important than a sense of accomplishment, usefulness, and feeling valued; having a sense of stability, order, and a daily rhythm; and developing interesting social contacts (Katzenbach 2003; Brooks 2007). Functionalists also maintain that wage inequities spur people to persevere and to set higher goals. Low-paying jobs, for example, can motivate people to work harder or obtain further education to move up the economic ladder (see Chapter 8).

### Critical Evaluation

Critics of functionalism point out that work often leads to stress and myriad health problems. Many people

## TABLE 11.2

# Sociological Explanations of Work and the Economy

| THEORETICAL PERSPECTIVE | LEVEL OF ANALYSIS | KEY POINTS |
| --- | --- | --- |
| **Functionalist** | Macro | Capitalism benefits society; work provides an income, structures people's lives, and gives them a sense of accomplishment. |
| **Conflict** | Macro | Capitalism enables the rich to exploit other groups; most jobs are low-paying, monotonous, and alienating; productivity isn't always rewarded. |
| **Feminist** | Macro and micro | Gender roles structure women's and men's work experiences differently and inequitably. |
| **Symbolic Interactionist** | Micro | How people define and experience work in their everyday lives affects their workplace behavior and relationships with coworkers and employers. |

© Cengage Learning

have a McJob, an unstimulating and low-paid job that requires repetitive and routine tasks, instead of feeling connected to a product or a group (see the material on McDonaldization in Chapter 6).

Functionalists can also be faulted for minimizing many U.S. corporations' disinterest in workers' well-being. As many as 43 percent of full-time private sector workers in the United States don't have paid sick days (Roan 2008). Even when business improves, many employers convert full-time jobs to part-time or temporary positions, and many have been cutting health benefits (Schultz 2011).

## 11-4b Conflict Theory: The Economy Can Be Hazardous to Your Health

Whereas functionalists emphasize the economy's benefits, conflict theorists argue that capitalism creates social problems. They contend, for example, that globalization leads to job insecurity; a handful of transnational conglomerates have enormous power; and many workers have been laid off, especially because of offshoring.

### Key Characteristics

Conflict theorists assert that low wages alienate employees rather than motivate them to work harder, as functionalists claim, because many workers realize that their labor benefits the wealthy and not themselves. Conflict theorists contend that many people are stuck in low-paying, monotonous jobs and don't always get promotions regardless of their skills, attitudes, perseverance, or productivity (Clements 2012).

Conflict theorists argue that capitalism enables the rich and powerful to exploit other groups, resulting in huge wealth disparities. In most industrialized countries, CEOs earn 10 to 25 times more than the average worker. In the United States, CEOs—including those responsible for the Great Recession's financial crisis—earn 350 times more than the typical worker. Since the mid-1970s, CEO compensation has increased by 1,200 percent compared with 308 percent for workers (Lazonick 2011; Pearlstein 2011; Ferguson 2012).

### Critical Evaluation

Are conflict theorists too quick to blame capitalism for economic problems? Critics fault conflict theorists for emphasizing economic constraints rather than choices. For example, middle-class Americans, especially, are spending more than they're earning and getting deeper into unmanageable debt. Even before the Great Recession, 35 percent of Americans criticized other people

AP Photo/Connecticut Post, Ned Gerard

Many unemployed men—including recent Asian and Latino immigrants—are joining the military because it provides steady work, college education benefits, and an opportunity for career advancement.

Is one of these people a millionaire? Between 2005 and 2009, jobless millionaires collected more than $74 million in taxpayer-financed federal and state unemployment benefits (Steinhauer 2011).

for living beyond their means—buying luxury products, taking expensive vacations, and often eating at restaurants (Center for American Progress 2006).

Also, deindustrialization and globalization have led to layoffs in assembly-line jobs, but some manufacturers complain that they have turned down business contracts because of a shortage of skilled workers—such as welders, electricians, and machinists—despite offering decent pay (up to $25 an hour for a beginner), full health benefits, matching retirement funds, and annual bonuses (Hagenbaugh 2006; see also Trumbull 2011). Thus, functionalists might criticize conflict theorists for denouncing capitalism instead of showing how it can benefit many people with low education levels and few marketable skills.

## 11-4c Feminist Theories: The Economy Creates and Reinforces Sex Inequality in the Workplace

Like functionalists, feminist scholars believe that capitalism benefits society. They also agree with conflict theorists that there is widespread inequality in the workplace, but they see gender as a critical factor in explaining the inequality.

> Are conflict theorists too quick to blame capitalism for economic problems?

### Key Characteristics

At all income levels and in all occupations, women—especially Latinas and black women—earn less than men (see *Figure 11.3* on p. 208). Women rarely penetrate the top ranks, often regardless of ability, experience, or educational attainment.

For college-educated women, on average, the gender wage gap emerges quickly. Just one year after graduation, women earn only 80 percent of what their male counterparts make, even if they went to the same college, got the same grades, had the same major, are in the same kind of job, and have the same demographic characteristics such as marital status, race-ethnicity, and number of children. Ten years after graduation, women earn only about 69 percent of what men earn. Over a 40-year career, women can receive up to $1 million less in pay (Boushey et al. 2010; AAUW 2011; Hegewisch et al. 2011).

When the recession began in 2007, men suffered about 70 percent of the job losses. Those hit hardest included men in higher paying occupations and well-paid union jobs that were offshored or cut to decrease manufacturing costs. More women survived because they were in lower paying jobs in the health care and service industries that can't be offshored, but women experienced 70 percent of all public sector job losses (National Women's Law Center 2011; Rivers and Barnett 2011).

### Critical Evaluation

Critics of feminist perspectives note that emphasizing men's domination of women in the workplace ignores social class in the many situations where high-status women control low-status men and women. Second, many feminist scholars maintain that capitalism exploits women by crowding them into lower paying occupations. However, there is considerable economic gender inequality in socialist, communist, and mixed economies in both industrialized and developing nations (Cudd and Holstrom 2010).

Third, some people question whether the economy reinforces sex discrimination or simply reflects cultural sexism that's engrained in other institutions. Many families, for example, have gendered expectations that require employed women, but not men, to curtail their economic ambitions and activities because women are expected to raise children, perform most of the domestic tasks, and care for older family members. For example, 42 percent of Americans, including women, say that it's best for the mother not to be employed, even part-time (Kramer 2005; Parker 2009). In effect, then, patriarchy (see Chapter 7) may be a more critical factor than discrimination in explaining many women's inequality in the economy.

Each of the four theoretical perspectives focuses on different aspects, but taken together, they provide us with a broad lens for understanding work and the economy.

## 11-4d Symbolic Interaction Theories: We Learn Work Roles

Symbolic interactionists rely on micro-level approaches to explain the day-to-day meaning of work. They are especially interested in the formal and informal rules that develop in workplaces.

### Key Characteristics

Interactionists have provided numerous insights on how people define and experience work. Many low-paid workers who feel disengaged from their jobs tolerate their situations because they have few options. They endure drug tests, constant surveillance, rat- and

AP Photo/Donna Abu-Nasr

For symbolic interactionists, cultural expectations shape women's and men's work, as well as determining whether women should even be employed. In Saudi Arabia, one of the world's most patriarchal societies, women are still not permitted to vote, travel abroad, work without the permission of a male relative, or drive. Here, female employees work at the first car showroom in Saudi Arabia where women sell cars, but only to female buyers.

cockroach-infested buildings, physical pain, and work-related occupational injuries. Regardless of the occupation, coworkers may develop close relationships that become friendships outside of work.

Interactionists have also studied professionals to determine how they are socialized into their jobs. In medicine, for example, surgeons teach their interns and residents that it's normal to make some mistakes, but that carelessness and continued errors are unprofessional (Bosk 1979). Assembly-line workers often control coworkers, especially verbally, who overproduce or underproduce by punishing or rewarding them (see Chapter 6).

### Critical Evaluation

The most common criticism is that interactionism, although providing in-depth analyses, sacrifices scope. For example, sociologist Deirdre Royster (2003) studied young men in Baltimore who had graduated from the same vocational high school at about the same time. She found that white male teachers tended to provide white students, but not black students, with active assistance such as information about job vacancies and job references in trade occupations. We don't know, however, whether these findings are applicable to other cities or types of jobs because the study was based on a small and nonrepresentative sample of young men.

## 11-5 Global Political Systems

Every society has a **government**, a formal organization that has the authority to make and enforce laws. Governments are expected to maintain order, provide social services, regulate the economy, establish educational systems, create armed forces to discourage (real or imagined) attacks by other countries, and ensure the inhabitants' safety from internal attacks that range from individual crimes to groups that seek to overthrow the government.

Worldwide, governments vary from democratic to totalitarian, but there are also authoritarian governments and monarchies. Let's begin by looking at democracy.

### 11-5a Democracy

A **democracy** is a political system in which, ideally, citizens have control over the state and its actions.

> **government** a formal organization that has the authority to make and enforce laws.
>
> **democracy** a political system in which, ideally, citizens have control over the state and its actions.

Democracies are based on several principles:

- Individuals are the best judges of their own interests and participate in governmental decisions.
- Citizens select leaders who are responsive to the wishes of the majority of the people.
- Suffrage (the right to vote) is universal, and elections are frequent, free, fair, and secret.
- The government recognizes individual rights, such as freedom of speech (including dissent), press, and assembly, and the right to organize political parties whose members compete for public office.

In practice, democracy doesn't ensure equality and respect for human rights. For example, the U.S. Constitution initially limited voting to white male landowners, African American men got the right to vote in 1870, and U.S. women gained the vote only in 1920. Today, as you'll see shortly, U.S. democratic values are endangered because special interest groups have unprecedented influence on the political system.

Worldwide, 35 percent of the world's population (almost 2.5 billion people) lives in countries with repressive governments that deny basic political and civil rights. People in sub-Saharan and North Africa, the Middle East, China, and Russia are the least likely to have political freedom (see *Figure 11.4*). Despite the Arab Spring (political uprisings in the Arab world that began in 2010), democracy declined around the world in 2011 for the sixth consecutive year (Kurlantzick 2012; Puddington 2012).

## 11-5b Totalitarianism and Dictatorships

At the opposite end of the continuum from democracy is **totalitarianism**, a political system in which the government controls every aspect of people's lives. Totalitarianism has several distinctive characteristics (Taylor 1993; Tormey 1995; Arendt 2004):

- A pervasive ideology that legitimizes state control and instructs people how to act in their public and private lives.
- A single political party controlled by one person, a *dictator*—a

supreme, sometimes idolized leader—who stays in office indefinitely.

- A system of terror that relies on secret police and the military to intimidate people into conformity and to punish dissenters.
- Total control by the government over other institutions, including the military, education, family, religion, economy, media, and all cultural activities, including the arts and sports.

Contemporary examples of the most repressive totalitarian governments include China, Burma (Myanmar), Cuba, Iran, Libya, North Korea, Somalia, Sudan, Syria, Tibet, Turkmenistan, and Uzbekistan. Within these countries and territories, state control over daily life is pervasive and wide-ranging, independent organizations and political opposition are banned or suppressed, and those who criticize the government are imprisoned, tortured, or killed (Puddington 2012).

## 11-5c Authoritarianism and Monarchies

Many nations have some version of democracy or totalitarianism. However, a number of countries are characterized by **authoritarianism**, a political system in which the state controls the lives of citizens but permits some degree of individual freedom. In the Middle East, for example, the ruler of Qatar has absolute power and discourages public criticism of his policies but has implemented a constitution that specifies that two-thirds of the officeholders must be elected (rather than appointed) and has supported numerous progressive reforms, such as women's active participation in politics (Harman 2007).

A **monarchy**, the oldest type of authoritarian regime, is a political system in which power is allocated solely on the basis of heredity and passes from generation to generation. In this form of government, a member of a royal family, usually a king or queen, reigns over a kingdom. A monarch's power and authority are legitimized by tradition and religion (the right to govern is bestowed on the monarch by God).

There are more than 40 monarchies around the world. In some countries—especially in the Middle

## FIGURE 11.4
## Political Freedom Around the World, 2011

This map shows the degree of political freedom, measured by political rights and civil liberties, in 194 countries around the world. *Political rights* include free and fair elections, competitive parties, and no discrimination against minorities. *Civil liberties* include freedom of speech and the press; freedom to practice one's religion; and the freedom of teachers, professors, and students to discuss political issues without fear of physical violence or intimidation by the government.

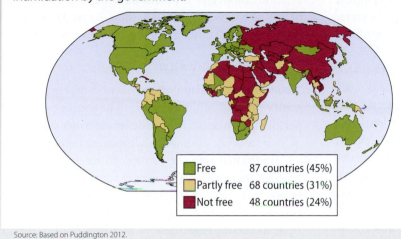

| | Free | 87 countries (45%) |
| --- | --- | --- |
| | Partly free | 68 countries (31%) |
| | Not free | 48 countries (24%) |

Source: Based on Puddington 2012.

East and parts of Africa—monarchs have absolute control. Those in Norway, Denmark, Belgium, Britain, Japan, and many other countries have little political power because they are limited by democratic constitutions, serving primarily ceremonial roles (Glauber 2011).

We see, then, that political systems vary worldwide. Despite the variation, political leaders wield considerable power and authority.

# 11-6 Politics, Power, and Authority

**C**harles de Gaulle, one of France's past presidents, once remarked, "I have come to the conclusion that politics are too serious a matter to be left

In 2007, pro-democracy activists led the initial demonstrations in Rangoon, Burma, when the military-run government increased the price of fuel, which hiked the costs of staples such as rice and cooking oil. Tens of thousands of Buddhist monks joined the protests, denouncing the government as "the enemy of the people." The monks' participation was significant because they are highly revered and influential.

to the politicians." Despite such tongue-in-cheek comments, politics is a serious matter because it can improve or diminish the quality of our everyday lives.

## 11-6a Power

Politics encompasses two important concepts, power and authority (Lasswell 1936). **Power** is the ability of a person or group to affect the behavior of others despite resistance and opposition. Examples of political power include a government's quashing a peaceful protest or, conversely, protecting people's right to conduct such a protest.

U.S. political power includes *earmarks* (known commonly as "pork")—funding requests by lawmakers in Congress to provide federal money to companies, projects, groups, and organizations in their district. Earmarks are inserted into the annual

Reuters/Landov

spending bills and don't require external competitive bidding. Despite calls to ban earmarks, pork spending has grown considerably—from $10 billion in 1995 to $17 billion in 2010. Some examples of recent pork barrel projects include $200 million for corporations to promote their products overseas and $4.9 million for 11 states to determine how to use wood (Citizens Against Government Waste 2011).

Both Democrats and Republicans maintain that earmarks create jobs and pay for pressing needs, such as maintaining local museums. Critics contend that earmarks decrease funding for important federal programs; influence elections; and finance *lobbyists*, people who try to influence legislation on behalf of a special interest. Earmarks subsidize projects for the lawmakers' personal gain and go to companies that, in turn, make generous campaign contributions (O'Harrow 2008; Fallis et al. 2012; Higham et al. 2012; Luo and McIntire 2012). President Obama pledged to limit lobbyists' influence, but, every day, there's a "steady stream" of lobbyists at the White House (Farnam 2012).

Whether based on persuasion or coercion, power—especially political power—is about controlling others. Because people may revolt against sheer force, many governments depend on authority to establish order, shape people's attitudes, and control their behavior.

## 11-6b Authority

**Authority**, the legitimate use of power, has three characteristics. First, people *consent* to authority because they believe that their obedience is for the greater good (e.g., following traffic rules). Second, people see the authority as *legitimate*—valid, justifiable, and necessary (e.g., paying taxes to support public education, increase national security, and remove trash and snow). Third, people accept authority because it is *institutionalized* in organizations (e.g., police departments and government agencies).

Max Weber (1925/1978) described three ideal types of legitimate authority: traditional, charismatic, and rational-legal (see *Table 11.3*). Remember that ideal types are models that describe the basic characteristics of any phenomenon (see Chapters 1 and 5). In reality, these types of authority often overlap.

### Traditional Authority

**Traditional authority** is power based on customs that justify the position of the ruler. The source of power is personal because the ruler inherits authority on the basis of long-standing customs, traditions, or religious beliefs.

Traditional authority is most common in nonindustrialized societies where power resides in kinship groups, tribes, and clans. In the past, kings and emperors ruled because of heredity and the belief that they had a divine right to power, regardless of intelligence or ability. In many African, Middle Eastern, and Asian countries today, power is passed down among men within a family line (Tétreault 2001; see also Chapter 8).

### Charismatic Authority

**Charismatic authority** is power based on exceptional individual abilities and characteristics that inspire devotion, trust, and obedience. Like traditional authority, charismatic authority is personal and reflects extraordinary deeds or even a belief that a leader has been chosen by God, but the leaders don't pass their power down to their offspring.

Charismatic leaders can inspire loyalty and passion whether they are heroes or tyrants. Examples

---

### TABLE 11.3

# Weber's Three Types of Authority

| TYPE OF AUTHORITY | DESCRIPTION | SOURCE OF POWER | EXAMPLES |
|---|---|---|---|
| Traditional | Power is based on customs, traditions, and/or religious beliefs. | Personal | Medieval kings and queens, emperors, tribal chiefs |
| Charismatic | Power is based on exceptional personal abilities or a calling. | Personal | Adolf Hitler, Gandhi, Martin Luther King, Jr. |
| Rational-legal | Power is based on the rules and laws that are inherent in an elected or appointed office. | Formal | U.S. presidents, members of Congress, state officials, police, judges |

© Cengage Learning

Joan of Arc (1412–1431) is a well-known example of a charismatic woman who defied the status quo. She led the French to victory over the English in many battles. During the 1960s, California's César Chávez (1927–1993), another charismatic leader, organized strikes to call attention to farm workers' low wages and poor working conditions.

In some countries, such as Japan and England, emperors, kings, and queens have traditional authority and perform symbolic state functions such as attending religious ceremonies and granting titles (like that of Sir Elton John, a singer). Both of these countries also have parliaments that exercise legal-rational authority in determining laws and policies.

of the latter include historical figures such as Adolf Hitler, Napoleon, Ayatollah Khomeini, and Fidel Castro. These and other dictators have been spellbinding orators who radiated magnetism, dynamism, and tremendous self-confidence and promised to improve a nation's future (Taylor 1993).

### Rational-Legal Authority

**Rational-legal authority** is power based on the belief that laws and appointed or elected political leaders are legitimate. Unlike traditional and charismatic authority, rational-legal authority comes from rules and regulations that pertain to an office rather than to a person. For example, anyone running for mayor must have specific qualifications (such as U.S. citizenship). When a new mayor (or governor or other politician) is elected, the rules don't change, because power is vested in the office rather than the person currently holding the office.

### Mixed Authority Forms

People may have more than one type of authority. For example, some historians believe that presidents Abraham Lincoln, Theodore Roosevelt, John F. Kennedy, and Ronald Reagan enjoyed charismatic appeal beyond their rational-legal authority.

# 11-7 Politics and Power in U.S. Society

Politics plays a critical role in our everyday lives. Two important facets of the political system are political parties and voting.

## 11-7a Political Parties

A **political party** is an organization that tries to influence and control government by recruiting,

In 1995, at age 3, Oyo Nyimba Kabamba Iguru became the world's youngest monarch when he was crowned as king of the Ugandan Kingdom of Toro. At age 18, he officially took control of the kingdom.

nominating, and electing its members to public office. Political parties provide a way for citizens to shape public policy at the local, state, and national levels.

## Functions of U.S. Political Parties

Political parties engage in a wide variety of activities—everything from stuffing envelopes and calling voters to drafting laws if a candidate is elected. Through their activities, parties perform a number of vital political functions, especially recruiting candidates for public office, organizing and running elections, and if elected, running the government (Schmidt et al. 2001).

Ideally, that's how political parties *should* work. In reality, parties perform some of these functions more effectively than others. For example, parties typically concentrate on winning elections rather than making promised changes once candidates are in office.

## The Two-Party System

Many democracies around the world have a number of major political parties: 3 in Canada, 5 in Germany, 9 in Italy, more than 20 in Israel, and about 1,050 in India (Yardley 2009). Thus, compared with many other countries, the two-party system of Democrats and Republicans in the United States is uncommon.

Political parties base their actions on an *ideology*, a set of ideas that constitute a person's or group's beliefs, goals, expectations, and actions. Those who identify with the Democratic Party typically believe that the government should provide social programs, especially for the poor; most of those in the Republican Party believe that the federal government should be involved in few social programs. Table 11.4 offers some examples of other major ideological differences between Democrats and Republicans.

Within a political party, Americans also identify themselves as conservative, moderate, or liberal on social and economic issues. In 2011, for example, 51 percent of Republicans and 16 percent of Democrats described their political views as conservative (Saad 2012). Moreover, 9 percent of Americans, "libertarians," express fairly liberal views on social issues and conservative views on economic issues, but lean to the Republican side (Keeter 2011). Thus, our two-party system is more complex than appears on the surface.

## Sociology in Your Life

Where do you fit on the American political spectrum? Take the quiz by clicking on the link at CengageBrain.com.

What about the Tea Party and independents? Analysts view the Tea Party as a social movement rather than a political party. What unites Tea Party supporters is their firm conviction that the federal government has gotten too big and powerful and that taxes are too high. Because of these views, 62 percent of Tea Party followers are ideologically similar to conservative Republicans (Newport 2010; Page and Jagoda 2011).

In the past, about equal percentages of Americans identified themselves as either Republicans or

## TABLE 11.4

# How Do Democrats and Republicans Differ?

| ISSUE | MANY DEMOCRATS BELIEVE THAT . . . | MANY REPUBLICANS BELIEVE THAT . . . |
|---|---|---|
| **Family** | Government programs should implement universal health insurance for all families, especially those with low incomes. | Government should cut all welfare benefits to unwed mothers, stigmatize divorce, and emphasize "traditional" family values. |
| **Abortion** | Women should have the right to choose abortion in practically all cases; the government should pay for abortions for low-income women. | The unborn should be protected in all cases; there should be no public money for abortions. |
| **Gay Rights** | Gay men and lesbians should have the right to marry and to adopt children. They should be fully accepted in the military and other institutions. | Marriage and adoption should be limited to heterosexuals. Inclusion of gay men and lesbians in the armed forces creates interpersonal problems. |
| **Education** | The government should strengthen public schools by raising salaries for teachers and decreasing classroom sizes. It should not use tax dollars to help students attend private schools. | The government should increase state and local control of schools. It should use tax dollars to fund students to attend private schools of their choice if their local school is underperforming. |

Sources: Based on Benokraitis 2000; Welch et al. 2004; and "Modest Rise in Number Saying . . ." 2011.

Democrats. In 2011, however, 27 percent identified themselves as Republicans, 31 percent said that they were Democrats, and a record 40 percent identified themselves as independent (Jones 2012). Because there's no national Independent Party, independents usually vote for Democrats, Republicans, or not at all.

Other parties (such as the Green Party and Reform Party) have emerged in the United States, but they've had little impact. Most third-party candidates have little chance of being elected because the media don't take them seriously and give them only passing, if any, coverage. Campaign contributors and voters pay little attention to third-party candidates because they believe that their donations and ballots will be wasted.

## 11-7b Who Votes, Who Doesn't, and Why

Many Americans are disillusioned with politics. In late 2011 and early 2012, for example, only 10 percent of Americans approved of the job Congress was doing, a historic low (Newport 2012).

One might expect that such dissatisfaction would result in higher voter turnout, but this isn't the case. Of 163 countries around the world that hold democratic elections, the United States ranks 140th in voter turnout. In 34 nations, at least 80 percent of the eligible population votes, and more than 90 percent does so in Austria, Italy, and Luxembourg—compared, on average, with about 50 percent of U.S. citizens (International Institute for Democracy and Electoral Assistance 2007; Thompson 2010).

Some of the high turnout rates in other countries may be partly due to *compulsory voting*: The government demands explanations or imposes fines for not voting, may make it difficult for nonvoters to get benefits such as public assistance, or requires evidence of voting to get a passport or a driver's license. Generally, these kinds of sanctions send the message that voting is not a privilege but a civic responsibility (Holder 2006; International Institute for Democracy and Electoral Assistance 2007).

In 2008, 64 percent of Americans voted in the presidential election. This was a larger than usual percentage, but lower than the numbers who voted in presidential elections during the 1950s and 1960s (File and Crissey 2010). Who votes, who doesn't, and why reflect demographic characteristics, attitudes about politics, and situational and structural factors.

### Demographic Factors

Many demographic factors affect registration and voting, but the most important are age, marital status, social class, race and ethnicity, and religion. Sex isn't a major variable because both men and women turn out in roughly equal proportions (File and Crissey 2010).

*Age.* Historically, the voting rate is higher among older than younger people; it generally increases with age but drops off at about age 75, probably because of poor health. Why the age difference? A key factor is registration. Only about half of young people (ages 18–34) are registered to vote, compared with 75 percent of those age 55 and older. Young people have lower registration rates because they are more mobile than older people and are less likely to reregister after a move. They may also be preoccupied with major life events such as going to college and finding jobs (Schachter 2004). In effect, the low participation rate of young voters means that they have little impact on political processes.

*Marital Status.* Married people are more likely to vote than those who are widowed, divorced, or have never married. People who have never married tend to be younger, which influences their voting rate. Married couples are more likely to be registered voters, to be homeowners, to live in established neighborhoods, to be parents of young children in school, and to be employed (Day and Holder 2004). This combination of characteristics suggests that married people, compared with never-married singles, have a bigger stake in society and, as a result, are more likely to vote.

*Social Class.* Social class has a significant impact on voting behavior. At each successive level of educational attainment, the voting rate increases. For example, the voting rate of those with a college degree (77 percent) is almost twice as high as that of people who have not completed high school (39 percent), and those with advanced degrees are the most likely to vote (83 percent). People with the highest educational levels are usually more informed about and interested in the political process and more likely to believe that their vote counts ("Who Votes . . ." 2006; File and Crissey 2010).

Voting rates also increase with income levels. In 2008, the voting rate of people with annual family incomes of $100,000 or more was 92 percent, compared with 56 percent of those with annual incomes

Theresa McCracken/Cartoonstock

"Now that we've agreed on the loopholes,
what should the tax laws be?"

under $30,000. People with higher incomes are more likely to be employed and to have assets (such as houses and stock) and, therefore, are more likely to be aware of the costs of not voting to protect or increase their resources. In contrast, people who are unemployed usually have few assets and may be too disillusioned with the political system to vote ("Who Votes . . ." 2006; File and Crissey 2010).

### Race and Ethnicity

Whites are typically the most likely to vote. Here again, a key to voter turnout is registration. Across all racial-ethnic groups—including both native-born and naturalized citizens—the higher the voter registration, the higher the voting rate (File and Crissey 2010).

Low voter registration rates among racial-ethnic groups are associated with other demographic and social characteristics. For example, compared with white citizens, minorities tend to be younger and poorer, are more likely to be unemployed, and have lower education levels. Because they are less likely to be homeowners or married, they may be less invested in election results. Minorities are also more likely than whites to be pessimistic about government and politics (Frey 2008).

Despite such constraints, during the 2008 presidential election, minorities played an important role in the outcome. Nearly all (95 percent) black voters cast their ballots for Barack Obama, as did 67 percent of Latinos and 62 percent of Asian voters. In contrast, white voters supported John McCain (55 percent) over Obama (43 percent).

### Religion

Only 14 percent of Americans say that their religious beliefs influence their politics, but there are strong links between these variables. Among Jews and mainline Protestants (e.g., Lutherans and Methodists), more than 80 percent are registered to vote, compared with 69 percent of Catholics, 48 percent of Muslims, 42 percent of Hindus, and only 13 percent of Jehovah's Witnesses. The low voter registration rates of Muslims and Hindus are primarily due to the high number of recent immigrants who aren't eligible to vote (Pew Forum on Religion & Public Life 2008).

Religion also shapes political affiliation. Mormons, mainline Protestants, and members of evangelical churches (including born-again Christians) tend to describe themselves as conservative and support the Republican Party. Jews, Buddhists, Hindus, and those not affiliated with a religious group describe their political beliefs as liberal and favor the Democratic Party. The connection between religious affiliation and political opinion appears to be especially strong for certain issues such as abortion and homosexuality (Pew Forum on Religion & Public Life 2008; see also *Table 11.4* on p. 216).

## Attitudes

Millions of Americans have a low opinion of the federal government and Congress. Many simply don't trust politicians: "They all lie," "I've never met an honest politician," "They're all corrupt," and so on. When Gallup asked people to rate 22 occupations according to honesty and integrity, members of Congress and stockbrokers were among the lowest, at 9 percent each (Saad 2009). Whether such cynicism is warranted or not, many people don't vote because they believe that their lives won't improve regardless of who's elected.

## Situational and Structural Factors

Situational and structural factors also discourage voting. For example, illness, disability, out-of-town travel,

# How Would *You* Describe Congress?

In 2010, the Pew Research Center asked a nationally representative sample of Americans to provide one word that described Congress. The responses were put into a "word cloud." The larger the word, the more times it was used (Auxier 2010). What one word would *you* use to describe Congress?

Word cloud graphic, created using http://wordle.net, from "Congress in a Wordle", Mar. 22, 2010, The Pew Research Center For the People & the Press, a project of the Pew Research Center

and transportation or registration problems can all impede voter turnout (File and Crissey 2010).

Because a key to voter turnout is registration, some states have removed structural barriers to registering. In some states—Idaho, Maine, Minnesota, New Hampshire, Wisconsin, and Wyoming—people can vote the same day that they register. In Oregon, elections have been conducted by mail since 2000. In all of these states, the voting rates are much higher than the national average, sometimes well above 72 percent (Welch et al. 2004).

In many European and other countries, where voter turnout is typically high, voters are registered automatically when they pay taxes or receive public services, and elections are held on Saturday. In the United States, in contrast, the government does little to help people register to vote, elections are held on a weekday (usually Tuesday), and those with long job commutes can't get to the polls before they close at 8:00 P.M. (Harder and Krosnick 2008).

Instead of making voting easier, since 2011, 34 states (particularly those controlled by Republican legislatures) have passed or are considering laws that will make it more difficult for at least 5 million people to register or to vote. Some of the new rules require government-issued photo identification (such as a driver's license or passport) and a birth certificate or proof of citizenship, and some states have reduced early voting. State legislators contend that the new laws

will crack down on voter fraud in future elections, but opponents argue that the new regulations will stifle voter turnout, especially among students, the poor, and minorities, who tend to vote for Democrats (Weiser and Norden 2011; Jonsson 2012).

Because many Americans don't vote, who has the most power? And do government leaders represent the average citizen? Such questions have generated considerable debate among sociologists and other social scientists.

> **pluralism** a political system in which power is distributed among a variety of competing groups in a society.

# 11-8 Sociological Perspectives on Politics and Power

According to symbolic interactionism, power is socially constructed through interactions with others; people learn to be loyal to a political system—whether it's a democracy or a monarchy—and to show respect for its symbols and leaders. For the most part, however, sociologists rely on macro-level theories of functionalists, conflict theorists, and feminists to explain political power. *Table 11.5* summarizes these perspectives.

## 11-8a Functionalism: A Pluralist Model

For functionalists, the people rule through **pluralism**, a political system in which power is distributed among a variety of competing groups in a society (Riesman 1953; Polsby 1959; Dahl 1961).

### Key Characteristics

In the pluralist model, individuals have little direct power over political decision making but can influence government policies through special interest groups—trade unions, professional organizations, and so on (see *Figure 11.5*). The various groups rarely join ranks because they concentrate on single issues such as health care, pollution, or education. This focus on different issues fragments groups, but also results in a broad representation of a variety of interests and a distribution of power.

Because there are a number of single-issue groups, functionalists maintain, there are multiple leaderships (Dye and Ziegler 2003). As a result, many leaders, not

TABLE 11.5

# Sociological Explanations of Political Power

| | **FUNCTIONALISM: A PLURALIST MODEL** | **CONFLICT THEORY: A POWER ELITE MODEL** | **FEMINIST THEORIES: A PATRIARCHAL MODEL** |
|---|---|---|---|
| **Who has political power?** | The people | Rich upper-class people—especially those at top levels in business, government, and the military | White men in Western countries; most men in traditional societies |
| **How is power distributed?** | Very broadly | Very narrowly | Very narrowly |
| **What is the source of political power?** | Citizens' participation | Wealthy people in government, business corporations, the military, and the media | Being white, male, and very rich |
| **Does one group dominate politics?** | No | Yes | Yes |
| **Do political leaders represent the average person?** | Yes, the leaders speak for a majority of the people. | No, the leaders are most concerned with keeping or increasing their personal wealth and power. | No, the leaders are rarely women who have decision-making power. |

© Cengage Learning

just a few, can shape decisions that represent many groups and issues. Pluralists note that people also have power outside of interest groups: They can vote, run for office, contact lawmakers, and collect signatures to place specific issues on a ballot. Therefore, there are continuous checks and balances as individuals and groups vie for power and try to influence laws and policies.

## Critical Evaluation

Does pluralism work as democratically as functionalists maintain? Critics argue that interest groups have unequal resources. The poor and disadvantaged rarely have the skills and educational backgrounds to organize or promote their interests. In contrast, wealthy individuals and organizations can influence

## FIGURE 11.5

# Pluralist and Power Elite Perspectives of Political Power

### PLURALIST MODEL

Power is dispersed among multiple groups that influence the government. For example,

- government employees
- victims' rights groups
- labor unions
- banks and other financial institutions
- realtors and home builders
- teachers
- environmental groups
- women's rights groups

### POWER ELITE MODEL

Power is concentrated in a very small group of people who make all the key decisions. For example,

**Top level** (1 percent)—CEOs of large corporations, high-ranking lawmakers in the executive branch, and top military leaders

**Middle level** (8 percent)—most members of Congress, lobbyists, entrepreneurs of small businesses, leaders of labor unions and other interest groups (in law, education, medicine, etc.), and influential media commentators

**The masses** (91 percent)—people who are unorganized and exploited and either don't know or don't care about what's going on in government

© Cengage Learning

government through political contributions and personal connections.

Critics also maintain that pluralists aren't realistic about the emergence of multiple leaderships because the affluent control the government. In 2010, about 1 percent of all Americans were millionaires, but 50 percent of those in Congress were. Members' average wealth was at least $7 million in the House and $14 million in the Senate. All the Supreme Court Justices and top executive branch members are also multimillionaires (Center for Responsive Politics 2011). The vast wealth disparity between the representatives and the represented means that most lawmakers are unlikely to understand or worry about the economic pressures many Americans face in keeping their families financially afloat (Whoriskey 2011).

## 11-8b Conflict Theory: A Power Elite Model

Unlike functionalists, conflict theorists contend that the United States is ruled by a **power elite**, a small group of influential people who make the nation's major political decisions. Sociologist C. Wright Mills (1956) coined the term *power elite* to describe a pyramid of power that he believed characterized American democracy (see *Figure 11.5* on p. 220).

### Key Characteristics

According to Mills, the power elite is made up of three small but influential groups of people at the top level who run the country: political leaders (specifically the president and his top aides), business heads (the corporate rich, who have enormous wealth), and military chiefs (who govern the Pentagon). Practically all of the members of these groups are white, Anglo-Saxon, Protestant men who form an inner circle of power.

**power elite** a small group of influential people who make the nation's major political decisions.

For Mills, those at the middle level of power include members of Congress, lobbyists, influential media commentators, leaders of labor unions and other interest groups, and the heads of state and local governments. The bottom level, and the largest and least powerful group, the masses, is composed of everyone else (see *Figure 11.5*). The power elite tolerates the masses—including their elections and laws—but in the end simply does what it wants, such as declaring wars, decreasing taxes for the rich, and cutting off benefits for the poor (e.g., see Kivel 2004; Domhoff 2006).

Many contemporary power elite theorists maintain, like Mills, that the United States is not a democracy because of the close ties between politics, business, and the military. For example, President Obama's top administration includes recent high-ranking employees from corporations such as AT&T, IBM, Allstate Insurance, and Goldman Sachs (a financial corporation that enjoyed a bailout at taxpayers' expense in 2009) (Myerson 2011).

Since 2000, a total of 130,000 people have been either lawmakers who then became lobbyists or former lobbyists who went on to hold important positions as congressional staff members. In effect, then,

One of the reasons many Americans are disillusioned with government and politics is that dishonest lawmakers are rarely punished. A recent exception is Rod R. Blagojevich, former governor of Illinois, whom the FBI indicted on 19 felony charges, accusing him of using his office for the financial and political benefit of himself, his family, and friends (Davey and Saulny 2009). In 2011, he was sentenced to 14 years in prison.

UPI/Brian Kersey/Landov

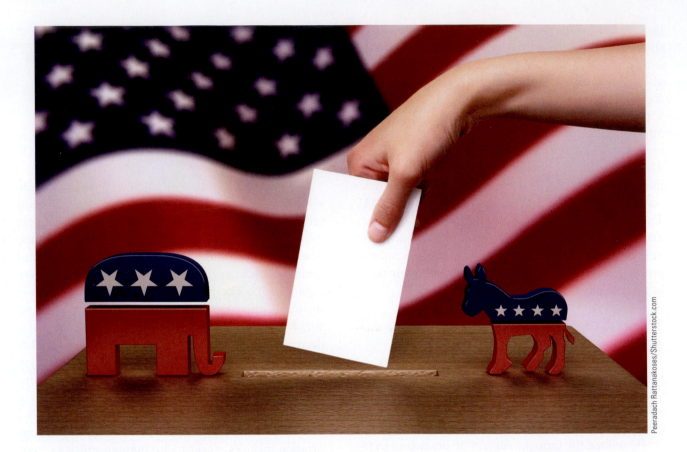

Peeradach Rattanakoses/Shutterstock.com

there's a "revolving door" among the power elite (LegiStorm 2011).

According to conflict theorists, the upper classes have enormous power over political institutions because they finance political campaigns and get elected or appointed to political offices. Because many lawmakers are wealthy, they have the power and resources to craft legislation that promotes their own interests (Blumgart 2011; Rivlin 2011; Schweizer 2011; Berger 2012).

### Critical Evaluation

Functionalists accuse conflict theorists of exaggerating the power elite's importance. They point out, for example, that "the masses" aren't just puppets but are aware of the power elite's influence. As a result, they vote, mobilize, support particular interest groups, and protest current administrative and corporate policies (e.g., the Tea Party and Occupy Wall Street movements).

Also, according to critics, the power elite perspective assumes, incorrectly, that the wealthy people at the top are unified in their interests and goals. For example, because the Democratic and Republican political parties and their top officials endorse very different agendas, the power elite are rarely unified.

Also, critics maintain, the wealthy can't always buy political offices. In 2010, for example, Meg Whitman (the founder of eBay) spent almost $142 million of her personal fortune in her bid for California governor. She lost to Jerry Brown, who contributed only $24,000 of his own money to his $4 million campaign funds (Crowley 2010; Palmeri and Green 2010).

## 11-8c Feminist Theories: A Patriarchal Model

Feminist theorists emphasize that women are generally excluded from the most important political positions. Women's status in the Western world has improved considerably during the last 100 years, but as you saw in Chapter 9, relatively few women, especially minority women, serve in Congress and at the highest government levels.

### Key Characteristics

Of the 197 world leaders who are presidents or prime ministers, only 11 percent are women, and only 7 percent of the world's monarchs are women (Institute for Women's Leadership 2011). Of 188 countries worldwide,

In 2008, Ann E. Dunwoody became the first woman to achieve the rank of four-star general in the U.S. Army. Many hailed the appointment as a major step forward for women. Feminist scholars ask why it's taken so long to break "the brass ceiling" and why women are still highly underrepresented at top military levels.

women occupy only 20 percent of the positions in decision-making bodies that are comparable with our Congress. The United States ranks 71st in women's political leadership, well below many African, European, and Asian countries, and even below most of the Arab countries that many Westerners view as repressing women (Inter-Parliamentary Union 2011).

Women make up 51 percent of the U.S. population but only 17 percent of Congress (see Chapter 9). Today, half of all national governments have legal requirements to include women in political positions. As a result, women make up at least 30 percent of the governing bodies in Argentina, Cuba, Finland, India, Sweden, Rwanda, and other countries (Inter-Parliamentary Union 2011). The United States doesn't have such mandates because it ignores political gender inequity (Schmitz 2010).

Feminist theorists maintain that most American women have been shut out of the political process because the United States is still a patriarchal society. Women are successful fund-raisers, vote in higher numbers than men, and often run for office. Compared with men, however, women get considerably less media coverage of their political campaigns or receive coverage that is more disparaging (Zurbriggen and Sherman 2010).

### Critical Evaluation

Feminist scholars go further than conflict theorists in showing that most members of the powerful elite are men. Despite such contributions, critics reject several feminist explanations of political power. They claim, for example, that most women's organizations tend to focus on single issues such as abortion, child care, or domestic violence. Such splintering dilutes women's mobilization within a political party and decreases their chances of winning office.

One might also question many feminists' contention that patriarchy is the root of political power differences between women and men. Some of the most patriarchal societies—those in much of Africa, Latin America, and many Arab countries, for example—have many more women in high-ranking positions in political bodies than does the United States, which professes equality between the sexes. Thus, patriarchy may not be as critical in explaining political gender inequality as many feminist scholars claim.

Some critics also point out that American women have made considerable progress in achieving high-level positions. Contemporary examples include Secretary of State Hillary Rodham Clinton; Supreme Court Justices Ruth Bader Ginsburg, Sonia Sotomayor, and Elena Kagan; Secretary of Homeland Security Janet Napolitano; and ambassador to the United Nations Susan Rice.

## Study Tools

Ready to study? Go to **Sociology CourseMate** at **www.cengagebrain.com** to complete practice quizzes, review flashcards, watch videos, and more.

The family is an important institution in all societies.

Masterfile

# Families and Aging

In 2007, 95-year-old Nola Ochs received a college diploma. The mother of 4, grandmother of 13, and great-grandmother of 15 began taking classes after her husband died in 1972. Instead of viewing aging and her husband's death as setbacks, Ochs forged ahead. She had always yearned for a college education, but was too busy raising their children on the family's farm. Ochs may not be typical, but her determination shows that, throughout the life course, we have choices and can overcome constraints.

> **family** an intimate group consisting of two or more people who (1) live together in a committed relationship, (2) care for one another and any children, and (3) share close emotional ties and functions.

In this chapter, we'll examine how families have changed, their diversity, family conflict and violence, and our aging society. Let's begin by considering what sociologists mean by *family*.

## Key Topics

In this chapter, we'll explore the following topics:

| | |
|---|---|
| **12-1** | What Is a Family? |
| **12-2** | How U.S. Families Are Changing |
| **12-3** | Diversity in American Families |
| **12-4** | Family Conflict and Violence |
| **12-5** | Our Aging Society |
| **12-6** | Sociological Explanations of Family and Aging |

## 12-1 What Is a Family?

Ask five of your friends to define *family*. Their definitions will probably differ not only from each other's but from yours. For our purposes, a **family** is an intimate group consisting of two or more people who (1) live together in a committed relationship, (2) care for one another and any children, and (3) share close emotional ties and functions. This definition includes households (such as those of foster families and same-sex couples) whose members aren't related by birth, marriage, or adoption.

Contemporary household arrangements are complex; family structures vary across cultures and have changed over time. In some societies, a family includes uncles, aunts, and other relatives. In others, only parents and their children are viewed as a family.

### 12-1a How Families Are Similar

The family, a social institution, exists in some form in all societies. Worldwide, families are similar in fulfilling some functions, encouraging marriage, and trying to ensure that people select appropriate mates.

#### Family Functions

Families vary considerably in the United States and globally but fulfill at least five important functions that ensure a society's survival (Parsons and Bales 1955):

- *Sexual regulation.* Every society has norms regarding who may engage in sexual relations, with whom, and under what circumstances. In the

## What do you think?

It should be much harder to get a divorce.

| 1 | 2 | 3 | 4 | 5 | 6 | 7 |
|---|---|---|---|---|---|---|

strongly agree　　　strongly disagree

**incest taboo** cultural norms and laws that forbid sexual intercourse between close blood relatives.

**marriage** a socially approved mating relationship that people expect to be stable and enduring.

**endogamy** cultural practice of marrying within one's group.

**exogamy** cultural practice of marrying outside one's group.

United States, having sexual intercourse with someone younger than 18 (or 16 in some states) is a crime, but some societies permit marriage with girls as young as 8. One of the oldest rules that regulate sexual behavior is the **incest taboo**, a set of cultural norms and laws that forbid sexual intercourse between close blood relatives, such as brother and sister, father and daughter, or uncle and niece.

- *Reproduction and socialization.* Reproduction replenishes a country's population. Through socialization, children acquire language; absorb the accumulated knowledge, attitudes, beliefs, and values of their culture; and learn the social and interpersonal skills needed to function effectively in society (see Chapter 4).

- *Economic security.* Families provide food, shelter, clothing, and other material resources for their members.

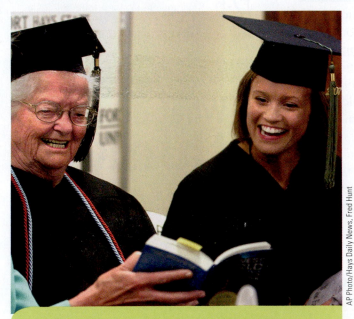

Nola Ochs, 95, and her 21-year-old granddaughter were in the same graduating class at Fort Hays State University in Kansas. When she was handed her degree, the crowd gave Nola Ochs a standing ovation, breaking a rule against applauding until the names of all 2,176 graduates had been read.

- *Emotional support.* Families supply the nurturance, love, and emotional sustenance that people need to be happy, healthy, and secure. Our friends may come and go, but our family is usually our emotional anchor.

- *Social placement.* Initially, our social position is based on our parents' social class. Family resources affect children's ability to pursue opportunities such as higher education, but we can move up or down the social hierarchy in adulthood (see Chapter 8).

Some sociologists include *recreation* as a basic U.S. family function. Although not critical for survival, since the 1950s many parents spend much more time with their children on leisure activities, such as visiting amusements parks and playing video games together (see Coontz 2005).

### Marriage

**Marriage**, a socially approved mating relationship that people expect to be stable and enduring, is also universal. Countries vary in their specific norms and laws dictating who can marry whom and at what age, but marriage everywhere is an important rite of passage that marks adulthood and its related responsibilities, especially providing for a family.

### Endogamy and Exogamy

All societies have rules, formal or informal, defining an acceptable marriage partner. **Endogamy** is a cultural practice of marrying within one's group. The partners are similar in religion, race, ethnicity, social class, and/or age. Across the Arab world and in some African nations, about half of married couples are first or second cousins because such unions reinforce kinship ties and increase a family's resources (Aizenman 2005; Bobroff-Hajal 2006).

**Exogamy** is a cultural practice of marrying outside one's group. In the United States, for example, 24 states prohibit marriage between first cousins, even though violations are rarely prosecuted. In some parts of India, where most people still follow strict caste rules, the government is encouraging exogamy by offering a $1,250 cash award to anyone who marries someone from a lower caste. This is a hefty sum in areas where the annual income is less than half that amount (Chu 2007; see also discussion of castes in Chapter 8).

### 12-1b How Families Differ

Families also differ around the world. Some variations affect the family's structure, whereas others regulate where people reside and who has the most household power and authority.

In China's Himalayas, the Mosuo may be a matriarchal society. For the majority of Mosuo, a family is a household consisting of a woman, her children, and the daughters' offspring. An adult man will join a lover for the night and then return to his mother's or grandmother's house in the morning. Any children resulting from these unions belong to the female, and it is she and her relatives who raise them (Barnes 2006).

## Nuclear and Extended Families

In Western societies, the typical family form is the **nuclear family** made up of married parents and their biological or adopted children. In much of the world, however, the most common family form is the **extended family**, composed of parents and children plus other kin, such as uncles and aunts, nieces and nephews, cousins, and grandparents.

As the number of single-parent families increases in industrialized countries, extended families are more common. By helping out with household tasks and child care, other adult members make it easier for a single parent to work outside the home. Many Americans assume that the nuclear family is the most common arrangement, but such families have declined—from about 40 percent of all U.S. households in 1970 to 20 percent in 2010 (Kreider and Elliott 2009; Tavernise 2011).

## Residence Patterns

How do families differ in their living arrangements? In a **patrilocal residence pattern**, newly married couples live with the husband's family; in a **matrilocal residence pattern**, they live with the wife's family; and in a **neolocal residence pattern**, the couple sets up its own residence.

Around the world, the most common residence pattern is patrilocal. In industrialized societies, married couples are typically neolocal. Since the early 1990s, however, the tendency has increased for young married adults to live with the parents of either the wife or the husband, or sometimes with the grandparents of one of the partners. Such "doubled-up" U.S. households have always existed, but escalated during the Great Recession for economic reasons (DeNavas-Walt et al. 2011; see also Chapter 11). As a result, a recent phenomenon is the **boomerang generation**, young adults who move back into their parents' home after living independently for a while or who never leave home in the first place. Parents try to launch their children into the adult world, but like boomerangs, some keep coming back. Some U.S. journalists call this group *adultolescents* because they're still "mooching off their parents" instead of living on their own.

## Authority and Power

Residence patterns often reflect who has authority and power in the family. In a **matriarchal family system**, the oldest women (usually grandmothers and mothers) control cultural, political, and economic resources and, consequently, have power over males. Some American Indian tribes were matriarchal, and in some African countries, the oldest women have considerable authority and influence. For the most part, however, matriarchal societies are rare.

A more widespread pattern is a **patriarchal family system**, in which the oldest men (grandfathers, fathers, and uncles) control cultural, political, and economic resources and, consequently, have power over women. In some patriarchal societies, women have few rights within or outside the family; they may not be permitted to work outside the home or attend college. In other patriarchal societies, women may have considerable decision-making power in the home but few legal or political rights, such as getting a divorce or running for political office (see Chapter 9).

**nuclear family** a family form composed of married parents and their biological or adopted children.

**extended family** a family form composed of parents and children, as well as other kin.

**patrilocal residence pattern** newly married couples live with the husband's family.

**matrilocal residence pattern** newly married couples live with the wife's family.

**neolocal residence pattern** a newly married couple sets up its own residence.

**boomerang generation** young adults who move back into their parents' home after living independently for a while or who never leave it in the first place.

**matriarchal family system** the oldest women control cultural, political, and economic resources and, consequently, have power over males.

**patriarchal family system** the oldest men control cultural, political, and economic resources and, consequently, have power over females.

In an **egalitarian family system**, both partners share power and authority fairly equally. Many Americans believe they have egalitarian families, but patriarchal families are much more common. For example, employed women shoulder almost twice as much housework and child care as men (see Chapter 9). As you'll see shortly, women are considerably more likely than men to experience intimate partner violence and economic hardship after a divorce. Also, employed women are more likely than men to provide caregiving to aging family members (Cynkar and Mendes 2011).

## Courtship and Mate Selection

Sociologists often describe U.S. dating as a **marriage market**, a courtship process in which prospective spouses compare the assets and liabilities of eligible partners, and choose the best available mate. Marriage markets don't sound very romantic, but such open dating fulfills several important functions: recreation and companionship; a socially acceptable way of pursuing love; opportunities for sexual intimacy and experimentation; and finding a spouse (see Benokraitis 2011).

Many societies, in contrast, discourage open dating because marriage is a family rather than an individual decision that increases solidarity between families and preserves endogamy. Children may have veto power, but they believe that if partners are compatible, love will result. In much of India, marriages are usually carefully arranged to ensure a union that the family deems acceptable. In some of India's urban areas, arranged marriages also rely on nontraditional methods

Does this marriage depict endogamy, exogamy, homogamy, or heterogamy? Why?

Stephen Coburn/Shutterstock.com

(such as online dating services) to find prospective spouses (Cullen and Masters 2008).

Guided by endogamy, in both open and arranged mate selection systems, most people practice **homogamy**, marrying someone who is similar in characteristics such as race, ethnicity, age, education, social class, or religion. *Heterogamy* refers to marrying someone who is different in social class, race, ethnicity, age, or other characteristics. Although not widespread, heterogamy is increasing in some societies. For example, U.S. interracial marriages rose from 3 percent of all marriages in 1980 to 10 percent by 2010, but the spouses are usually similar in social class and age (Wang 2012). Thus, whom we marry is not completely a matter of choice but is limited by cultural mate selection norms and values.

## Monogamy and Polygamy

There are several types of marriages worldwide. In **monogamy**, one person is married exclusively to another person. Where divorce and remarriage rates are high, as in the United States, people engage in **serial monogamy**. That is, they marry several people but one at a time—they marry, divorce, remarry, divorce, and so on. In a classic and often cited study, a well-known anthropologist concluded that only about 20 percent of societies are strictly monogamous; the others permit either polygamy or combinations of monogamy and polygamy (Murdock 1967).

**Polygamy** is a form of marriage in which a man or woman has two or more spouses. There are two types of polygamy: In *polygyny*, one man is married to two or more women; in polyandry, one woman is married to two or more men. Although rare, polyandry exists. Among the Pimbwe people in western Tanzania, Africa, for example, some women have several husbands because of a shortage of women and the difficulty of one man to provide for a family (Borgerhoff Mulder 2009).

Polygyny is common in many societies, especially in some regions of Africa, South America, and the Middle East. Osama bin Laden—who orchestrated the 9/11 terrorist attacks—had 4 wives and 10 children; his father had 11 wives and 54 children (Coll 2008; Grant 2008). Western and industrialized societies forbid polygamy, but there are pockets of isolated polygynous groups in the United States, Europe, and Canada. The Church of Jesus Christ of Latter-Day Saints (Mormons) banned polygamy in 1890 and excommunicates members who follow such beliefs. Still, an estimated 300,000 families in Texas,

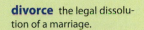
## HOW MUCH DO YOU KNOW ABOUT CONTEMPORARY U.S. FAMILIES AND AGING?

**True or False?**

1. Out-of-wedlock births to teenagers have increased over the past 20 years.
2. Cohabitation (unmarried couples living together) decreases the likelihood of divorce.
3. Almost half of all children live in one-parent households.
4. Having children increases marital satisfaction.
5. About 15 percent of all Americans never marry.
6. Women and men are equally likely to experience violence by an intimate partner.
7. Baby boomers (those born between 1946 and 1964) are the fastest growing segment of the population.
8. Most people age 65 and older are isolated and lonely.

See p. 230 for answers.

© Cengage Learning

Arizona, Utah, and Canada are headed by males of the Fundamentalist Church of Jesus Christ of Latter Day Saints (FLDS), a polygynous sect that broke away off from the mainstream Mormon church more than a century ago. Wives who have escaped from these groups have reported forced marriage between men in their 60s and girls as young as 10 years old, who have experienced sexual abuse and incest (Janofsky 2003; Madigan 2003).

# 12-2 How U.S. Families Are Changing

The American family has changed dramatically over the past half century. Noteworthy changes include divorce, singlehood, cohabitation, unmarried parents, and two-income families.

## 12-2a Divorce

Couples of all ages experience **divorce**, the legal dissolution of a marriage, but how long are marriages lasting? On average, first marriages last about 8 years. Among women who divorced in 2010, 42 percent did so within the first 9 years of marriage, compared with 1 percent who divorced after 50 or more years of marriage (Kreider and Ellis 2011; Payne and Gibbs 2011).

### Trends

Before the twentieth century, when life spans were much shorter, marriages typically ended with the death of one of the spouses. Now, over a lifetime, about 46 percent of American marriages end in divorce (Schoen and Canudas-Romo 2006). The U.S. divorce rate rose steadily during the twentieth century, peaked during the early 1980s, and then declined (see *Figure 12.1*). Thus, divorce rates are *lower* today than they were between 1980 and 2009.

## FIGURE 12.1
## Divorce Rates in the United States, 1870–2010

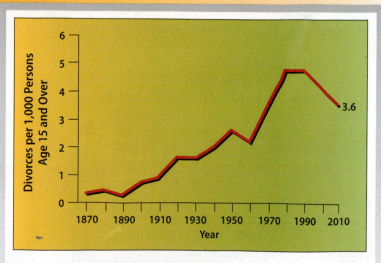

Note: The *divorce rate* is the number of divorces that occur in a given year for every 1,000 marriages for women aged 15 and older.
Sources: Based on Plateris 1973; Tejada-Vera and Sutton 2009, Table A; U.S. Census Bureau 2010, Table 78.

## Why Do People Divorce?

There are many *macro-level reasons* for divorce. It's easier to get a divorce than in the past because all states have enacted **no-fault divorce** laws, meaning that neither partner needs to establish guilt or wrongdoing on the part of the other. Legally valid reasons for divorce range from infidelity and abuse to simple incompatibility. Technological advances, such as the Internet, have also made divorce more accessible. Some do-it-yourself divorce kits available online cost as little as $50 for all the necessary court forms and documents.

The economy also affects divorce rates. Divorces fell during the Great Depression, but the Great Recession may have increased marital stability (or postponed separations) because it was too costly to live alone, one spouse might have lost health benefits, divorce itself can be expensive and so forth. On the other hand, high levels of credit card debt, unemployment, and foreclosures can lead to more conflict, unhappiness, and divorce (Dew 2009; Cohen 2012). Until there is more research, the link between divorce rates and the recent recession is speculative.

*Demographic variables* also help explain divorce rates. Marrying at an early age—especially younger than 18—is one of the strongest predictors of divorce. For example, 48 percent of first marriages of females younger than 18 dissolve within 10 years, compared with 24 percent of first marriages of women who are at least 25 at the time of the wedding (Bramlett and Mosher 2002). In most cases, young couples aren't prepared, emotionally or financially, to handle marital and parental responsibilities.

### HOW MUCH DO YOU KNOW ABOUT CONTEMPORARY U.S. FAMILIES AND AGING?

**If you answered true to any of the statements, you're wrong. Based on material in this chapter, all of the statements are FALSE.**

1. Out-of-wedlock births to teenagers have decreased over the past 20 years, especially since the early 2000s.
2. Couples who cohabit before marriage have a higher divorce rate than those who don't.
3. A majority of U.S. children (70 percent) live in homes with married parents.
4. Generally, child-rearing decreases marital satisfaction for both partners.
5. At least 90 percent of Americans will marry at least once during their lifetime.
6. Women are almost three times more likely than men to experience violence by an intimate partner.
7. People age 85 and older are the fastest growing segment of the U.S. population.
8. Most people age 65 and older visit friends and relatives daily, and two-thirds work or engage in numerous volunteer activities.

© Cengage Learning

Americans without college degrees are three times more likely than those with a college education to divorce in the first 10 years of marriage (Wilcox and Marquardt 2011). The latter have lower divorce rates not because they're smarter but because going to college postpones marriage. As a result, better educated couples are often more mature and capable of dealing with personal crises. They also have higher incomes and better health care benefits, which lessen marital stress over financial problems (Glenn 2005; Kreider and Ellis 2011).

There are also *micro-level (interpersonal) reasons* for divorce. The most common are infidelity, communication and financial problems, substance abuse, and spousal abuse. For example, among those age 40 and older, 23 percent of women, compared with only 8 percent of men, say that verbal, physical, or emotional abuse was "the most significant reason for the divorce" (Montenegro 2004; Glenn 2005).

These and other reasons have led to a greater acceptance of divorce. In 2008, 70 percent of Americans said that divorce is "morally acceptable" (up

Mike Kemp/Rubberball/Jupiter Images

from 56 percent in 2001). In another national survey, 67 percent of Americans said that children are better off if their unhappy parents get a divorce rather than remain married (Taylor et al. 2007; Saad 2008).

The major positive outcome of divorce is that it provides an escape for people in miserable marriages. Despite the high divorce rate, most people aren't disillusioned about marriage. Indeed, nearly 85 percent of Americans who divorce remarry, half of them within 4 years (Kreider and Ellis 2011). Through remarriage, many couples form a **stepfamily**, a household in which two adults who are biological or adoptive parents, with a child from a prior relationship, marry or cohabit. In 2010, of the almost 89 million American children living in families, 93 percent were biological children, 5 percent were stepchildren, and 2 percent were adopted (Lofquist et al. 2012).

## 12-2b Singlehood and Postponing Marriage

In 2011, one of the most publicized media stories was that only 51 percent of U.S. adults were married, compared with 72 percent in 1960. In 2010, 40 percent of Americans believed that marriage is obsolete compared with 28 percent in 1978. And, 28 percent of all households now consist of just one person—the highest level in U.S. history (Cohn et al. 2011; Klinenberg 2012).

Some syndicated columnists bemoaned such statistics and wrote articles on how to revive marriage in America (see McManus 2012). Is such hand-wringing warranted?

### Trends

The number of single Americans increased from 38 million in 1970 to 100 million in 2010, comprising 44 percent of those age 18 and older. Singles include people who are divorced and widowed, but those who have never married make up the largest and fastest growing segment of the single population—27 percent in 2010, up from 22 percent in 1960 (U.S. Census Bureau 2012).

Are Americans giving up on marriage? No, because about 93 percent will marry at least once during their lifetime. Instead, many people are postponing marriage. In 1960, the median age at first marriage was 20 for

## Sociology in Your Life

How have U.S. marriage rates, age at first marriage, and cohabitation trends changed since 1960? Find out at the Pew Research Center link at CengageBrain.com.

women and 23 for men. By 2011, these ages had increased to 27 for women and 29 for men, the oldest average ages for first marriages ever recorded by the U.S. Census Bureau (U.S. Census Bureau News 2012).

### Why Are Singles Postponing Marriage?

There are many *micro-level (individual) reasons* for postponing marriage. A survey of Generation Y (those born between 1980 and 2000) found that many believe in marriage and lifelong commitment, but are more relaxed than previous generations about sex, hooking up, "stayovers" (staying together several nights a week), and living together. Because many of these young adults grew up in divorced families, 63 percent believe that most marriages aren't happy, and they're postponing marriage and children until they feel they can do a better job that their parents did (Jamison and Ganong 2011; Zimmerman 2012). Thus, marriage and parenthood—once seen as prerequisites for adulthood—are now viewed more as "lifestyle choices."

*Demographic factors* also help explain the large number of single people. The longer one waits to marry,  the smaller the pool of eligible partners because more of one's peers have paired off. The likelihood of singlehood also varies by education level. For example, 85 percent of unmarried people age 25 and older are high school graduates, compared with 25 percent who have at least a bachelor's degree (U.S. Census Bureau News 2011). More education often means more income, and more income reduces financial barriers to marriage.

*Macro-level variables* have also increased the number of single people. For example, because it's difficult to juggle a job and a family, many women have chosen to advance their education and careers before marrying and having children (Kent 2011). For men, the well-paid blue-collar jobs that once enabled high school graduates to support families are mostly gone, and unemployment rates are high. Generally, the less money a male has, the less likely he is to marry (Taylor 2010). Such factors delay marriage and encourage cohabitation.

### 12-2c Cohabitation

Another major change affecting U.S. families is an increase in **cohabitation**, an arrangement in which two

**stepfamily** a household in which two adults who are biological or adoptive parents, with a child from a prior relationship, marry or cohabit.

**cohabitation** an arrangement in which two unrelated people aren't married but live together and have a sexual relationship.

People live together rather than marry for a variety of reasons. Do you think that women or men benefit more from cohabitation? Why?

unrelated people aren't married but live together and have a sexual relationship (shacking up, in plain English). Because it's based on emotional rather than legal ties, "cohabitation is a distinct family form, neither singlehood nor marriage" (Brown 2005: 33).

## Trends

In 2011, married couples comprised 48 percent of all households, a sharp decline from 78 percent in 1950 (U.S. Census Bureau Factfinder 2011). Some of the decline was due to a small percentage of people who'll probably never marry and those who haven't remarried yet, but most of the increase was due to cohabitation.

The number of unmarried U.S. heterosexual couples has increased more than 17-fold—from 0.4 million in 1960 to almost 8 million in 2010. This number rises by at least another 551,000 if we include same-sex unmarried partners. However, only about 10 percent of the population is cohabiting at any given time (Lofquist 2011; Wilcox and Marquardt 2011).

## Why Do People Cohabit?

People who cohabit are diverse in age. Only 20 percent are 24 or younger, whereas more than half of those ages 30 to 49 have cohabited at some point in their lives (Martinez et al. 2006; Taylor 2010). Older people sometimes cohabit rather than marry because of financial reasons. A 72-year-old woman who lives with her 78-year-old partner, for example, has no intention of getting married because she'd forfeit her late husband's pension: "My income would be cut by

## TABLE 12.1

# Some Benefits and Costs of Cohabitation

| BENEFITS | COSTS |
|---|---|
| • Couples have the emotional security of an intimate relationship but can also maintain their independence by spending time with their friends separately and visiting family members alone (McRae 1999). | • Cohabitants have fewer legal rights than married couples. For example, there's no automatic inheritance if a partner dies without a will, and it's more difficult to collect child support from a cohabiting partner than an ex-spouse (Silverman 2003; Grall 2011). |
| • Couples can save money by sharing living expenses, dissolve the relationship without legal problems, and leave the relationship more easily if it becomes abusive (DeMaris 2001; Silverman 2003). | • Women in cohabiting relationships do more of the cooking and other household tasks than do wives (Coley 2002; Ciabattari 2004). There's also no evidence that cohabiting women can leave easily if the man is abusive. |
| • Couples find out how much they really care about each other when they have to cope with unpleasant realities, such as a partner who doesn't pay bills or has low hygiene standards. | • People who cohabit before marriage tend to have fewer problem-solving skills (such as controlling anger) than those who don't cohabit. Cohabitors are also more likely than biological parents to abuse their children sexually, physically, and emotionally (Cohan and Kleinbaum 2002; Popenoe and Whitehead 2002). |
| • Children in cohabiting households can reap economic advantages from living with two adult earners and enjoy a stable home life (Kalil 2002; Cabrera et al. 2008). | • Among cohabitors who have a college degree, the annual median income is $5,000 lower than for married couples (Fry and Cohn 2011). Because cohabiting parents are more than twice as likely as married parents to break up, children's emotional and financial insecurity often increases (Wilcox and Marquardt 2011). |

© Cengage Learning

## FIGURE 12.2
## Percentages of Births to Unmarried Women, by Race and Ethnicity, 2009

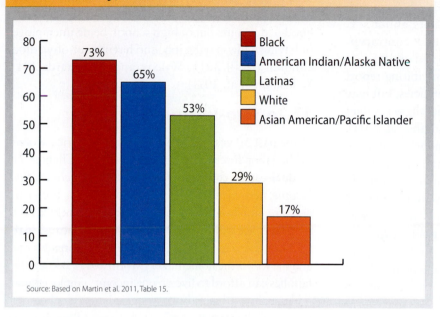

Legend:
- Black
- American Indian/Alaska Native
- Latinas
- White
- Asian American/Pacific Islander

Bar values: Black 73%, American Indian/Alaska Native 65%, Latinas 53%, White 29%, Asian American/Pacific Islander 17%

Source: Based on Martin et al. 2011, Table 15.

2011). Nonmarital childbearing has increased significantly since 1970, but most of the research still focuses on women.

### Trends

In 1950, only 3 percent of all U.S. births were to unmarried women. In 2009, there were almost 1.7 million such births, accounting for 41 percent of all U.S. births. Thus, 4 in 10 American babies are now born outside of marriage, a new record (Hamilton and Ventura 2012).

Births to unmarried women vary widely across racial-ethnic groups. White women have more out-of-wedlock babies than do other groups. Proportionately, however, nonmarital birth rates are highest for black women and lowest for Asian American women (see *Figure 12.2*).

Unmarried teenage births peaked in the early 1990s, have declined steadily since then, and are now at their lowest level since 1968, with 34 births per 1,000 women ages 15 to 19 (Kost and Henshaw 2012; Hamilton and Ventura 2012). Teenagers account for 21 percent of all nonmarital births, compared with 72 percent for women ages 20 to 34. Thus, the number of births to unmarried women ages 20 to 34 is almost four times greater than for teenagers (Martin et al. 2011).

Still, 87 percent of teenage births are nonmarital, compared with 62 percent for women in their 20s, and 20 percent for those ages 30 to 34 (Martin et al. 2011). Because teenage mothers are less likely than their older counterparts to have the resources, parenting skills, and maturity to raise healthy children, their offspring are especially vulnerable to poverty and neglect (Carlson et al. 2005).

$500 a month if I got married, and we can't afford that" (Silverman 2003).

Does cohabitation lead to a better marriage? No. The probability of a first marriage ending in separation or divorce within 5 years is 20 percent for couples who didn't cohabit before marriage, but 49 percent for those who did. Thus, living together before marriage often increases a couple's risk for divorce (Bramlett and Mosher 2002; Lichter et al. 2006; Stevenson and Wolfers 2007).

Why do cohabiters generally experience higher divorce rates? Some are unsuccessful marriage partners because of drug abuse or mental health problems, chronic unemployment, and sexual infidelity. Cohabitants are less likely than noncohabitants to put effort into the relationship and to compromise, more likely to dissolve a relationship than work on it, and to have poorer communication skills—characteristics that tend to lead to divorce (Dush et al. 2003; Tach and Halpern-Meekin 2009). Because of the higher divorce rate it leads to, some social scientists are adamantly opposed to cohabitation, but others see benefits (see *Table 12.1*).

### 12-2d Nonmarital Childbearing

Nearly half of American men ages 15 to 44 report that at least one of their children was born outside of marriage, and 31 percent say that all of their children were born outside of marriage (Livingston and Parker

> The number of births to unmarried women ages 20 to 34 is almost four times greater than for teenagers.

## Why Has Nonmarital Childbearing Increased?

On a *micro (individual) level,* many teens don't communicate with their parents or other adults about sex, condoms, and contraception. About 20 percent of married women and 65 percent of those who aren't married or cohabiting report having unintended or unwanted pregnancies, but may not get abortions because of personal or religious values ("About Teen Pregnancy" 2011; Wildsmith et al. 2011).

*Demographic variables* also affect nonmarital childbearing. Social class, especially education, is a key factor in out-of-wedlock births. For example, 63 percent of these births are to women who have not graduated from high school, compared with only 6 percent to women who have either a college or graduate/professional degree. Educated women usually have higher incomes to pay for birth control, more awareness of family planning, and more decision-making power in their relationships, such as insisting that men use condoms (Chandra et al. 2005; Guttmacher Institute 2009).

Many researchers also attribute rising nonmarital childbearing rates to *macro-level variables,* especially the economy. You saw earlier that many people postpone marriage for financial reasons and cohabit instead. More than half of all nonmarital births occur to cohabiting couples. Because many of these unions are short-lived and most fathers don't pay child support, mothers and children often experience poverty. Poverty increases the likelihood of dropping out of high school, being unemployed or having a low-paying job, and having out-of-wedlock children (Grall 2011; Wilcox and Marquardt 2011; Wildsmith et al. 2011).

### 12-2e Two-Income Families

In the past 50 years, the proportion of married women in the labor force has almost tripled (see Chapter 11). In **dual-earner couples** (also called *dual-income, two-income, two-earner,* or *dual-worker couples*), both partners are employed outside the home. The best kind of marriage, according to 30 percent of Americans, is the traditional one in which the husband is the breadwinner and the wife is the homemaker (Morin 2011). Because few families can afford to live on one income, employed married couples with children younger than 18 make up 66 percent of all married couples (U.S. Census Bureau 2012).

For several decades, a number of sociologists described employed women, especially mothers, as working a *second shift*—having to perform housework and child care tasks after coming home from a job. These days, however, *both* employed partners often experience a second shift. Employed mothers shoulder twice as much child care and housework as their spouses, but employed fathers often work much longer hours than do mothers. Thus, the second shift can be equally stressful for both parents (Hochschild 1989; Bianchi et al. 2006).

Every year, the media feature and applaud stay-at-home dads, but their numbers are negligible. In 2010, 176,000 fathers cared for children younger than 15 while their wives worked outside the home. Among all married two-parent families, only about 3 percent of fathers are stay-at-home dads during a given year (U.S. Census Bureau News 2012). The role is usually temporary and due to unemployment or health problems.

© Getty Images/Comstock/Jupiter Images

## 12-3 Diversity in American Families

*Family diversity* refers to "the variety of ways that families are structured and function to meet the needs of those defined as family members" (Stewart and Goldfarb 2007: 4). Three major sources of diversity are social class, race and ethnicity, and gay and lesbian households.

### 12-3a Social Class Variations

Social class is a critical factor in determining a family's financial security, and, consequently, its members' opportunities and life outcomes. The 28 percent of U.S. children fortunate enough to be born to higher income parents (those who earn at least $131,000 a year) have a good chance of enjoying good health, achieving a high educational level, and suffering few economic hardships. In contrast, in 2011 about 19 percent of families (up from 10 percent in 1996) lived in extreme poverty, surviving on $2 or less per person per day (Federal Interagency Forum on Child and Family Statistics 2011; Shaefer and Edin 2012).

Several factors predict children's likelihood of experiencing poverty and extreme poverty. In particular, women, single-parent families (especially those headed by women), African Americans and Latinos, and adults with low levels of educational attainment and/or limited work experience are the most likely to be poor (Redd et al. 2011).

The most disadvantaged parents include those who have children with more than one partner. Among unmarried parents, 63 percent have multiple partners. In these relationships, most of the fathers invest little time and money in their children because the fathers are young, have low education levels, or have been incarcerated. The fathers don't live with their children, know little about them, and move from one sexual relationship to another. Children whose parents have offspring with multiple partners are likely not only to be poor, but also to be depressed and to show more "externalizing behaviors" such as acting out and aggression, dropping out of high school, and being delinquent (Scommegna 2011b; see also Dodson and Luttrell 2011).

## 12-3b Racial and Ethnic Families

You saw in Chapter 10 that there is considerable variation across racial-ethnic groups. Among Latino and Asian American families in particular, family structures vary depending on the members' country of origin, time of arrival, status as either immigrants or refugees, and socioeconomic status. Latino and Asian families are more likely to be poor if they are recent immigrants, don't speak English, and have low education levels (DeNavas Walt et al. 2011; Kochhar et al. 2011; Lopez and Velasco 2011). In contrast, at least 41 percent of Middle Eastern parents have at least a college degree college, compared with 30 percent of the general U.S. adult population, providing academically successful role models for their children (Brittingham and de la Cruz 2005; U.S. Census Bureau 2012).

Until 1980, married couple families were the norm in African American families. Since then, black children have been more likely than children in other

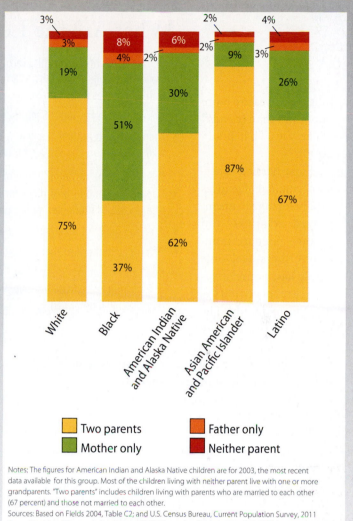

### FIGURE 12.3
### Where American Children Live, 2011

Legend:
- Two parents (yellow)
- Mother only (green)
- Father only (orange)
- Neither parent (red)

Notes: The figures for American Indian and Alaska Native children are for 2003, the most recent data available for this group. Most of the children living with neither parent live with one or more grandparents. "Two parents" includes children living with parents who are married to each other (67 percent) and those not married to each other.

Sources: Based on Fields 2004, Table C2; and U.S. Census Bureau, Current Population Survey, 2011 Annual Social and Economic Supplement 2011, November. Retrieved March 3, 2012 (www.census.gov).

racial-ethnic groups to grow up with only one parent, usually the mother (see *Figure 12.3*). About 67 percent of Latino children now live in two-parent families, down from 78 percent in 1970 (Lugaila 1998). These shifts reflect a number of social and economic factors: postponement of marriage, high divorce and separation rates, low remarriage rates, male unemployment, and out-of-wedlock births, especially among teenagers.

Extended families are especially common in Latino, black, American Indian, and Asian families. For many Latinos, *familism*—the belief that family relationships take precedence over individual decisions—and the strength of the extended family have traditionally provided emotional and economic support. In a national poll, for example, 82 percent of Latinos, compared with 67 percent of the general U.S. population, said that "relatives are more important than friends" ("The Ties That Bind" 2000).

Some African American nuclear families expand to take in unemployed relatives and become extended families. Others welcome **fictive kin**, nonrelatives who are accepted as part of the family. The ties with fictive kin may be as strong as those formed by blood or marriage because these household members provide support—such as caring for children—when parents are employed or negligent (Billingsley 1992; Dilworth-Anderson et al. 1993).

Parents in all social classes want their children to be successful. Low-income black parents often have few contacts with any community institutions beyond church and school, and they depend on both to help their children succeed (Willie and Reddick 2003). Many Asian American parents, regardless of social class, teach their children to excel in school and endure hardships—even selling their house—to ensure the best college opportunities for their children (Chan 1997; Fong 2002).

Besides these racial-ethnic family variations, ethnically and racially mixed marriages have increased (Wang 2012). This means that more children than ever before identify with more than one racial-ethnic

## TABLE 12.2

# What Are the Legal Benefits of Marriage?

Some states, counties, and cities recognize civil unions and domestic partnerships (see Chapter 9). However, there are more than 1,128 federal benefits and protections tied to marriage that are denied to unmarried couples (U.S. General Accounting Office 2004). For example, partners who aren't married:

- Aren't eligible for Social Security benefits if a partner dies
- Aren't entitled automatically to a share of the property if there is no will, and even if there is a will, a court can appoint a stranger to administer the estate
- Are unlikely to be required to pay child support or to have visitation rights
- Can't collect a one-time payment of $100,000 in line-of-duty benefits paid to a surviving spouse of a deceased firefighter, public prosecutor, police officer, or corrections officer
- Aren't eligible for veterans' compensation if a partner is killed or disabled in action
- Often pay higher federal and state income taxes because they can't file a joint return
- Don't have automatic privileges for hospital visits, access to intensive care, transferring a partner to a different health care facility, making decisions about organ donations, or deciding where the deceased will be buried

© Cengage Learning

group; how successful they'll be depends largely on their parents' social class.

## 12-3c Gay and Lesbian Families

Nationally, of the 594,000 same-sex households in 2010, 19 percent reported having children (Lofquist 2011). Gay and lesbian parents are also raising 4 percent of all adopted and 3 percent of all foster children (Gates and Ost 2004; Gates et al. 2007). Thus, large numbers of lesbians and gay men, single or partnered, are parents.

In most respects, lesbian and gay families are like heterosexual families: The parents must make a living, may disagree about finances, and must develop problem-solving strategies. There are three major differences, however. One is that same-sex parents receive less social support from their relatives than heterosexual parents do. Second, gay and lesbian parents face the added burden of raising children who may experience discrimination because of their parents' sexual orientation (Kurdek 2004). A third difference is that banning same-sex marriage denies gay and lesbian parents and their children numerous benefits that married couples and their children enjoy (see *Table 12.2*).

© MBI/Alamy

Many Middle Eastern American children (84 percent) live with both parents, compared with 67 percent of all U.S. children. Nuclear families are the norm, but extended family ties are important.

# 12-4 Family Conflict and Violence

Conflict is a normal part of family life, but violence is *not* normal. Over a lifetime, we are much more likely to be assaulted or killed by a family member than by a stranger (Truman 2011).

## 12-4a Intimate Partner Violence and Abuse

**Intimate partner violence (IPV)** occurs between people in a close relationship. The term *intimate partner* refers to current and former spouses, couples who live together, and current and former boyfriends or girlfriends.

### Trends

IPV is pervasive in U.S. society. In a recent national study, 36 percent of women and 29 percent of men said that they have been victims of IPV at some time in their lives (Black et al. 2011). Nearly half of all U.S. women and men have experienced psychological aggression by an intimate partner, but female victims outnumber males in other forms of violence (see *Figure 12.4*). These numbers are conservative because many people, both women and men, are too ashamed to report the victimization, believe that no one can help, or fear reprisal.

Each year, IPV results in about 1,200 deaths and 2 million injuries among women, compared with 330 deaths and nearly 600,000 injuries among men, and is the leading cause of death for women ages 15 to 44 (Catalano 2007; Black and Breiding 2008; Tessier 2008). Women are much more likely than men to experience IPV over a lifetime regardless of age, race or ethnicity, annual household income, or education level (Fox and Zawitz 2007; Black et al. 2011).

**intimate partner violence (IPV)** abuse that occurs between people in a close relationship.

## FIGURE 12.4

### Lifetime Prevalence of Experiencing Intimate Partner Violence, by Sex, 2010

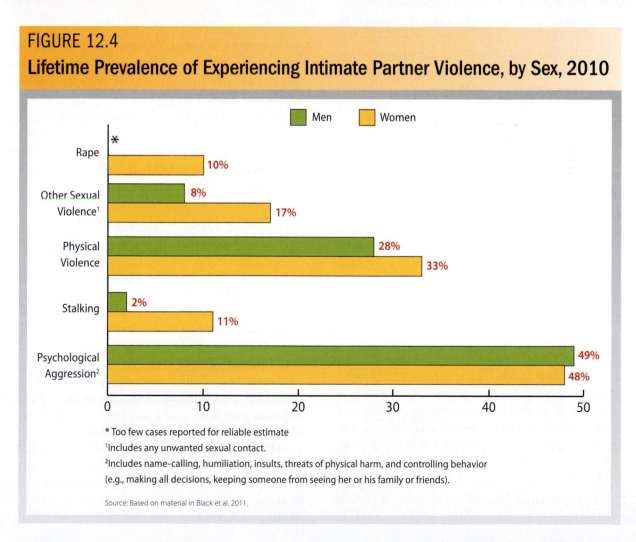

* Too few cases reported for reliable estimate
[1] Includes any unwanted sexual contact.
[2] Includes name-calling, humiliation, insults, threats of physical harm, and controlling behavior (e.g., making all decisions, keeping someone from seeing her or his family or friends).

Source: Based on material in Black et al. 2011.

**child maltreatment** (also called *child abuse*) a broad range of behaviors that place a child at serious risk, including physical and sexual abuse, neglect, and emotional mistreatment.

## Why Does Intimate Partner Violence and Abuse Occur?

Aside from being male, there is no "typical" batterer, but some characteristics are common to abusers. Some reflect *macro-level* influences, such as unemployment and poverty. For example, although IPV cuts across all social classes, women living in households with an annual income of less than $7,500 are nearly seven times more likely to be abused than those living in households with an annual income of $75,000 or more, because poverty and unemployment increase the likelihood of stress and violence (Thompson et al. 2006).

Cultural factors, such as a strong orientation toward family and community, especially common among Latino and Asian American families, may increase the likelihood of violence against women who don't "obey" their husbands. Women in Latino or Asian families, especially those who don't speak English well, may not report marital violence because they fear deportation or being ostracized by their community, are dependent on batterers for economic survival, or don't know about or trust social service organizations that provide help (National Latino Alliance . . . 2005; Gorman 2010).

*Micro-level* variables also increase the likelihood of IPV. For example, violence escalates if one or both partners abuse drugs, if they have more children than they can afford (which intensifies financial problems), or if either partner has been raised in a violent household where abuse was common in resolving conflict (Benson and Fox 2004). Couples fight about a variety of things, but the most common disagreements have four sources that can escalate into abuse and violence: gender role expectations (such as who does what housework); money (saving and spending); children (especially discipline); and infidelity, both personally and online (Anderson and Sabatelli 2007; Benokraitis 2011).

## 12-4b Child Maltreatment

**Child maltreatment** (also called *child abuse*) includes a broad range of behaviors that place a child at serious risk or result in serious harm, including physical and sexual abuse, neglect, and emotional mistreatment. The victims often experience several types of maltreatment.

### Trends

In 2010, an estimated 754,000 children were victims of child maltreatment, representing 10 per 1,000 U.S.

The movie *Precious* (2009) received many awards, but some critics contended that the film stereotyped black female teenagers—especially those living in low-income households—as obese, illiterate, and living with abusive mothers (Amusa 2010). If you've seen the movie, do you agree with such criticism or not? Why?

children. Because only a fraction of the total number of child victimization cases is reported, however, federal agencies assume that millions of American children experience abuse and neglect on a daily basis (U.S. Department of Health and Human Services 2011).

Almost 82 percent of the perpetrators are one or both parents, 6 percent are relatives, and another 5 percent are a parent's girlfriend or boyfriend. Nearly 80 percent of the children who die of abuse and neglect are younger than 3 years old; of these, almost half are infants (U.S. Department of Health and Human Services 2011).

### Why Does Child Maltreatment Occur?

The reasons for both IPV and child maltreatment are similar. On a *macro level*, for example, economic hardship increases the likelihood of stress that leads to child abuse, including infant deaths due to severe head traumas (Berger et al. 2011). On a *micro level*, child mistreatment is most common in single-parent and stepparent households, those with parental substance abuse and mental illness, if there is adult and sibling violence, and if the child has emotional or developmental problems (Finkelhor et al. 2011).

Almost 26 percent of all children live in homes where parents or other adults engage in violence (Hamby et al. 2011). Whether children are targets of abuse or see it, a growing body of research has linked childhood experience of violence with lifelong developmental problems, including depression, delinquency, suicide, alcoholism, low academic achievement, unemployment, and medical problems in adulthood (Currie and Tekin 2006; McDonald et al. 2006; Putnam 2006; Finkelhor et al. 2011).

## 12-4c Elder Abuse and Neglect

**Elder abuse** (sometimes called *elder mistreatment*) is any knowing, intentional, or negligent act by a caregiver or other person that causes harm to people age 65 or older. This term includes physical, psychological, and sexual abuse; neglect; isolation from family and friends; deprivation of basic necessities, such as food and heat; and not providing needed medications.

### Trends

Family members and acquaintances mistreat an estimated 5 percent of people age 65 and older every year. Some researchers call elder abuse "the hidden iceberg" because about 93 percent of cases aren't reported to police or other protective agencies (National Center on Elder Abuse 2005).

Sylvia Bouchard/Shutterstock.com

Who are the victims? About 83 percent are white; the average age is 76; 76 percent are women; 84 percent live in their own homes; 86 percent have a chronic disease or other health condition; 57 percent are married or cohabiting; 53 percent have not graduated from high school; 50 percent suffer from dementia, Alzheimer's, or other mental illness; 46 percent feel socially isolated; and the average combined household income is less than $35,000 a year (Acierno et al. 2009; Jackson and Hafemeister 2011). These data suggest that the most likely victims of elder abuse are those who are the most vulnerable physically, mentally, socially, and financially.

Who are the perpetrators? Most are adult children, spouses or cohabiting partners, or other family members. Less than a third are acquaintances, neighbors, or nonfamily service providers. The average age of the abuser if 45; 77 percent are white; 61 percent are males; 82 percent have a high school diploma or less; 50 percent abuse alcohol and/or other drugs; 46 percent have a criminal record; 42 percent are financially dependent on the elder; 37 percent live with the elder; 29 percent are chronically unemployed; and 25 percent have mental health problems (Jackson and Hafemeister 2011).

### Why Does Elder Abuse and Neglect Occur?

Family members neglect or abuse older people for a variety of reasons. On a *micro level*, for example, abuse of alcohol and other drugs is more than twice as likely among family caregivers who abuse elders as among those who don't. Both victims and offenders also report a childhood history of witnessing or experiencing family violence, poor family relationships in the past and currently, and communication problems.

On a *macro level*, a shared residence is a major risk factor for elder mistreatment because the caregiver(s) may depend on the older person for housing whereas the elder is dependent on the caregiver(s) for physical help. These situations increase the likelihood of everyday tensions and conflict. Also, fewer resources, such as education and income, increase the risk of experiencing or inflicting abuse.

> **elder abuse** (sometimes called *elder mistreatment*) any knowing, intentional, or negligent act by a caregiver or other person that causes harm to people age 65 or older.

# 12-5 Our Aging Society

**M**any older Americans are vigorous and productive, but others experience physical and mental limitations as they age. (Researchers often use the terms *elderly, aged,* and *older people* interchangeably.) Let's begin by considering what we mean by "old."

## 12-5a When Is "Old"?

What images come to mind when you hear the word *old*? My mother once remarked, "It's very strange. When I look in the mirror, I see an old woman, but I don't recognize her. I know I'm 86, but I feel at least 30 years younger." She knew her health was failing, yet my mother's identity, like that of many older people, came from within, despite her chronological age.

"Old" is a social construction. In many African nations, where people rarely live past 50 because of diseases, 40 is old. In industrialized societies, where the average person lives to at least 75, 40 is considered young. Still, regardless of how we feel physically, society usually defines old in terms of chronological age. In the United States, for instance, people are

Between the ages of 65 and 91, avid mountaineer Hulda Crooks scaled Mount Whitney (the highest mountain in the continental United States) 23 times. She died at the age of 101 in 1997.

AP Photo/Itsuo Inouye

71 years in 1970 (Administration on Aging 2011; U.S. Census Bureau 2012).

The number of Americans age 65 and older is booming. Almost 41 million (13 percent of the total population) are 65 or older, a dramatic increase from just 4 percent in 1900. One of the fastest growing groups is the oldest-old, whose numbers increased from 100,000 in 1900 to 4.2 million in 2000. By 2030, this group will comprise almost 3 percent of the population. By 2020, about 214,000 Americans will be *centenarians*, age 100 or older. If we continue to decrease mortality, some predict, many children born since 2000 will celebrate their 100th birthdays in the twenty-second century (He and Muenchrath 2011; Scommegna 2011b; U.S. Census Bureau 2012).

**gerontologists** scientists who study the biological, psychological, and social aspects of aging.

typically deemed old at age 65, 66, or 67 because they can retire and become eligible for Medicare and Social Security benefits.

**Gerontologists**—scientists who study the biological, psychological, and social aspects of aging—emphasize that older people shouldn't be lumped into one group. Instead, there are significant differences between the *young-old* (65–74 years old), the *old-old* (75–84 years old), and the *oldest-old* (85 years and older) in their ability to live independently or to work and in their health needs. Generally, for example, a 65-year-old woman is much less likely than an 85-year-old woman to need caregiving from family and friends.

## 12-5b Life Expectancy and Multigenerational Families

In 1800, the average person's chance of living to 100 was roughly 1 in 20 million; today it's 1 in 500 (Jeune and Vaupel 1995 and author's calculations). More people are reaching age 65 than ever before. For example, an American child born in 2009 has a *life expectancy*—the average length of time people of the same age will live—of 78 years (75 for men and almost 81 for women), compared with 47 years in 1900 and

## Sociology in Your Life

The world population is growing and aging. How will these changes affect industrialized and developing nations in 2050? To find out, see the world population clocks at CengageBrain.com.

### FIGURE 12.5
### The Young and the Old in the United States, 1900–2030

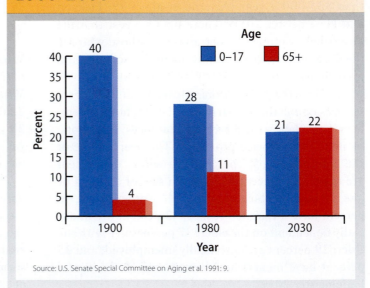

Source: U.S. Senate Special Committee on Aging et al. 1991: 9.

© iStockphoto.com/David Freund

Today, you have a 1 in 500 chance of living to be 100.

The number of older Americans has increased, whereas the proportion of young people has decreased. By 2030, there will be more older than young people in the United States (see *Figure 12.5*). One result of longer life spans is the increase in multigenerational families. Many young children enjoy relationships not only with their grandparents, but with their great-grandparents and even great-great-grandparents. On the other hand, millions of adults spend more years caring for frail and elderly parents, grandparents, and other relatives. These adults are often referred to as the **sandwich generation** because they are between two other generations, caring for their own children and their aging parents.

### 12-5c Who Will Care for Our Graying Population?

In 2002, 35 percent of Americans worried about not having enough money for retirement. In 2011, across all income groups, 77 percent of Americans between the ages of 30 and 49, and 70 percent of those ages 50 to 64, worried about retirement funds (Mendes 2011).

There are three reasons why such concerns are warranted. First, more than 1 in 6 Americans (16 percent of men and 20 percent of women), including 16 percent of those age 65 and older, care for an older family member or relative. Doing so means losing wages for an average of seven workdays per year, with costs to the U.S. economy of $25.2 billion a year in lost productivity (Cynkar and Mendes 2012; Mendes 2012; Witters 2012).

The caregiving may be a labor of love, but caregivers often report emotional stress, income loss, and physical exhaustion. There's also a question about the availability of future caregiving for aging parents because of high divorce rates, the rise of single-parent families, and cohabitation. The people in these groups are less likely to form strong enough bonds with their parents to care for them when the parents are elderly (Jacobsen et al. 2011).

A second reason why Americans should worry about retirement, regardless of age, is that the old-age dependency ratio is decreasing. The **old-age dependency ratio** (sometimes called the *elderly support ratio*) refers to the number of working-age adults ages 18 to 64 for every person age 65 and older who's not in the labor force. In effect, this ratio is an indicator of the burden on the working-age population of supporting those 65 and older who aren't employed.

As *Figure 12.6* shows, the old-age dependency ratio has decreased considerably. By 2050, fewer than 3 working-age people will be supporting each older person, compared with 14 workers per older person in 1900. Thus, many people will have to pay much higher federal and state taxes to support our graying population.

A third reason for many Americans' justifiably worrying about retirement is that two of our major entitlement programs (Social Security and Medicare) consume about 15 percent of the country's gross domestic product, compared with only 4 percent in 1970. Medicare is expected to run out of money in 2024, and Social Security funds may be exhausted in 2037 (The Board of Trustees . . . 2011). So, unless you're wealthy or expect a huge inheritance, you'll probably work into your 70s.

## 12-6 Sociological Explanations of Family and Aging

The four sociological perspectives are useful in understanding families and aging. *Table 12.3* summarizes the key points of these theories.

### 12-6a Functionalism

We began this chapter by looking at the vital functions that families perform—such as procreation and socialization—that promote societal stability and

**sandwich generation** people in a middle generation who care for their own children and their aging parents.

**old-age dependency ratio** (sometimes called the *elderly support ratio*) the number of working-age adults ages 18 to 64 for every person age 65 and older who's not in the labor force.

## FIGURE 12.6

## Old-Age Dependency Ratios in the United States, 1900 to 2050

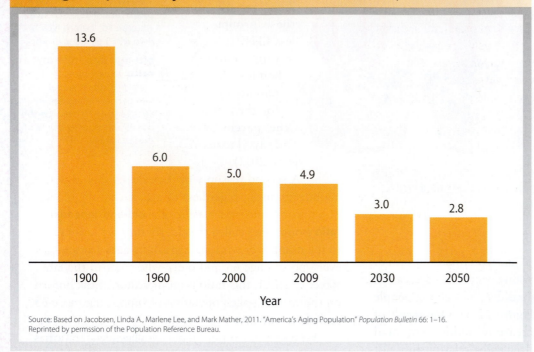

Source: Based on Jacobsen, Linda A., Marlene Lee, and Mark Mather, 2011. "America's Aging Population" *Population Bulletin* 66: 1–16. Reprinted by permssion of the Population Reference Bureau.

### Stability and Activity

Rapid global aging suggests that many countries will face debates over health care, Social Security, and other costs. From a functionalist perspective, there's also the question of how much a rapidly graying population can contribute to the tasks that are critical for a society's survival.

In the late 1950s, functionalists maintained that as people age, they withdraw from their social roles and become isolated. This *disengagement theory* was abandoned when many gerontologists found that disengagement was due to age discrimination, not choice, and was limited to older people with disabilities and poor health. Functionalists replaced disengagement theory with **activity theory**, which proposes that many older people remain engaged in numerous roles and activities, including work, and that those who do so adjust better to aging and are more satisfied with their lives (see Atchley and Barusch 2004 for a summary of some of this research).

**activity theory** proposes that many older people remain engaged in numerous roles and activities, including work.

individual well-being. Functionalists recognize that families differ in structure (e.g., nuclear vs. extended) but believe that their similar functions ensure a society's continuity. Problems arise, for instance, when parents can't or don't provide their children with the necessary financial and emotional support.

## TABLE 12.3

# Sociological Perspectives on Families and Aging

| THEORETICAL PERSPECTIVE | LEVEL OF ANALYSIS | KEY POINTS |
|---|---|---|
| **Functionalist** | Macro | • Families are important in maintaining societal stability and meeting family members' needs.<br>• Older people who are active and engaged are more satisfied with life. |
| **Conflict** | Macro | • Families promote social inequality because of social class differences.<br>• Many corporations view older workers as disposable. |
| **Feminist** | Macro and Micro | • Families both mirror and perpetuate patriarchy and gender inequality.<br>• Women have an unequal burden in caring for children as well as older family members and relatives. |
| **Symbolic Interactionist** | Micro | • Families construct their everyday lives through interaction and subjective interpretations of family roles.<br>• Many older family members adapt to aging and often maintain previous activities. |

## Critical Evaluation

For many functionalists, marriage, followed by procreation, is critical in promoting social order and stability. Critics maintain, however, that such stances have several weaknesses. First, functionalism seems to support traditional family arrangements, such as the nuclear family, but largely devalues other family forms, such as single-parent and same-sex families and unmarried parents. Also, married couples who choose to be child free contribute to social order and stability by working, paying taxes, and playing important kinship roles (such as uncle and aunt).

Second, it's questionable whether some family functions are as universal or necessary as functionalists claim. Procreation often occurs outside of marriage, and the government has assumed some of the family's functions, such as caring for some children (as in foster homes) and the aged (through Medicare, for example) (Lundberg and Pollak 2007).

Third, is activity theory as representative of older people as some functionalists claim? Many people continue to work even in their 80s, not because they choose to do so but because they can't afford to retire, even though they are in poor health and unhappy in their low-income jobs. As health deteriorates, many older people become less active and more isolated (Kinsella and Phillips 2005; Lee 2009).

## 12-6b Conflict Theory

Conflict theorists agree that families serve important functions but point out that some groups benefit more than others. Families are sources of social inequality that mirror the larger society, and the inequities persist from birth to old age (Cruikshank 2009).

### Inequality, Social Class, and Power

For conflict theorists, families perpetuate social inequality. Those in high income brackets have the greatest share of capital, including wealth, that they can pass down to the next generation. This inheritance reduces the likelihood that all families can compete for resources such as education, decent housing, and health care (see Chapter 8).

There's also a question of whether U.S. society cherishes children as much as it professes. For example, 30 countries—including Cuba, Slovakia, and most of Europe—have lower infant mortality rates than the United States. And of 16 industrialized countries, the United States has the highest child poverty rates and the lowest rates of spending on children's services (Children's Defense Fund 2008; Haub and Kaneda 2011).

"In return for an increase in my allowance, I can offer you free unlimited in-home computer tech support."

Aaron Bacall/Cartoonstock

Unequal access to resources continues into old age. Most employers insist that they value older workers' loyalty, work ethic, reliability, and experience, but many are less likely to hire or retain older people because they are usually more expensive than younger workers. Because many large companies have cut their pension plans, numerous older workers must work long after they expected to retire (Roscigno 2010; Jacobsen et al. 2011).

## Critical Evaluation

Conflict theory is limited for several reasons. First, conflict theorists tend to overlook the fact that most older people, especially those in the middle and upper classes, don't have to struggle for resources. Between 1960 and 2007, for example, the U.S. government's spending on children declined from 20 percent to 15 percent, but spending on people age 65 and older, regardless of social class, grew from 22 percent to almost 46 percent because of programs such as Social Security and Medicare (Carasso et al. 2008). A second weakness is that conflict theory links family inequality to capitalism and social class, but there is also considerable family inequality in countries that aren't capitalist (see Chapters 8 and 11).

## 12-6c Feminist Theories

Feminist scholars agree with conflict theorists that there's considerable inequality between low-income and wealthy families in accessing necessary resources. However, feminist theorists emphasize the inequality of gender roles in families, especially in patriarchal societies (including the United States).

### Gender Roles and Patriarchy

Feminist scholars view the patriarchal family as a major reason for women's inequality. Because employed mothers do almost twice as much housework and child care

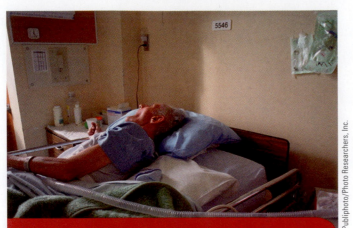

The top U.S. medical centers spend anywhere from $30,000 to almost $94,000 a year per older patient in his or her last two years of life, and all of the costs are covered by Medicare or Medicaid (Wennberg et al. 2008). Is this a good investment of our resources?

battery to take his wife, the victim, to Red Lobster for dinner and then bowling (McEwan 2012). Thus, some judges still view domestic violence charges as frivolous.

Feminist sociologists have for some time challenged popular myths that children of single and employed mothers have emotional, cognitive, and behavioral problems. They have also documented the ways that sexism, racism, and classism intersect to oppress women and children. For example, the poorest older adults are most likely to be minority women. Caregivers, who are predominantly women, must often leave their jobs or work only part-time to care for children and aging parents (Allen and Beitin 2007; Houser 2007).

### Critical Evaluation

Feminist theories have noted the impact of gender and patriarchy on family relationships, but some critics question whether the perspectives overstate women's oppression. For example, patriarchal family structures limit women's rights, but they also often provide important economic resources for women and children (Benokraitis 2011).

Some critics believe that feminist scholars have a tendency to view full-time homemakers as victims rather than as individuals who choose the role. Thus, some maintain, feminist scholars are "in danger of refusing to listen to a multiplicity of women's voices" (Johnson and Lloyd 2004: 160).

Some critics also contend that feminist scholars exaggerate many women's subordinate role in the family. For example, 29 percent of wives earn more than their husbands (*Women in the Labor Force . . .* 2011). Many of these women still shoulder a larger share of the workload at home. However, women's higher incomes also mean that they have more domestic decision-making power, don't have to rely on men financially, and are able to pay for housekeeping, child care, and elder care services.

## 12-6d Symbolic Interaction

For symbolic interactionists, people create subjective meanings of what a family is and does. Thus, people learn, through interaction with others, how to act as a parent, a grandparent, a teenager, a stepchild, and so on throughout the life course.

### Learning Family and Aging Roles

Interactionists often use exchange theory to explain mate selection and family roles. The fundamental premise of **exchange theory** is that people seek through their

**exchange theory** posits that people seek through their social interactions with others to maximize their rewards and minimize their costs.

as men, many men benefit from women's unpaid labor at home. In most countries, males determine laws about property and inheritance rights, marriage and divorce, and many other regulations that give men authority over women. For example, the United States is similar to most other nations in that men (particularly lawmakers and Catholic church officials) control women's decisions about reproduction and access to abortion (see Chapters 9 and 13).

Men hold most of the power, resources, and privilege, and many feel free to use women and children as targets of sexual and physical abuse. In many cases, legal, political, and religious institutions don't take violence against women and children as seriously as they should (Lindsey 2005). Recently, for instance, a judge in Florida "sentenced" a man charged with domestic

Because employed mothers do twice as much housework and child care as men, many men benefit from women's unpaid labor at home.

In most Latino communities, the quinceañera (pronounced "keen-say-ah-NYAIR-ah") is an important coming-of-age ritual that celebrates a girl's entrance into adulthood on her fifteenth birthday. The quinceañera is an elaborate and dignified religious and social event that reaffirms strong ties with family, relatives, and friends.

social interactions to maximize their rewards and minimize their costs. In mate selection, people trade their resources—such as wealth, intelligence, good looks, youth, and/or status—for more, better, or different assets.

Many people stay in unhappy marriages and other intimate relationships because the rewards seem equal to the costs. For example, many women tolerate an abusive relationship because they fear loneliness or losing the economic benefits that a man provides (Choice and Lamke 1997).

A well-known psychologist who interviewed more than 200 couples over a 20-year period found that the difference between lasting marriages and those that split up was a "magic ratio" of 5 to 1. That is, if there are five positive interactions between partners for every negative one, the marriage is likely to be stable over time (Gottman 1994). Thus, we can improve our family relationships by learning to interact in more positive ways.

Stereotypes about older people are deeply rooted in U.S. society. For example, 84 percent of Americans age 60 and older report one or more incidents of **ageism**, or discrimination against older people, including insulting jokes, disrespect, patronizing behavior, and assumptions about being frail or unhealthy (Roscigno 2010). **Continuity theory** posits that older adults can substitute satisfying new roles for those they've lost (Atchley and Barusch 2004). For example, a retired music teacher can offer private lessons. Thus, developing new roles may lessen some of the emotional distress due to ageism.

## Critical Evaluation

A common criticism is that interactionism, a micro-level perspective, doesn't address macro-level constraints. For example, families living in poverty, especially single mothers, are often stigmatized for not living up to middle-class parenting norms such as raising their children in safe neighborhoods or participating in school activities. Doing so is nearly impossible, however, because low-income parents can't move to safer communities or take time off from work to attend school functions (Dodson and Luttrell 2011).

Second, exchange theory is useful in explaining why people behave as they do, but individuals don't always calculate the potential costs and rewards of every decision. In the case of women who care for older family members, for example, genuine love and concern can override cost-benefit decisions, especially in many traditional Asian, Latino, and Middle Eastern families, where culturally defined kinship duties take precedence over individual rights (Hurh 1998; Do 1999; see also Chapter 10).

Third, continuity theory ignores societal obstacles that discourage many older people from engaging in satisfying new roles. After retirement, for example, many people must get by on low, fixed incomes and can't afford recreational activities that they enjoyed in the past. Also, older people who experience illnesses because of low-quality health care over a lifetime may become too sick to participate in family and community activities or pursue hobbies.

**ageism** discrimination against older people.

**continuity theory** posits that older adults can substitute satisfying new roles for those they've lost.

## Study Tools

Ready to study? Go to **Sociology CourseMate** at **www.cengagebrain.com** to complete practice quizzes, review flashcards, watch videos, and more.

Education and religion have a
powerful impact on our lives.

Courtesy of the Town of
Stoddard, NH/Stoddard
Historical Society

# Education and Religion

Education and religion are influential social institutions. Both impart values, shape our attitudes, maintain traditions, bring people together, exert control over our behavior, and grapple with social change. Let's begin with education.

> **education** a social institution that transmits attitudes, knowledge, beliefs, values, norms, and skills to its members through formal, systematic training.
>
> **schooling** formal training and instruction provided in a classroom setting.

## 13-1 What Is Education?

**E**ducation is a social institution that transmits attitudes, knowledge, beliefs, values, norms, and skills to its members through formal, systematic training. **Schooling**, a narrower term, is formal training and instruction provided in a classroom setting.

U.S. education and schooling have undergone four significant transitions since the beginning of the twentieth century: Universal education has expanded, community colleges have flourished, public higher education has burgeoned, and student diversity has increased as more women and racial-ethnic groups enrolled at colleges and universities. As a result of these and other changes, 88 percent of Americans 25 or older have completed high school, and 30 percent have obtained a bachelor's or higher degree (see *Figure 13.1*).

## 13-2 Sociological Perspectives on Education

**S**ociologists offer several explanations regarding the purpose of education. See *Table 13.1* on p. 249 for a summary of the four major perspectives.

### 13-2a Functionalism: What Are the Benefits of Education?

For functionalists, education contributes to society's stability, solidarity, and well-being and provides people with an opportunity for upward mobility. In analyzing the value of education, functionalists distinguish between manifest and latent functions.

#### Manifest Functions of Education

Some of the functions of education are *manifest*; that is, they are open, intended, and visible. The major *manifest functions* of education include the following:

## Key Topics

In this chapter, we'll explore the following topics:

## What do you think?

People freely decide whether to go to college.

| 1 | 2 | 3 | 4 | 5 | 6 | 7 |
|---|---|---|---|---|---|---|

strongly agree     strongly disagree

## FIGURE 13.1
## U.S. Educational Attainment, 1940–2011

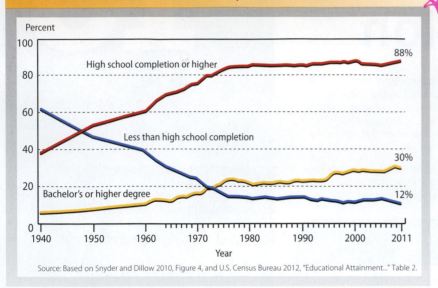

Source: Based on Snyder and Dillow 2010, Figure 4, and U.S. Census Bureau 2012, "Educational Attainment…" Table 2.

- Schools are *socialization agencies* that teach children how to get along with others and prepare them for adult economic roles (Durkheim 1898/1956; Parsons 1959).

- Education *transmits knowledge and culture*. Schools teach skills such as reading, writing, and counting; they also instill cultural values that encourage competition, achievement, patriotism, and democracy (see Chapter 3).

- Similar values increase *cultural integration*, the social bonds that people have with each other and with the community at large, and *societal cohesion*.

- Education promotes *cultural innovation*. For example, faculty at research universities receive billions of dollars every year to develop computer technology, treatments for diseases, and programs to address social problems such as domestic violence.

- Education *benefits taxpayers* because more highly educated people tend to pay higher taxes and are less likely to use public assistance programs and to commit crimes (Carroll and Erkut 2009).

Almost all parents (94 percent) want their children to get a college degree. Many Americans (69 percent) agree with functionalists that a college degree is important in getting a good job. In a recent national survey, 74 percent of college graduates said their education was very useful in helping them grow intellectually, and 55 percent said it was useful in preparing them for a job or career (English 2011; Parker et al. 2011).

Education increases earnings. Over a lifetime, education affects earnings five times more than other demographic factors such as sex, race, ethnicity, and age (Julian and Kominski 2011). By age 55, college graduates bring home 90 percent more than people with only a high school diploma. Lifetime earnings are even higher for people with advanced degrees (see *Figure 13.2*). During economic downturns, the more educated people are, the less likely they are to be out of work or experience poverty.

Those who are better educated—regardless of race, ethnicity, and marital status—live longer and report better physical and emotional health. They are more informed about the benefits of healthful behaviors and are better able to handle stress because of their enhanced economic and social position (Jemal et al. 2008; Loucks et al. 2011; Rheault and McGeeney 2011).

## Sociology in Your Life

Are you optimistic or pessimistic about getting a well-paid job when you finish college? To see how the rest of the Millennial generation feels, check out the Pew Research data at CengageBrain.com.

### Latent Functions of Education

In addition to manifest functions, education has *latent functions*—hidden, unstated, and sometimes unintended consequences. For example:

- Schools *provide child care*, particularly after-school programs, for the growing number of single-parent and two-income families.

- High schools and colleges are *matchmaking institutions* that bring together unmarried people.

- Education *decreases job competition*; the more time that young adults spend in school, the longer the jobs of older workers are safe.

## TABLE 13.1

# Major Sociological Perspectives on Education

| THEORETICAL PERSPECTIVE | LEVEL OF ANALYSIS | VIEW OF EDUCATION | SOME MAJOR QUESTIONS |
|---|---|---|---|
| Functionalist | Macro | Contributes to society's stability, solidarity, and cohesion and provides opportunities for upward mobility | What are the manifest and latent functions of education? |
| Conflict | Macro | Reproduces and reinforces inequality and maintains a rigid social class structure | How does education limit equal opportunity? |
| Feminist | Macro and Micro | Produces inequality based on gender | What are the gender gaps in education? |
| Symbolic Interactionist | Micro | Teaches roles and values through everyday face-to-face interaction and practices | How do tracking, labeling, self-fulfilling prophecies, and engagement affect students' educational experiences? |

© Cengage Learning

- Educational institutions *create social networks* that can lead to jobs or business opportunities.

- Education is *good for business.* Thousands of companies offer services that tutor and test students and produce textbooks and related materials.

## Critical Evaluation

Do functionalists exaggerate education's benefits? Many people in trade occupations—construction workers and electricians, for instance—have higher lifetime earnings than some people with college degrees ("Usual Weekly Earnings . . ." 2012). Among the college educated, what people study heavily affects their financial payoff. The median income of full-time, full-year workers with bachelor's degrees varies dramatically—from $29,000 for counseling psychology majors to $120,000 for petroleum engineering majors (Carnevale et al. 2011b). Many millionaires don't have college degrees, nor do 25 percent of state legislators (Ellsberg 2011; Smallwood and Richards 2011). Earning potential isn't the only issue a student considers when selecting a major, but critics contend that the higher education establishment greatly overstates the economic value of a college degree (see Marsh 2011).

Also, according to some critics, functionalists tend to gloss over education's dysfunctions. Almost half of Americans say they can't afford to go to college. About 40 percent of undergraduates leave college with no debt, but those with loans owe, on average, almost $38,000, and the outstanding student debt has surpassed $1 trillion, primarily because of rising college costs. Even Americans age 60 and older still owe

## FIGURE 13.2

### Education Pays Off

Lifetime earnings for full-time, year-round workers (in millions of dollars)

| | |
|---|---|
| Doctoral degree | $3.3 |
| Professional degree | $3.6 |
| Master's degree | $2.7 |
| Bachelor's degree | $2.3 |
| Associate's degree | $1.7 |
| Some college | $1.5 |
| High school graduate | $1.3 |
| Not high school graduate | Almost $1.0 |

Source: Carnevale et al. 2011a, Figure 1.

Comstock/Jupiter Images

FIGURE 13.3

## Attainment of College Degree or Higher, by Race and Ethnicity, 1970 and 2010

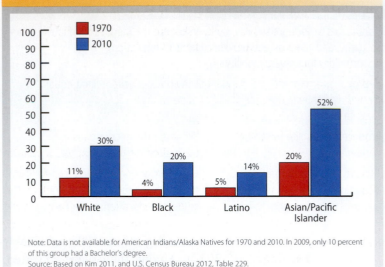

Legend:
- 1970
- 2010

| Race/Ethnicity | 1970 | 2010 |
|---|---|---|
| White | 11% | 30% |
| Black | 4% | 20% |
| Latino | 5% | 14% |
| Asian/Pacific Islander | 20% | 52% |

Note: Data is not available for American Indians/Alaska Natives for 1970 and 2010. In 2009, only 10 percent of this group had a Bachelor's degree.

Source: Based on Kim 2011, and U.S. Census Bureau 2012, Table 229.

about $36 billion in student loans (Dell 2011; Mitchell and Jackson-Randall 2012; Brown et al. 2012).

Depending on the major, the lifetime earnings return on an associate's or bachelor's degree far outweighs the debt (Parker et al. 2011). Critics claim, however, that a large portion of soaring college costs subsidizes "supersized bureaucracies" that spend much of their budget not on education but on administrative salaries, athletics, student recreational facilities, and sometimes faculty who spend most of their time doing research rather than teaching (Gillen et al. 2011; Hacker and Dreifus 2011).

## 13-2b Conflict Theory: Does Education Perpetuate Social Inequality?

Conflict theorists ask why education benefits some people more than others. From preschool to graduate school, conflict theorists maintain, education creates and perpetuates social inequality based on social class, race, and ethnicity. Schools also use standardized tests and social control to maintain the status quo.

### Social Class, Race, and Ethnicity

College enrollments have surged since 1990, especially among minority groups, but who graduates? As *Figure 13.3* shows, Asian Americans/Pacific Islanders have made the largest gains. Of the 48 groups in this population, 87 percent of Asian Indians have at least a bachelor's degree, compared with only 21 percent of Samoans (Teranishi 2011; see also Fry 2011). Such variations, as with other minority subgroups, are due to many factors, including English language barriers, but the best predictor of educational attainment is social class.

Fifty years ago, the largest academic achievement gap was between blacks and whites; today it's between social classes. The achievement gap between children from high- and low-income families has grown by about 40 percent since the 1960s (Pell Institute 2011; Reardon 2011).

Wealthy parents have the greatest amount of *economic capital* (income and other monetary assets), *cultural capital* (advanced degrees and positive attitudes toward education), and *social capital* (social networks that provide support and information about schooling). Differential access to all three types of capital reinforces and reproduces the existing class structure from one generation to the next (Bourdieu 1984; Dickert-Conlin and Rubenstein 2007).

Students from lower socioeconomic backgrounds—many of whom are minorities—are less likely to attend high schools that offer high-level courses in mathematics and science. This can mean the difference between passing or failing required courses during the first few

"That concludes the list of students with outstanding grades. And now for those of you with outstanding student loan payments..."

Loren Fishman/Cartoonstock

years of college. Many college youths from low-income and middle class families must also work part-time or full-time and, consequently, have more stress and less time for academic work (Adelman 2006; Lopez 2009).

## Gatekeeping and Standardized Testing

An **intelligence quotient (IQ)** is an index of an individual's performance on a standardized test relative to the performance of others of the same age. IQ and other standardized tests are examples of *gatekeeping*, controlling the access to education or jobs of people from lower socioeconomic levels. Schools typically use IQ scores to place students into different ability groups at an early age.

Many scholars have challenged the validity of these tests because, they argue, social and environmental factors have at least as much influence on IQ scores as inherited factors. For example, raising a child in an upper-middle-class environment versus a lower-class environment can increase IQ scores by 12 to 18 points. Better schooling (small class sizes, computer access, and teachers' skills) also boosts IQ scores, especially for poor and minority children (Nisbett 2009).

What about other standardized tests? The SAT, administered by the College Board, is required by many colleges as part of the application process. SAT prep courses, which cost $1,000 or more, increase scores by an average of only 30 points. The increase is negligible but may bring a student's score above a college's cutoff for admission (Briggs 2009). Consequently, such coaching gives an advantage to students from higher income families.

SAT defenders claim that the test scores are accurate predictors of how well a high school graduate will perform in college (Caperton 2009). Critics contend that the SAT measures social class rather than ability or intelligence because students at lower socioeconomic levels have less economic, cultural, and social capital. For example, mastering math or writing rules is more likely at affluent high schools that enjoy generous funding, up-to-date textbooks and laboratories, small classes, experienced teachers, and a high-quality curriculum ("Quality Counts . . ." 2007; Sternberg 2010).

Jaimie Duplass/Shutterstock.com

Advanced Placement (AP) courses and exams are also gatekeeping tools. In 2011, 30 percent of the 3 million students who graduated from high school took at least one AP exam. Only 18 percent scored high enough to get credit for one or more introductory college courses in 37 subjects (College Board 2012). Still, many high school teachers encourage students to take AP courses and exams because they impress college admissions officers, increase the chances of getting a scholarship, and may enable students to graduate a semester or two early, reducing the cost of a college education. As a result, the number of students taking AP courses has doubled since 2001. Students—including those from low-income urban high schools—who do well on AP exams are more likely to attend college, earn a degree, be employed after graduation, and earn higher wages. However, many low-income students don't have access to AP courses (Klopfenstein and Thomas 2005; Jackson 2012).

## Education and Social Control

Functionalists see education as an avenue for upward mobility. Conflict theorists maintain that mobility is restricted because of a hidden curriculum, credentialism, and privilege.

*Hidden Curriculum.* Every school has a formal curriculum that includes reading, writing, and learning other skills. Schools also have a **hidden curriculum**, practices that transmit nonacademic knowledge, values, attitudes, norms, and beliefs that tend to legitimize "economic inequality and the staffing of unequal work roles" (Bowles and Gintis 1977: 108).

Schools in low-income and working-class neighborhoods tend to stress obedience, following directions, and punctuality so that students can fill jobs (as in restaurants, retail stores, and hospitals) that require these characteristics. Instead of preparing students for college or careers, for example, many low-income high schools have curricula such as "medical careers and health professions" that focus on training nursing aides, health assistants, and other low-paid personnel for hospitals and nursing homes (Kozol 2005).

Schools in middle-class neighborhoods tend to emphasize proper behavior and appearance, cooperation, following rules, and deference to authority

**intelligence quotient (IQ)** an index of an individual's performance on a standardized test relative to the performance of others of the same age.

**hidden curriculum** school practices that transmit nonacademic knowledge, values, attitudes, norms, and beliefs that legitimize economic inequality and fill unequal work roles.

**credentialism** an emphasis on certificates or degrees to show that people have certain skills, educational attainment levels, or job qualifications.

because many of these students will be working in middle-level bureaucracies that require such attributes. Routine is the norm because it's good training for many bureaucratic jobs (Hedges 2011). In contrast, selective private schools encourage leadership, creativity, independence, and people skills—all prized characteristics in elite circles (see Chapter 8). In effect, then, the hidden curriculum reproduces the existing class structure and provides workers for jobs and occupations in the stratification hierarchy.

*Credentialism.* Have you noticed that doctors', lawyers', and dentists' offices are usually wallpapered with framed degrees? Faculty members, similarly, have letterheads and name plates that include their titles to signal their educational achievement. Whether the tangible symbols of people's achievements are framed, hung on office doors, or embossed on business cards, all of them reflect **credentialism**, an emphasis on certificates or degrees to show that people have certain skills, educational attainment levels, or job qualifications. Besides conferring social status, credentials connote knowledge or expertise in an area.

Functionalists maintain that credentialism rewards people for their accomplishments, sorts out those who are the most qualified for jobs, and stimulates upward social mobility. Conflict theorists, however, contend that for many positions, people can gain skills on the job with a few weeks of training or succeed because of ability or other factors. For example, 7 percent of high school dropouts make as much as or more than someone with a bachelor's degree (Carnevale et al. 2011a).

Because of a large supply of high school graduates, employers can demand higher levels of education even though some jobs (such as sales and law enforcement) don't require a college degree for competent performance, a process called *credential inflation.* As more people obtain a college degree, its value diminishes, and students from low-income families, who are the least likely to have access to a college education, fall further behind (Collins 2002; Bollag 2007).

*Privilege.* Many colleges say that they welcome low-income and minority students. In fact, in 2010–2011, colleges and universities awarded $5.3 billion in financial aid to students from high-income families. Among students with a college GPA of 3.5 or higher on a 4.0 scale, white students received more than three times as much in merit-based scholarships as minority students (Baum and Payea 2011; Kantrowitz 2011).

Another privileged group is *legacies*, the children of alumni who have "reserved seats" regardless of their accomplishments or ability. For example, President George W. Bush—who had average high school grades and standardized test scores—was admitted as a legacy at Yale University, which his father and grandfather had attended (Golden 2006; Wickenden 2006). Legacies also include students with inferior academic records who are the children of wealthy people, including celebrities, who make million-dollar donations to a school (Massey 2007). Among selective public and private colleges and universities, almost 75 percent employ legacy preferences, which account for up to 25 percent of the student population (Hurwitz 2011).

## Critical Evaluation

Despite their contributions, conflict theorists have been criticized for ignoring the gains that many low-income and minority students have made, including graduating from college and graduate school (Marks 2007). Conflict theorists downplay the importance of education

Graduating from a selective college or university has benefits that include a network of useful contacts and better job offers. However, a student's motivation, ability, creativity, and ambition may be more important factors than his or her alma mater. Steven Spielberg, the well-known movie producer and director, graduated from California State University at Long Beach after being rejected by the more prestigious University of Southern California and University of California at Los Angeles film schools.

in upward mobility, but as you saw earlier, there is a strong association between educational achievement and economic success. Also, conflict theorists may denounce credentialism and credential inflation, but both are a fact of life. Increasingly, employers are requiring college degrees for most jobs, including those that were filled in the past by people with only a high school diploma (Carnevale et al. 2010).

## 13-2c Feminist Theories: Is There a Gender Gap in Education?

Because women across all racial and ethnic groups are more likely than men to finish college, some researchers are describing this phenomenon as "the feminization of higher education" (McCormack 2011; Pollard 2011). A closer look at the data shows that higher education isn't nearly as feminized as some analysts claim. For example, the percentage of female college and university presidents rose from 2006 to 2011, but women still comprise only 26 percent of this group (Cook and Kim 2012). Is there also a gender gap among students?

### Degrees Earned

As *Figure 13.4* shows, equal numbers of girls and boys are high school graduates. Women sail past men in earning associate's, bachelor's, and master's degrees, but their percentage of professional and doctoral degrees drops considerably. Such data contradict the description of higher education as feminized, but why do women achieve more degrees than men up to the master's degree?

In both high school and college, women spend more time studying than men, earn better grades, hold more leadership posts, and are more involved in student clubs and community volunteer work than men. Women (81 percent) are also more likely than men (67 percent) to have a positive view of the value of higher education (Goldin et al. 2006; Wang and Parker 2011).

Women may be more motivated than men and take school more seriously because they realize they need more credentials to succeed. In fact, the general public is more likely to say that a college education is more important for women to get ahead than it is for men (77 percent and 68 percent, respectively) (Wang and Parker 2011). On average, for example, women have to have a Ph.D. to make as much as a man with a bachelor's degree (Carnevale et al. 2011a). Regardless of the reasons for women's success, many colleges have been giving men preferential treatment

The Taliban, a conservative religious group in Afghanistan, outlawed all education for girls in 1996, maintaining that educating girls violates Islam. The group was ousted from national power, and today 6 million students, boys and girls, attend school, but there's no space for 5.3 million children. About half of Afghanistan's 12,000 schools have no permanent structure, with classes held in tents or in the open air (McCanna 2009).

in admissions to avoid large gender imbalances in their student bodies (Gewertz 2009; Miners 2009; Kahlenberg 2010).

### Majors and Careers

A major gender gap is the underrepresentation of women in the fields of science, technology, engineering, and mathematics (STEM). In elementary, middle, and high school, girls and boys take math and science courses in roughly equal numbers, and about the same numbers leave high school planning to pursue STEM majors in college. After a few years, however, men outnumber women in nearly every STEM field, and in some—such as physics, engineering, and computer science—women earn only 20 percent of the bachelor's degrees. Their numbers decline further at the graduate level and yet again in the workplace (Hill et al. 2010).

Why is there a gradual attrition? As early as middle school, many teachers and parents expect boys, not girls, to be interested in math and science. Less encouragement may reduce girls' self-confidence and their interest in taking STEM courses and AP exams in high school. In college, female students are initially as persistent as men in a STEM major and get higher

## FIGURE 13.4

## Educational Attainment, 18 Years and Older, by Sex, 2011

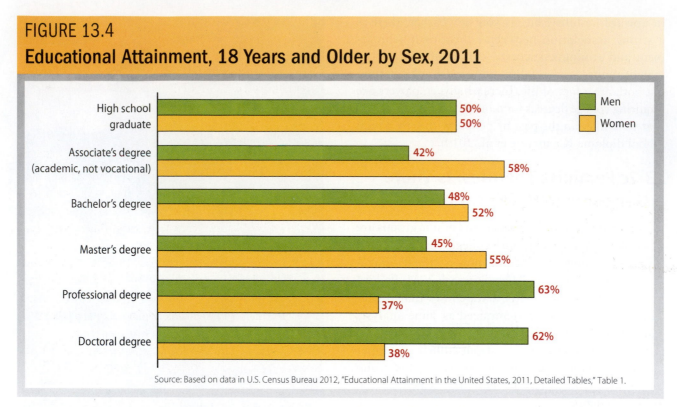

**Men** / **Women**

| | Men | Women |
|---|---|---|
| High school graduate | 50% | 50% |
| Associate's degree (academic, not vocational) | 42% | 58% |
| Bachelor's degree | 48% | 52% |
| Master's degree | 45% | 55% |
| Professional degree | 63% | 37% |
| Doctoral degree | 62% | 38% |

Source: Based on data in U.S. Census Bureau 2012, "Educational Attainment in the United States, 2011, Detailed Tables," Table 1.

grades, but they are less satisfied than men with the core courses and more likely to doubt their ability to succeed in a male-dominated discipline. The exit from a STEM major is also associated with having few female faculty role models and mentors and with women's expectations that they won't be paid or promoted equally with men (Karukstis 2009; Lord et al. 2009; Hunt 2010; Williams and Ceci 2012).

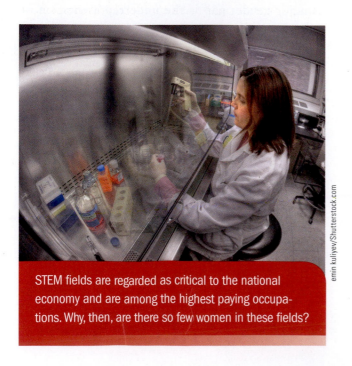

STEM fields are regarded as critical to the national economy and are among the highest paying occupations. Why, then, are there so few women in these fields?

emin kuliyev/Shutterstock.com

### Critical Evaluation

A common criticism is that feminist scholars tend to ignore women's tremendous gains in higher education. Some fault feminist theorists, who support gender equality, for being more interested in women's than men's gender gaps, especially at the bachelor's and master's levels (Baenninger 2011). It's also not clear why feminist scholars devote little attention to issues such as why boys are more disruptive in elementary school, which leads to suspensions and decreases the chance of attending college by at least 16 percent for each suspension; whether the general public's image of men as underachievers affects men's motivation; and why parents are more likely to pay for their daughters' (40 percent) than their sons' (29 percent) college education (Bertrand and Pan 2011; Smith 2011; Wang and Parker 2011).

### 13-2d Symbolic Interactionism: How Do Social Contexts Affect Education?

None of us is born a student or a teacher. Instead, these roles, like others, are socially constructed (see Chapters 3–5). For symbolic interactionists, education is an active *process* that includes students, teachers, peers, and parents and involves tracking, labeling, and student engagement.

## Tracking

Beginning in kindergarten, practically all schools sort students by aptitude. Such sorting results in **tracking** (also called *streaming* or *ability grouping*), assigning students to specific educational programs and classes on the basis of test scores, previous grades, or perceived ability.

Some educators believe that tracking is beneficial because students learn better in groups with others like themselves, and it allows teachers to develop curricula for students with similar ability. Many symbolic interactionists maintain, however, that tracking creates and reinforces inequality.

- High-track students take classes that involve critical thinking, problem solving, and creativity that high-status occupations require. Low-track students take classes that are limited to simple skills, such as punctuality and conformity, that usually characterize lower status jobs.

- High-track students have more homework, better quality instruction, and more enthusiastic teachers. One result is that high-track students are more likely to see themselves as "bright," whereas low-track students see themselves as "dumb" or "slow."

- The effects of tracking are usually cumulative and lasting. Teachers tend to have low expectations for low-track students, who therefore fall further behind every year in reading, mathematics, and interaction skills (Oakes 1985; Riordan 1997; Hanushek and Woessman 2005).

Middle school and high school become even more stratified as high-track students are sorted into gifted, honors, and AP programs and courses. In college, students continue to be tracked and sorted into honors' programs and accelerated undergraduate courses.

## Labeling

Tracking often leads to labeling, a serious problem because "there is a widespread culture of disbelief in the learning capacities of many of our children, especially children of color and the economically disadvantaged" (Howard 2003: 83). Labeling, in turn, can result in a *self-fulfilling prophecy*, whereby students live up or down to teachers' expectations and evaluations that are influenced by a student's social class, skin color, hygiene, accent, and test scores (see Chapter 5).

Almost all parents would like their children to get a college degree, but parents affect their children's academic outcomes through conscious or unintentional labeling. Only about half of low-income parents (those with annual incomes of $25,000 or less) expect their children to attain a bachelor's degree or higher, compared with 87 percent of parents earning $75,000 or more. Because of high expectations, more educated and higher income parents read to their preschool children; take them to libraries, plays, historical sites, zoos, and aquariums; and encourage them to persist despite any difficulties they encounter (Child Trends 2010). Higher income parents have fewer budget constraints than their lower income counterparts, and they "program" their youngsters to succeed regardless of the children's innate abilities (Kagan 2011).

> **tracking** (also called *streaming* or *ability grouping*) assigning students to specific educational programs and classes on the basis of test scores, previous grades, or perceived ability.

## Student Engagement

Many U.S. schools have been assessing performance not only through tests but also in terms of *student engagement*, how involved students are in their own learning. A team of researchers who spent thousands of hours in more than 2,500 first-, third-, and fifth-grade classrooms concluded that the typical U.S. child has only a 1 in 14 chance of being in a school that encourages her or his engagement. Because of standardized tests, teachers in public elementary schools spend most of the school day on basic reading and math drills and little time on problem solving, reasoning, science, and social studies. As a result, the study found, classrooms are often "dull and bleak places" where kids don't get a lot of teacher feedback or face-to-face interaction with their peers (Pianta et al. 2007).

U.S. high school students aren't as engaged in their education as they could be. For example, 26 percent admit that they usually don't do their homework, and 10 percent drop out. The students who are most likely to be disengaged are from poor or low-income families, are African American or Latino, don't live with both biological parents, attend financially strapped urban public high schools that are typically overcrowded and understaffed, and have fewer resources such as computers and even textbooks (Finn 2006; Planty et al. 2007).

In college, the typical undergraduate spends more time partying, socializing, and working for pay than studying. On average, even full-time college students

FIGURE 13.5
## Who Studies the Least in College?

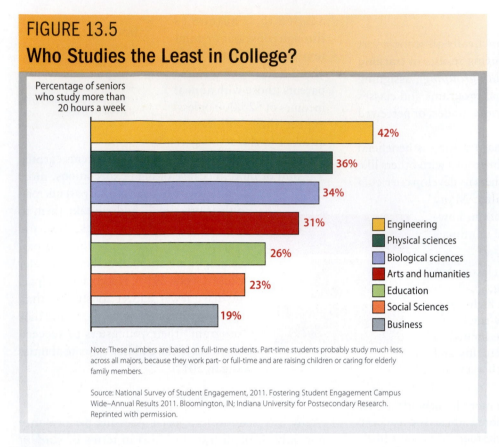

Percentage of seniors who study more than 20 hours a week

- Engineering — 42%
- Physical sciences — 36%
- Biological sciences — 34%
- Arts and humanities — 31%
- Education — 26%
- Social Sciences — 23%
- Business — 19%

**Legend:**
- Engineering
- Physical sciences
- Biological sciences
- Arts and humanities
- Education
- Social Sciences
- Business

Note: These numbers are based on full-time students. Part-time students probably study much less, across all majors, because they work part- or full-time and are raising children or caring for elderly family members.

Source: National Survey of Student Engagement, 2011. Fostering Student Engagement Campus Wide–Annual Results 2011. Bloomington, IN; Indiana University for Postsecondary Research. Reprinted with permission.

study only 14 hours a week compared with 24 hours a week in 1961 (Babcock and Marks 2011). This is well below the 24 to 30 hours faculty members say students should be spending on class preparation if they're taking three courses. As *Figure 13.5* shows, the amount of studying varies by major, with business majors studying the least. Even among engineering students, who study the most, 25 percent say that they don't read the required material before going to class (National Survey of Student Engagement 2011).

According to one college instructor, "Education is the only business in which the clients want the least for their money" (Perlmutter 2001: B1). However, some analysts also blame professors for students' plummeting amount of studying. Many faculty have watered down their required readings, accept that some students don't even buy the textbook, and they often give easy exams to avoid complaints about low grades. They may view students as "consumers" who increase enrollment figures and ensure their jobs, rather than as adults who should be expanding their knowledge. When faculty dilute their course requirements and tests, some students don't study because they're disenchanted with unchallenging courses that are boring and don't stimulate them intellectually (Benton 2011; Flaherty 2011; Glenn 2011).

## Critical Evaluation

One weakness is that interactionists deemphasize the power of individuals to change the course of events, including overcoming the effects of labeling. For example, low-income Latino parents in Chicago, New York, and Los Angeles who were recent immigrants said that they contacted school personnel on a regular basis, despite the language barrier. Other parents said that they went to college open houses to speak to college representatives, with their children translating the questions and answers (Tornatzky et al. 2002).

Another limitation is that symbolic interactionists, because of their microlevel analysis, neglect the macro-level constraints that are built into society. As conflict and feminist theories show, U.S. education is affected by gatekeeping and control of resources by privileged and powerful groups, which often shape what teachers teach, what students learn, and who will and won't have access to higher education and to particular colleges (Sacks 2007). Such constraints have created numerous problems in U.S. education.

# 13-3 Some Problems With U.S. Education

Only 34 percent of Americans have "a lot" of confidence in our public schools (Morales 2011). Despite the many opportunities that it offers, the U.S. educational system suffers from problems including inadequate schooling, high dropout rates (in both high school and college), and widespread grade inflation and cheating.

## 13-3a Quality and Quantity of Schooling

The U.S. educational system ranks 26th in the world, well behind Finland, Germany, Singapore, Canada,

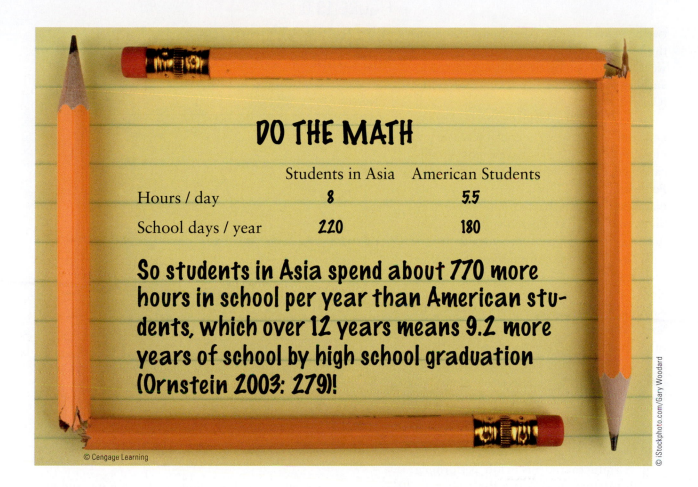

# DO THE MATH

|  | Students in Asia | American Students |
|---|---|---|
| Hours / day | 8 | 5.5 |
| School days / year | 220 | 180 |

So students in Asia spend about 770 more hours in school per year than American students, which over 12 years means 9.2 more years of school by high school graduation (Ornstein 2003: 279)!

© Cengage Learning

© iStockphoto.com/Gary Woodard

and South Korea. Among 64 countries, U.S. 15-year-olds rank 30th in math, 23rd in science, and 17th in reading (Fleischman et al. 2010; OECD 2011). In *Academically Adrift*, a book that received considerable media coverage, sociologists Richard Arum and Josipa Roksa (2011) found that 36 percent of U.S. college seniors were no better in writing and reasoning skills than they were in their first semester.

No single factor explains why many U.S. students are doing so poorly. Some believe that the typical U.S. curriculum, compared with that of other countries, covers "too many topics too superficially" (Grant and Murray 1999: 25). Besides not studying very much, American students also spend less time in the classroom. Students in Asia, for example, have longer and more school days than their American counterparts (see the "Do the Math" box).

Some policy analysts maintain that many U.S. students are performing poorly mainly because of inadequate school funding. On average, states spend almost three times as much on each prisoner as they do on each public school student (see Chapter 7). However, high funding doesn't guarantee high outcomes because teachers play a critical role.

## 13-3b Teachers' Effectiveness

Among first-year American college students, 36 percent need remedial coursework (Aud et al. 2011). Thus, college faculty, in contrast to many high school teachers, see students as unprepared for college. For example, only 33 percent of college instructors, compared with 76 percent of high school teachers, say that students have the necessary writing skills for college-level work (ACT 2007).

A combination of teacher-related factors helps explain why U.S. high school students need remedial coursework in college, and why many students have much lower performance scores than those in other countries. First, the top-performing countries accept only the top applicants for education programs. Finland, Singapore, and South Korea recruit the top 5 percent to 30 percent of high school graduates, but accept only 10 percent who have high standardized test scores in science, math, and reading (Auguste et al. 2010; Sawchuck 2012).

In top-performing countries, 100 percent of teachers are in the top third of their graduating classes. In the United States, 47 percent of kindergarten through twelfth grade teachers comes from the bottom third (Auguste et al. 2010). "In other words, we hire lots of our lowest performers to teach, and then we scream when our kids don't excel" (Cloud 2010: 48).

Second, compared with teachers in Europe and elsewhere, many American teachers are out of field; that is, they have neither certification nor a major in the subject they teach. For example, only 8 percent of fourth-grade math teachers in the United States majored or minored in math, compared with 48 percent in Singapore (Levine 2006; Crowe 2010).

Third, the top-performing countries offer their teachers competitive salaries. In South Korea and Singapore, teachers on average earn more than lawyers and engineers. In South Korea, Singapore, Finland, and many other countries, teachers receive enormous respect and enjoy the same prestige as physicians and other high-status professions and have lifelong careers, resulting in very low teacher attrition. In contrast, about 30 percent of new U.S. teachers leave the profession after five years. High turnover rates are especially harmful to students in low-performing schools (Auguste et al. 2010; Ronfeldt et al. 2011).

About 72 percent of U.S. principals say that tenure policies and teachers' unions prevent them from firing bad teachers (U.S. Chamber of Commerce et al. 2009). Such claims may be warranted, but 40 percent of American teachers, especially those working in low-income schools, are disheartened because principals don't support them, the schools are decrepit, they can't control misbehaving students, and they have little input on the curriculum (Yarrow 2009).

## 13-3c Dropping Out

About 8 percent of young Americans ages 16 to 24 are high school dropouts, down from 15 percent in 1972. Boys are more likely to drop out than girls. The dropout rate is highest among Latinos, but the gap between that group and others has decreased slightly since the early 1990s (see *Figure 13.6*). Foreign-born youths, many of whom were already behind in school when they arrived in the United States, make up 20 percent of the nation's high school dropouts (Fry 2010).

Overall, 22 percent of children who have lived in poverty don't graduate from high school, compared with 6 who have never been poor. Poverty increases the likelihood of experiencing stressful events such

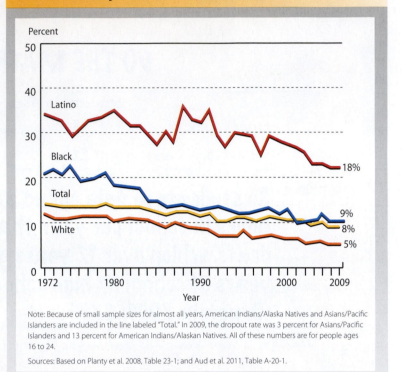

### FIGURE 13.6
### High School Dropout Rates, by Race and Ethnicity: 1972–2009

Note: Because of small sample sizes for almost all years, American Indians/Alaska Natives and Asians/Pacific Islanders are included in the line labeled "Total." In 2009, the dropout rate was 3 percent for Asians/Pacific Islanders and 13 percent for American Indians/Alaskan Natives. All of these numbers are for people ages 16 to 24.

Sources: Based on Planty et al. 2008, Table 23-1; and Aud et al. 2011, Table A-20-1.

as crime, attending a low-quality school, and falling behind in reading skills (Hernandez 2011).

The U.S. college dropout rate, 40 percent, is the highest in the industrialized world. Only 57 percent of students in 4-year colleges complete a bachelor's degree within 6 years, and only 30 percent of students at 2-year colleges earn a certificate or associate's degree within 3 years (Aud et al. 2011).

A growing number of policymakers and researchers attribute the high high school and college dropout rates to a disconnect between education and opportunities in the labor market. Unlike the United States, between 40 and 70 percent of high school students in Austria, Denmark, Finland, the Netherlands, Norway, and Switzerland have programs that combine classroom and workplace learning, including internships that give students "currency" in the labor market ("Pathways to Prosperity . . ." 2011; see also Tabarrok 2012).

The major reason U.S. students give for leaving college is having to work and go to school at the same time. Among students in 4-year schools, 45 percent work more than 20 hours a week. Among those attending community colleges, 60 percent work more than 20 hours a week, and 26 percent work more than 35 hours a week. Moreover, 23 percent of college students have dependent children, making it difficult to juggle academic and domestic responsibilities (Johnson et al. 2009). Despite these constraints, the high dropout rates of U.S. students are somewhat surprising because there is considerable grade inflation at both the high school and college levels.

## 13-3d Grade Inflation

Has an A replaced the C of the 1970s? In 1973, only 20 percent of high school students had an A average, compared with 47 percent in 2003. However, standardized tests show that between 1992 and 2007, the percentage of twelfth-graders who performed at or above a basic level in reading decreased from 80 to 73 percent, and those performing at or above a proficient level declined from 40 to 35 percent (Ewers 2004; Lee et al. 2007; Planty et al. 2008). Such data suggest that grades are rising, but learning is lagging.

A professor who tracks GPAs says that grade inflation "has gone wild: When students walk into a classroom knowing that they can go through the motions and get a B+ or better, that's what they tend to do, give minimal effort. Our college classrooms are filled with students who do not prepare for class" (Rojstaczer 2009: 9). According to another critic, "The demanding professor is close to being extinct" because "professors are under pressure to accommodate students" (Murray 2008: 32).

Grade inflation, especially in college, is due to a number of factors, both institutional and individual. In 1971, only 26 percent of first-year college students expected to earn at least a B average in college, compared with 52 percent in 1995, and a whopping 70 percent in 2010 (Pryor 2011) even though, as you saw earlier, students are studying less than in the past.

Many faculty members give high grades because it decreases student complaints, involves less time and thought in grading exams and papers, and reduces the chances of students' challenging a grade. Some faculty believe that they can get favorable course evaluations from students by handing out high grades, and others accept students' view of high grades as a reward for simply showing up in class. Inflating grades decreases student attrition and satisfies administrators, especially when state legislators base funding on graduation rates (Carroll 2002; Kamber and Biggs 2002; Halfond 2004; Bartlett and Wasley 2008).

Most recently, some law schools, including those that are prestigious, have retroactively inflated transcript grades to make their graduates more competitive in the labor market (Rampell 2010). Among other problems, such as lying, grade inflation gives students

Technology has made cheating easier. Students often use cell phones to store data, to search for answers on the Internet, and to send text messages to friends for answers.

an exaggerated and unrealistic sense of their ability and accomplishments. One result is a sense of entitlement in job searches because "I'm an A student!"

## 13-3e Cheating

Despite grade inflation, cheating is common in high school, college, and graduate school. Recently, an investigation of 56 Atlanta public schools found that teachers and principals in 78 percent of the schools inflated the scores of standardized tests to meet specific score targets (Strauss 2011). New York authorities charged 20 students from an affluent Long Island community with paying impersonators up to $3,600 per SAT to increase their chances of getting into a good school (Winter 2011).

Some 59 percent of high school students admit to cheating on tests, and 33 percent have plagiarized from the Internet. Between 50 and 75 percent of college students say that they have cheated at least once on tests and assignments. Dozens of studies have uncovered cheating at the Air Force Academy and in numerous graduate programs, including those in business administration and dentistry (McCabe et al. 2006; Frosch 2007; Wasley 2007; Gabriel 2010; Josephson Institute 2011).

An employee of a "paper mill" company that churns out custom student papers says that "business is booming." The staff of 50 writers isn't large enough to meet

Would you want a military officer, a financial planner, or a dentist who cheated her or his way through college or graduate school?

**religion** a social institution that involves shared beliefs, values, and practices related to the supernatural that unites believers into a community.

**sacred** anything that people see as mysterious, awe-inspiring, extraordinary and powerful, holy, and not part of the natural world.

**secular** anything that is not related to religion.

**religiosity** the ways people demonstrate their religious beliefs.

**cult** a religious group that is devoted to beliefs and practices that are outside of those accepted in mainstream society.

**new religious movement (NRM)** term used instead of *cult* by most sociologists.

the demands for plagiarized papers at bachelor's, master's, and doctorate levels on behalf of students who "couldn't write a convincing grocery list." One of his recent assignments included writing a graduate thesis on business ethics (Dante 2010: B1).

Education and religion are often entwined. At all educational levels, many private schools are operated by religious organizations but receive considerable funding from the government and public taxes. Religion also affects the family, politics, and other social institutions.

## 13-4 What Is Religion?

For sociologists, **religion** is a social institution that involves shared beliefs, values, and practices related to the supernatural that unites believers into a community. Belonging to a community is important because groups have different beliefs, values, and practices. For example, being Catholic involves confessing (telling one's sins to a priest), a practice not followed by Protestants, Jews, Muslims, and other religious groups.

### 13-4a The Sacred and the Secular

Every society distinguishes between the sacred and the secular. **Sacred** refers to anything that people see as mysterious, awe-inspiring, extraordinary and powerful, holy, and not part of the natural world. **Secular** refers to anything that is not related to religion. Sociologists also differentiate religion from religiosity and spirituality.

### 13-4b Religion, Religiosity, and Spirituality

*Religion* refers to a community of people who share a faith, but the frequency and intensity of religious expression can vary. When sociologists examine **religiosity**, the ways people demonstrate their religious beliefs, they often find that religion and religiosity differ. For example, 72 percent of Americans say that they are "deeply religious," but only 44 percent attend worship services once a week. About 40 percent of college students say that they're religious, but a majority of them report engaging in casual sex, watching pornography, and cheating on exams (Astin and Astin 2005; Lyons 2005; Freitas 2008).

*Spirituality* is a personal quest to feel connected to a reality greater than oneself. About 25 percent of Americans who never go to church consider themselves spiritual: They feel united with the people around them; have made personal sacrifices, such as working with others to decrease poverty; and often believe in miracles (Adler 2005; Pew Forum on Religion & Public Life 2008).

People who see themselves as spiritual may not be religious, however. For example, 77 percent of students at one college described themselves as "spiritual," but only 16 percent participated in religious activities on campus (Denton-Borhaug 2004; see also Astin et al. 2010 and Winograd and Hais 2011). Thus, belonging to a specific religious group, religiosity, and spirituality can each involve different behaviors.

## 13-5 Types of Religious Organization and Some Major World Religions

Religion is important in all known societies, but there's considerable diversity in its expression. People manifest their religious beliefs most commonly through organized groups, including cults, sects, denominations, and churches. The major world religions also differ in their membership and beliefs.

### 13-5a Cults (New Religious Movements)

A **cult** is a religious group that is devoted to beliefs and practices that are outside of those accepted in mainstream society. Many sociologists use **new religious movement (NRM)** rather than *cult* because the media have used the latter term in pejorative ways to describe any unfamiliar, new, or seemingly bizarre religious movement (Roberts 2004).

iStockphoto.com/Shelly Au

The U.S. Supreme Court has ruled that a 6-foot-high monument containing the Ten Commandments outside the Texas state capitol is constitutional, even though it's on government property. How do such rulings reflect the overlap between the sacred and the secular?

**charismatic leader** a religious leader whom followers see as having exceptional or superhuman powers and qualities.

**sect** a religious group that has broken away from an established religion.

**denomination** a subgroup within a religion that shares its name and traditions and is generally on good terms with the main group.

**church** a large established religious group that has strong ties to mainstream society.

NRMs usually organize around a **charismatic leader** (like Jesus) whom followers see as having exceptional or superhuman powers and qualities. Some NRMs have become established religions. The early Christians were a renegade group that broke away from Judaism. Islam, the second largest religion in the world today, began as a cult organized around Muhammad, and the Church of Jesus Christ of Latter-Day Saints (Mormons) began as a cult around Joseph Smith. Most contemporary cults are fragmentary, loosely organized, and temporary, but others have developed into groups that are more lasting, organized, and highly bureaucratic (such as the Church of Scientology).

## 13-5b Sects

A **sect** is a religious group that has broken away from an established religion. Those who begin sects are usually dissatisfied members who believe that the parent religion has become too secular and has abandoned key original doctrines. Like cults, some sects are small and disappear after a time, whereas others become established and persist. Examples of sects that have persisted include the Amish, the Jewish Hassidim, Jehovah's Witnesses, Quakers, Seventh-Day Adventists, and the Fundamentalist Church of Jesus Christ of Latter-Day Saints (FLDS) (Finke and Stark 1992; Bainbridge 1997). Some sects develop into denominations.

## 13-5c Denominations

A **denomination** is a subgroup within a religion that shares its name and traditions and is generally on good terms with the main group. Denominations can form slowly or develop rapidly, depending on factors such as geography, immigration, and culture. Some scholars describe a denomination as somewhere between a sect and a church. Like sects, denominations have a professional ministry. Unlike sects, however, denominations view other religious groups as valid and don't make claims that only they possess the truth (Niebuhr 1929; Hamilton 2001).

Denominations typically accommodate themselves to the larger society instead of trying to dominate or change it. As a result, people may belong to the same denomination as did their grandparents or even great-grandparents. Denominations exist in all religions, including Christianity, Judaism, and Islam. In the United States, for example, the many Protestant denominations include Episcopalians, Baptists, Lutherans, Presbyterians, Methodists, and Evangelicals.

## 13-5d Churches

A **church** is a large established religious group that has strong ties to mainstream society. Because leadership is attached to an office rather than a specific leader, new generations of believers replace previous ones, and members follow tradition or authority rather than a charismatic leader. As in a denomination, people are usually born into a church, even though they may decide to withdraw later.

Churches (e.g., Roman Catholic Church, Anglican Church in North America) are typically bureaucratically organized, have formal worship services and trained clergy, and often maintain some degree of control over political or educational institutions (see Chapter 11). Because churches are an integral part of the social order, they often become dependent on, rather than critical of, the ruling classes (Hamilton 1995).

## 13-5e Some Major World Religions

Worldwide, the largest religious group is Christians, followed by Muslims. If the world's population is represented as an imaginary village of 100 people, it has about

- 33 Christians
- 21 Muslims
- 16 nonreligious individuals (people who may believe in God but are not affiliated with a religious group) or atheists (people who don't believe in God)
- 14 Hindus
- 6 Buddhists
- 6 Chinese Universalists (followers of a complex set of beliefs and practices that combines ancestor cults, Confucianism, Buddhism, Taoism, folk religion, and goddess worship)
- 4 believers in another religion (including Judaism and Sikhism) (based on "Major Religions of the World . . ." 2007)

Note that no religious group comes close to being a global majority, that the third largest group consists of the nonreligious and nonbelievers, and that non-Christians outnumber Christians 2 to 1. Five religious groups in particular have had a worldwide impact on economic, political, and social issues. *Table 13.2* provides a brief overview of these groups.

## Sociology in Your Life

How much do you know about religion? And how do you compare with the average American? Find out at CengageBrain.com.

## 13-6 Religion in the United States

Among adult Americans, 92 percent believe in God, 7 percent don't believe in God, and 1 percent have no opinion on this issue (Newport 2011). For sociologists, religiosity is a better measure of being religious than simply asking people whether they believe in God (or a universal spirit) and which religion they follow. Religiosity includes a number of variables, but the most common are religious belief, affiliation, and attendance at services.

### 13-6a Religious Belief

Some 55 percent of Americans say that religion is very important in their lives, down from 75 percent in 1952 (Newport 2011). Not surprisingly, religion is not very important for those who are *agnostics* (say that it's impossible to know whether there is a God), *atheists* (believe that there is no God), or others who are skeptics. Nonetheless, about 25 percent of Americans, including some atheists and agnostics, embrace the tenets of some Eastern religions or elements of New Age spirituality such as reincarnation, meditation, astrology, and the evil eye (casting of harmful curses and spells) (Pew Forum on Religion & Public Life 2009b).

### 13-6b Religious Affiliation

About 15 percent of Americans say that they have no religious preference or affiliation (see *Table 13.3* on page 264), up from 8 percent in 1990 (Kosmin and Keysar 2009). About half of U.S. adults—especially those who were raised as Catholics or Protestants—have changed their religious affiliation since childhood, some have done so more than once, and 15 percent have opted for no religion at all. The most important reasons for becoming unaffiliated include not believing the teachings, seeing many religious people as hypocritical or judgmental, and losing respect for religious leaders who focus on power and money (Pew Forum on Religion & Public Life 2009a).

### 13-6c Religious Participation

Turning to the third measure of religiosity, religious participation, about 40 percent of Americans attend religious services at least once a week; 27 percent seldom or never attend. Thus, many people are more likely to believe in a religion than to practice it by attending services regularly. Mormons (77 percent), evangelical Protestants (58 percent), and members of historically black churches (59 percent) have higher attendance rates at religious services than do Catholics (42 percent), Muslims (40 percent), or Jews (16 percent) (Pew Forum on Religion & Public Life 2008; "Mormons in America" 2012).

## TABLE 13.2
# Characteristics of Five Major World Religions

| RELIGION | DATE OF ORIGIN | FOUNDER | NUMBER OF FOLLOWERS | CORE BELIEFS |
|---|---|---|---|---|
| Christianity <br> ©iStockphoto.com/Sebastien Cote | 0 C.E. | Jesus Christ | 2.2 billion | Jesus, the son of God, sacrificed his life to redeem humankind. Those who follow Christ's teachings and live a moral life will enter the Kingdom of Heaven. Sinners who don't repent will burn in hell for eternity. |
| Islam <br> ©iStockphoto.com/Kenneth C. Zirkel | 600 C.E. | Muhammad | 1.6 billion | God is creator of the universe, omnipotent, omniscient, just, forgiving, and merciful. Those who sincerely repent and submit (the literal meaning of *islam*) to God will attain salvation, while the wicked will burn in hell. |
| Hinduism <br> ©iStockphoto.com/Ferenc Cegledi | Between 4000 and 1500 B.C.E. | No specific founder | 887 million | Life in all its forms is an aspect of the divine. The aim of every Hindu is to use pure acts, thoughts, and devotion to escape a cycle of birth and rebirth (*samsara*) determined by the purity or impurity of past deeds (*karma*). |
| Buddhism <br> ©iStockphoto.com/Pablo Salgado Barrientos | 525 B.C.E. | Siddhartha Gautama | 386 million | Life is misery and decay with no ultimate reality. Meditation and good deeds will end the cycle of endless birth and rebirth, and the person will achieve *nirvana*, a state of liberation and bliss. |
| Judaism <br> ©iStockphoto.com/Howard Sandler | 2000 T | Abraham | 15 million | God is the creator and the absolute ruler of the universe. God established a particular relationship with the Hebrew people. By obeying the divine law God gave them, Jews bear special witness to God's mercy and justice. |

Note: C.E. (Common Era) is the nondenominational abbreviation for A.D. (Anno Domini, Latin for "In the year of our Lord)" and B.C.E. (Before the Common Era) is the nondenominational abbreviation for B.C. (Before Christ).

Sources: Based on a number of sources including the Center for the Study of Global Christianity 2007, "Religions of the World . . ." 2007, and "Global Christianity . . ." 2011.

## 13-6d Some Characteristics of Religious Participants

Americans differ in their beliefs and affiliations. Religious participation also varies by sex, age, race and ethnicity, and social class.

### Sex

Across all age and faith groups in 145 countries, women tend to be more religious than men in believing in God, praying, attending services, and saying that religion is very important in their lives (Pew Forum on Religion & Public Life 2008; Deaton 2009; Kosmin and Keysar 2009; Taylor et al. 2009). It may be that women are expected to be more pious and spiritual because, especially as nurturers, they transmit religious values to their children (see Chapter 9). Another reason may be men's greater involvement in public life (e.g., employment, politics) that demands more of their time and energy.

### Age

Generally, Americans age 65 and older are more likely than younger people to describe themselves as religious, to say that religion is very important in their lives, and to attend services at least weekly. These age-related differences may reflect several factors: Older Americans grew up decades ago when church attendance was higher

## TABLE 13.3

## Religious Affiliation in the United States, 2011

| CHRISTIAN | 78% |
|---|---|
| Protestant/Other Christian | 52.5% |
| Catholic | 23.6% |
| Mormon | 1.9% |
| **OTHER RELIGIONS** | **4.5%** |
| Jewish | 1.6% |
| Muslim | 0.5% |
| Other non-Christian religion | 2.4% |
| **UNAFFILIATED** (None, Atheist, Agnostic) | **15%** |

Note: "Other non-Christian religion" includes groups such as Buddhists and Hindus. The numbers don't sum to 100% because 2.5% didn't respond.

Source: Based on Newport, "Christianity Remains Dominant . . . ." 2011.

Megachurches are Christian congregations that have a regular weekly attendance of more than 2,000 and tend to be evangelical. They represent only 0.5 percent of all U.S. churches, but their number has more than doubled (to more than 1,200) since 2000 (Thumma et al. 2005). Pictured here, the Lakewood Church in Houston, Texas, has the largest congregation in the United States, averaging more than 43,000 in attendance per week.

**secularization** a process of removing institutions such as education and government from the dominance or influence of religion.

for all Americans, they seek spiritual comfort as elderly friends and relatives die, they want to lessen a sense of isolation or loneliness, and they are preparing for death (Taylor et al. 2009).

"Millennial" young adults (those born in 1981 and later) are less religious than previous generations: They are less likely to pray, to regularly attend worship services, or to identify themselves with a religious group. It appears, however, that religiosity increases with age. For example, only 40 percent of Millennials say religion is very important compared with 60 percent of Boomers (those born between 1946 and 1964). In the late 1970s, however, when Boomers were the same age as Millennials, only 39 percent said that religion was very important (Pond et al. 2010).

### Race and Ethnicity

In the last decade or so in the United States, the fastest growing group has been those who describe themselves as nonreligious. Within this population, 39 percent are Asian, 34 percent are white, 22 percent are Latino, and 13 percent are black (McCauley 2012).

Some Catholic and evangelical Protestant churches have been especially successful in attracting recent Latino immigrants because they are more likely to offer services in Spanish, their services are expressive rather than formal, and the clergy are more responsive to their members' social and economic needs (Campo-Flores 2005; Wolfe 2008). Asian Americans may be the most diverse religious group in America: "Asian

Buddhist and Hindu temples, Muslim mosques, and Catholic and Protestant churches have mushroomed in Los Angeles, New York, and other major cities over the past thirty years" (Min 2002: 5).

### Social Class

People with less education are generally more religious than those who have attained higher educational levels. For example, 60 percent of people with a high school diploma or less say that religion is very important in their daily lives, compared with 50 percent of college graduates (Pew Forum on Religion & Public Life 2008). Because education and income are highly correlated, as income increases, the importance of religion generally decreases (see *Figure 13.7*).

## 13-6e Secularization: Is Religion Declining?

Industrialized nations have been experiencing **secularization**, a process of removing institutions such as education and government from the dominance or influence of religion. Some sociologists maintain that secularization is increasing rapidly in the United States, but others contend that this claim is greatly exaggerated.

### Is Secularization Increasing?

There is evidence of increased secularization in the United States. For example:

## FIGURE 13.7
## Religion and Household Income

**fundamentalism** the belief in the literal meaning of a sacred text.

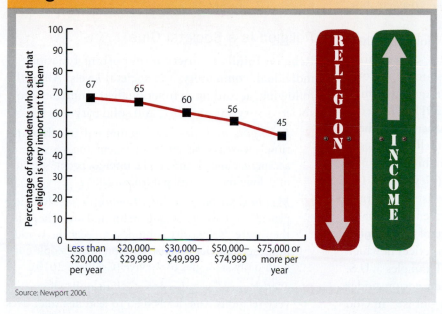

Source: Newport 2006.

recently banned women's full-face Islamic veils, and Syria has done the same in both public and private universities ("Face Veil Banned in Universities" 2010; "After Much Debate . . ." 2011).

### Is the Extent of Secularization Exaggerated?

Many sociologists contend that the extent of secularization has been greatly overstated. Some point out that **fundamentalism**, the belief in the literal meaning of a sacred text

- The number of Americans who don't identify with any religion (15 percent) has doubled since 1990 (Zuckerman 2011).

- Attendance at religious services has decreased, and fewer Americans say that religion is "very important" in their lives (see pp. 263–264).

- In a recent poll, 70 percent of Americans—compared with only 14 percent in 1957—said that religion is losing influence on U.S. life (Newport 2010).

- A majority (52 percent) of Americans say that religious groups should keep out of political matters, up from 43 percent in the mid-1990s ("Public Views . . ." 2012).

Many Americans believe that, through its financial support of religious organizations, the government is escalating a "campaign" to trample their religious freedom. Catholic Charities, one of the nation's most extensive social service networks, receives about 62 percent of its annual revenue from the government. Some states now require adoptive agencies that receive state money to comply with nondiscrimination laws. Roman Catholic bishops in several states have discontinued their adoptive services because they refuse to consider same-sex couples as potential adoptive parents (Goodstein 2011).

There is also evidence of greater "forced secularization" worldwide. Nearly a third of the world's 7 billion people live in countries where government restrictions on religion have increased since 2006. Some of the laws and policies constrict religious beliefs or practices; others allow acts of religious hostility by individuals, organizations, and social groups ("Rising Restrictions . . ." 2011). For example, some European countries have

(such as the Christian Bible, the Muslim Qur'an, or the Jewish Torah), has increased in the United States and worldwide (Woolf 2004; Jones 2011). There is also considerable evidence that religion has an enormous impact on many Americans' lives. For example:

- Increasing numbers of pharmacists have refused to fill prescriptions for oral contraceptives because doing so violates their religious beliefs (Bergquist 2006).

- The number of organizations engaged in religious lobbying in Washington has increased almost fivefold—from fewer than 40 in 1970 to 211 in 2010 ("Lobbying for the Faithful . . ." 2011).

- In a recent poll, 61 percent of Americans said that it's important for members of Congress to have strong religious beliefs ("Growing Number of Americans . . ." 2010).

- Another poll reported that 83 percent support Christmas displays on government property (Pew Research Center for the People & the Press 2010).

Over the years, one of the most divisive secularization issues has been the clash over which explanations for human origins should be taught in public schools. Scientists claim that human beings (and other creatures) evolved from earlier animals through a process known as "natural selection" over several million years. In contrast, creationism, based on a fundamentalist interpretation of the Bible, argues that God created humans in their present form about 10,000 years ago. Intelligent design (also referred to as "theistic evolution") is a watered-down version of creationism: It doesn't reject evolution outright but holds that the universe is so complex that it must have been created by a supreme being (National Academy of Sciences 2008; Robinson 2011a).

Many biology teachers worry that conservative state legislators are coming closer to passing laws that will require teaching creationism in science classes (Boston 2011). Their concerns may be justified: 40 percent of Americans believe in creationism, 38 percent believe in intelligent design, and only 16 percent believe in evolution through natural selection (up from 9 percent in 2000) (Newport 2010).

Sociologists (and other social scientists) who argue that secularization isn't increasing in the United States and elsewhere point to the prevalence of **civil religion** (sometimes called *secular religion*), practices in which citizenship takes on religious aspects. Examples of U.S. civil religion include the phrase "one nation under God" in the Pledge of Allegiance, the phrase "In God We Trust" on U.S. coins, and a legally mandated National Day of Prayer, an annual day of observance held on the first Thursday of May.

# 13-7 Sociological Perspectives on Religion

You've seen that religion plays an important role in many people's lives. Why? How does society affect religion? And how does religion affect society? *Table 13.4* summarizes the major sociological perspectives in answering these questions.

## 13-7a Functionalism: Religion Benefits Society

Durkheim (1961: 13, 43) described religion as an "essential and permanent aspect of humanity" whose "essential task is to maintain, in a positive manner,

© Pascal Deloche/Godong/Corbis

the normal course of life." Religion can also be dysfunctional, but first let's consider its positive influence.

### Religion Is a Societal Glue

Religion fulfills a variety of important functions at individual, community, and societal levels. All of the following, according to functionalists, contribute to a society's survival, stability, and solidarity:

- *Belonging and identity.* Communal worship and rituals increases social contacts, enhances a sense of acceptance and identity, and reinforces people's feeling of belonging to a group (Gruber 2005).

- *Meaning, purpose, and emotional comfort.* Religion provides a sense of meaning in life and offers hope for the future. Religiosity is highest among the world's poorest countries because religion helps people cope with daily struggles to survive (Crabtree 2010).

- *Well-being.* Several recent Gallup polls show that very religious Americans tend to lead healthier lives and experience less depression and worry (Newport et al. 2010, 2012). Also, couples who share religious affiliations, beliefs, and practices report being happier than those who don't (Ellison et al. 2010).

- *Social service.* Religions provide numerous social services that benefit their members and others. For example, religious organizations raise millions of dollars for the victims of natural disasters, distribute food and clothing, and find shelter for displaced families (Brown 2005). Also, Americans who are active in a church or other religious or spiritual organization are more likely than those who are not religious to engage in civic and community projects (Jansen 1011).

- *Social control.* Because people who are religious internalize rules about right and wrong and fear damnation in the afterlife, they try to practice self-control, which encourages social conformity and discourages deviant behavior (Durkheim 1961).

### Religion Promotes Social Change

For functionalists, religion can also spearhead social change. Mohandas Gandhi (1869–1948), a spiritual leader in India, worked for his country's independence from Great Britain through nonviolence and peaceful negotiations. In the United States, religious leaders, especially the Reverend Martin Luther King, Jr., were at the forefront of the civil rights movement during the late 1960s. More recently, after 33 years of debate, the Presbyterian Church has changed its constitution to allow openly gay people in same-sex relationships to be ordained as ministers, elders, and deacons, and numerous Lutheran churches have welcomed gay pastors (Goodstein 2010).

## TABLE 13.4
# Sociological Perspectives on Religion

| THEORETICAL PERSPECTIVE | LEVEL OF ANALYSIS | VIEW OF RELIGION | SOME MAJOR QUESTIONS |
|---|---|---|---|
| **Functionalist** | Macro | Religion benefits society by providing a sense of belonging, identity, meaning, emotional comfort, and social control over deviant behavior. | How does religion contribute to social cohesion? |
| **Conflict** | Macro | Religion promotes and legitimates social inequality, condones strife and violence between groups, and justifies oppression of poor people. | How does religion control and oppress people, especially those at lower socioeconomic levels? |
| **Feminist** | Macro and Micro | Religion subordinates women, excludes them from decision-making positions, and legitimizes patriarchal control of society. | How is religion patriarchal and sexist? |
| **Symbolic Interactionist** | Micro | Religion provides meaning and sustenance in everyday life through symbols, rituals, and beliefs and binds people together in a physical and spiritual community. | How does religion differ within and across societies? |

© iStockphoto.com/PaulMaguire

© Cengage Learning

Max Weber (1920) asserted that religion sparks economic development. His study of Calvinism, a Christian sect that arose in Europe during the sixteenth century, led to his coining the term **Protestant ethic**, which he described as a belief that hard work, diligence, self-denial, frugality, and economic success would lead to salvation in the afterlife. According to Weber, the harder the early Calvinists worked and saved, the more likely they were to accumulate money, to become successful, and to drive the growth of capitalism.

Was Weber right about the relationship between the Protestant ethic and the rise of capitalism? The data are mixed. Some studies show that people are diligent not because of religious beliefs but simply because amassing savings provides resources when disaster strikes (as when droughts wipe out crops). On the other hand, a study of 59 industrialized and developing countries found that religious beliefs that encourage hard work and thrift spur economic growth (Cohen 2002; Barro and McCleary 2003).

## How Is Religion Dysfunctional?

Functionalists emphasize the benefits of religion, but they also recognize that religion can be dysfunctional when it harms individuals, communities, and societies. For example, religious intolerance can lead to conflict between groups (including vandalizing churches,

mosques, and synagogues) and attacks on religious minorities. The United States has a long history, beginning with the Pilgrims in 1620, of people discriminating against other religious groups even though they themselves sought religious freedom (Davis 2010).

> **Protestant ethic** a belief that hard work, diligence, self-denial, frugality, and economic success will lead to salvation in the afterlife.

## Critical Evaluation

Critics accuse functionalists of emphasizing benefits and glossing over numerous dysfunctional aspects of religion that create and maintain conflict. For example, disagreements over whose deity is the "real God" have led to wars, terrorism, and genocide (see Chapter 10).

Critics also note that functionalists, by emphasizing the needs that religion fulfills, imply that religion is indispensable to leading a good life. However, people who aren't religious do many good things for society, whereas some religious people commit heinous acts, such as the molesting of young children by some Catholic priests. It's also not clear why Jews, who are one of the least religious U.S. groups, report the highest rates of well-being, and why the nonaffiliated groups, such as atheists, report higher rates of well-being than religious groups (see Newport et al. 2012).

In 2006, a dairy truck driver burst into an Amish school-house in Lancaster County, Pennsylvania. He killed five girls and wounded three others before shooting himself because, according to his suicide note, he was angry with life and at God. Pictured here, an Amish funeral procession escorts the victims to a burial ground. The Amish community, in which violence is unthinkable, accepted the tragedy as the will of God and started a charity fund to help not only the victims' families but also the gunman's widow and children. From a function-alist perspective, why does their religion help the Amish and other religious people cope with tragedy?

## 13-7b Conflict Theory: Religion Promotes Social Inequality

From a conflict perspective, religion promotes and reinforces social inequality. Throughout history and currently, religion has created discord and divisiveness within groups, between groups, and in the larger society.

### "The Opium of the People"

Much conflict theory reflects the work of Karl Marx (1845/1972), who described religion as "the sigh of the oppressed creature" and "the opium of the people" because it encouraged passivity and acceptance of class inequality. For Marx, religion taught people to endure suffering and deprivation instead of revolting against injustice.

Marx viewed religion as a form of **false conscious-ness**, an acceptance of a system of beliefs that prevents people from protesting oppression. Contemporary conflict theorists don't view religion as an opiate, but they agree with Marx that religion legitimizes social inequality and sometimes leads to social disruption and violence.

### A Source of Social Disruption and Violence

For conflict theorists, religion tends to promote strife because, typically, religious groups differentiate between "we" and "they" ("We're right and they're wrong."). Such distinctions spark numerous dis-putes within and across societies. For example, ter-rorist attacks by militant Muslims on Christians in Africa, the Middle East, and Asia increased 309 per-cent between 2003 and 2010 (Ali 2012). The schism between Iraq's Sunnis and Shiites dates back more than 1,300 years, and there are long-standing conflicts between Muslims and Jews in Israel and Palestine (Ghosh 2007; Robinson 2011b).

For thousands of years, many governments and religious leaders have condoned or perpetrated wide-spread violence in the name of religion. Terrorists who targeted the World Trade Center on 9/11 didn't see themselves as "crazed terrorists" but as "true believers" who were carrying out "God's will" by imposing their religious beliefs on others or destroying their "religious enemies" (Juergensmeyer 2003).

Many conflict theorists point out that religion can be a major source of mutual hatred, discrimination, and stereotypes. For example, many non-Muslims in the West and many Muslims in the Middle East and Asia generally describe people on the other side as immoral, sinful, or fanatical in their religious views and practices (Pew Global Attitudes Project 2006).

### A Legitimation of Social Inequality

Conflict theorists also see religion as a tool dominant groups use to control society, protect their own inter-ests, and derail social change. U.S. Roman Catholic bishops have a long history of helping the poor and fostering social justice. Most recently, however, most have opposed comprehensive health care because it would provide taxpayer funding for abortion, have publicly chastised pro-choice Catholic politicians, and have campaigned against same-sex marriage—all of which fuel social inequality and conflict (Doyle 2011).

Conflict theorists also point out that private evan-gelical schools, colleges, and universities receive mas-sive state and federal financial aid that diverts much-needed tax revenues from public schools and colleges. Many of the schools' textbooks describe homosexuals and abortion rights supporters as "evil," and are hos-tile toward other religions, including nonevangelical Protestants, Jews, and Catholics (Berkowitz 2011; Tabachnick 2011).

### Critical Evaluation

Functionalists may overemphasize consensus and har-mony, but conflict theorists, according to some critics, often ignore the role that religion plays in creating

>SIGH< WHERE WOULD I BE WITHOUT YOU?

RELIGION

WAR

Don Addis/Tampa Bay Times

social cohesion and cooperation. Also, the altruism of many religious people counters Marx's notion of false consciousness. For example, many religious people run and support charities and volunteer their time in low-income neighborhoods rather than seeking economic rewards at work or enjoying recreational activities (Roberts 2004). Compared with most other countries, the United States is religiously diverse, and many Americans are religiously tolerant. Such characteristics are powerful sources of societal stability that decrease religious conflict (Putnam and Campbell 2012).

## 13-7c Feminist Theories: Religion Subordinates and Excludes Women

Feminist theorists agree with conflict theorists that religion can create violence and maintain inequality. They go further, however, by criticizing organized religions for being sexist and patriarchal and for shutting women out of leadership positions.

### Sexism and Patriarchy

From a feminist perspective, most religions are patriarchal because they emphasize men's experiences and a male point of view and see women as subordinate to men. According to the apostle Paul, for example, "Wives should submit to their husbands in everything" (*Ephesians* 5: 24). The idea that Eve was created out of Adam's rib is often used to justify men's domination of women. Such beliefs are propagated in many religious teachings and institutions today (Mananzan 2002).

In Orthodox Judaism, a man's daily prayers include this line: "Blessed art thou, O Lord, our God, King of the Universe, that I was not born a woman." The Qur'an tells Muslims that men are in charge of women and that women should obey men. Almost all contemporary religions worship a male God, and none of the major world religions treat women and men equally (Gross 1996; Jeffreys 2011).

Some feminist scholars charge that the Bible is often interpreted in a patriarchal manner. For example, Jesus encouraged women's intellectual pursuits when it was not the norm, and there are numerous passages in the Bible about women spreading Jesus' teachings. Muslim feminists, similarly, note that men have interpreted sacred Islamic texts to ensure male dominance and control. Women were among some of Muhammad's earliest converts, and the Qur'an has numerous passages that establish women's equal rights in inheritance and family roles (Menissi 1991, 1996; Gross 1996). Thus, women's subordination in many Islamic societies is due not to religious tenets but to the men who have interpreted the tenets to maintain their power and privilege (Smith 1994).

> **false consciousness** an acceptance of a system of beliefs that prevents people from protesting oppression.

### Exclusion of Women From Leadership Positions

Women earned 33 percent of theology degrees in 2009, a dramatic increase from only 2 percent in 1970. Still, women make up only 13 percent of the nation's clergy, and only 3 percent of female clergy lead large congregations (those with more than 350 people) (Winseman 2004; U.S. Census Bureau 2012).

Neither Roman Catholicism nor Orthodox Judaism allows women to be ordained because both groups

In Iraq and Afghanistan, Muslims serving in the U.S. military have faced a double-edged sword. They are prized for their language skills and cultural knowledge by commanders, but may be singled out by insurgents as traitors (Dreazen 2009).

AP Photo/The Dallas Morning News, Sonya N. Hebert

**What Is a Hajj?** Every year, more than 2 million Muslims engage in an important ritual, the Hajj, by making a pilgrimage to Mecca (in Saudi Arabia). The Hajj is one of the five pillars of the Muslim faith that demonstrate the solidarity of Muslims and their submission to God. Every able-bodied follower who can afford it is expected to perform the pilgrimage at least once in a lifetime. The Hajj occurs from the 8th to the 12th days of the last month of the Islamic year (roughly in the November to January period of the Western calendar).

believe that women should serve and not lead. In the United States, Catholic women contribute $6 billion a year during Sunday Masses, but "the presence of women anywhere within the institutional power structure is virtually nil" (Miller 2010: 39; see also Gibson 2012).

Some Protestant denominations justify women's exclusion from leadership positions based on biblical passages such as "I permit no woman to teach or have authority over men; she is to keep silent" (I Timothy 2: 11–12). Many Protestant groups—including Southern Baptists and especially evangelical groups (such as born-again Christians)—interpret this passage and similar ones to mean that women should never, under any circumstances, instruct men, within or outside of religious institutions. Thus, Southern Baptist seminaries accept women for theological studies, but rarely hire them as

> "Children's lives would be harmed if the nation had a female president. . . . [It is not] God's desire to have a woman rule the institutions of the family, the church, and the state."
>
> —Email sent to Rick Santorum, a candidate for the 2012 Republican Party's presidential nomination, by Jamie Johnson, one of Santorum's staff and a pastor at a central Iowa church (Diamond 2012)

faculty members. Even in liberal Protestant congregations, female clergy tend to be relegated to specialized ministries with responsibilities for music, youth, or Bible studies (Banerjee 2006; Bartlett 2007).

### Critical Evaluation

Some feminist Muslim scholars have criticized Western feminists for misreading and misinterpreting Islamic and other sacred scriptures, such as reducing practically all discussions on gender to the *hijab* (a veil or scarf that Muslim women wear) instead of focusing on justice for both women and men in marriage, employment, and other areas (Choudhury et al. 2006; Esposito and Mogahed 2007; Fakhraie 2009).

Another criticism is that religious women aren't as oppressed as many feminist scholars claim. Even in the most conservative religions, women don't blindly submit to religious dogma. Instead, they challenge existing doctrines, stop volunteering at church, don't attend weekly services, and establish their own parishes (Avishai 2010; Padgett 2010; Miller 2012).

### 13-7d Symbolic Interactionism: Religion Is Socially Constructed

For symbolic interactionists, religion isn't innate but learned. Symbols, rituals, and beliefs are three of the most common vehicles for learning and internalizing religion.

### Symbols

A *symbol* is anything that stands for or represents something else to which people attach meaning (see Chapter

## FIGURE 13.8
## Religious Symbols

Top row from left to right: Buddhist, Christian, Hindu, Indigenous and Ancient Religions, Islam and Judaism.

Bottom row from left to right: Spirituality, Lutheran, Confucian, Baha'i, Scientology

© Cengage Learning

**ritual** (sometimes called a *rite*) a formal and repeated behavior in which the members of a group regularly engage.

3). Many religious symbols are objects (a cross, a steeple, a Bible), but they also include behaviors (kneeling and bowing one's head), words ("Holy Father," "Allah," "the Prophet"), and physical appearance (the wearing of head scarves, skull caps, turbans, clerical collars).

Religious symbols, like all symbols, are shorthand communication tools (see *Figure 13.8*). Some symbolic interactionists define a religion as "a system of symbols" because it's a community that's unified by its symbols (Berger and Luckmann 1966; Geertz 1966).

### Rituals

A **ritual** (sometimes called a *rite*) is a formal and repeated behavior in which the members of a group regularly engage. Religious rituals, like secular ones, strengthen a participant's self-identity (Reiss 2004). Religious rites of passage—such as the *bat mitzvah* for girls and the *bar mitzvah* for boys in the Jewish community and the first communion and confirmation for Catholic children—reinforce the individual's sense of belonging to a particular religious group. A group's rituals symbolize its spiritual beliefs and include a wide range of practices such as praying, chanting, fasting, singing, dancing, and offering sacrifices.

All religions have rituals that mark significant life events, such as birth and marriage (see Chapters 3, 5, and 12). Death rituals are probably the most elaborate and sacred worldwide. They vary across religious groups and societies, but all of them comfort the living and show respect for the dead.

### Beliefs

Rituals and symbols come from *beliefs*, convictions about what people think is true. Religious beliefs can be passive (believing in God but never attending formal services) or active (participating in rituals and ceremonies). Beliefs bind people together into a spiritual community.

One of the strongest beliefs worldwide is that prayer is important. Islam requires prayer five times a day. In the United States, 58 percent of U.S. adults say that they pray more than once a day and do so for a variety of reasons, such as feeling close to God, as well as requesting better health, more money, and cures for sick pets (Pew Forum on Religion & Public Life 2008; Wicker 2009). Prayer offers psychological and spiritual benefits such as comfort and a sense of unity among those who pray together, but depending on God to make decisions may also dampen people's motivation to actively shape their own lives (Schieman 2010).

### Critical Evaluation

A common criticism is that interactionists' focus on micro-level behavior ignores the ways that religion promotes social inequality at the macro level. Conflict theorists and feminist scholars, especially, maintain that people often use religion to justify violence and women's subordination. Some critics also wonder if interactionists paint too rosy a picture of religion even on a micro level. After all, people's attachment to their religious symbols, rituals, and beliefs can create considerable conflict, such as the violence and protests that ensued in the spring of 2012 after American troops in Afghanistan burned copies of the Qur'an containing extremist statements in the margins written by imprisoned militants and terrorists.

 **Study Tools**

Ready to study? Go to **Sociology CourseMate** at **www.cengagebrain.com** to complete practice quizzes, review flashcards, watch videos, and more.

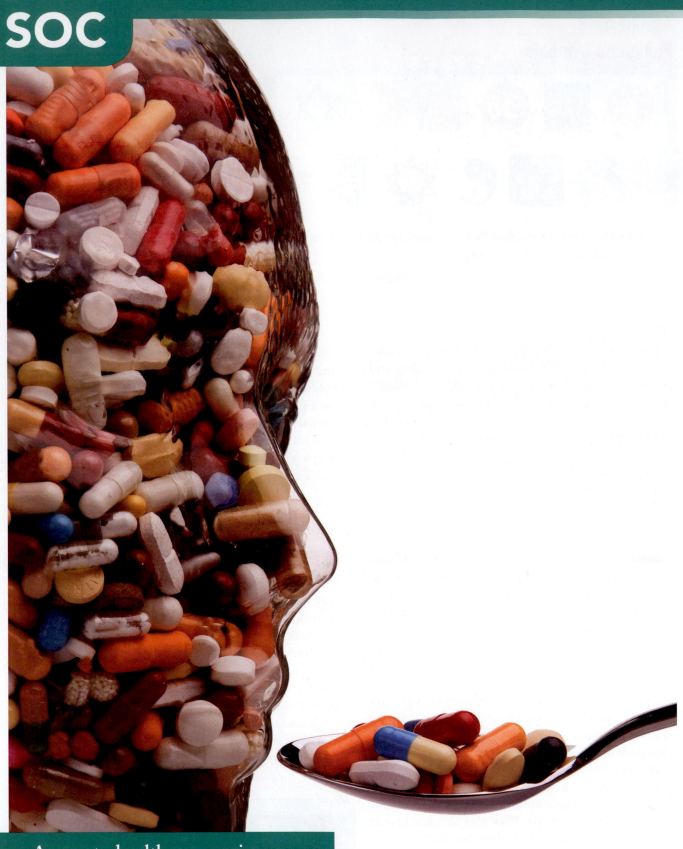

# SOC

Access to health care varies considerably in the United States and worldwide.

# Health and Medicine

Access to good health is distributed unequally in our society and many others. Before you read further, take the short quiz on the next page to see how much you know about U.S. health and medicine.

**Health** is the state of physical, mental, and social well-being. We can determine physical health, to a great extent, by using objective measures such as weight and blood pressure. Mental and social well-being are more difficult to gauge because they're more subjective, as you'll see shortly; they're influenced by culture and new technologies (which change over time) and by how authority figures, especially physicians, define health.

Health varies among individuals and societies, but all people experience **disease**, an alteration of the normal physical and/or mental structures of the body or mind. When people feel sick, most seek **health care**, any activity that improves a person's well-being. A vital part of health care is **medicine**, a system of individuals, organizations, and institutions that provide scientific diagnosis, treatment, and prevention of illness, injury, and other health impairments.

**health** the state of physical, mental, and social well-being.

**disease** an alteration of the normal physical and/or mental structures of the body or mind.

**health care** any activity that improves a person's well-being.

**medicine** a system of individuals, organizations, and institutions that provide scientific diagnosis, treatment, and prevention of illness, injury, and other health impairments.

**disability** physical or mental impairments that limit a person's ability to perform an important activity.

## Key Topics

In this chapter, we'll explore the following topics:

**14-1** Health and Illness in the United States

**14-2** Health Care: The United States and Around the World

**14-3** Sociological Perspectives on Health and Medicine

## What do you think?

Prescription pills are more dangerous than using illegal drugs, including marijuana.

| 1 | 2 | 3 | 4 | 5 | 6 | 7 |

strongly agree    strongly disagree

## 14-1 Health and Illness in the United States

Americans' life expectancy has increased (see Chapter 12), but today's children are in danger of becoming the first generation in U.S. history to live shorter and less healthy lives than their parents (Levi et al. 2011). Many will experience **disability**, physical or mental impairments that limit a person's ability to perform an important activity. Others will die at an earlier age than expected for a variety of macro- and micro-level reasons.

The question of why some people are healthier than others is important for all countries. In addressing these issues, health and medical practitioners, researchers, and sociologists examine epidemiology and what affects illness.

## HOW MUCH DO YOU KNOW ABOUT U.S. HEALTH AND MEDICINE?

**True or False?**

1. Genes determine how long people live.

2. Women have higher depression rates than men.

3. White women are more likely to die of breast cancer than women of other racial-ethnic groups.

4. The most important preventable health hazard is drug abuse.

5. Chronically ill people are responsible for almost 50 percent of all U.S. health care spending.

6. Compared with other groups, Latinos are the most likely to die from illegal drugs and prescription drug abuse.

7. Most women have better health care coverage than men.

8. Because of healthier eating habits and exercise, medical researchers predict that adult obesity rates will decline in the future.

The answers to 2 and 5 are true. The others are false. You'll see why as you read this chapter.

© Cengage Learning

Sociologists incorporate epidemiology in their analyses, but they're more interested in understanding and explaining *why* people vary so much in the causes and distribution of disease.

## 14-1b Some Reasons for Contemporary Illness and Early Death

Genetics affects all of us, but sociologists are most interested in the *social* sources of health, illness, and early death. The three most important types of interlocking reasons are environmental, demographic, and lifestyle factors.

## 14-1a Epidemiology

**epidemiology** the study of the causes and distribution of disease within a population.

**Epidemiology** is the study of the causes and distribution of disease within a population. Why, for example, do people live much longer in some countries than in others? And what are the causes of heart disease? In answering such questions, epidemiologists look at two factors. One is *incidence*, the number of new cases of a health problem that occur in a given population during a given time period (such as the number of people ages 45 to 55 who experienced health problems in 2013). The other measure of epidemiology is *prevalence*, the total number of cases (extent) of an illness or health problem within a population or at a particular point in time (such as the number of Americans age 70 and older who, in 2013, experienced problems walking).

### Environmental Factors

People who live in a high-income country, such as the United States, experience fewer health problems than their counterparts in middle- and low-income countries (World Health Organization 2012). Nonetheless, a nation's wealth doesn't guarantee good health because of macro-level variables, including environmental hazards.

## Sociology in Your Life

Have you searched for online health advice but wondered if the sites were trustworthy? And what are some of the most common searches about health? For an encyclopedia of more than 1,600 health topics from A to Z, go to CengageBrain.com.

**"IT'S WHEAT-FREE, DAIRY-FREE, FAT-FREE, NUT-FREE, SUGAR-FREE AND SALT-FREE...ENJOY!"**

Brian Fray/Cartoonstock

## FIGURE 14.1
## U.S. Death Rates by Age, 2010

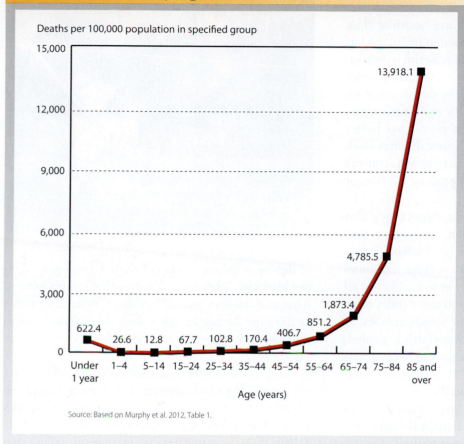

Deaths per 100,000 population in specified group

Source: Based on Murphy et al. 2012, Table 1.

preventable medical errors, such as doctors operating on the wrong patient or wrong body part (Chamberlain et al. 2012). Medication errors—mistakes such as misdiagnosing an illness, prescribing the wrong drugs, or giving patients drugs that interact dangerously—injure about 1.3 million Americans and kill about 7,000 every year (Brown and McGann 2011).

### Demographic Factors

Religion, family size, urban-rural residence, and other variables can affect health. Four of the most important demographic factors, however, are age, sex, race and ethnicity, and social class.

*Age.* Not surprisingly, age is the single most important predictor of illness and death. Death rates drop sharply shortly after birth, begin to rise at about age 45, and escalate with age (see *Figure 14.1*). After age 65, chronic rather than acute diseases comprise the majority of health problems, including disability. **Chronic diseases** (such as asthma) are long-term or lifelong illnesses that develop gradually or are present from birth. In contrast, **acute diseases** (such as chicken pox) are illnesses that strike suddenly and often disappear rapidly but can cause incapacitation and sometimes death.

Chronic diseases increase as people age. For example, 8 percent of Americans ages 18 to 29 have high blood pressure, compared with 16 percent of those ages 30 to 44, 37 percent of 45- to 64-year-olds, and 58 percent of people age 65 and older (Witters 2012). The age differences may be due to a combination of factors,

Insulated and energy-efficient buildings prevent heat loss but also increase health hazards because of poor ventilation, dampness that generates mold, and indoor pollutants (Institute of Medicine 2011). Numerous toxic chemicals in everyday products such as dishwashing detergents, shampoos, shaving products, and makeup have been linked to asthma and some types of cancer (Rochman 2011; Dodson et al. 2012).

Millions of Americans also suffer from foodborne illness because of disease-causing bacteria such as *E. coli* and salmonella (Cowell and Kanter 2011; Zimmer 2011). Each year an estimated 115,000 Americans experience salmonella food poisoning, resulting in about 42 deaths (Kramer 2011). Salmonella bacteria are found in food products such as raw poultry, eggs, and beef, sometimes on unwashed fruit, and in polluted water. Food prepared on surfaces that have previously been in contact with raw meat products can become contaminated with the bacteria.

Access to health care can prolong life, but an estimated 98,000 Americans die each year because of including obesity, lifelong smoking, and inadequate medical coverage.

Both younger and older people are healthier than they used to be. In the latter group, health is deteriorating less rapidly than in the past, even among those age 80 and older. One reason may be that people with the most severe health problems die at an earlier age. Other reasons may be that older people are better informed about how to manage chronic diseases (such as arthritis) and have greater access to new treatments (such as cancer surgery) that can prolong life (Cutler and Landrum 2011).

*Sex.* The gender gap in life expectancy has decreased since 1975, but women, on average, still live longer than men (see Chapter 12). Men of all ages and across all racial-ethnic groups are about four times more likely than women to die by suicide. Men of all racial-ethnic groups are also two to three times more likely than women to die in motor vehicle crashes, to be victims of homicide, to smoke and drink alcohol, to abuse drugs, and to work in dangerous occupations such as mining, construction, and public safety (Frieden 2011; see also Chapters 7 and 9).

Although women tend to live longer than men, this doesn't mean that they're necessarily healthier. Compared with men, women have more chronic diseases such as arthritis, asthma, cancer, and mental illness. Some of these problems may be due to obesity because, across all ages and most racial-ethnic groups, women are more likely than men to be overweight. However, women ages 18 to 64 are much less likely than men to have health insurance. As a result, twice as many men as women have regular access to health care, which means that women are less likely than men to seek preventive care and to receive early diagnoses of health problems (White House Council on Women and Girls 2011).

Many people assume that women are emotionally healthier than men because they talk more, express their feelings, and have more friends and family members with whom they chat to relieve stress and anxiety (see Chapter 9). Some of these assumptions are correct because many women have strong social networks comprised of close female friends and relatives, but from age 12 to 60 and older, women report experiencing higher depression rates than men (White House Council on Women and Girls 2011).

*Race and Ethnicity.* A good measure of a population's health is its **infant mortality rate**, the number

Dangerous occupations, such as mining, can decrease life expectancy because of accidents, disease, and a higher risk of death.

of deaths of infants (younger than 1 year) per 1,000 live births in a population. Nationally, infants born to African American women are more than three times as likely to die before age 1 than are infants born to Latinas and women who are white and Asian (see *Figure 14.2*).

Low birth weight and socioeconomic disparities are among the most important factors in explaining infant mortality rates, but at any age and every SES level, African Americans mothers are the most likely to lose their infants. Some researchers speculate that black women are more prone to suffer from some conditions—such as high blood pressure, diabetes, and obesity—that have negative birth outcomes (Shore and Shore 2009; Williams 2011).

Lower infant mortality rates increase longevity, but not for all racial-ethnic groups. As *Figure 14.3* on p. 278 shows, Latinos have higher life expectancy rates than African Americans and white males. This finding is a paradox because, on average, as you saw in prior chapters, many Latinos have high poverty rates and low socioeconomic status.

Some researchers have proposed that Latinos have higher life expectancy rates because (1) they are more likely than other groups to be undercounted, therefore leading to a false appearance of greater longevity; (2) healthy foreign-born people are more likely to migrate and to stay in the United States; and (3) Latinos tend to have strong family structures and social networks that protect them from hardships and low socioeconomic conditions. There has been considerable research about these and other causes, but the explanations are still tentative (Arias 2010).

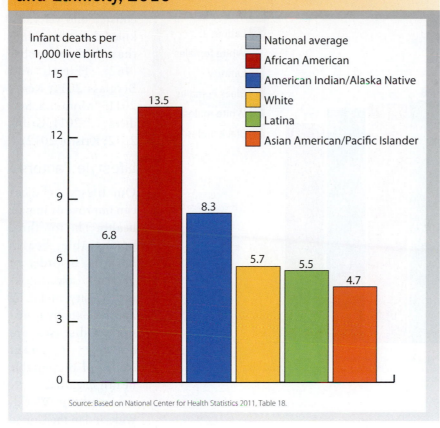

## FIGURE 14.2
## U.S. Infant Mortality Rates by Mother's Race and Ethnicity, 2010

Infant deaths per 1,000 live births

- National average
- African American
- American Indian/Alaska Native
- White
- Latina
- Asian American/Pacific Islander

National average: 6.8
African American: 13.5
American Indian/Alaska Native: 8.3
White: 5.7
Latina: 5.5
Asian American/Pacific Islander: 4.7

Source: Based on National Center for Health Statistics 2011, Table 18.

*Social Class.* Social class has a strong influence on health. Numerous studies show that higher SES provides an array of resources, such as money and knowledge, that increase good health. Americans with the least education are more than twice as likely to die from cancer as those with the most education, mainly because of risky behaviors such as smoking and drug abuse, as well as less access to health care. As early as age 45, the lower the family income, the greater the likelihood that adults will experience two or more chronic diseases (National Center for Health Statistics 2012).

SES also affects the likelihood of having Internet access, which facilitates searching for health information online, coping with a chronic condition, and improving one's health. More educated people are not only healthier than the less educated, but they also have a positive impact on their children's education and health. Living in poor neighborhoods increases the likelihood of experiencing stress (which affects diet, smoking, and alcohol/drug usage), having a sedentary lifestyle because of limited recreational facilities, and having less access to health care (Phelan et al. 2010; Williams and Sternthal 2010; American Cancer Society 2011; Beckles and Truman 2011; Dubowitz et al. 2011; Keenan and Rosendorf 2011; Murphey et al. 2011; Tu 2011; Mocan and Altindag 2012).

People in lower social classes are more likely to have jobs that endanger their lives. In 2010, for example, more than 4,500 U.S. workers died as a result of work-related injuries, and almost 1.2 million experienced workplace injuries or illnesses caused, among other things, by falls, overexertion, and accidents using equipment ("Injuries, Illnesses, and Fatalities" 2011).

Many military recruits come primarily from lower social classes. The soldiers, not officers, are typically on a war's front lines. Between 2003 and mid-2012, almost 6,500 U.S. soldiers were killed in Iraq and Afghanistan, thousands more returned home with serious physical disabilities, and many more suffered traumatic brain injuries or posttraumatic stress disorders. Being a recent

There are other health gaps among racial and ethnic groups. For example, the number of people diagnosed with asthma grew by more than 4 million from 2001 to 2009. Scientists don't know why asthma rates are rising, but they increased the most (almost 50 percent) among black children ("Asthma in the US" 2011).

Among racial-ethnic groups, white women are the most likely to get breast cancer, but black women are the most likely to die from this disease (Whitman et al. 2012). It could be that black women are diagnosed at later stages of the disease or that doctors aren't as aggressive in screening and treating black women with cancer, but no one knows for sure. Compared with other groups, Latinos are the least likely to die from illicit drugs and prescription drug abuse (Frieden 2011). One reason may be that Latinos, especially recent immigrants, have strong family and kinship ties that decrease the likelihood of engaging in unhealthy behavior, including drug abuse (see Chapters 7 and 10).

## FIGURE 14.3
## U.S. Life Expectancy at Birth by Race, Ethnicity, and Sex, 2010

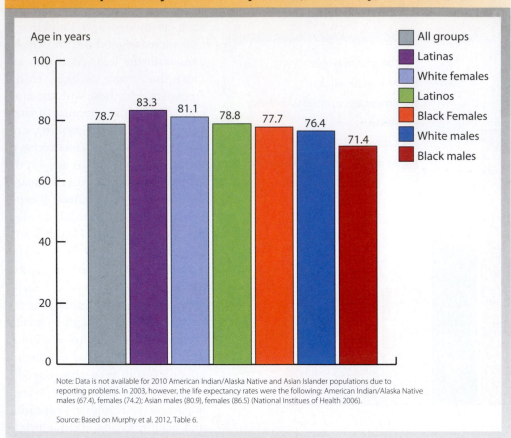

Age in years

Legend:
- All groups
- Latinas
- White females
- Latinos
- Black Females
- White males
- Black males

Bar values: 78.7, 83.3, 81.1, 78.8, 77.7, 76.4, 71.4

Note: Data is not available for 2010 American Indian/Alaska Native and Asian Islander populations due to reporting problems. In 2003, however, the life expectancy rates were the following: American Indian/Alaska Native males (67.4), females (74.2); Asian males (80.9), females (86.5) (National Institues of Health 2006).

Source: Based on Murphy et al. 2012, Table 6.

ZUMA Press/Newscom

veteran also doubles one's risk of suicide; for the second year in a row, in 2010 more U.S. soldiers, both enlisted and veterans, killed themselves (468) than died in combat (462) (Harrell and Berglass 2011; Keyes 2011; "More U.S. Soldiers . . ." 2011; Griffis 2012; Kristof 2012).

### Lifestyle Factors

Our lifestyle choices can improve or impair health. The top three preventable health hazards, in order of priority, are smoking, obesity, and drug abuse. Sexually transmitted diseases are another important source of preventable health hazards.

*Smoking.* Worldwide and in the United States, tobacco use is the leading cause of preventable death and disability. In the United States, tobacco use is responsible for about 1 in 5 deaths annually, and, on average, smokers die 13 to 14 years earlier than nonsmokers. Smoking can cause cancer, heart disease, stroke, and lung diseases (including emphysema and bronchitis) and costs more than $193 billion each year in lost productivity and health care expenditures ("Smoking and Tobacco Use" 2012).

In 1975, Minnesota was the first state to restrict smoking in most public spaces. It was only in 1995 that some states started passing laws that banned smoking at home because of the dangers of secondhand smoke, especially for children (Zhang et al. 2011). After that, smoking bans were extended to offices, restaurants, other public spaces, and cars where children were passengers.

Cigarette smoking by adolescents and adults has decreased (see *Figure 14.4*), but is still high. Each day, more than 3,800 American teenagers smoke their first cigarette, and more than 1,000 of them become daily cigarette smokers. The vast majority of Americans who smoke every day during adolescence are addicted to nicotine by young adulthood. Nearly all adults who smoke every day started smoking when they were 26 or younger (U.S. Department of Health and Human Services 2012).

Prevention is difficult because the tobacco industry spends almost $10 billion a year to market its products, half of all movies for children under 13 contain scenes of tobacco use, half of our states continue to allow smoking in public places, and images and messages normalize tobacco use in magazines, on the Internet, and at retail stores frequented by youth. In 2011, states collected more than $25 billion from tobacco taxes and legal settlements but spent only 2 percent of this money on tobacco control programs (U.S. Department of Health and Human Services 2012).

Obesity is one of the top three causes of death and disability, but Americans are more likely to disapprove of someone who smokes (25 percent) than someone who's overweight (12 percent) (Saad 2011). Almost 85 percent of Americans don't believe it's right for companies to refuse to hire smokers and obese people, but 60 percent say that smokers should pay higher health insurance rates. Only 42 percent feel the same way about people who are obese (Mendes 2011), perhaps because more of us are obese than smokers.

*Obesity.* Americans continue to get fatter: 36 percent of adults and 17 percent of children are obese. In 1990, not a single state had an obesity rate above 15 percent. Now all states have obesity rates above 20 percent, and 12 states have obesity rates equal to or greater than 30 percent. Since 1988, the prevalence of obesity has increased among adults at all income and education levels (Ogden et al. 2010, 2012). At the current rate, some medical researchers predict that

## FIGURE 14.4
## Cigarette Smoking by U.S. High School Students and Adults, 1965–2010

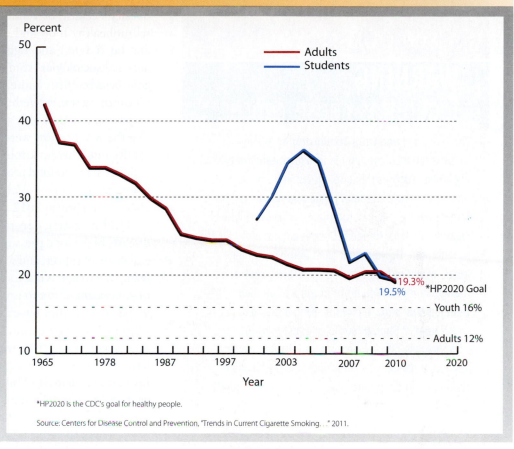

*HP2020 is the CDC's goal for healthy people.

Source: Centers for Disease Control and Prevention, "Trends in Current Cigarette Smoking..." 2011.

Should smoking be banned in all public places, including outside of office buildings?

1-800-QUIT-NOW

© U.S. HHS

FDA via Getty Images

**WARNING:**
**Cigarettes cause cancer.**

The U.S. Food and Drug Administration is seeking to force tobacco companies to use graphic warning labels, such as this one, on cigarette packs and ads. A Washington federal judge has ruled that this requirement is unconstitutional because it violates cigarette makers' freedom of speech under the First Amendment. Do you think such graphic warning labels would reduce tobacco use? Should alcohol manufacturers be forced to put labels on their products that show a cirrhotic liver, a failing heart, or a DUI death? Should food corporations, whose products increase obesity, be forced to include images—such as a person with an amputated leg because of diabetes—on doughnuts, ice cream, and junk food?

## FIGURE 14.5
## U.S. Adult Obesity Over Time

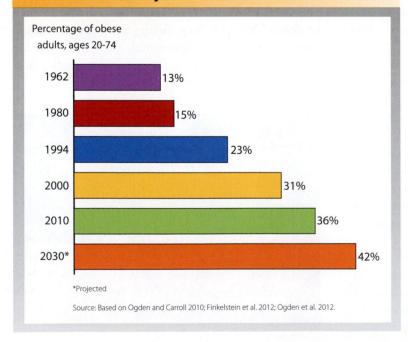

Percentage of obese adults, ages 20-74

| Year | Percentage |
|------|-----------|
| 1962 | 13% |
| 1980 | 15% |
| 1994 | 23% |
| 2000 | 31% |
| 2010 | 36% |
| 2030* | 42% |

*Projected

Source: Based on Ogden and Carroll 2010; Finkelstein et al. 2012; Ogden et al. 2012.

42 percent of American adults will be obese in 2030 (see *Figure 14.5*), and 25 percent of that group will be severely obese (Finkelstein et al. 2012).

For years, many Americans, especially nutritionists and parents, have blamed the food industry for children's obesity because of vending machines at schools that sell unhealthy food and drinks that are high in sugar and fat (Layton and Eggen 2011). However, a recent national sociological study found that children's weight gain between fifth and eighth grades doesn't vary significantly by sex, race/ethnicity, or family SES. The most surprising result, highly publicized by the media, was that the weight gains weren't due to buying soft drinks, candy, and other junk food at school because children's eating habits and food preferences are firmly established before the fifth grade (Van Hook and Altman 2012; see also An and Sturm 2012 for similar research findings).

Healthy eating can reduce people's risk for heart disease, high blood pressure, diabetes, osteoporosis, several types of cancer, and weight. Many people claim that being overweight is due to their genes, but more than half of Americans admit to unhealthful eating, even though 91 percent say they have access to affordable fruits and vegetables in their communities (Cochrane 2012). The 86 percent of full-time employees who are overweight have at least one chronic disease and high absenteeism rates, resulting in $153 billion in lost productivity every year (Witters and Angrawal 2011).

*Drug Abuse.* Excessive alcohol use and other drug abuse is the third leading lifestyle-related cause of death in the nation. Approximately 79,000 Americans die each year because of excessive drinking, more than 1.6 million are hospitalized, and more than 4 million visit emergency rooms for alcohol-related problems ("Alcohol Use and Health" 2011).

What is excessive alcohol use? *Binge drinking* is defined as drinking five or more drinks on the same occasion on at least 1 day in the past 30 days; *heavy drinking* is defined as drinking five or more drinks on the same occasion on 5 or more days in the past 30 days. Between 1993 and 2009, the prevalence of binge drinking in the United States increased from 14 percent to 16 percent, and the prevalence of heavy drinking increased from 3 percent to 5 percent ("Alcohol and Public Health" 2011).

Excessive alcohol use results in myriad immediate and long-term health risks.

They include unintentional injuries, violence, risky sexual behaviors, miscarriage and stillbirth, physical and mental birth defects, alcohol poisoning, unemployment, psychiatric problems, heart disease, several types of cancer, and liver disease ("Alcohol Use and Health" 2011).

*Illicit* (illegal) *drugs* include marijuana/hashish, cocaine and crack, heroin, hallucinogens (such as LSD and PCP), inhalants, and any prescription-type psychotherapeutic drug (such as stimulants and sedatives) used nonmedically. Illicit drug use is more common among men (10 percent) than women (6 percent), most common among those ages 16 to 35, and least common among Asian Americans (4 percent) and Latinos (6 percent) (National Center for Health Statistics 2011).

Illicit drugs have the same health risks as excessive alcohol use but can also have immediate life-threatening consequences. In addition, illicit drug users have higher rates of mental illness, suicidal thought and behavior, and major depressive episodes that last at least 2 weeks and impair a person's ability to function at school, home, or work (Substance Abuse and Mental Health Services Administration 2012).

***Sexually Transmitted Diseases.*** People who engage in unprotected sex can transmit or contract one or more of 50 sexually transmitted diseases (STDs) such as chlamydia, gonorrhea, syphilis, and genital herpes. People who are infected with STDs are at least two to five times more likely than uninfected individuals to contract HIV, the virus that causes AIDS. More than 1 million Americans are living with HIV, and 1 in 5 aren't aware of their infection. The population most severely affected by HIV is men (whether gay, bisexual, or straight) who have sex with men. Another high-risk group is people who inject drugs. Although HIV and genital herpes are incurable, these and other STDs can be treated if diagnosed early. Nonetheless, almost 18,000 Americans died of AIDS in 2009 ("The Role of STD Prevention . . ." 2010; "HIV in the United States . . ." 2012).

We've looked at some of the macro and micro reasons that help explain why some people are healthier than others. Who gets health care when it's needed? And who pays for it?

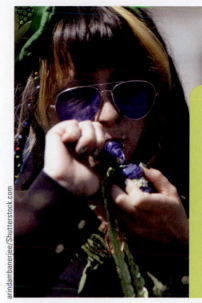

Some medical researchers believe that smoking marijuana is a gateway to using harder drugs such as cocaine and heroin. Others disagree and claim that smoking marijuana is less addictive than smoking cigarettes.

arindambanerjee/Shutterstock.com

## 14-2 Health Care: The United States and Around the World

The United States is one the richest nations in the world, but only the very wealthy don't have to worry about receiving and paying for the best medical care available. In this section, we'll examine why many Americans are worse off than their counterparts in other high-income countries, especially in Europe. Let's begin by looking at health care coverage in the United States.

### 14-2a U.S. Health Care Coverage and Who Pays for Medical Care

You'll see shortly that U.S. health care costs are skyrocketing. So, how many Americans have health insurance? And who pays for medical care?

#### How Many Americans Have Health Insurance?

In 2011, 17 percent of Americans had no health insurance, up from 13 percent in 1987. People who are the least likely to be insured are Latinos, those ages 19 to 34, and people with household incomes of less than $25,000 a year (DeNavas-Walt et al. 2011; Mendes 2012).

## Sociology in Your Life

To find out more about the prevalence and neurological effects of marijuana usage, go to CengageBrain.com.

## TABLE 14.1

## Health Insurance Coverage in the United States, 2011

| Percentage of Americans who receive health insurance from . . . | |
| --- | --- |
| Employer-based programs | 45 |
| Government health insurance | 25* |
| Other private health insurance | 12 |
| Not insured | 17 |

*Government health insurance includes Medicare, Medicaid, and military/veterans' benefits.

Source: Based on Mendes 2012, "Fewer Americans . . ."

**health maintenance organization (HMO)** a business organization that provides medical care to subscribers for a fixed fee.

There are also considerable variations across states. In 2011, for example, Massachusetts had the smallest number of uninsured residents (5 percent), compared with a high of 28 percent in Texas (Jones 2012). A major reason for such differences is that Massachusetts has enacted an insurance program that provides health care to most residents, and other states have government programs that provide more health coverage than does Texas.

### Who Pays for Medical Care?

About 57 percent of Americans receive their health coverage through *private insurance*, primarily employer-based programs that cover most or all costs for employees and their family members, but also through private insurance not provided by employers. Some 25 percent of Americans have government health insurance; others have no insurance (see *Table 14.1*).

Employer-based health insurance coverage has deteriorated. Between 2000 and 2011, such coverage declined from 69 percent to 45 percent as millions of full-time workers ages 18 to 64 lost their jobs (Gould 2012). Nearly half of small business owners report that they aren't hiring additional employees they need because they can't afford the rising health insurance costs (Jacobe 2012).

Instead of not hiring or dropping health coverage entirely, many employers are passing on more of the costs to workers. Since 2006, workers at both large and small firms have been making higher contributions to premiums and paying higher deductibles or copayments (Claxton et al. 2011; Mercer 2011).

*Medicare*, a government program, pays many of the medical costs of Americans age 65 and over, regardless of income. Thus, even billionaires are eligible for Medicare. *Medicaid*, another government program, provides medical care for the poor, those living below the poverty level (see Chapter 8).

Whether their health insurance is private or government sponsored, many Americans belong to a **health maintenance organization (HMO)**, a business organization that provides medical care to subscribers for a fixed fee. HMOs require members to use only HMO doctors. They reduce health care costs by providing preventive care (which helps reduce the costs of more extensive medical care in the future) and monitoring doctors' fees to economize on the cost of medical professionals (Weitz 2013).

U.S. health care costs have risen steadily—per person, nationally, and as a percentage of the gross domestic product. By 2020, health care expenses will consume $1 of every $5 in the economy (see *Table 14.2*). Chronically ill patients, comprising only 5 percent of the population, are responsible for almost 50 percent of all U.S. health care spending (Schoenman and Chockley 2011).

## 14-2b The United States Compared With Other Countries

The United States spends more on health care than any other nation in the world; no other country spends more than 12 percent of its total economy (GDP) on health care. Compared with 14 other high-income countries, U.S. employers and the government spend more per person on health care, have increased their spending, and have had one of the highest growth rates in health care spending since 1980 (Kaiser Family Foundation 2011).

Although we spend more on health care than other countries, we cover a smaller percentage of the total population and have poorer health outcomes. For example, the United States ranks only 37th worldwide in life expectancy, and life expectancy has declined, particularly in the South, because of growing racial-ethnic and income inequality (Kulkarni et al. 2011). We also have fewer and shorter doctor visits and fewer days of inpatient hospital care (Weitz 2013).

In Canada, Germany, Great Britain, France, and Sweden, the government picks up most of the health care bill. Because there is some version of national insurance in these and other countries, patients and health care workers don't have to submit bills to several insurance providers, resulting in considerable administrative savings. People are insured by the government,

TABLE 14.2

# The Increasing Cost of U.S. Health Care, 1980–2020

| | 1980 | 1990 | 2000 | 2010 | 2020 |
|---|---|---|---|---|---|
| Average cost per person | $1,100 | $2,864 | $4,878 | $8,402 | $13,708 |
| National health expenditure | $256 billion | $724 billion | $1.4 trillion | $2.6 trillion | $4.7 trillion |
| Percent of gross domestic product (GDP) | 9% | 13% | 14% | 18% | 20% |

Notes: The "average cost per person" includes medical care, supplies, drugs, and health insurance. All numbers are in 2010 dollars; the numbers for 2020 are projected.

Sources: Based on Centers for Medicare & Medicaid Services 2010, Table 1; and 2012, Table 1.

nonprofit organizations, or large groups (such as cities or industries) that have considerable clout in keeping down drug costs and setting prices for health care providers and services. There are also private hospitals, but most are nonprofit.

Compared with doctors in many other countries, those in the United States are more likely to adopt new, expensive, and often unproven technologies (such as bone marrow transplants) that benefit very small populations. These treatments might advance medical technology and provide help for people who are dying, but they are expensive for patients and insurers and highly profitable for both physicians and hospitals. Because many high-income and middle-income countries provide almost universal health care across all social classes, wealth isn't a major factor in accessing quality health care (Holland 2011; Weitz 2013).

## 14-2c What Kind of Health Reform Should Americans Endorse?

In 2010, Congress passed health care reform legislation—the Patient Protection and Affordable Care Act (which goes by the acronym ACA)—that is designed to be implemented fully by 2020. Republicans challenged the constitutionality of the ACA, especially because of its *individual mandate*, which requires all Americans to either obtain health insurance or pay a fine. In 2012, however, the U.S. Supreme Court ruled (in a 5-4 decision) that the individual mandate is legal.

Some key features of the new law that took effect immediately prohibit health care insurers from denying coverage to children with a preexisting condition, imposing lifetime spending limits, and denying health claims without giving people a chance to appeal the

Would you be better off living outside the United States when it comes to health care costs? Here's how the United States measured up against some other countries in 2011 in charging patients for some routine medical procedures.

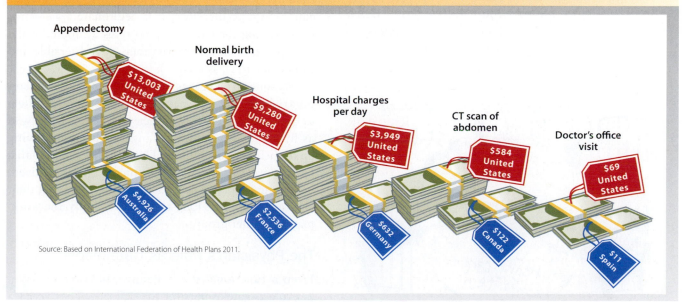

Appendectomy
$13,003 United States
$4,926 Australia

Normal birth delivery
$9,280 United States
$2,536 France

Hospital charges per day
$3,949 United States
$632 Germany

CT scan of abdomen
$584 United States
$122 Canada

Doctor's office visit
$69 United States
$11 Spain

Source: Based on International Federation of Health Plans 2011.

insurer's decision. The ACA also requires insurers to provide free preventive health services, such as tests for diabetes, mammograms, and colonoscopies, and routine vaccinations; allows young adults to be insured under their parents' policies until age 26; permits people to choose a primary care doctor outside of the health plan's provider network; and allows people to visit the nearest emergency room outside of a plan's network without penalty (Roller 2010; Tumulty et al. 2010).

Some have hailed the ACA as an important milestone in health care reform. The law extends health coverage to 30 million uninsured Americans and brings the country a step closer to achieving universal health insurance. ACA also moves away from the traditional fee-for-service model in which doctors are paid for each test, treatment, or X-ray they provide. Instead, many insurance companies will offer doctors higher payments for better quality care, such as longer office visits that result in better diagnoses and help prevent illness (Eibner and Price 2012; Lynch 2012).

Others argue that government has no right to require people to purchase health insurance, the primary basis of the Supreme Court case. Some critics also claim that the ACA does little to control health care costs. According to some policy analysts, national health spending will increase even more because more people will be covered by Medicaid and Medicare (with the latter patients demanding more health services), the newly insured will consume more prescriptions because of substantially lower out-of-pocket requirements, and implementing and operating the various ACA provisions will mean higher administrative costs (Hsieh 2011; Keehan et al. 2011). So far, however, opponents of the ACA haven't proposed alternatives.

# 14-3 Sociological Perspectives on Health and Medicine

Sociologists focus on different aspects of health, health care, and medicine. Taken together, the four perspectives offer valuable insights on understanding these issues. *Table 14.3* summarizes these perspectives.

## 14-3a Functionalism: Good Health and Medicine Benefit Society

For functionalists, good health and medicine are critical for a society's survival and stability. As you saw earlier in this chapter, the medical profession has increased life expectancy and has provided many Americans with health care that increases our quality of life.

One of the functionalists' contributions has been in addressing the idea of disease and illness as dysfunctional because they prevent people from performing their everyday roles in the family, the economy, and other institutions. In these analyses, functionalists have emphasized the importance of understanding the sick role and the physician's role as gatekeeper.

### The Sick Role

Talcott Parsons, an influential sociologist, introduced the concept of the *sick role* and saw illness as deviant (Parsons 1951). The **sick role** is a pattern of behavior accepted as appropriate for people who are ill. In Parsons' model, sick people aren't responsible for their condition and, therefore, have legitimate reasons for not fulfilling their usual social roles at home, in the workplace, or in other situations ("I feel terrible and can't come to class today").

Parsons also emphasized that the sick role is temporary, and that sick people must seek medical help to hasten their recovery. Otherwise, they'll be viewed as hypochondriacs and malingers who aren't living up to their responsibilities. Thus, from a functionalist perspective, the sick role is legitimate if it's short-lived, but dysfunctional if people feign illness to avoid their role responsibilities long-term in the family, the workplace, or other groups.

### The Physician's Role as Gatekeeper

From a functionalist perspective, physicians play a key role as gatekeepers for the sick role. They verify a

Ed Fischer/Cartoonstock

## TABLE 14.3

# Sociological Perspectives on Health and Medicine

| PERSPECTIVE | LEVEL OF ANALYSIS | KEY POINTS |
|---|---|---|
| **Functionalist** | Macro | • Health and medicine are critical in ensuring a society's survival and preserving social order.<br>• Illness is dysfunctional because it prevents people from performing expected roles.<br>• Sick people are expected to seek professional help and get well. |
| **Conflict** | Macro | • There are gross inequities in the health care system.<br>• The medical establishment is a powerful social control agent.<br>• A drive for profit ignores people's health needs. |
| **Feminist** | Macro and Micro | • Women are less likely than men to receive high-quality health care.<br>• Gender stratification in medicine and the health care industry reduces women's earnings.<br>• Men control women's health. |
| **Symbolic Interactionist** | Micro | • Illness and disease are socially constructed.<br>• Labeling people as ill increases the likelihood of being stigmatized.<br>• Medicalization has increased the power of medical associations, parents, and mental health advocates and the profits of pharmaceutical companies. |

How sick is too sick to miss work or class? Are menstrual cramps enough? What about a severe cold? A hangover? And when does being sick become dysfunctional—after a day, a week, a month?

Iakov Filimonov /Shutterstock.com

person's condition as sick and provide an excuse for temporarily not performing necessary roles. They also designate a patient as "recovered" and ready to meet societal role expectations again. Doctors' specialized knowledge gives them considerable authority in defining health and illness that is unmatched by other health care providers, such as nurses and physical therapists.

## Critical Evaluation

Functionalist theories are limited in explaining illness-related behavior for several reasons. First, those with chronic illnesses (such as arthritis and diabetes) are usually viewed as healthy enough to perform their expected roles regardless of how they feel. Second, many people can't assume even a short-lived sick role. Some don't have the resources to see physicians or may lose a portion of their earnings if they miss workdays. Others, such as full-time moms and stay-at-home dads, don't have the luxury of exiting parenting roles unless the illness is life-threatening. Third, illness and sick roles aren't necessarily dysfunctional. Disease and illness generate jobs, and a serious illness can provide an opportunity to reorder one's priorities or change an unhealthy lifestyle.

## 14-3b Conflict Theory: Health Care and Medicine Benefit Some More Than Others

For conflict theorists, medicine and the health care industry benefit some groups much more than others.

Unlike functionalists, conflict theorists argue that the medical system doesn't always benefit society because it reinforces social inequality, exerts social control to maintain the status quo, and is often driven by a profit motive rather than a concern for people's well-being.

## Social Inequality in Health Care

You saw earlier that illness and death rates vary considerably, especially by social class. Generally, the wealthy can afford health insurance and high-quality medical care. In contrast, low-income groups, who are often minorities, receive poorer treatment or none at all because they don't have the resources to pay for it.

## Social Control of Medicine and Health Care

One of the central concepts of conflict theory is that medicine is an institution in which those at the top of the economic hierarchy exert social control and maintain the status quo. Physicians, for example, have almost absolute power in diagnosing an illness, providing treatment, and deciding on medical procedures.

Since 1847, the powerful American Medical Association (AMA) has successfully fought off the efforts of other health practitioners (such as midwives, acupuncturists, chiropractors, and pharmacists) to be recognized as medical professionals (Weitz 2013). One of the results is that the services of these groups aren't covered by employer-sponsored or government (Medicare and Medicaid) insurance programs.

According to conflict theorists, the medical profession and health care industry maintain the status quo because of the **medical-industrial complex**, a network of business enterprises that influences medicine and health care. The medical-industrial complex includes many groups—doctors, nurses, hospitals, lawyers, HMOs, nursing homes and hospices, insurance companies, drug manufacturers, consulting firms, accountants, banks, and real estate and construction businesses.

Who benefits from the medical-industrial complex? All of those involved, especially physicians and drug companies. When pharmaceutical companies seek FDA approval of a new drug, they don't submit negative findings, only the data they want to get published in medical journals. The companies pay physicians to write the articles, say nothing about therapies that work better, and employ influential doctors to promote a drug and to serve on advisory panels (Hyman 2011).

Direct-to-consumer prescription drug marketing, which began in the 1980s, increases a drug's sales ("Ask your doctor about . . .") (Holmes 2011). To treat chronic diseases and control symptoms, the average American now takes about 12 medications annually, compared with 7 medications 20 years. For the first time, Americans are now more likely to die from prescription pills than auto accidents, the leading cause of death in the past. Most of the deaths are due to unintentional overdoses of opiate-based pain relievers such as Vicodin and Oxycontin. Some asthma drugs cause more deaths than the disease itself (Rosenberg 2011; Warner et al. 2011).

## Profit Is More Important Than Health

From a conflict perspective, health care is big business. In 2011, CEOs' pay packages boomed. Of the top 10 earners, 3 were in the health care industry (Rushe 2011). Since the ACA's passage, large insurers' revenues jumped from 36 to 42 percent. To cut expenditures, government policymakers have outsourced components of the Medicare and Medicaid programs to private insurers, increasing the latter's profits (Frier 2012).

Some analysts accuse drug companies and some doctors of fueling addiction. A study of Oxycontin patients found that less than half of the patients had regular meetings with their doctors to check for symptoms of addiction, and more than a quarter of patients received multiple early refills. At some pain

Acupuncture is a medical treatment to relieve pain. It originated in China in the second century. Recently, acupuncture has gained tentative endorsement by the U.S. Institutes of Health and the World Health Organization, but most American physicians still dismiss this treatment as nonsense.

Yuri Arcurs/Shutterstock.com

management centers, doctors prescribe powerful narcotics without performing physical exams or requiring evidence of injury (Gwynne 2011).

Surgeons are among the highest paid doctors, but how much of the surgery is actually necessary? A study of 1.8 million Medicare recipients age 65 and older who died in 2008 found that a third had surgery in the last year of life. Nearly 1 in 5 had surgery in the last month of life, and nearly 1 in 10 had surgery in the last week of life. Among those undergoing end-of-life surgery, almost 60 percent were 80 and older (Kwok et al. 2011). Surgery and other treatments can relieve pain and prolong life, but conflict theorists contend that financial profit has turned many doctors, hospitals, and the pharmaceutical industry into multibillion-dollar medical-industrial corporations that aren't using limited resources as wisely as they should.

### Critical Evaluation

The most common criticism is that conflict theorists often ignore the contributions of medical and health care systems. Without them, people would suffer more, die at a younger age, and have a lower quality of life.

Second, not all medical providers are motivated by profit. For example, doctors in family practices earn, on average, only about half as much ($189,400) as surgeons (Bureau of Labor Statistics 2012, *Occupational Outlook Handbook . . .*).

Third, medical scientists and some medical associations, such as the American Academy of Family Physicians, have been among the most vocal critics of unneeded surgery, tests, and procedures such as cardiac stress tests, antibiotic prescriptions for sinusitis that usually resolves itself within a few weeks, colonoscopies, and routine prostate cancer screening for men because the risks can outweigh the benefits (Chou et al. 2011; Agnvall 2012; Shelton and Deardorff 2012).

## 14-3c Feminist Theories: Health and Medicine Benefit Men More Than Women

Feminist scholars agree with conflict theorists on practically all points. They go further, however, by addressing issues such as the health costs of being a woman, gender stratification in medicine and health care, and men's control over women's choices.

### The High Health Costs of Being a Woman

Women, compared with men, have greater health care needs, especially during their reproductive years. They

maron/age fotostock

### IS BEING FASHIONABLE HAZARDOUS TO WOMEN'S HEALTH?

Squeezing into skinny jeans and tight pants can cause nerve compression, numbness, and digestive problems. Narrow-toed shoes and high heels wreak havoc on unsuspecting feet; blisters, bunions, hammer toes, nerve damage, bone death, stress fractures, and ankle sprains are all potential consequences. Earrings can tear through an earlobe, an injury that takes a long time to repair and heal. Thongs can cause chafing and small breaks in sensitive skin, increasing the likelihood of fungal growth and bacterial infections (Sifferlin 2012).

Many cosmetics contain chemicals that can trigger skin problems such as itching, redness, and acne. Others—contaminated with bacteria, yeasts, or molds—can lead to a range of problems from simple rashes to serious infections that can cause swelling and breathing problems. Tainted cosmetics, especially eye makeup, can also cause serious infections. Some products, including shampoos and perfumes, are suspected of causing long-term health problems such as asthma and cancer (Malkan 2007; Dahl 2011).

Except for some hair color additives, the FDA doesn't require companies to test products for safety or list toxic ingredients. More than 500 cosmetic products sold in the United States contain ingredients that are banned in Japan, Canada, and the European Union (Rano and Houlihan 2012). Skin Deep (www.ewg.org /skindeep) is an excellent site that, among other things, ranks the safety of a range of cosmetic products.

## TABLE 14.4

## Median Annual Earnings of Full-Time Health Care Workers by Sex, Selected Occupations, 2011

| | MEN | WOMEN | PERCENTAGE OF WOMEN IN THIS OCCUPATION |
|---|---|---|---|
| **All health care practitioners** | **$58,708** | **$50,180** | **75** |
| Pharmacists | $103,896 | $98,696 | 56 |
| Physicians and surgeons | $100,620 | $79,404 | 36 |
| Physical therapists | $79,144 | $63,232 | 60 |
| Registered nurses | $56,212 | $53,768 | 90 |
| Emergency medical technicians and paramedics | $39,364 | $33,748 | 32 |
| Support technologists and technicians | $35,464 | $32,864 | 76 |

Source: Based on Bureau of Labor Statistics, "Median Weekly Earnings of Full-Time Wage and Salary Workers by Detailed Occupation and Sex," Table 39. Accessed April 23, 2012 (www.bls.gov).

have also historically played a central role in coordinating the health care needs of multiple generations of family members, including children, spouses, and aging parents (see Chapter 12).

You saw earlier that, on average, women live longer than men but are more likely than men, particularly at later ages, to suffer from stress, obesity, hypertension, and chronic illnesses. Women are also less likely than men to receive high-intensity treatments such as organ transplants and coronary bypasses (Weitz 2013).

Social class affects every person's health, but women experience higher health costs simply because of their sex. Women earn approximately 77 to 81 cents for every dollar men earn (see Chapter 11) but use more health care services than men do. Less income, greater usage of health care services, and rising medical costs jeopardize many women's health. In 2010, for example, 48 percent of adult women (compared with 34 percent in 2001) reported that because of cost they didn't fill a prescription; skipped a recommended test, treatment, or follow-up; had a medical problem for which they didn't visit a doctor; or didn't see a specialist when needed. In addition, 44 percent of women, compared with 35 percent of men, had problems paying their medical bills or paying off medical debts (Robertson and Collins 2011).

*Gender rating*, the practice of charging women more than men for identical health care plans, results in considerably higher costs for women. In states that haven't banned such discrimination, 92 percent of the best-selling plans rate by gender. Excluding maternity coverage, one company charges 25-year-old women 85 percent more than men with the same coverage. In fact, 56 percent of the best-selling plans charge a 40-year-old woman who doesn't smoke more than a 40-year-old male smoker (Garrett 2012).

The ACA bans gender rating for plans offered in both the individual and small-group markets (organizations employing 100 or fewer people). However, as part of a compromise, the new law allows insurance companies to continue this discriminatory practice in plans offered in the large-group market (Pearsall 2011).

### Gender Stratification in Medicine and Health Care

According to U.S. Labor Department projections, job growth in health care occupations will surge until at least 2020 because of a graying population and corporations' inability to outsource medical services (see Kolet and Chandra 2012). Many of these jobs, however, such as home health aides and dental assistants (not dental hygienists), are low-paying and dominated by women.

In the higher-paying health care jobs, and in all of these occupations, men consistently earn more than women (see *Table 14.4*). Among physical therapists, for example, males earn almost $79,200 a year compared with $63,200 for females. Thus, in only 15 years, male therapists earn $240,000 more than their female counterparts, even when both have the same level of education, years of experience, and work responsibilities, and the occupation has more women than men (see also Chapters 8 and 11).

Many feminist scholars attribute much of the gender wage gap to male doctors' gatekeeping. Because many registered nurses, pharmacists, and physical therapists (occupations that have more women than men) now receive doctoral degrees, they want to use the honorific title of "doctor." Doing so would win more respect from patients, help women land top administrative jobs that pay more, provide more autonomy in treating patients, and bring higher fees from health insurers. Physicians are fighting back because they don't want to lose control of the doctor title for several reasons: They want to maintain their status and prestige in the health care industry; they treat patients first, whereas pharmacists, nurses, and physical therapists play only secondary roles; and they have considerably more education and training than others to diagnose and treat illness and disease (Harris 2011).

### Men's Control of Women's Health

Feminist scholars maintain that in patriarchal societies, including ours, men control many aspects of women's health. Catholic bishops, all of whom are men, condemn contraceptives as sinful, even though 98 percent of U.S. Catholic women have used contraception (Jones and Dreweke 2011). Male politicians have passed laws in many states to decrease women's ability to get abortions (see Chapter 9). Most recently, after numerous protests by women, Virginia legislators backed off a new law that would have required women to undergo an intrusive and painful vaginal ultrasound exam to discourage them from having an abortion (Mandell 2012). There are no similar exams for men.

In some states, conservative lawmakers are trying to restrict or block usage of RU-486, an abortion-inducing drug. They aren't proposing similar measures to limit the availability of Viagra and other erectile dysfunction medications for men because "Viagra is a wonderful drug" (Beadle 2012).

### Critical Evaluation

Critics have questioned feminist theories for four reasons. First, feminist scholars sometimes ignore the fact that social class, rather than sex, determines whether men and women are healthy and receive health care services. Second, it's not entirely clear that men can be blamed for gatekeeping because 56 percent of physicians and surgeons are women. Third, most feminist scholars have been mute in the recent political debates over women's reproductive rights, including abortion. Thus, according to some observers, feminist scholars don't press for women's rights on these and other issues (Reimer 2012; Sullivan 2012).

Finally, perhaps the gatekeeping by physicians (men or women) is justified because a doctoral degree in nursing, pharmacy, or physical therapy requires considerably less schooling and training than getting a Doctor of Medicine (M.D.) degree. It's doubtful, for example, that patients could be diagnosed as well and given treatments and prescriptions by nurses, pharmacists, and physical therapists as they are by doctors (Harris 2011).

## 14-3d Symbolic Interactionism: Health, Illness, and Medicine Are Socially Constructed

Symbolic interactionists focus on how we define and construct views about health, illness, and medicine and then implement these definitions in everyday life. People's social constructions include labeling, stigmatizing behavior, and medicalizing attitudes and behaviors as normal or sick.

### The Social Construction of Illness

Medical models assume that illness is an objective label given to anything that deviates from normal biological functioning and that each illness has specific features that any doctor can recognize. In contrast,

In 2012, when President Obama proposed that the full cost of contraceptives for women should be covered by health insurance, all of the panelists in the congressional hearings were men. Where are the women in political decisions that affect their reproductive rights?

AP Photo/Carolyn Kaster

interactionists view illness and medical knowledge as social constructions, not always medical facts, that can change over time. That is, medical discourse influences people's behavior, affects their subjective experiences, shapes their self-identities, and legitimates medical intervention (Foucault 1975). For example, drunkenness used to be attributed to a lack of moral character, an absence of self-discipline, and individual choice. The medical model now defines alcoholism as a preventable and treatable disease.

Doctors sometimes diagnose disease differently because of the patient's age and sex. For example, despite clear symptoms, they are more likely to diagnose men and older patients with coronary heart disease (Lutfey et al. 2010). Social class also affects social constructions of illness. For example, as educational level rises, people report poorer health, presumably because they're more aware of medical knowledge and evaluate their own physical well-being more critically (Schnittker 2009).

## Labeling and Stigma

Conditions and behaviors that are diagnosed and labeled as illness or disease change over time, differ among groups, and vary across countries. A record-high 50 percent of Americans favor legalizing marijuana use, up from only 12 percent in 1970, and 70 percent want to make it legal nationally for doctors to prescribe marijuana for medical reasons, particularly to reduce pain and suffering (Newport 2011). Despite a growing consensus that medical marijuana should be legal, only 16 states have allowed medical marijuana and stopped labeling its users as deviant.

The American Psychiatric Association (APA) is the world's most powerful agent in creating mental illness and disorder labels. First published by the APA in 1952, *The Diagnostic and Statistical Manual of Mental Disorders*, or *DSM* for short, has become a "bible of mental illness." The *DSM* relies on subjective definitions and descriptions, but it's very influential in labeling (or unlabeling) mental illness.

TDC Photography/Shutterstock.com

The *DSM*'s diagnoses and labels have changed over the years. For example, the APA once classified homosexuality as a psychological disorder, dropping sexual orientation from its roster of mental illnesses only in 1973. The number of mental illnesses surged from 106 in 1952 to more than 300 in 2000. The new revision, which is to be published in 2013, is expected to add new mental illnesses such as binge-eating disorder, Internet addiction, and hypersexual disorder. The *DSM* affects patients, doctors, insurers, pharmaceutical companies, and taxpayers because psychiatrists and other physicians use its definitions to bill insurance companies.

Stigma is a major consequence of labeling (see Chapter 7). In the United States, for example, obese women have higher rates of gynecological cancers than nonobese women. They avoid routine exams, however, because of the stigma of obesity and the possible negative attitudes of health care professions (Amy et al. 2006). Western stigmas about fat people seem to be spreading to other nations. A study of people in 10 countries and territories found that some societies that had traditionally viewed plumper bodies as attractive or associated with being wealthy now stigmatize being overweight as a disease. Some of the results of stigmatizing include ridicule, criticism, low self-esteem, and negative self-identities (Brewis et al. 2011). Labeling and stigma also lead to the medicalization of illness.

## The Medicalization of Illness

Another subjective component of socially constructing illness is **medicalization**, a process that defines and treats a nonmedical condition or behavior as an illness, disorder, or disease that requires a medical solution. For example:

- The normal cramps, bloating, and headaches that some women experience before a menstrual cycle now comprise a premenstrual "syndrome" (PMS) that can be treated with medications and vitamins.

- Male impotence, which commonly increases as men age, is now called "erectile dysfunction" and can be treated with drugs, regardless of age, to enhance a man's sexual experience (e.g., "Cialis is ready when you are").

Digital Vision/Jupiterimages

As you read earlier, a growing number of Americans are obese. Do you think that, as a result, stigmatizing overweight people will decrease in the future?

- Children with attention and behavior problems are now often defined and treated with drugs for "deficit hyperactivity disorder" (ADHD).

- Aging is normal, but women, especially, undergo cosmetic surgery and take anti-aging pills to avoid looking their age.

Medicalization of normal behavior is a lucrative business. Pharmaceutical corporations have reaped enormous profits because everyday normal anxieties, discomforts, and stresses (such as frustration with traffic, boredom with routine housekeeping chores, and feeling sad or personally insecure) can be "fixed" by popping a Prozac or other pill (Herzberg 2009). Since the 1990s, there's been a 900 percent increase in prescriptions for Ritalin, a drug for ADHD (Waters 2011).

Physicians benefit from medicalization because they can achieve higher incomes by charging insurance companies for more diagnoses and treatments of illnesses and diseases. And the more behaviors that the *DSM* defines as mental illness, the more likely psychiatrists are to increase their number of patients and charge insurance companies for treating mental disorders.

Mental rights advocates and parents also benefit from medicalization. For example, both groups worry that that if the *DSM* places autism in a larger category, such as autism spectrum disorder (ASD), a high-functioning autistic child may not get coverage from insurance plans or receive special treatment, such as more attention from teachers and health specialists who now devote much of their time to AHDH children (Conrad and Barker 2010; Rochman 2012).

## Critical Evaluation

Despite their contributions, symbolic interaction theories are limited for several reasons. First, interaction theories don't address structural factors, such as social class, that can affect people's physical and emotional health. For example, among Americans age 65 and older, those who are at least middle-class, still working part-time, have Social Security benefits, and have fewer out-of-pocket medical expenses report higher physical and emotional health than those in lower classes who don't have such resources (Lantz et al. 2007; Rheualt and McGeeney 2011).

Second, not all people accept medical definitions of health and illness. For example, many deaf people have ignored medical recommendations to get cochlear implants (devices that can increase hearing) because they don't see deafness as a medical disability, but a social reality that helps them form a community and identify with other deaf people (Conrad and Barker 2010).

Third, the prevalence of some developmental disorders, such as autism, has increased in the United States from 1 per 1,000 children in the 1970s to 1 per 88 children today (Baio 2012). Interactionists may be correct that such increases might be largely due to medicalization that exaggerates the prevalence of such illnesses, but no one knows for sure.

Finally, are interactionists exaggerating the importance of socially constructed illnesses and diseases? For example, viruses cause measles, and numerous STDs can result in HIV/AIDS and other diseases. Thus, many medical diagnoses and treatments are critical in ensuring a society's continuation.

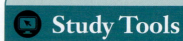 **Study Tools**

Ready to study? Go to **Sociology CourseMate** at **www.cengagebrain.com** to complete practice quizzes, review flashcards, watch videos, and more.

**Population growth**
and urbanization are changing
our environment.

Scott Olson/Getty Images

# Population, Urbanization, and the Environment

In 2011, the world's population surpassed 7 billion people. By 2025, India is expected to overtake China as the world's largest country ("China's Population . . ." 2012). Our planet has more inhabitants than ever before, and many are living longer than ever before. This chapter examines how such changes affect the population, urbanization, and the environment in the United States and worldwide. Let's begin with population.

> **demography** the scientific study of human populations.
>
> **population** a group of people who share a geographic territory.

## Key Topics

In this chapter, we'll explore the following topics:

**15-1** Population Dynamics

**15-2** Urbanization

**15-3** Environmental Issues

## What do you think?

Individuals can do little to prevent global warming.

| 1 | 2 | 3 | 4 | 5 | 6 | 7 |
|---|---|---|---|---|---|---|

strongly agree          strongly disagree

## 15-1 Population Dynamics

Population growth was one of the most significant changes of the twentieth century. Since 1900, the world's population has more than tripled. In fact, the world adds almost 383,000 people each day (Haub and Kaneda 2011). Most of the births are in other countries, but the world's population growth affects all Americans—now and in the future.

Information about population growth comes from **demography**, the scientific study of human populations. Demographers analyze populations in terms of size, composition, distribution, and why they change. A **population** is a group of people who share a geographic territory. A territory can be as small as a town or as vast as the planet, depending on a researcher's focus. Demographers also study personal data such as when and where you were born, your probability of getting married or divorced, the kind of job you'll probably have, how many times you'll move, and how long you'll probably live. According to one demographer, "If people are not interested in demographic phenomena, they are not interested in themselves" (McFalls 2007: 3).

### 15-1a Why Populations Change

Global population has grown rapidly since 1800 (see *Figure 15.1*). It reached 1 billion in 1804, 5 billion in 1987, 6.5 billion in 2005, and is expected to reach 9.4 billion by 2050 (U.S. Census Bureau

2010). When demographers examine population changes, they look at the interplay among three key factors: how many people are born (fertility), how many die (mortality), and how many move from one area to another (migration).

## Fertility: Adding New People

The study of population changes begins with **fertility**, the number of babies born during a specified period in a particular society. There are several ways to measure fertility, but one of the most commonly used is the **crude birth rate**, also known as the *birth rate*, the number of live births for every 1,000 people in a population in a given year.

"Crude" implies that the rate is a general measure of a society's childbearing because it's based on the total population rather than more specific measures such as

a woman's age or marital status. However, the crude birth rate allows comparisons for a given year across populations or countries. In 2011, for example, that rate was 20 worldwide, 36 for Africa, 13 for the United States, and 11 for Europe (Haub and Kaneda 2011).

Birth rates also vary within a country, but are much higher in the least developed nations (35) than in the developed nations (11) (Haub and Kaneda 2011). In the United States and other countries, the more affluent and those with more education, regardless of race and ethnicity, have fewer children than the poor because they are more likely to use contraceptives and to postpone childbearing, which decreases the number of children they will have over a lifetime (Zirulnick 2011; see also Chapters 9 and 12).

## Mortality: Subtracting People

The second factor in population change is **mortality**, the number of deaths during a specified period in a population. Demographers typically measure mortality by the **crude death rate** (also called the *death rate*), the number of deaths per 1,000 people in a population in a given year. In 2011, for example, the crude death rate was 8 worldwide and for the United States, 11 for Europe, and 15 or higher for some African nations (Haub and Kaneda 2011).

A death rate isn't necessarily the best measure of a population's health, however. Death rates are high in developed countries—even though these nations have better medical services, better nutrition, and healthier environments than most developing countries—because industrialized nations also have large proportions of people who are 65 and older.

A better measure of a population's health is the *infant mortality rate*, the number of deaths of infants younger than 1 year per 1,000 live births. Generally, as the standard of living improves—meaning better access to clean water, adequate sanitation, and medical care—the infant mortality rate decreases (see Chapter 14). In 2011, the infant mortality rate was 6 in Europe and the United States, 18 in South America and the Caribbean, and as high as 131 in Afghanistan (Haub and Kaneda 2011).

Lower infant mortality greatly raises *life expectancy*, the average number of years that people who were born at about the same time can expect to live. Worldwide, in 2011, the average life expectancy was 70 years—68 for men and 72 for women. Again, however, there are considerable variations across countries—from a high of 83 in Japan and 82 in Sweden and France to a low of 44 in Afghanistan. The United

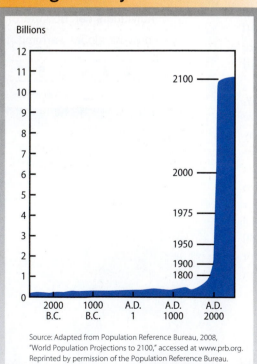

## FIGURE 15.1
## World Population Growth Through History

Billions

Source: Adapted from Population Reference Bureau, 2008, "World Population Projections to 2100," accessed at www.prb.org. Reprinted by permission of the Population Reference Bureau.

Because of high population growth, the roads in many of India's largest cities, such as this one in New Delhi, are chaotic: "Cars, trucks, buses, motorcycles, taxis, rickshaws, cows, donkeys, and dogs jostle for every inch of the roadway as horns blare and brakes squeal. Drivers run red lights and jam their vehicles into available space, ignoring pedestrians" (Hamm 2007: 49–50).

MANAN VATSYAYANA/AFP/Getty Images

States, with a life expectancy of 78, ranks below at least 25 other industrialized countries and only slightly higher than less developed countries such as Cuba, the Czech Republic, and Uruguay (Haub and Kaneda 2011; see also Chapter 14).

In general, life expectancy has been increasing worldwide. Much of the increase is due to medical advances that have wiped out many diseases and the availability of clean drinking water, sanitation, immunization, and antibiotics, all of which tend to prolong life. However, civil wars, genocide, and deaths caused by AIDS have devastated many African countries (Ashford 2006). In almost every society, people with a higher socioeconomic status live longer and healthier lives. They are more aware of the benefits of nutrition, work in jobs that are relatively safe, and have the resources to access medical services.

## Migration: Adding and Subtracting People

The third demographic factor in population change is **migration**, the movement of people into or out of a specific geographic area. Migration is the product of both push and pull factors.

*Push factors* encourage or force people to leave a residence. These factors include war, political or religious persecution, unemployment, high crime rates,

and natural disasters. After Hurricanes Katrina and Rita devastated much of the Louisiana-Texas border in 2005, for example, those who were the most disadvantaged—up to 75 percent—moved to other states because they didn't have home insurance policies that covered the cost of the damages (Myers et al. 2010).

*Pull factors* attract people to a new location. Some of these factors include religious freedom, better schools, lower crime rates, and, especially, economic opportunities. For example, the United States is the destination of most Mexicans (both legal and undocumented) because they see America as a land of opportunity. Such optimism helps explain why 10 percent of Mexicans born in their homeland now live in the United States (Passel et al. 2012).

There are two types of migration: international and internal. *International migration* is movement to another country. Such migration includes *emigrants* (people who are moving out of a country) and *immigrants* (people who are moving into a country). You may be a product of international migration, for example, if your great-great-grandparents emigrated from Ireland and immigrated to the United States.

International migration is often in the news, but only about 3 percent of the world's population migrates to a different country and ends up staying for a year or longer. Most emigrants move to a neighboring country (from Mexico to the United States, for example, rather than from Mexico to Canada). International migration is relatively uncommon both because most people have no desire to leave their family and friends and because governments try to regulate border crossings. However, 20 percent of international migrants live in the United States, more than in any other county in the world (Martin and Zürcher 2008; Bremner et al. 2009).

Of the 700 million adults worldwide who say that they would like to relocate permanently to another country if they could, 24 percent would like to move to

**migration** the movement of people into or out of a specific geographic area.

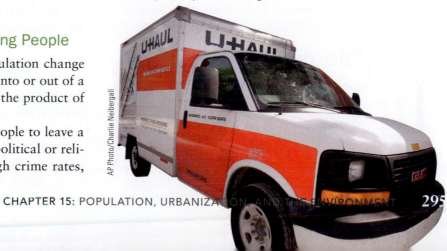

AP Photo/Charlie Neibergall

the United States to reunite with family members living there, to find jobs, or to provide better lives for their children (Esipova et al. 2010). Thus, the reasons for international migration aren't very different today from those of immigrants during the twentieth century. Because of some states' tougher immigration laws and economic recession, however, many emigrants, especially Mexicans, have been going home or simply not coming to the United States (see Chapter 10).

The second type of migration is *internal migration*, movement within a country. About 40 million Americans move within the United States every year, many of them to the South and West because of better economic opportunities. In 1910, the West made up just 7 percent of the U.S. population compared with 25 percent today. Most internal migrants are single and college-educated, and are more likely to relocate to central cities than to suburbs or rural areas because of job prospects (Franklin 2003; Tavernise and Zelony 2010).

## 15-1b Population Composition and Structure

Demographers examine age and sex to understand a population's composition and structure. Two of the most common measures are *sex ratios* and *population pyramids*.

### Sex Ratios

The proportion of men to women in a population is its **sex ratio**. A sex ratio of 100 means that there are equal numbers of men and women; a sex ratio of 95 means that there are 95 men for every 100 women (fewer males than females).

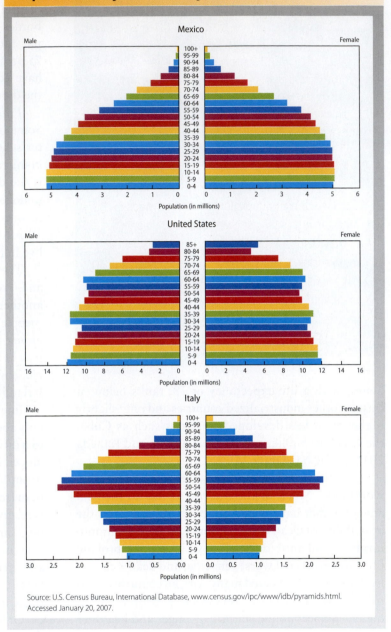

**FIGURE 15.2**

**Population Pyramid Projections, 2025**

Source: U.S. Census Bureau, International Database, www.census.gov/ipc/www/idb/pyramids.html. Accessed January 20, 2007.

Sex ratios are important because they affect the availability of marriageable partners, marriage rates, and childbearing (see Chapter 12).

Worldwide, 105 boys are born for every 100 girls (Haub and Gribble 2011). This newborn sex ratio decreases, however, as males experience higher rates of mortality throughout life. Although it's not clear why, male fetuses die in miscarriages at a higher rate than female fetuses (Christenson et al. 2004). However, sex ratios are skewed in favor of males in some countries, especially China and India, because of the practice of *female infanticide* (sometimes called *gendercide*)—the intentional killing of female infants because male offspring bring higher social status (Almond et al. 2009; Zhu et al. 2009; Hvistendahl 2012).

In India, where the sex ratio is 112, the abortion of female fetuses increased from 27 percent in 2001 to 56 percent in 2011. Most female infanticide occurs among educated or richer households that could afford ultrasound and abortion services (Haub and Gribble 2011; Jha et al. 2011). The government has tried to discourage female infanticide, especially in rural villages with the highest rates, by offering about $2,200 for producing more girl than boy births. The program has had limited success, however, because of a cultural preference for males (Arnoldy 2011).

## Population Pyramids

A **population pyramid** is a visual representation of the makeup of a population in terms of the age and sex of its members at a given point in time. As *Figure 15.2* shows, Mexico is a young country: much of its population is under age 45 (which also means that many women are in their childbearing years), and there are relatively few people 65 years and older. In contrast, Italy is an old country, and the United States is somewhere in the middle.

The shape of the population pyramid (a triangle for Mexico, a rectangle for the United States, and a diamond for Italy) has future implications for young and old countries. Italy, for example, has a relatively small number of women ages 15 to 44 (in their reproductive years) and a bulge of people ages 45 to 79. This suggests that there may not be enough workers to support an aging population in the future and that there will be a greater need for social services for older people than for children and adolescents (see Chapters 12 and 14). Thus, population pyramids give us a snapshot of a country's demographic profile and suggest some of the problems that people are likely to experience in the future.

## 15-1c Population Growth: A Ticking Bomb?

Eight of the countries with the largest populations, many of them in the developing world, will grow even more by 2050 (see *Table 15.1*). So, has population growth gotten out of hand? There are many views on this question, but two of the most influential have been Malthusian theory (which argues that the world can't sustain its unprecedented population surge) and demographic transition theory (which maintains that population growth is slowing).

### Malthusian Theory

For many demographers, population growth is a ticking bomb. They subscribe to the **Malthusian theory,** the idea that the population is growing faster than the food supply needed to sustain it. This theory is named after Thomas Malthus (1766–1834), an English economist, clergyman, and college professor who maintained that humans are multiplying faster than the ability of the earth to produce sufficient food.

According to Malthus (1798/1965), population grows at a *geometric rate* (2, 4, 8, and so on), whereas the food supply grows at

**population pyramid** a visual representation of the makeup of a population in terms of the age and sex of its members at a given point in time.

**Malthusian theory** the idea that population is growing faster than the food supply needed to sustain it.

### TABLE 15.1

## The World's Largest Countries, 2011 and 2050

| COUNTRY | 2011 POPULATION (IN MILLIONS) | ESTIMATED 2050 POPULATION (IN MILLIONS) |
|---|---|---|
| China | 1,346 | 1,313 |
| India | 1,241 | 1,692 |
| United States | 312 | 423 |
| Indonesia | 238 | 309 |
| Brazil | 197 | 223 |
| Pakistan | 177 | 314 |
| Nigeria | 162 | 433 |
| Bangladesh | 151 | 226 |

Note: Remember to add six zeros to these numbers. For example, China's population in 2011 was about 1,346,000,000, or 1.3 billion people.

Source: Based on Haub and Kaneda 2011.

Hulton Archive/Getty Images

Thomas Malthus

**demographic transition theory** the idea that population growth is kept in check and stabilizes as countries experience economic and technological development.

an *arithmetic rate* (1, 2, 3, 4, and so on). That is, two parents can have 4 children and 16 grandchildren within 50 years. The available number of acres of land, farm animals, and other sources of food can increase in that time period, but certainly not quadruple. In effect, then, the food supply will not keep up with population growth. Because there are millions of parents, the results could be catastrophic, with masses of people living in poverty or dying of starvation.

Malthus first posited that only war, famine, and disease acted as *preventive checks* on population growth. In later essays, he also included "moral restraint" as a necessary preventive check, especially for the lower classes. The lack of moral restraint characterized people who married at an early age and didn't practice sexual abstinence before and outside of marriage. Such behavior resulted in large families and out-of-wedlock children that the working men couldn't save from "rags and squalid poverty" (Malthus 1872/1991).

Except for the notions about moral restraint, Malthusian theory has had a lasting influence. *Neo-Malthusians* (or New Malthusians) agree that the population is exploding beyond food supplies. The world's population reached its first billion in 1800. In the 200 years that followed, the world added another 5 billion people (see *Figure 15.1* on p. 296). As a result of this growth, according to some influential neo-Malthusians, the earth has become a "dying planet"—a world with insufficient food and a rapidly expanding population that pollutes the environment (Ehrlich 1971; Ehrlich and Ehrlich 2008).

The number of hungry people in the world, primarily in sub-Saharan Africa and South Asia, increased from 825 million in 1995 to 1.02 billion in 2009. The resources and technical knowledge are available to increase food production by 70 percent by 2050, but poverty and difficult growing conditions plague the countries that need food the most (Food and Agriculture Organization of the United Nations 2009).

## Demographic Transition Theory

Some demographers are more optimistic than neo-Malthusians. **Demographic transition theory** maintains that population growth is kept in check and stabilizes as countries experience economic and technological development, which, in turn, affects birth and death rates. According to this theory, population growth changes as societies undergo industrialization, modernization, technological progress, and urbanization. During these processes, a nation goes through four stages (see *Figure 15.3*), from high birth and death rates to low birth and death rates.

■ *Stage 1: Preindustrial society.* In this initial stage, there is little population growth. The birth rate is high because people rarely use birth control and they want as many children as possible to provide unpaid agricultural labor and support parents in old age. However, a high death rate offsets the high birth rate. Many children don't survive infancy, and mortality is high at all ages because of diseases and minimal access to health care.

■ *Stage 2: Early industrial society.* There is significant population growth because the birth rate is higher than the death rate. The birth rate may even increase over what it was in Stage 1 because mothers and their children enjoy improved health care. Couples may still have large numbers of children because they fear that many of them will die, but the death rate declines because of better sanitation, better nutrition, and medical advances (e.g., immunizations and antibiotics). Most of the world's poorest countries are currently in Stage 2.

■ *Stage 3: Advanced industrial society.* As the infant mortality rate declines, parents have fewer children.

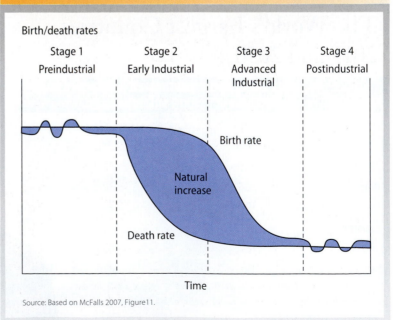

## FIGURE 15.3
## The Classical Demographic Transition Model

Birth/death rates

| Stage 1 Preindustrial | Stage 2 Early Industrial | Stage 3 Advanced Industrial | Stage 4 Postindustrial |

Birth rate

Natural increase

Death rate

Time

Source: Based on McFalls 2007, Figure 11.

Effective birth control reduces family size. The decrease in child care responsibilities, in turn, enables women to work outside the home. China and many countries in Latin America are currently in Stage 3 (Gelbard et al. 1999; Brea 2003).

■ *Stage 4: Postindustrial society.* In this stage, the demographic transition is complete, and the society has low birth and death rates. Women tend to be well educated and to have full-time jobs or careers. If there is little immigration, the population may even decrease because the birth rate is low. This is the case today in Canada, Japan, Singapore, Hong Kong, Australia, New Zealand, the United States, and many European countries, including Italy and Scotland.

## Critical Evaluation

The dire predictions of Malthus and his successors that global population growth would lead to worldwide famine, disease, and poverty haven't come true. Still, today more than 20 percent of people live in abject poverty, subsisting on less than $1 a day (World Bank 2008; see also Chapter 8).

Despite neo-Malthusians' fears, global fertility is half of what it was in 1972. The population of some industrialized countries is declining because people aren't having enough babies to replace themselves. These countries are experiencing **zero population growth (ZPG)**, a stable population level that occurs when each woman has no more than two children.

Future population growth is difficult to predict because there are many unknowns. Some low-birth nations with fertility rates below ZPG are paying women to have more children because there won't be enough young workers to pay for social security systems and the rising cost of health care for aging populations. For example, Russia gives mothers with one child $9,000 for each additional baby; Japan has expanded its day care facilities and offers families a monthly allowance of $145 per child younger than 15; and China is now encouraging newly married couples to have two children, easing the one-child policy introduced in 1978 (Ford 2010; Haub 2010; Yamazaki and Ito 2010).

Some neo-Malthusians maintain, however, that it's irresponsible for *any* country to encourage higher fertility rates. These demographers worry about the consequences of adding 3 billion more inhabitants to the planet in less than 50 years, especially for many developing countries "with desperate economic outlooks" (Sachs 2005; Shorto 2008).

The birth rate in the United States has declined, but the country is third in the world in population growth, behind India and China, largely because of high immigration rates, and many of the immigrants are young women with high fertility rates. Thus, the U.S. population "is growing by more than all other developed countries *combined*" (Ryerson 2004: 21).

One result of population growth is urban growth. Cities attract both immigrants and native-born residents because of jobs and cultural activities, but the population growth of urban areas also creates numerous problems.

# 15-2 Urbanization

If you've flown over the United States, you've probably noticed that people tend to cluster in and around cities. After sunset, some areas glow with lights, whereas others are engulfed in darkness. The average person, in the United States and worldwide, is more likely to live in a city than a rural area, and this trend is rising.

A **city** is a geographic area where a large number of people live relatively permanently and make a living primarily through nonagricultural activities. **Urbanization**, which increases the size of cities, is the movement of people from rural to urban areas. Most of this discussion focuses on U.S. cities, but let's begin with a brief look at urbanization globally.

## 15-2a Urbanization: A Global View

In 2008, for the first time in history, a majority of the world's population lived in urban areas. By 2050, urban

North Wind Picture Archives via AP Images

## TABLE 15.2

## Urbanization Around the World

| PERCENTAGE OF PEOPLE LIVING IN URBAN AREAS | | | |
|---|---|---|---|
| REGION | 1950 | 2011 | 2030 (PROJECTED) |
| World | 29 | 51 | 60 |
| Africa | 14 | 40 | 48 |
| Asia | 18 | 45 | 56 |
| Latin America and the Caribbean | 41 | 79 | 83 |
| Northern America | 64 | 82 | 86 |
| Europe | 51 | 73 | 77 |
| Oceania | 62 | 71 | 71 |

Source: Based on United Nations Department of Economic and Social Affairs 2012, Tables 2 and 6.

**megacities** metropolitan areas with at least 10 million inhabitants.

### Origin and Growth of Cities

Cities are one of the most striking features of modern life, but they have existed for centuries. About 7,000 years ago, for example, people built small cities in the Middle East and Latin America to protect themselves from attackers and to increase trade. By 1800, 56 cities in Western Europe had a population of 40,000 or more (Chandler and Fox 1974; Flanagan 1990; De Long and Shleifer 1992).

Before the Industrial Revolution, which began in the late eighteenth century, urban settlements in Europe, India, and China developed largely because people figured out how to use natural resources, such as mining coal and transporting water efficiently for irrigation and consumption. The Industrial Revolution spurred ever-increasing numbers of people to move to cities in search of jobs, schooling, and improved living conditions. As a result, the urban population surged—from 3 percent of the world's population in 1800 to 14 percent in 1900 (Sjoberg 1960; Mumford 1961).

### World Urbanization Trends

As industrialization advanced, urbanization increased. Between 1920 and 2007, the world's urban population increased from 270 million to 3.3 billion, and is expected to rise to 6.3 billion by 2050. The pace of urbanization is most rapid in the less developed regions of the world, especially Asia, Africa, and Latin America (see *Table 15.2*). By 2050, most of the world's urban population will be concentrated in Asia and Africa (United Nations Department of Economic and Social Affairs 2012).

Many of the world's largest cities are becoming **megacities**, metropolitan areas with at least 10 million inhabitants. In 1950, the world had only two megacities: Tokyo (11.3 million) and New York–Newark (12.3 million). By 2025, there will be 37 megacities, but only three of them in the United States (New York–Newark,

dwellers will make up 67 percent of the world's population (United Nations Department of Economic and Social Affairs 2012). Why is urbanization increasing? And where is most of it taking place?

© Shawn Baldwin/The New York Times/Redux Pictures

With a population of almost 20 million in 2012, Cairo, Egypt, is one of the world's largest cities. Cairo is the cultural center of the Arab world, but millions of Egyptians, including this fisherman, live in dire poverty and don't experience any of the city's cultural benefits. This man sleeps in his boat, makes tea from the water of the Nile River (which is infested with life-threatening parasites), often smiles and waves dutifully as tour boats motor up the river with tourists snapping his picture, and, on a good day, earns a few dollars (Slackman 2007).

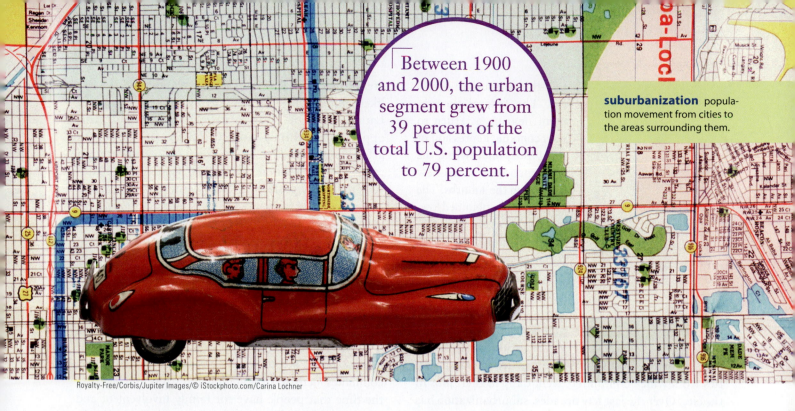

Between 1900 and 2000, the urban segment grew from 39 percent of the total U.S. population to 79 percent.

**suburbanization** population movement from cities to the areas surrounding them.

Royalty-Free/Corbis/Jupiter Images/© iStockphoto.com/Carina Lochner

Chicago, and Los Angeles–Long Beach–Santa Ana). Besides Tokyo—which will probably remain the most populous city in the world, with nearly 39 million inhabitants in 2025—there will be 22 megacities in Asia, 6 in Latin America, and 3 each in Africa and Europe. Most of the megacities in Asia will be at least twice as large as those in the United States and Europe (United Nations Department of Economic and Social Affairs 2012).

Should the explosive growth of cities and megacities concern us? Generally, cities provide jobs, offer better health services, and have more educational opportunities, but not everyone benefits from such advantages. The urban poor are often crowded into slums, where children are less likely to be enrolled in school, sanitation is inadequate, and the economic gap between haves and have-nots is widening (Bruinius 2010; Laneri 2011).

## 15-2b Urbanization in the United States

Like many other countries, the United States is becoming more urban. During the Industrial Revolution, millions of Americans in agricultural areas migrated to cities to find jobs. As a result, between 1900 and 2000, the U.S. rural population shrank from 61 percent to 21 percent of the total population, whereas the urban segment increased from 39 percent to 79 percent (Riche 2000; U.S. Census Bureau 2010).

### Shifts in Urban and Rural Populations

Despite rapid population growth in parts of the South and West, 45 percent of all U.S. counties (1,346 counties)

shrank in population between 2000 and 2007, and 85 percent were rural. Many rural communities, particularly in the Midwest, have been losing inhabitants for decades and "are on the brink of extinction" (Mather 2008).

The fastest growing counties are located near large metropolitan areas, such as those around Atlanta, Chicago, Dallas, Houston, Los Angeles, Miami, New York, and Washington, D.C. How, specifically, has urban America been changing?

### How Urban America Is Changing

U.S. cities have changed quite a bit, especially since 1980, resulting in "blurring the lines that have long separated cities and suburbs" (Berube 2011). We'll look at suburbs shortly, but there are several ways that U.S. urbanization has been changing.

First, many middle-class blacks have moved to suburbs, leaving low-income African Americans behind in inner cities (Keen 2011). Second, in metropolitan areas, the average white American lives in a more diverse neighborhood where her/his neighbors aren't white (23 percent in 2010 compared with 12 percent in 1980; Turner and McDade 2011). Third, suburbs are attracting fewer young adults, especially those with young children. This means that many communities no longer have the taxes that support schools, retail stores, services for older people, and public recreational facilities (Frey 2011; El Nasser and Overberg 2011).

Urban growth sparked **suburbanization**, population movement from cities to the areas surrounding them. During the 1950s, suburbs mushroomed, attracting two-thirds of urban dwellers. The federal government, fearing a return to the economic depression of the 1930s, underwrote the construction of much new housing in the suburbs. The general public obtained low-interest mortgages, veterans were offered the added incentive of being able to purchase a home with a $1 down payment, and massive highway construction programs enabled commuting by car (Rothman 1978).

Originally, most suburbs were bedroom communities from which commuters went daily to their jobs in the city. Over the last few decades, suburbanization has generated **edge cities**, business centers that are within or close to suburban residential areas and include offices, schools, shopping, entertainment, malls, hotels, and medical facilities.

People have also created **exurbs**, areas of new development beyond the suburbs that are more rural but on the fringe of urbanized areas. About 6 percent of Americans live in exurbs. The average exurbanite is white, a middle-income earner, married with children, a "super commuter" (one who travels two or more hours a day for work), and owns a large house outside of an expensive metropolitan suburb (Berube et al. 2006; Lalasz 2006).

## Some Consequences of Urbanization and Suburbanization

Cities offer many benefits. Among other advantages, people can often walk, bicycle, or take a bus or subway to work; they are surrounded by a vast array of restaurants and shops; and they have easy access to numerous cultural activities, such as museums and theaters. Urbanization also creates problems, such as urban sprawl, increased traffic congestion, a scarcity of affordable housing, and racial segregation.

*Urban sprawl.* **Urban sprawl**—the rapid, unplanned, and uncontrolled spread of development into regions adjacent to cities—is widespread. Between 1995 and 2002, New Jersey, the nation's most densely populated state, lost 29 percent of its farmland, forests, wildlife habitats, and open recreational areas to urban sprawl (Lathrop and Hasse 2007).

Urban sprawl has created rapid *job sprawl*, which occurs when companies move jobs from metropolitan areas to suburbs. The more distant the suburb, the less likely minorities—especially African Americans and Latinos—and the poor are to hear about employment opportunities through informal networks, be able to afford houses in these areas, and have transportation to the jobs (Kneebone 2009; Raphael and Stoll 2010).

*Traffic congestion.* In most cases, the only way to get around in urban sprawl areas is by automobile. This means that most suburban households face the costs of buying, fueling, insuring, and maintaining multiple cars. The U.S. population has increased by 23 percent over the last 25 years, but total highway miles have increased by only about 5 percent, resulting in greater traffic congestion within and outside cities. More than 3 million Americans (about 3 percent of workers) now travel 90 minutes or more to work every day, a proportion that has increased by 95 percent since 1990. Traffic snarls and long commutes increase air pollution and stress and decrease the time that people have for family involvement and leisure pursuits (Sullivan 2007; U.S. Census Bureau 2010).

*Lack of affordable housing.* In some cases, the poor are pushed out by **gentrification**, a process in which upper-middle-class and affluent people buy and renovate houses and stores in downtown urban neighborhoods. Governments in many older cities encourage gentrification to increase dwindling populations, to revitalize urban areas, and to augment tax revenues. However, rent increases have displaced many low-income residents and small businesses.

*Residential segregation.* Residential segregation has been declining slowly since 1980. Among large metropolitan areas with a population of 500,000 or more, the least segregated are in the South and West, and the most segregated are mainly in the Northeast and Midwest (Scommegna 2011).

Gentrification improves old city neighborhoods and increases property values, but also displaces low-income residents.

The nation's 20 most multiethnic metropolitan regions are now "global neighborhoods" that have substantial numbers of whites, blacks, Latinos, and Asians (Logan and Zhang 2011). However, the average black or Latino household lives in a poorer neighborhood than the average white household (Reardon and Bischoff 2011).

For the first time, a majority of all racial/ethnic groups in large metropolitan areas live in suburbs. In 2010, more than a third of all 13.3 million new

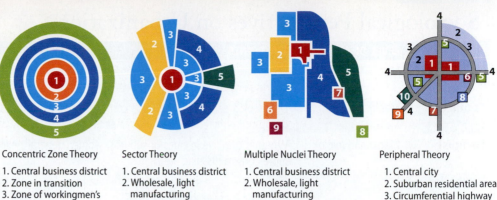

## FIGURE 15.4
## Four Models of City Growth and Change

**Concentric Zone Theory**
1. Central business district
2. Zone in transition
3. Zone of workingmen's homes
4. Residential zone
5. Commuters' zone

**Sector Theory**
1. Central business district
2. Wholesale, light manufacturing
3. Lower-class residential
4. Middle-class residential
5. Upper-class residential

**Multiple Nuclei Theory**
1. Central business district
2. Wholesale, light manufacturing
3. Lower-class residential
4. Middle-class residential
5. Upper-class residential
6. Heavy manufacturing
7. Outlying business district
8. Residential suburb
9. Industrial suburb

**Peripheral Theory**
1. Central city
2. Suburban residential area
3. Circumferential highway
4. Radial highway
5. Shopping mall
6. Industrial district
7. Office park
8. Service center
9. Airport complex
10. Combined employment and shopping center

Sources: Based on Park and Burgess 1921, Hoyt 1939, Harris and Ullman 1945, and Harris 1997.

suburbanites were Latinos, who are following jobs to suburban areas (Tavernise and Gebeloff 2010; Frey 2011). Between 1999 and 2008, the suburban poor population grew by 25 percent. As in cities, blacks and Latinos make up a disproportionate share of the poor residents in suburbs (Kneebone and Garr 2011).

Some communities, particularly on the West coast, are becoming "ethnoburbs." These are suburban ethnic clusters of residential areas and business districts in large metropolitan areas that are multiracial, multiethnic, multicultural, multilingual, and often multinational. One ethnic minority group has a significant concentration, but doesn't necessarily constitute a majority (Li 2009).

## 15-2c Sociological Explanations of Urbanization

How and why do cities change? And how do these changes affect their populations? In answering these and other questions, functionalists provide insights on urban development, conflict theorists emphasize the impact of capitalism and big business, feminist scholars focus on gender roles and space, and symbolic interactionists examine the quality of city life (*Table 15.3* on p. 306 summarizes these perspectives).

### Functionalism: How and Why Cities Change

In the 1920s and 1930s, sociologists at the University of Chicago developed theories of **urban ecology**, the

study of the relationships between people and urban environments. Initially, these sociologists based their theories on Chicago, the city where their university was located, but social scientists later revised the descriptions (see *Figure 15.4*).

Sociologists Robert Park and Ernest Burgess (1921) proposed *concentric zone theory* to explain the distribution of social groups within urban areas. According to this model, a city grows outward from a central point in a series of rings. The innermost ring, the central business district, is surrounded by a zone of transition, which contains industry and poor-quality housing. The third and fourth rings have housing for the working and middle classes. The outermost ring is occupied by people who live in the suburbs and commute daily to work in the central business district.

In developing *sector theory*, economist Homer Hoyt (1939) refined concentric zone theory. He proposed that cities, including Chicago, develop in sectors instead of rings. Pie-shaped wedges radiate from the central business district, their orientation depending on transportation routes (such as rail lines and highways) and various economic and social activities. Thus, some sectors are predominantly industrial, some contain stores and offices, and others, generally farther away from

**urban ecology** the study of the relationships between people and urban environments.

## TABLE 15.3

## Sociological Perspectives on Urbanization

| PERSPECTIVE | LEVEL OF ANALYSIS | KEY POINTS |
|---|---|---|
| **Functionalist** | Macro | People create urban growth by moving to cities to find jobs and to suburbs to enhance their quality of life. |
| **Conflict** | Macro | Driven by greed and profit, large corporations, banks, developers, and other capitalist groups determine the growth of cities and suburbs. |
| **Feminist** | Macro and micro | Whether they live in cities or suburbs, women generally experience fewer choices and more constraints than men. |
| **Symbolic Interactionist** | Micro | City people are more tolerant of different lifestyles, but they tend to interact superficially and are generally socially isolated. |

© Cengage Learning

**new urban sociology**
the view that urban changes are largely the result of decisions made by powerful capitalists and high-income groups.

the central business district, are middle- and upper-class residential areas.

Geographers Chauncey Harris and Edward Ullman (1945) developed another influential model, *multiple-nuclei theory*, which proposed that a city contains more than one center around which activities revolve. For example, a "minicenter" often includes an outlying business district with stores and offices that are accessible to middle- and upper-class residential neighborhoods, whereas airports typically attract hotels and warehouses. Thus, heavy industry and high-income housing are rarely in the same part of the city.

As cities grew after World War II, these models no longer described urban spaces. Thus, Chauncey Harris (1997) proposed a *peripheral* theory of urban growth, which emphasized the development of suburbs around a city but away from its center. According to this model, as suburbs and edge cities burgeon, highways that link the city's central business district to outlying areas and beltways that loop around the city provide relatively easy access to airports, the downtown, and surrounding areas.

### Conflict Theory: The Impact of Capitalism and Big Business

For functionalists, people's choices shape urban changes. In contrast, **new urban sociology**, a perspective heavily influenced by conflict theory, views urban changes as largely the result of decisions made by powerful capitalists and high-income groups. That is, economic and political factors favorable to the rich, not ordinary citizens, determine urban growth or decline. For example, when a local government

wants to rejuvenate parts of the inner city, it typically offers tax breaks, changes zoning laws, and allows real estate, construction, and banking industries to seek profits with little regard for the needs of low-income households or the homeless (Feagin and Parker 1990; Macionis and Parrillo 2007).

For conflict theorists, urban space is a commodity that is bought and sold for profit. It is not the average American, they argue, but bankers, corporate executives, developers, politicians, and influential businesspeople who determine how urban space is used. Increasing the value of some property is a higher priority than respecting community values, considering neighborhood needs, or maintaining a livable city. As a result, poor and low-income people are crowded into dilapidated neighborhoods (Logan and Molotch 1987; Gottdiener and Hutchison 2000).

### Feminist Theories: Gender Roles, Space, and Safety

Feminist theories emphasize gender-related constraints. Whether they live in cities or suburbs, women generally experience more problems than men because living spaces are usually designed by men who have tended to ignore women's changing roles.

Many women fear the city, especially urban public spaces such as streets, parks, and public transportation (Domosh and Seager 2001). They see these places as risky for their physical safety, despite the fact that most violence against women occurs at home. A few cities provide public transportation, such as minivans that operate seven nights a week, to prevent crimes against women, usually minorities, who must travel to work after 8:00 P.M. and return home before dawn. For the most part, however, such services are rare (Saegert and Winkel 1981; Hayden 2002).

Electronics City, India's version of California's Silicon Valley, is an industrial park that spans more than 330 acres and houses more than a hundred businesses, such as Hewlett-Packard, Motorola, and Infosys, as well as a premier graduate school that focuses on information technology (Hamm 2007). If governments and corporations can find the capital to build such facilities, ask conflict theorists, why can't they provide affordable housing for people in surrounding neighborhoods?

In the suburbs, both women's and men's physical mobility is limited because of the scarcity of public transportation systems. Consequently, many suburban households have two or more cars, but women living in the suburbs usually experience greater problems than men. For example, there may be fewer job opportunities (especially if women have domestic responsibilities such as being at home when children return from school), and women often spend much of their time maintaining a single-family home, which decreases their time for educational or leisure activities (Cichocki 1981; Hayden 2002).

### Symbolic Interactionism: How People Experience City Life

Symbolic interactionists are most interested in the impact of urban life on city residents. In a classic essay, sociologist Louis Wirth (1938: 14) described the city as a place where "our physical contacts are close, but our social contacts are distant."

Wirth defined the city as a large, dense, and socially and culturally diverse area. These characteristics produce *urbanism*, a way of life that differs from that of rural dwellers. Wirth saw urbanites as more tolerant of a variety of lifestyles, religious practices, and attitudes than residents of small towns or rural areas.

He also believed that urbanism has negative consequences, such as alienation, friction because of physical congestion, pursuit of self-interest, impersonal relationships, and a disintegration of kinship and friendship ties. Some studies have supported Wirth's theory of urbanism (see Guterman 1969), but others have challenged his views. In a recent national study, for example, people living in large metropolitan areas scored higher than those living in small towns and rural areas on characteristics such as physical and emotional health, access to basic necessities, and being satisfied with life (Witters 2010).

### Critical Evaluation

Conflict theorists point out that functionalists tend to overlook the negative political and economic impact, especially when profit and greed guide urban planning. Conflict theory seems to assume that residents are helpless victims as developers and corporations raze low-income houses. In fact, environmental groups have had considerable success in pushing through legislation to maintain and even increase open public spaces and build energy-saving homes in low-income neighborhoods (Moore 2008).

Feminist sociologists have made important contributions through studies of the everyday lives of low-income women, especially in central cities (see Chapter 11), but urbanization has received much less attention. According to critics, urbanites are more diverse than some symbolic interactionists claim. People living in cities aren't necessarily more self-centered or isolated than those in small towns or rural areas. Instead, many have close family bonds, friends, and satisfying relationships with coworkers (Gans 1962; Crothers 1979; Wilson 1993). Also, the symbolic interactionist perspective doesn't show how political, educational, religious, and economic factors shape people's experiences of city life (Hutter 2007).

You've seen that the world's population is growing rapidly and is becoming more urbanized; both are taxing the planet's limited resources.

## 15-3 Environmental Issues

Consider the following:

- Most of us have at least 116 toxic chemicals in our bodies that did not exist in the environment (much less in humans) just 75 years ago ("Second National Report . . ." 2003).

■ Every American now produces, on average, 5 pounds of garbage a day, compared with 2.7 pounds a day in 1960 (EPA 2012).

■ Commercial logging—spurred by high U.S. demand for hardwoods such as teak, mahogany, and rosewood—destroys 50,000 species every year, including plants that produce life-saving medicines (Raintree Nutrition, Inc. 2008).

■ As many as 7 million Americans get sick every year from swimming in water contaminated with bacteria, viruses, or parasites that cause a wide range of diseases, including ear, nose, and eye infections, hepatitis, skin rashes, and respiratory illnesses (Natural Resources Defense Council 2007).

Such environmental problems threaten our **ecosystem**, an area in which all forms of life live in relation to one another and a shared physical environment. Plants, animals, and humans depend on each other for survival. Because the ecosystem is interconnected worldwide, what happens in one country affects others. Let's look more closely at clean water, air pollution, and global warming—three interrelated environmental issues that are endangering the U.S. and global ecosystem.

## 15-3a Water

An expanding world population, extreme weather patterns, and industrial pollution are making water a scarce commodity. The introduction of water filtration and chlorination in major U.S. cities between 1900 and 1940 was responsible for about 43 percent of the total decline in mortality over that period (Scommegna 2005). Few Americans die because of contaminated water, but as many as 19 million become sick each year because of the parasites, viruses, and bacteria in drinking water (Duhigg 2009b).

More than 1 billion people worldwide don't have clean water, and nearly half the world's population drinks contaminated water. Water-related diseases cause 50 percent of illnesses and deaths worldwide every year, and diarrheal diseases kill more than 3 million children every year (Water Quality & Health Council 2005; United Nations World Water . . . 2006). In some developing countries, families often spend up to 25 percent of their income to purchase water, and many women and children spend up to 6 hours a day carrying it home (United Nations Development Programme 2006).

### Availability and Consumption

Just 3 percent of the earth's water is fresh, and two-thirds of that water is locked up in the ground, glaciers, and ice caps. That leaves about 1 percent of the earth's water for the world's more than 7 billion people (Schirber 2007). The world's demand for water has tripled over the last half century. By 2030, total global water supply will meet just 60 percent of the demand (Hamilton 2011).

Some refer to water as "blue gold" because it's becoming one of the earth's most precious and scarce commodities. Water shortages—rather than oil or diamonds—are behind conflicts and even wars in a number of countries. According to a past Secretary-General of the United Nations, "Too often, where we need water, we find guns . . ." (World Water Assessment Programme 2009: 20).

Industrialized nations not only have greater access to clean water than the developing world, they use more and pay less for it. The average person in the United States uses about 151 gallons of water per day compared with 101 gallons in Italy, 23 in China, and less than 3 in Mozambique. Among people living in industrialized countries, Americans pay the lowest rate for water ($2.49 a gallon), whereas Danes and Germans pay the most (almost $9.00 a gallon). In the developing world, people typically pay five times as much as Europeans (United Nations Development Programme 2006; Lavelle 2007). Thus, in many countries, clean water is a luxury rather than a basic human right.

In much of India, residents don't know when the next water delivery will arrive. They spend days waiting for, and often fighting over, the shipments.

Ruth Fremson/The New York Times/Redux Pictures

## Sociology in Your Life

To find information on water and air quality, legislation, and cleanup initiatives in your state, go to CengageBrain.com to find the latest updates from the EPA.

# How Much Water Does It Take to Make. . .

Both U.S. droughts and population growth have increased. In 1950, Americans used 150 billion gallons of water every day, compared with 400 billion gallons today (Walsh 2011). It takes water to make everything. For example, it takes:

**2,600 gallons to make a pair of blue jeans**

Jacob Kearns/Shutterstock.com

**713 gallons to make a cotton shirt**

Stocksnapper/Shutterstock.com

Christopher Elwell/Shutterstock.com

**634 gallons to make an average hamburger**

magicover/Shutterstock.com

**53 gallons to produce a cup of to-go latte coffee**

**1.5 gallons to produce an average 18 ounces of bottled water**

Evgeny Karandaev/Shutterstock.com

Sources: Based on Connell 2011; Postel 2012.

## Threats to Water Supplies

Precipitation (in the form of rain, snow, sleet, or hail) is the main source of water for the ecosystem. Clean water has been depleted for many reasons, including pollution, privatization, and mismanagement.

*Pollution.* Toxins from cities, factories, and farms are spoiling U.S. freshwater supplies. Every year, more than 860 billion gallons of sewage, pesticides, fertilizers, automotive chemicals, and trash enter the country's rivers (Gurwitt 2005).

Since 2004, the 1972 Clean Water Act has been violated more than 500,000 times by more than 23,000 companies. About 60 percent of the polluters have dumped chemicals that can contribute to mental retardation and cancer. Fewer than 3 percent of the violations resulted in fines or other punishment by state officials (Duhigg 2009a).

The 1974 Safe Water Drinking Act requires communities to deliver safe tap water to local residents. Since 2004, however, the water provided to more than 49 million Americans has contained illegal concentrations of chemicals (e.g., arsenic) or radioactive substances (e.g., uranium). Fewer than 6 percent of the violators were ever fined or punished by state or federal officials (Duhigg 2009b).

*Privatization.* Water is a big business because of *privatization*, transferring some or all of the assets or operations of public systems into private hands. Perrier, Evian, Coca-Cola, PepsiCo—and particularly the French giants Vivendi and Suez—have been buying the rights to extract water in the United States and other countries at will from aquifers (underground layers of rock that hold water), then bottling and selling it around the world.

About 70 percent of Americans say they drink bottled water, almost 9 billion gallons in 2008 (Mui 2009). Bottling water is lucrative for corporations, but there are many environmental drawbacks. For example:

■ It depletes local water supplies, whether the water comes from municipal sources (40 percent) or local springs (60 percent) (Velasquez-Manoff 2009).

■ About 86 percent of the empty plastic bottles in the United States clog landfills instead of being recycled (Food & Water Watch 2007).

Damon Winter/The New York Times/Redux Pictures

Many of this 7-year-old boy's teeth are capped. Dentists say that pollutants in drinking water have damaged many Americans' teeth. The boy's mother, an accountant at a large state bank, asks "How can we get digital cable and Internet in our homes, but not clean water?" (Duhigg 2009a: A1).

*Mismanagement.* Most water pollution problems are due to human mismanagement, not nature. In China and India, for example, many government officials support economic growth and rarely punish local or international polluters who dump chemicals and waste into rivers and lakes (Carmichael 2007; Ford 2007).

Of all available water worldwide, agriculture consumes about 70 percent, industry uses 20 percent, and 10 percent is residential (World Water Assessment Programme 2009). In agriculture, many irrigation systems are inefficient, farmers often grow water-hungry crops such as cotton and sugarcane in arid areas, and pesticide and chemical fertilizer runoff from fields pollutes streams, rivers, and lakes. Some Americans are trying to conserve water, but they're probably a minority. For example, nearly 75 percent of residential water use in California—which has experienced numerous water emergencies—goes to outdoor purposes, mostly landscaping (Goodale 2009; Clayton 2010).

In the United States, a significant water pipe bursts, on average, every 2 minutes somewhere in the country (Duhigg 2010). The Environmental Protection Agency (EPA) estimates that it would cost from $17 billion to $23 billion per year for the next 20 years to replace the country's substandard pipes, especially in cities along the eastern seaboard. Most of the pipes in those cities are nearly 200 years old, and some are made of wood. A water pipe that leaks or bursts wastes huge amounts of drinking water, damages streets and homes, and seeps dangerous pollutants into drinking water (American Rivers 2005; Lavelle 2007). Some environmentalists wonder why the federal government spent more than $700 billion to bail out bankers (see Chapter 11) but spends less than $2 billion a year to replace aging water and sewage pipes that affect millions of Americans.

## 15-3b Air Pollution and Global Warming

Global warming is a serious environmental problem. Let's begin by looking at air pollution, the major cause of global warming.

### Air Pollution: Some Sources and Causes

More than 188 hazardous air pollutants can have negative effects on human health or the environment ("Hazardous Air Pollution . . ." 2005). There are many reasons for air pollution, but four are among the most common. First, a major source of air pollution is the burning of *fossil fuels*, substances obtained from the earth, including coal, petroleum, and natural gas. The exhaust gases of cars, trucks, and buses contain poisons—sulfur dioxide, nitrogen oxide, carbon dioxide ($CO_2$), and carbon monoxide. Power plants that produce electricity by burning coal or oil also spew pollutants.

Second, manufacturing plants that produce consumer goods pour pollutants into the air. Formaldehyde-based vapors that can lead to cancer and respiratory problems are emitted by many household and personal care products: pressed wood (often used in furniture), permanent-press clothes, plastic grocery bags, waxed paper, latex paints, detergents, nail polish, cosmetics, shampoos, and hair conditioners ("Formaldehyde" 2004).

Third, winds blow contaminants in the air across borders and oceans. For example, air pollution from Asia, especially China, has affected the air quality in the Sequoia and Kings Canyon national parks in California. Air pollution originating in Europe has been tracked to Asia, the Arctic, and the United States (Spotts 2004; Lamb 2009). Many of the smog-producing gases that come from Asia have also increased pollution and life-threatening health problems, such as lung and heart disease, in rural areas in the western United States where there is little industry or automobile traffic (Cooper et al. 2010).

Finally, government policies that affect air pollution vary from one administration to another. Between 2002 and 2006, for example, Justice Department lawsuits against polluters declined by 70 percent, and criminal and civil fines for polluting decreased by more than half, compared with the period from 1996 to 2000. The Bush administration blocked the efforts of 18 states to cut emissions from cars and trucks because it believed that tougher regulations would hurt the U.S. economy (Environmental Integrity Project 2007; Pelton 2007; "No Action on Greenhouse Gases" 2008). The Obama administration promised to decrease air pollution, but has "caved in" to influential petroleum, chemical, utility, and coal company lobbyists and affluent potential campaign contributors (Banerjee 2011: 8; Lerner and Bitetti 2011; Hertsgaard 2012).

© iStockphoto.com/Günay Mutlu

## Global Warming and the Greenhouse Effect

**Global warming** is an increase in the average temperature of Earth's atmosphere. The warming has resulted from numerous factors. Some are natural, such as changes in solar radiation, the Earth's orbit, and the frequency or intensity of volcanic activity. Most of the factors underlying global warming, however, are due to people, not nature.

Global warming begins with the **greenhouse effect**, the heating of Earth's atmosphere because of the presence of certain atmospheric gases. When heat from the sun enters the atmosphere, some of it is absorbed by Earth's surface, and some of it is reflected back to space. The presence of greenhouse gases in the atmosphere traps some of this heat. Heat is necessary to support life, but when greenhouse gases increase in the atmosphere, Earth becomes warmer than it would be otherwise, endangering public health and the welfare of current and future generations (Fagan 2008; U.S. Environmental Protection Agency 2009).

Air pollutants, especially $CO_2$, ignite the greenhouse effect. Every year, humans release at least 1 billion tons of $CO_2$ into the atmosphere, primarily from coal-fired power plants. In mid-2007, the United States, which has only 5 percent of the world's population, was responsible for 25 percent of all $CO_2$ emissions. Since then, China's $CO_2$ emissions have exceeded those of the United States by 8 percent (Netherlands Environmental Assessment Agency 2007).

In the United States, agriculture accounts for almost 14 percent of global greenhouse gas emissions. About 80 percent of all greenhouse gases come from industrial coal-fired power plants such as Southern Company and its subsidiaries in Georgia, and plants in Alabama, Indiana, Ohio, and West Virginia owned by America Electric Power. At least 10 other states also have high-polluting power plants (Food and Agriculture Organization of the United Nations 2011; Cappiello 2012).

## Some Effects of Climate Change

**Climate change** is a change in overall temperatures and weather conditions over time. Global warming probably began thousands of years ago when "humans began changing the planet [by] clearing land to grow food by cutting down and burning forests" (Fischman 2009: B11). Such *deforestation*, clearing massive amounts of forests, affects global climate changes because forests recycle carbon dioxide into oxygen.

Regardless of when climate changes began, humans have greatly accelerated global warming. Since such record keeping began in 1850, the years 2001 to 2012 have been the warmest. Scientists predict that extreme weather, especially heat waves, will produce heavier rainfall in some regions, more floods, stronger hurricanes and tornadoes, landslides, wildfires, and more intense droughts around the world (Intergovernmental Panel on Climate Change 2011; NOAA National Climactic Data Center 2011; Rice 2012).

Changes have already occurred because of global warming and climate change. For example:

- Higher $CO_2$ levels are turning oceans more acidic, resulting in some shellfish (e.g., clams and crabs) dying or not developing (Spotts 2009).

- In Syria, more than 800,000 people have lost their livelihoods and had to abandon their homes because of a 4-year dry spell, increased by climate change, and can't afford rising food prices (Akkad 2009).

- The current Arctic sea ice decline has been unprecedented since 1,450 years ago (Kinnard et al. 2011). On Alaska's coasts, ice shelves that acted as shields against storms and tidal forces are melting. Waters are advancing 80 feet a year because of coastal erosion, and more than 180 Alaskan villages are in danger of being engulfed unless they relocate (Tizon 2008).

> **global warming** the increase in the average temperature of Earth's atmosphere.
>
> **greenhouse effect** the heating of Earth's atmosphere because of the presence of certain atmospheric gases.
>
> **climate change** a change in overall temperatures and weather conditions over time.

Photo courtesy of American Hydrotech, Inc.

> In 2007, the prestigious American Institute of Architects chose a Seattle branch of a public library as one of its "Top Ten Green Projects." The library has 4 inches of soil planted with native grasses and succulent ground covers designed to absorb and filter rainwater, remove $CO_2$ from the atmosphere, and insulate the building in both summer and winter. The roof also lasts longer than a traditional hard roof (Adler 2007).

**sustainable development** economic activities that meet the needs of the present without threatening the environmental legacy of future generations.

Such data show that our planet is ailing. Nonetheless, people can change the situation, in the United States and globally, through better environmental policies and practices. One promising practice is sustainable development.

## 15-3c Is Sustainable Development Possible?

**Sustainable development** refers to economic activities that meet the needs of the present without threatening the environmental legacy of future generations. Is there an inherent contradiction between "sustainable" and "development"? There are reasons to be both pessimistic and optimistic about achieving sustainable development.

### Reasons to Be Pessimistic

Which country is the greenest? A recent study of environmental performance ranked 149 countries on factors such as maintaining and improving air and water quality, cooperating with other countries on environmental problems, overfishing, and emitting greenhouse gases. Switzerland, Sweden, Norway, Finland, and Costa Rica were the top five. The United States ranked 39th, well below a number of developing countries, including Slovakia and Albania (Esty et al. 2008).

Why, compared with other industrialized and even some developing countries, does the United States rank so low in environmental performance? There are many reasons, but two are especially important. First, lawmakers often accommodate business. For example, oil and utilities industries have been successful in pressuring Congress to drop laws that would have required them to develop renewable energy sources (such as wind and solar) and to pay $13 billion more in taxes for using only fossil fuels (Broder 2007; see also Chapter 11 on the close ties between the government and corporations).

Second, environmental issues are a low priority for some administrations. Since 2000, for example, the EPA hasn't enforced many environmental laws because of funding cuts, which resulted in the agency's having fewer staff members to investigate polluters and to apply penalties (Stephenson 2007). Some environmentalists are especially dismayed by *greenwashers*, companies and other organizations that pollute the planet while presenting an environmentally responsible public image (Elgin 2008).

Oliver Hoffmann/shutterstock.com

Most Americans want fluffy and ultra-plush toilet paper. Environmentalists complain that such products are a menace because they're made by chopping down and grinding up the pulp of trees that are decades or even a century old, a threat to the world's trees. Big U.S. toilet-paper makers say that they've introduced "Earth-friendly" toilet paper, but "customers are unwavering in their desire for the softest paper possible" (Fahrenthold 2009: A1).

A third reason, and especially important, is that many Americans aren't willing to sacrifice for a cleaner environment. Recent surveys show that Americans today are no more environmentally friendly than they were in 2000. For example, 85 percent say that they recycle paper, glass, aluminum, or other items, but national data show that only 34 percent really do so (EPA 2012). Sales for hybrid cars have declined by 17 percent in 2011 (Ingraham 2012). And, as *Table 15.4* shows, since 1999 Americans' worries about global warming have waned and they see economic growth as more important than environmental protection.

### TABLE 15.4

## Percentage of Americans Who. . .

|  | WORRY ABOUT GLOBAL WARMING | BELIEVE THAT ECONOMIC GROWTH IS MORE IMPORTANT THAN PROTECTING THE ENVIRONMENT |
|---|---|---|
| 1999 | 68% | 28% |
| 2007 | 65% | 37% |
| 2011 | 51% | 49% |

Note: The numbers may not sum to 100 percent because of no responses and rounding.

Sources: Based on Jacobe 2012, and Newport 2012.

Many people are contributing to sustainable development. "Urban farmers" are using rooftops, backyards, and public patches of trash-strewn land to produce food for themselves, neighbors, and the community (left). In Vermont, a dairy farmer sells "poop power" by converting cow manure to electricity (right). Consumers pay about 4 percent more for the electricity, but the manure is abundant and endlessly renewable, and its conversion to power decreases air pollution and practically eliminates the smell of cow dung in rural areas (Moore 2006; Nordin 2011).

Americans say that they want cleaner energy, but most oppose wind farms, for example, because the 40-foot windmills would "disturb our views of the landscape" (Dunlap 2007). Thus, are Americans eco-friendly only when it's convenient?

### Reasons to Be Optimistic

There has been progress since 1970, when the United States celebrated its first Earth Day. Most cars no longer burn leaded gasoline, ozone-destroying chlorofluorocarbons (CFCs) have been generally phased out, and total emissions of the six major air pollutants declined by 54 percent during the same period as the U.S. population increased by 47 percent (Hayward and Kaleita 2007; Sperry 2008).

Since 2005, many companies have found that being green is good for their profits and image, as well as the environment. For example, Johnson & Johnson, the world's largest manufacturer of health care products, now relies on renewable energy sources, such as wind and solar, for 30 percent of its electricity (Cohn 2007). Method Products, recognized as a groundbreaking sustainable home care company, focuses on ridding homes of toxic cleaning chemicals (Kelley 2011).

Current efficiency standards for appliances, lighting, and other equipment that were implemented during the early 1980s will save the United States the equivalent of two years of energy use ($1.1 trillion by

## Are Americans eco-friendly only when it's convenient?

2035) and slash greenhouse gas emissions (Lowenberger et al. 2012; see also Fischer et al. 2012). And, unlike the federal government, local governments in southern California and some Texas cities are treating and recycling wastewater (Stepney 2011; Barringer 2012; National Academy of Sciences 2012).

In other cases, progress has been mixed. For example, Walmart says that it now reuses or recycles 80 percent of the waste in domestic stores, and hopes to achieve its goal of zero waste. Some environmentalists contend, however, that Walmart's manufacturing low-quality products increases landfills and accelerates consumer consumption (Clifford 2012).

### Study Tools

Ready to study? Go to **Sociology CourseMate** at **www.cengagebrain.com** to complete practice quizzes, review flashcards, watch videos, and more.

# SOC

Every society experiences
social change.

AP Photos/Ted S. Warren

# Social Change: Collective Behavior, Social Movements, and Technology

## Key Topics

In this chapter, we'll explore the following topics:

16-1 Collective Behavior
16-2 Social Movements } connection
16-3 Technology and Social Change

For the most part, as you've seen throughout this textbook, sociologists examine behavior and social processes that are relatively institutionalized, routine, stable, and highly predictable. This chapter examines **social change**, the transformations of societies and social institutions over time. Some collective behavior is short-lived with few long-term changes; others, particularly social movements, can have lasting effects. Technology has also played a critical role in sparking social change. Let's begin with collective behavior.

> **social change** the transformations of societies and social institutions over time.
>
> **collective behavior** the spontaneous and unstructured behavior of a large number of people.

## What do you think?

I sometimes feel overwhelmed in keeping up with texting, e-mail, and Facebook.

| 1 | 2 | 3 | 4 | 5 | 6 | 7 |
|---|---|---|---|---|---|---|
| strongly agree | | | | | | strongly disagree |

## 16-1 Collective Behavior

Do you have several boxes crammed with collectibles such as Barbie dolls or baseball cards? A tattoo? Have you ever signed a petition? Joined a club? Posted on Facebook or sent text messages? If so, you've engaged in collective behavior.

### 16-1a What Is Collective Behavior?

**Collective behavior** is the spontaneous and unstructured behavior of a large number of people. Collective behavior encompasses a wide range of actions, including riots, fads, fashion, panic, rumors, and responses to disasters.

Sociologists emphasize two important characteristics of collective behavior. First, it is an act rather than a state of mind. For example, you may *feel* panic when a tornado threatens your town, but you don't engage in collective behavior until you actually *leave* your home and head for a safer location.

Second, collective behavior varies in its degree of spontaneity and structure. Panic, the least structured form of collective behavior, is typically short-lived. Fads—like collecting baseball cards—are more structured. They may last several years, and many people carefully wrap and store their collectibles hoping to make money in the future. Other forms of collective behavior—such as pro- and anti-abortion groups—become highly institutionalized social movements that include a staff, budget, and lobbying.

## 16-1b When Does Collective Behavior Occur?

In 2004, Cindy Sheehan's 24-year-old son and six other soldiers were killed in Sadr City, almost a year after President Bush had declared the end of major combat operations in Iraq. Sheehan pitched a tent near the president's ranch in Crawford, Texas, protesting the continuing war and demanding that the president meet with her face-to-face, but he refused to do so.

Sheehan's stance elicited widespread national and international coverage. Gold Star Families for Peace, a coalition of military families whose relatives had died in the war, aired a television ad in which Sheehan accused the president of dishonesty about the weapons of mass destruction: "You lied to us and because of your lies, my son died" (Madigan 2005: 1D). Some situations, such as Sheehan's, are more likely to encourage collective behavior than others. Why?

In 2005, groups such as the Gold Star Families for Peace and the Vietnam Veterans for Peace supported Cindy Sheehan's protest of the Iraq war by pitching tents in front of President George W. Bush's ranch in Crawford, Texas.

AP Photos/J. Scott Applewhite

### Structural Strain Theory

According to sociologist Neil Smelser (1962), six macro-level conditions encourage or discourage collective behavior. These conditions are "value-added" in the sense that each condition leads to the next one, ending in an episode of collective behavior.

1. **Structural conduciveness.** Structural conduciveness refers to the social conditions that allow a particular kind of collective behavior to occur. When channels for expressing a grievance either aren't available or fail, like-minded people may resort to protests. In Sheehan's case, the families who lost members in the Iraq war supported Sheehan because they felt that the Bush administration was insensitive to their personal losses.

2. **Structural strain.** Structural strain occurs when an important aspect of a social system is seen as discriminatory or unjust, creates problems, or interferes with people's everyday lives. Many of the families that sided with Sheehan did so because they believed that the administration, by not moving to end the war in Iraq, increased the number of U.S. casualties.

3. **Growth and spread of a generalized belief.** People begin to see a situation as a widespread problem instead of just a personal experience. With the generalized belief that there is a problem comes a general recognition that something should be done about it. In the Sheehan case, many mothers, especially, supported Sheehan and became the most vocal leaders of a number of antiwar protests because they felt that their children were dying "for nothing" (Bumiller 2005).

4. **Precipitating factors.** Some incident or behavior triggers an event and inspires action. Sheehan's pitching a tent outside of President Bush's ranch spurred many other antiwar advocates to support her accusations on radio talk shows and in letters to the editor.

5. **Mobilizing people for action.** Mobilization often requires leaders who encourage agitation to change the status quo. Sheehan was the leader in agitating for peace in Iraq, but antiwar groups supported her efforts through television ads.

6. **Social control.** In this stage, opposing groups may try to prevent, interrupt, or repress those advocating social change. Government officials, the police, community and business leaders, courts, the mass media, and other social control agents—all of whom benefit from the status quo—may quash, ridicule, or challenge the emerging collective behavior. In Sheehan's case, the administration and right-wing political commentators on television and radio dismissed her as a "crackpot" and described her protests as "treasonous" (Madigan 2005).

## Critical Evaluation

Smelser's model has contributed to an understanding of the emergence and development of collective behavior by helping to predict when and where episodes of such behavior might break out. Structural strain theory also offers insights on why, at every stage, collective behavior may either fade or escalate (Locher 2002). If, for instance, the president had met with Sheehan to discuss the general anxiety surrounding the war (reducing structural strain), the initial criticisms of his administration would probably have died down (halting the spread of a generalized belief), and the antiwar groups would not have run the television ads (limiting the mobilization of participants).

Second, the model doesn't explain all forms of collective behavior. In the case of fads and rumors, as you'll see shortly, all six stages don't necessarily occur. Third, the sequence of the stages isn't necessarily the same as Smelser outlined (Berk 1974).

Fourth, the determinants don't always spark collective behavior. For example, many groups—such as college students who complain about the price of textbooks and tuition costs—experience structural strain and are free to protest and mobilize, but rarely engage in collective behavior to change a situation that they complain about.

## 16-1c Varieties of Collective Behavior

There are many types of collective behavior; some are more fleeting or harmful than others (Turner and Killian 1987). Let's begin with rumors, one of the most common types.

### Rumors

There were widespread rumors that on January 1, 2000, a glitch (Y2K) in operating systems would cause computers around the world to crash, leading to global power outages, banks losing all of their customers' statements, and even airplanes falling from the skies. None of this occurred.

A **rumor** is unfounded information that people spread quickly. Through modern communication technology, a rumor can spread to millions of people over the Internet within seconds, especially when the subject line says something like "THIS IS REALLY TRUE!" Rumors can incite riots, panic, or widespread anxiety. Because of the Y2K rumor, thousands of people built underground shelters, and millions of others stocked up on bottled water, canned food, batteries, and medical supplies.

Most rumors (that rock stars Elvis Presley and John Lennon are alive, for example) are harmless. Others can wreak considerable damage. After a woman claimed that she found part of a human finger in her cup of Wendy's beef chili, the restaurant's business dropped by half nationally and rumors warning people to stop eating fast food altogether spread over the Internet (Richtel and Barrionuevo 2005). Ultimately, the woman admitted that she had planted the finger to try to get a lucrative settlement. However, some customers are still leery about eating at fast-food restaurants.

Rumors are typically false, so why do so many people believe them? First, rumors often deal with an important subject about which—especially during uncertain economic times or natural disasters—people are anxious, insecure, or stressed. This makes people especially suggestible. In Hurricane Katrina's aftermath, for example, the media reported numerous rumors of carjacking, murders, thefts, and rapes, the overwhelming majority of which subsequently proved to be false.

Second, there is often little factual information to counter a rumor, or people distrust the sources of such information. During Y2K, the people who stockpiled groceries and so forth didn't believe computer scientists or federal officials who said that there would not be a major calamity. Third, rumors offer entertainment, diversion, and drama in otherwise mundane daily lives (Turner and Killian 1987; Marx and McAdam 1994; Campion-Vincent 2005; Heath 2005).

**rumor** unfounded information that people spread quickly.

During the healthcare reform debate, many opponents spread the rumor that the new bill would result in "death panels" that would deny care to older people and children born with birth defects. Neither was true (Snow et al. 2009).

AP Photo/Coeur d'Alene Press, Jerome A. Pollos

## Gossip

**Gossip** is the act of spreading rumors, often negative, about other people's personal lives. Someone once said that "no one gossips about other people's virtues." Because of its tendency to be derogatory, gossip makes us feel superior ("Did you know that Margie just had breast implants? Isn't she pathetic?"). Gossip is also interesting, entertaining, and exposes hypocrisy (Epstein 2011).

Sometimes, people gossip to control other people's behavior and to reinforce a community's moral standards. For example, comments about someone's drug abuse or marital infidelity reinforce notions of what's deviant or unacceptable (see Chapter 4). In other cases, people gossip because they resent or envy someone's success or accomplishments, or because the gossip reinforces what people think they already know (Sunstein 2009).

## Urban Legends

Another form of rumor is **urban legends** (also called *contemporary legends* and *modern legends*), stories— funny, horrifying, or just odd—that supposedly happened somewhere. Some of the most common and enduring urban legends, but with updated variations, deal with food contamination, such as the finger at Wendy's. Others have targeted politicians. We still hear, for example, false allegations that George W. Bush had the lowest IQ of any president in the past 50 years and that President Obama is a radical Muslim.

Why do urban legends persist much longer than gossip? First, they reflect contemporary anxieties and fears—about contaminated food, unscrupulous companies, and corrupt and unresponsive governments. Second, urban legends are cautionary tales that warn us to watch out in a dangerous world. For example, there have been tales that sunscreens cause blindness, and that women have died sniffing perfume samples sent to them in the mail. Third, we tend to believe urban legends because we hear them from people we trust—family members, coworkers, and friends. Finally, urban legends—such as the one about alligators living in New York City's sewer system—are fun to tell and "too beguiling to fade away" (Kapferer 1992; Brunvand 2001; Ellis 2005).

## Panic and Mass Hysteria

In 2003, an indoor fireworks display meant to kick off a heavy metal concert in West Warwick, Rhode Island, set off a fire that killed 100 people and injured 200 others. Panic ensued as thick black smoke poured through the audience and hundreds of patrons stampeded for the front door (even though there were three other exits), trampling and crushing those who had fallen beneath them.

Most of the deaths in West Warwick were due not to the fire but to **panic**—a collective flight, typically irrational, from a real or perceived danger. The danger seems so overwhelming that people desperately jam an escape route, jump from high buildings, leap from a sinking ship, or sell off their stock. Fear drives panic: "Each person's concern is with his [or her] own safety and personal security, whether the danger is physical, psychological, social, or financial" (Lang and Lang 1961: 83).

## Sociology in Your Life

For a taste of some popular urban legends still circulating today, check out CengageBrain.com.

© istockphoto.com/Kim Sohee;© istockphoto.com/Robert Blanchard

# BETTY CROCKER MAKEOVER

**1936**     **1955**     **1965**     **1968**

**1972**     **1980**     **1986**     **1996**

AP Photos/General Mills

In 1921, General Mills created Betty Crocker, a fictitious woman, to answer thousands of questions about baking that came in from consumers every year. As fashions and hairstyles changed, so did Betty Crocker's image. The original image of a stern, gray-haired older woman has morphed over the years so that, by 1996, she had a darker complexion and was dressed in casual attire. Can you think of other brands that have changed their image over the years to keep up with changing trends?

**mass hysteria** an intense, fearful, and anxious reaction to a real or imagined threat by large numbers of people.

**fashion** a standard of appearance that enjoys widespread but temporary acceptance within a society.

Panic can result in hundreds or thousands of casualties, but it's a relatively rare type of collective behavior (Smelser 1962). Most people try to rescue loved ones rather than fleeing a dangerous situation (such as a flood or hurricane), and many leave a life-threatening situation in an orderly fashion (as in the World Trade Center on 9/11).

Panic is similar to **mass hysteria**—an intense, fearful, and anxious reaction to a real or imagined threat by large numbers of people. An example is the scare over the H1N1 flu ("swine flu"). In mid-2009, the World Health Organization issued an alert that H1N1 would become a pandemic (worldwide epidemic) and result in numerous deaths. Millions of Americans stood in line for hours, sometimes overnight, because the vaccine was initially in short supply, but the swine flu never materialized on the scale that was predicted (Witters 2010). Unlike panic, which usually subsides quickly, mass hysteria may last longer because warnings—especially about health and food—reinforce our general fears and anxieties about life's dangers.

## Fashions

A **fashion** is a standard of appearance that enjoys widespread but temporary acceptance within a society. Whether they last for years or change after a few months, fashions are highly institutionalized products and styles that are popular among a large number of people (Smelser 1962; Blumer 1969). For example, black women's hairstyles have changed—from Afros in the late 1960s, to straightened hair during the 1980s, to braids, cornrows, dreadlocks, hair extensions, and coloring more recently. All reflect gender politics, racial solidarity, generational differences, identity, and changing images of beauty (Banks 2000; Desmond-Harris 2009).

Fashion also involves periodic changes in the popularity of clothes, furniture, music, language usage, books, automobiles, sports, recreational activities, the names parents give their children, and even the dogs that people own. Teenagers who want to be fashionable buy clothes with prominent labels, but what's fashionable changes from year to year.

Why do fashions, especially in clothes, change fairly quickly? One reason is that designers, manufacturers, and retailers must continuously create demand for new fashions to maintain their profits. Second, many people keep up with fashion because they don't want to seem different or they fear being perceived as out-of-date or dowdy. Third, shopping for new clothes and other products decreases the boredom of everyday routine. Also, clothes and other products are status symbols that signal being an insider: "Others . . . will admire me for

© Rachel Weill/FoodPix/Jupiterimages

**fad** a form of collective behavior that spreads rapidly and enthusiastically but lasts only a short time.

**craze** a fad that becomes an all-consuming passion for many people for a short time.

**disaster** an unexpected event that causes widespread damage, destruction, distress, and loss.

being the kind of person who makes stylish choices" (Best 2006: 85–86; for classic analyses of fashion and collective behavior, see Veblen 1899/1953; Barber and Lobel 1952; Packard 1959; and Bourdieu 1984).

## Fads

A **fad** is a form of collective behavior that spreads rapidly and enthusiastically but lasts only a short time (Turner and Killian 1987; Lofland 1993). Fads that have arisen and faded fairly quickly include *products* (such as bean bag chairs and Crocs), *activities* (such as disco dancing and a variety of diets), and widespread enthusiasm for *popular personalities and television characters* (such as the Lone Ranger during the 1950s, and more recently, the Kardashians).

Why do fads sprout? A major reason is profit. Because children and adolescents are especially likely to adopt fads, manufacturers create numerous products and activities that they hope will catch on. These products include toys, sportswear, and new cereals. The hottest fads are usually the must-have Christmas toys that children plead for every year. Parents may stand in line for hours, drive to nearby cities, and scour the Internet to get the product. A few months later, the toy may be thrown away or stuffed in the back of a closet.

Some people dismiss fads as "ridiculous" or "silly," but they serve several functions. In a mass society, where people often feel anonymous, a fad can develop strong in-group feelings and a sense of belonging, especially among people who share similar interests and attitudes. Fads can also be fun, promise to resolve a nagging problem (such as being overweight), and help us keep up with technological changes (Marx and McAdam 1994; Best 2006).

Most fads are soon forgotten, but some become established. For example, Pez candy dispensers, which originated in 1952 and cost 49 cents, are still inexpensive (about $1.49), are sold in more than 60 countries, and have been continuously updated to include popular television characters such as the Simpsons (Paul 2002). Other fads, such as streaking (running around nude in public places), reemerge from time to time ("Streaking" 2005).

*Crazes involving ways to make easy money are most common in capitalistic societies.*

## Crazes

Some fads are **crazes**, forms of collective behavior that become all-consuming passions for a short time. The Beanie Babies fad of the late 1990s turned into a craze. After an entrepreneur published a highly successful magazine on these stuffed toys, there was a mad rush to buy them. The creator limited the sales in many stores to generate demand and retired earlier models to make them more sought after by collectors. These strategies worked, and the entrepreneur became one of the richest people on Earth. What about the collectors? According to an owner of a large toy store, "[Beanie Babies] make great insulation if you stick them in the walls" (Mulligan 2004: 3C).

Crazes involving ways to make easy money are most common in capitalistic societies where many people want to make as much money as possible as quickly as possible. The problem with economic crazes is that they may end quite suddenly and with very disappointing results, as we saw during the recent home mortgage boom and collapse. Ten years earlier, most of the fast-growing dot-com businesses collapsed after 2000. Investors' portfolios fell by as much as 50 percent, thousands of workers were laid off, and almost all of the "hot" Internet sites disappeared nearly overnight (van Ginneken 2003; McCarthy 2004).

## Disasters

Whereas people choose to participate in fashion, fads, and crazes, a **disaster** is an unexpected event that causes widespread damage, destruction, distress, and loss. Some disasters are due to *social causes*, such as war, genocide, terrorist attacks, and civil strife. Some are due to *technological causes*, including oil spills, nuclear accidents, burst dams, building collapses, and mine explosions. Others are the result of *natural causes*, such as fires, floods, landslides, earthquakes, hurricanes, tsunamis, and volcanic eruptions (Marx and McAdam 1994).

Disasters often inspire organized behavior rather than chaos. Instead of panicking, most people are rational, cooperative, and altruistic. They often care for family members instead of fleeing, for example, and thousands of volunteers offer financial, medical, and other help.

## Publics, Public Opinion, and Propaganda

A **public** is a collection of people, not necessarily in direct contact with each other, who are interested in a particular issue. A public is not the same as the general public, which consists of everyone in a society.

There are as many publics as there are issues—abortion, gun control, pollution, health care, and gay marriage, to name just a few. Even within one organization or institution, there may be several publics that are concerned about entirely different issues. At a college, for instance, students may be most concerned about the cost of tuition, faculty may spend much time discussing instructional technology, and maintenance employees may be most interested in wages.

In most cases, the interaction within a public is carried on indirectly through the mass media (newspapers, books, radio, television, and motion pictures), social media, blogs, or professional journals. Because publics aren't organized groups with memberships, they're often transitory. Publics expand or contract as people lose or develop interest in an issue. For example, a public may surge during a highly publicized and controversial incident, such as removing a patient's life support, but then evaporate quickly. In some cases, however, publics organize and become enduring social movements (a topic we'll examine shortly).

Some publics express themselves through **public opinion**, widespread attitudes on a particular issue. Public opinion has three components: (1) it is a verbalization rather than an action; (2) about a matter that is of concern to many people; and (3) involves a controversial issue (Turner and Killian 1987). Like publics, public opinions wax and wane over time. People's interest in crime and education, for example, decreases when they are more concerned about pressing events such as an economic crisis.

Public opinion can be swayed by **propaganda**, the presentation of information to influence people's opinions or actions. Propaganda isn't a type of collective behavior, but it affects collective behavior in several important ways. First, it may create attitudes that will inspire collective outbursts such as strikes or riots. Second, propaganda may be used to try to prevent collective outbursts, as when corporations try to convince employees that the loss of jobs is due to the economy rather than to offshoring (see Chapter 11). Third, propaganda is often used to try to gain adherents for a cause, whatever it might be (Smelser 1962).

Propaganda is institutionalized in advertising, political campaign literature, and government policies. It is conveyed to people in many ways: the mass media, social media, political speeches, religious groups, rumor, and symbols (such as flags). Much propaganda presents misinformation intended to sway an audience toward a particular viewpoint, but propaganda isn't necessarily good or bad. Instead, it's a means of influencing people.

**public** a collection of people, not necessarily in direct contact with each other, who are interested in a particular issue.

**public opinion** widespread attitudes on a particular issue.

**propaganda** the presentation of information to influence people's opinions or actions.

Propaganda tries to influence people's attitudes and behavior. One of the best-known examples of propaganda in the United States is this Uncle Sam poster, designed in 1917 and used to recruit soldiers for World Wars I and II. Opponents of the Vietnam War used it as an anti-war poster during the 1960s and 1970s.

## Crowds

Much collective behavior is scattered geographically, but crowds are concentrated in a limited physical space. A **crowd** is a temporary gathering of people who share a common interest or participate in a particular event. Regardless of size—whether it's a few dozen people or millions—crowds come together for a specific reason, such as a religious leader's death, a concert, or a riot.

Crowds differ in their motives, interests, and emotional level:

- A *casual crowd* is a loose collection of people who have little in common except for being in the same place at the same time and participating in a common activity or event. There is little if any interaction, the gathering is temporary, and there is little emotion. Examples include people watching a street performer, spectators at the scene of a fire, and shoppers at a busy mall.

- A *conventional crowd* is a group of people that assembles for a specific purpose and follows established norms. Unlike casual crowds, conventional crowds are structured; their members may interact, and they conform to rules that are appropriate for the situation. Examples include people attending religious services, funerals, graduation ceremonies, and parades.

- An *expressive crowd* is a group of people who exhibit strong emotions toward some object or event. The feelings—which can range from joy to grief—pour out freely as the crowd reacts to a stimulus. Examples include attendees at religious revivals, revelers during Mardi Gras, and enthusiastic fans at a football game.

- An *acting crowd* is a group of people who have intense emotions and a single-minded purpose. The event may be planned, but acting crowds can also be spontaneous demonstrations or other focused group behavior. Examples of acting crowds include people fleeing a burning building, soccer fans storming a field, and college students having a water balloon fight.

- A *protest crowd* is a group of people who assemble in public to achieve a specific goal. Protest crowds demonstrate their support of or opposition to an idea or event. Most demonstrations—such as antiwar protests, civil rights marches, boycotts, and labor strikes—are usually peaceful. Peaceful protesters can become aggressive, however, resulting in destruction and violence (Blumer 1946; McPhail and Wohlstein 1983; Pell 2011).

One type of crowd can easily change into another. A conventional crowd at a nightclub can turn into an acting crowd if a fire causes people to panic and flee for safety. Any of the five types of crowds—from casual to protest—can become a mob or a riot.

## Mobs

A **mob** is a highly emotional and disorderly crowd that uses the threat of force, actual force, or violence against a specific target. The target can be a person, a group, or a property. In the Oakland neighborhood of Chicago, a van veered off the street and struck a group of people sitting on the stoop at home. Seven male spectators, ranging in age from 16 to 47, mobbed the van. They pulled the driver and passenger from the car, stomping and beating them to death with bricks and stones ("7 People Charged . . ." 2002).

Mobs often arise when people are demanding radical societal changes, such as the removal of corrupt government officials. In other cases, especially when authority breaks down, people who take advantage of a situation may engage in mob behavior. For example, after the earthquake in Port-au-Prince, Haiti, most of the city's 3 million survivors focused on clearing the streets of debris and pulling bodies out of the rubble. However, dozens of armed men—some wielding machetes, others with sharpened pieces of

In 2006 and 2010, hundreds of thousands of Latinos and their supporters participated in demonstrations and rallies in Washington, D.C. and other cities, calling on Congress to offer citizenship to illegal immigrants. Is this kind of collective behavior an example of an expressive, acting, or protest crowd?

REUTERS/Jason Reed /Landov

wood—"dodged from storefront to storefront, battering down doors and hauling away whatever they could carry. . ." (Romero and Lacey 2010: A1). After attacking, a mob tends to dissolve quickly.

### Riots

Compared with mobs, riots usually last longer. A **riot** is a violent crowd that directs its hostility at a wide and shifting range of targets. Unlike mobs, which usually have a specific target, rioters unpredictably attack whomever or whatever gets in their way during a rampage. Most riots arise out of long-standing anger, frustration, or dissatisfaction that may have smoldered for years or even decades. Some of these long-term tensions arise from discrimination, poverty, unemployment, economic deprivation, or other unaddressed grievances.

There are numerous protests in the United States every year, but race riots have been the most violent and destructive. More than 150 U.S. cities experienced race riots after the assassination of Martin Luther King Jr. in 1968. In 1992, riots broke out in 11 cities after the acquittal of four white police officers involved in the beating of Rodney King, a black motorist, in Los Angeles. The violence resulted in deaths and considerable property damage because of fires and looting.

Riots are usually expressions of deep-seated hostility, but this isn't always the case. Sports celebration riots, such as the one that followed the Los Angeles Lakers victory over the Boston Celtics in the 2010 Championship Series, occur because of extreme enthusiasm and excitement rather than anger or frustration: "Participants smash, trample, and knock things down to express their ecstasy. Celebration riots are an orgy of gleeful destruction" (Locher 2002: 95).

Much collective behavior, such as a mob or riot, is spontaneous and short-lived. Bringing about long-term social changes, especially at a macro level, requires collective behavior that is structured and enduring. Social movements are important vehicles for creating or suppressing societal changes.

## 16-2 Social Movements

There are hundreds of social movements in the United States alone. Why are they so widespread? And how important are they in changing society?

### 16-2a What Is a Social Movement?

A social movement is a large and organized activity to promote or resist a particular social change.

Examples of social movements include groups that focus on the rights of the disabled, crime victims, gun control, and drunk driving, to name just a few. "Social movements are as American as apple pie," notes sociologist Lynda Ann Ewen (1998: 81–82). "The abolition of slavery, women's right to vote, legal unions, open admissions to public colleges and student aid, and Head Start are all changes in our society that were won through social movements."

Unlike other forms of collective behavior, social movements are organized, goal-oriented, deliberate, structured, and can have a lasting impact on a society. And unlike many other forms of collective behavior (such as crowds, mobs, and riots), the people who make up a social movement are dispersed over time and space, and usually have little face-to-face interaction (Turner and Killian 1987; Lofland 1996).

Some U.S. social movements, such as those focusing on white supremacy, are relatively small. Others are large and have subgroups that appeal to different segments of the population. For example, the U.S. environmental movement has at least 50 subgroups, including Earth First!, Greenpeace, the National Audubon Society, the Union of Concerned Scientists, and the Wilderness Society.

### 16-2b Types of Social Movements

Sociologists generally classify social movements according to their goals (changing some aspect of society or resisting change) and the amount of change that they seek (limited or widespread) (Aberle 1982). As *Table 16.1* suggests, some social movements may be perceived as more threatening than others because they challenge the existing social order.

*Alternative social movements* focus on changing some people's attitudes or behavior in a specific way. These movements typically emphasize spirituality, self-improvement, or physical well-being. They are the least threatening to the status quo because they seek limited change and only for some people. For example, millions of non-Asian Americans, influenced by Asian religions, have embraced yoga, meditation, and healing practices such as acupuncture (Cadge and Bender 2004).

*Redemptive social movements* (also called *religious* or *expressive movements*) offer a dramatic change, but only in some people's lives. They are typically based on spiritual or supernatural beliefs, promising to renew

> **riot** a violent crowd that directs its hostility at a wide and shifting range of targets.
>
> **social movement** a large and organized activity to promote or resist a particular social change.

people from within and to guarantee some form of salvation or rebirth. Examples include any religious movements that actively seek converts, such as the Jehovah's Witnesses and certain Christian evangelical groups (see Chapter 13).

*Reformative social movements* want to change everyone, but only with regard to a particular topic or issue. These movements, the most common type in U.S. society, don't want to replace the existing economic, political, or social class arrangements, but to change society in some specific way. Examples include civil rights groups, gay rights activists, labor unions, and animal rights groups.

*Resistance social movements* (also called *reactionary movements*) try to preserve the status quo by blocking change or undoing change that has already occurred. Resistance movements are often called *countermovements* because they usually form immediately after an earlier movement has succeeded in creating change within a society. For example, anti-abortion groups that arose in the United States shortly after the Supreme Court decision in *Roe v. Wade* (1973), which legalized abortion, seek to reverse that decision.

*Revolutionary social movements* want to completely destroy the existing social order and replace it with a new one. These movements range from utopian groups that withdraw from society and try to create their own to radical terrorists who use violence and intimidation. Examples of the latter include militia groups in the United States that believe the federal government is evil and want to overthrow it. The French Revolution, the Communist Revolution in China, and Fidel Castro's socialist movement in Cuba all succeeded in replacing the existing social order with a new one.

## 16-2c Why Social Movements Emerge

A social movement is "an answer either to a threat or a hope" (Touraine 2002: 89). Yet not everyone who feels threatened or hopeful joins a social movement. Why not? Let's look at four explanations, beginning with the oldest.

## Mass Society Theory

Early on, sociologists believed that the people who formed social movements felt powerless, insignificant, and isolated in modern mass societies, which are impersonal, industrialized, and highly bureaucratized. Thus, according to *mass society theory*, social movements offer a sense of belonging to people who feel alienated and disconnected from others (Kornhauser 1959).

### Critical Evaluation

Mass society theory may explain why some people form extreme political movements, like Fascism and Nazism, but subsequent research has shown that movement organizers are typically not isolated but well-integrated into their families and communities. Also, historically, many political activists in the United States, such as those behind the civil rights and women's rights movements during the late 1960s, were not powerless but came from relatively privileged backgrounds (Davis 1991; McAdam and Paulsen 1994).

### Relative Deprivation Theory

Relative deprivation theory is broader than mass society theory. **Relative deprivation** is a gap between what people have and what they think they should have compared with others in a society. What matters is not what people actually have—whether it's money, social status, power, or privilege—but what they *think* they should have.

© istockphoto.com/Feng Yu

## TABLE 16.1

# Five Types of Social Movements

| MOVEMENT | GOAL | EXAMPLES |
|---|---|---|
| Alternative | Change some people in a specific way | Alcoholics Anonymous, transcendental meditation |
| Redemptive | Change some people, but completely | Jehovah's Witnesses, born-again Christians |
| Reformative | Change everyone, but in specific ways | Gay rights advocates, Mothers Against Drunk Driving (MADD) |
| Resistance | Preserve status quo by blocking or undoing change | Antiabortion groups, white supremacists |
| Revolutionary | Change everyone completely | Right-wing militia groups, Communism |

© Cengage Learning

Relative deprivation theorists note two other elements. First, people often feel that they *deserve* better than they have ("I've worked hard all my life"). Second, they believe that they *cannot attain their goals through conventional channels* ("I've written people in Congress, and they just ignore me"). Thus, shared beliefs combined with unfulfilled expectations can trigger change-oriented social movements, as witnessed by the gay rights and civil rights movements (Davies 1962, 1979; Morrison 1971).

## Critical Evaluation

Relative deprivation theory helps explain why some social movements emerge. Critics point out, however, that there is a certain amount of relative deprivation in all societies, but that people don't always react by forming social movements. Relative deprivation theory also doesn't explain why some people join movements even though they don't see themselves as deprived and don't expect to gain anything personally if the movement succeeds (Gurney and Tierney 1982; Johnson and Klandermans 1995; Orum 2001).

## Resource Mobilization Theory

It takes more than feeling alienated (mass society theory) or disadvantaged (relative deprivation theory) to sustain a social movement. Instead, according to *resource mobilization theory*, a social movement will succeed if it can put together (or mobilize) an organization and leadership dedicated to advancing its cause (Oberschall 1973, 1995; McCarthy and Zald 1977; Gamson 1990).

Other important resources include money, dedicated volunteers, paid staff, access to the media, effective communication systems, contacts, special technical or legal knowledge and skills, equipment, physical space, alliances with like-minded groups, and a positive public image. Many labor unions have fewer members than in the past, but those of teachers and police officers remain strong. Such unions have millions of members who pay dues, much of which goes to lobbyists who represent their interests in Congress (see Chapter 11). The unions are also well organized at state and national levels, and enjoy widespread support from the general public.

## Critical Evaluation

One of the major contributions of resource mobilization theory is its emphasis on structural factors (such as organization and leadership) in explaining why some social movements thrive whereas others disappear. However, a major criticism is that resource mobilization theory largely ignores the role of relative deprivation in the formation of a social movement. If there

aren't large numbers of dissatisfied people to initiate a movement, even plentiful resources will not be able to sustain it (Jenkins 1983; Klandermans 1984; Scott 1995; Buechler 2000).

## New Social Movements Theory

*New social movements theory*, which became prominent during the 1970s, emphasizes the linkages among culture, politics, and ideology. Unlike the earlier perspectives, new social movements theory proposes that many recent movements (such as those that work for peace and environmental protection) promote the rights and welfare of *all* people rather than specific groups in particular countries (Touraine 1981; Laraña et al. 1994; Melucci 1995). Thus, new social movements theory is especially interested in "the struggle to liberate the voices of the dispossessed" (Schehr 1997: 6).

For these theorists, recent social movements differ from older ones in two ways. First, they attract a disproportionate number of members who are well-educated and relatively affluent, represent a wide variety of professions (such as educators, scientists, actors,

The "Tea Party" is an emerging social movement in U.S. politics. It originated in early 2010, has an estimated 500 local groups, and has attracted more Republicans than Democrats. No single person leads the movement, so far, but its supporters are motivated by safeguarding individual liberty, cutting taxes, and ending bailouts for business while the American taxpayer gets burdened with more and more debt (Jonsson 2010; O'Hara 2010).

> ## If there aren't large numbers of dissatisfied people to initiate a movement, even plentiful resources will not be able to sustain it.

businesspeople, and political leaders), and share a broad goal—improving the quality of life for all people around the world. Second, recent social movements pursue goals or advance values that may be of no immediate personal benefit to the participants, such as eradicating measles or tuberculosis in developing countries (Obach 2004).

### Critical Evaluation

Unlike earlier perspectives, new social movements theory contributes to our understanding of collective behavior that crosses international boundaries. According to some critics, however, neither this perspective nor the groups that it examines are novel. For example, some social movements (like feminism and environmentalism) have been around for a long time and still focus on the same basic issues, such as women's second-class citizenship and population growth. In addition, some scholars point out that educated middle-class or wealthy activists were as common in the old social movements as in more recent ones (Rose 1997; Buechler 2000; Sutton 2000).

Critics also contend that new social movements theory often overstates people's altruistic motivations. For example, many people who join environmental groups do so for reasons referred to as NIMBY (not in my backyard). That is, they are concerned about some undesirable environmental condition in their own community but show little interest in environmental threats to people elsewhere (Obach 2004).

These four theories, despite their limitations, help us understand social movements because "no single theory is sufficient to explain the complexities of any social movement" (Blanchard 1994: 8). The next question is why some social movements thrive and others fail.

### 16-2d The Stages of Social Movements

Most social movements are short-lived. Some never really get off the ground; others meet their goals and disband. Social movements generally go through four stages: emergence, organization, institutionalization, and decline (see *Figure 16.1*) (King 1956; Mauss 1975; Spector and Kitsuse 1977; Tilly 1978).

### Emergence

During *emergence*, the first stage of a social movement, a number of people are upset about some condition and want to change it. One or more individuals, serving as agitators or prophets, emerge as leaders. They verbalize the feelings of the discontented, crystallize the issues, and push for action. If leaders don't get much support, the movement may die. If, on the other hand, the discontent resonates among growing numbers of people, public awareness increases and the movement attracts like-minded people. For example, many scholars attribute the rise of the women's movement in the 1960s to the popularity of *The Feminine Mystique*, a book in which Betty Friedan (1963), then a full-time housewife, questioned whether educated women should give up careers to be full-time homemakers.

### Organization

Once people's consciousness has been raised, the second stage is *organization*. The most active members form alliances, seek media coverage, develop strategies and tactics, recruit members, and acquire the necessary resources. A division of labor is established in which the leaders make policy decisions and the followers perform necessary tasks such as preparing mass mailings, developing websites, and responding to phone calls and emails. At this stage, the movement may develop chapters at local, regional, national, and international levels.

### Institutionalization

As a movement grows, it becomes *institutionalized* and more bureaucratic: The number of staff positions increases, members draw up governing by-laws, the organization may hire outsiders (such as writers, attorneys, and lobbyists) to handle some of the necessary tasks, and the charismatic leaders may spend more of their time on speaking tours, in media interviews, and at national or international meetings. As the social movement grows and becomes more bureaucratic and self-sufficient, the original leaders may move on to better paying and more influential positions in government or the private sector.

### Decline

All social movements end sooner or later. This *decline*, the last stage, may take a number of forms:

## FIGURE 16.1
## Typical Stages of a Social Movement

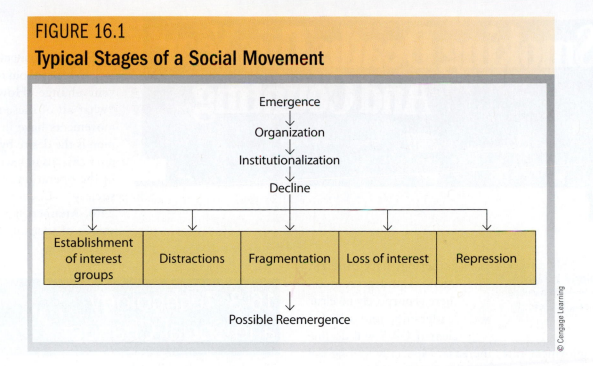

© Cengage Learning

- If a social movement is successful, it can become an *interest group* and a part of society's fabric. For example, a small antismoking movement that began in the mid-1970s now has the enthusiastic support of numerous prestigious organizations, such as the American Cancer Society, the American Heart Association, the American Medical Association, the World Health Organization, and governing bodies within and outside the United States (Wolfson 2001).

- Those involved in a social movement may become *distracted* because the group loses sight of its original goals, their enthusiasm diminishes, or both. For example, when Ralph Nader first criticized the automobile industry for car safety defects, he gained a large following. As Nader and his consumer rights groups expanded their focus to include environmental issues and corporate crimes, many of the initial followers lost interest.

- A social movement may experience *fragmentation* because the participants disagree about goals, strategies, or tactics. For example, the environmental movement has numerous groups that focus on different issues—air quality, marine life, land use, and global warming. Participants may also drift away from a movement because of time constraints, health problems, or similar reasons.

- Social movements may also decline as a result of *repression*. Many autocratic governments quash dissent. A government can crush an emerging social movement by arresting protestors and imprisoning or even executing leaders (see Chapter 11).

A social movement that declines can sometimes experience a later resurgence. For example, there have been several waves of the women's rights movement in the United States since the mid-nineteenth century. The first wave ensured women's right to vote in 1920; the second wave expanded employment and educational rights during the 1960s and 1970s; and the third wave, during the 1990s, focused on economic and other inequalities experienced by women of different social classes, sexual orientations, and nationalities (Kramer 2005).

## Is Occupy Wall Street a Social Movement?

In July, 2011 a few hundred demonstrators converged in the Wall Street financial district of New York City. The activists, who called themselves "Occupy Wall Street" (OWS), protested socioeconomic inequality, corporate greed, and corporate power over the U.S. government. The group's slogan, "We are the 99%," referred to the income inequality between the top 1 percent, who control over 43 percent of the income, and the rest of the population. By early October, similar demonstrations were held in 70 major U.S. cities and over 600 communities. Before long, similar "occupy" protests appeared in 900 cities worldwide (Moynihan 2011; Schneider 2011; Walters 2011).

Many analysts, including some sociologists, are referring to the OWS activism and the revolts in the Arab world (e.g., Egypt, Tunisia, Libya, Syria) as social movements (Barber 2012; Gould 2012; Milkman 2012). Some of the Arab protests ended quickly, whereas others erupted into civil wars. OWS is emerging as a social movement because its participants have organized protests of the status quo and seek to change some aspects of society. So far, however, the OWS

Social movements may seek to bring about or prevent change. However, "what all of these social movements have in common is the desire by ordinary citizens to have a say in the operation of their society" (Locher 2002: 246). Another important source of societal change is technology.

**technology** the application of scientific knowledge for practical purposes.

gatherings have different agendas and grievances, there is purposely no clear leadership, and it's not clear if OWS will decline if employment rates increase, especially among the college-educated.

Many people avoid getting involved in social movements because "they don't make any difference." Are these people right?

## 16-2e Why Social Movements Matter

As you've seen, social movements can either create or resist change. On an *individual level*, many of us enjoy a variety of rights as workers, consumers, voters, and even victims—rights that we owe to highly dedicated people who were determined to change inequitable laws and practices (e.g., see Jenness and Grattet 2001; Switzer 2003; Pell 2011).

On an *institutional level*, social movements can change general practices. For example, shopping for healthy food is much easier today than it was before the 1990s, when veggie burgers, tofu, and organic produce were practically nonexistent. Now, mainstream grocery stores have large organic food sections, a greater variety of fruits and vegetables, and breads and cereals that are made with whole grains, nuts, and less salt, sugar, and chemical additives. In effect, then, vegetarian and consumer groups have had a major impact on industries that control the choice of food products.

On a *societal level*, social movements have had a major effect in the United States and around the world. Most of the world's great religions began as protest movements (see Chapter 13). Also, democratic forms of government in the United States and other countries grew out of the activities of revolutionary groups that sought greater political and economic freedom (Giugni et al. 1999; della Porta and Diani 1999).

## 16-3 Technology and Social Change

This brief "history" of how to cure an earache has circulated on the Internet:

- 2000 B.C.—Here, eat this root.
- 1000 A.D.—That root is heathen, say this prayer.
- 1850 A.D.—That prayer is superstition, drink this potion.
- 1940 A.D.—That potion is snake oil, swallow this pill.
- 1985 A.D.—That pill is ineffective, take this antibiotic.
- 2000 A.D.—That antibiotic is artificial. Here, eat this root.

As this anecdote suggests, despite technological progress over the centuries, some of the old remedies are enjoying renewed popularity (see, for example, www.peoplespharmacy.com which offers home remedies for everything from arthritis to toenail fungus). Still, technological advances have brought enormous social changes, including benefits, costs, and ethical concerns.

## 16-3a Some Recent Technological Advances

**Technology**, the application of scientific knowledge for practical purposes, is a vital aspect of human life: "For good or ill, [technologies] are woven inextricably into the fabric of our lives, from birth to death, at home, in school, in paid work" (MacKenzie and Wajcman 1999: 3).

In the next 75 years, technology is likely to change our lives more dramatically than ever before. Several companies are working on "smart" pills designed to improve mental ability and restore brains that have been impaired by disease or injury. Our houses may be built with sensors that automatically test for carbon dioxide, anthrax, environmental contaminants, allergens, and radioactivity, and with devices that can defend against the release of chemical and biological agents (Rubin 2004; Murphy 2005). Whether such predictions will become reality is anyone's guess. In the meantime, technological advances continue to affect our lives. Let's look briefly at a few of the most influential.

## Computer Technology

In 1887, English mathematician Charles Babbage designed the first programmable computer. Since then, computers have gone through seven generations of evolution. One of the most practical aspects of computer technology is the development of *robots*, machines that are programmed to perform humanlike functions.

Robots are typically used for tasks that are too dull, dirty, or dangerous for humans—such as auto assembly, toxic waste cleanup, mining, minefield sweeping, and underwater and space exploration. Since 2001, American doctors have used a combination of computers, telecommunications, videoconferencing, and advanced robots to guide surgery that is carried out thousands of miles away (Rosen and Hannaford 2006).

In time, robots may be able to discuss stock market investment strategies, give you advice about a personal problem, read a book to you in any desired language or in a voice of either sex, and be emotionally savvy companions who cheer you up when you're unhappy (Breazeal 2002; Bar-Cohen and Breazeal 2003). On the negative side, computer scientists worry that the technology is already being used for criminal activities such as identity theft (see Chapter 7), and that robots will threaten even more jobs in the future (Markoff 2009).

## Biotechnology

*Biotechnology* is a broad term that applies to all practical uses of living organisms. It covers anything from the use of microorganisms (e.g., yeast) to ferment beer to *genetic engineering*, sophisticated techniques that can change the makeup of cells and move genes across species boundaries to produce new organisms.

One form of genetic engineering focuses on producing genetically modified (GM) crops to increase production and to make agriculture less costly, especially by developing new varieties of plants that can tolerate herbicides, resist pests (like the cotton bollworm), or both. The three main GM crops in the United States are corn (45 percent of all corn), soybeans (85 percent), and cotton (76 percent) (Pew Initiative on Food and Biotechnology 2004). Critics contend, however, that GM

"Freeganism" is a social movement that began in the mid-1960s but has gained momentum. In a practice that Americans call "dumpster diving," freegans get free, unspoiled food in the garbage of restaurants, grocery stores, and other food-related industries. Some freegans live on the food, whereas others give it to the poor. Such foraging, according to freegans, keeps perfectly edible food from adding to landfill clutter and unnecessarily depleting the environment.

crops can be toxic and less nutritious than their natural counterparts, increase pesticide use, harm soil quality, and disrupt ecosystems (Nestle 2012).

Another application of biotechnology that is more controversial than GM crops is stem cell research. A *stem cell* is a building block of the human body. It can replicate indefinitely and thus serve as a continuous source of new cells. These self-regenerating cells are found in embryos and umbilical cords, but also in parts of adult bodies, including the brain, blood, heart, skin, bone marrow, intestines, and other organs. Embryonic stem cells are more valuable in research because they can produce any cell type of the body, whereas adult stem cells are generally limited to the cell types of their tissue of origin such as a liver or kidney (National Institutes of Health 2006).

Opponents of stem cell research argue that all embryos deserve protection. Proponents maintain that hundreds of thousands of embryos that fertility clinics dispose of every year are a wasted resource that could be used to treat heart disease, leukemia and other cancers, diabetes, Parkinson's disease, and numerous other health problems (Slevin 2005; Weiss 2005). The controversy may be irrelevant in the future because scientists are using nanotechnology to develop stem cells that combine elements of human tissue with synthetic polymers ("Splicing Nanofiber, Stem Cells" 2012).

## Nanotechnology

Another promising recent technology is *nanotechnology*, the ability to build objects one atom or molecule at a time. The key characteristic of these objects is tiny size. Nanotechnology is based on structures measured in nanometers, a unit of measurement equal to 1 billionth of a meter, or 1/80,000th the width of a human hair.

Nanotechnology will probably result in computer chips that are barely visible to the human eye. In medicine, researchers foresee the day when thousands of nanotubes could potentially be packed into a hairlike capsule the size of a splinter that could be painlessly implanted under the skin to continuously monitor blood sugar, cholesterol, and hormone levels, and destroy tumors without also frying adjacent healthy cells, one of the negative effects of current chemotherapy and radiation treatments for cancer (Weiss 2004, 2005). Some scientists are also experimenting with nanoscale sensors that filter and purify water that contains bacteria, including E. coli (National Nanotechnology Coordination Office 2008).

These and other technological innovations promise longer and healthier lives in the future, but what about the present? What are some of the current benefits and costs of technological changes?

## 16-3b Some Benefits and Costs of Technology

The ever-accelerating pace of change means that most U.S. children born after 2000 are using technologies that didn't exist just a decade ago. Indeed, much of this "iGeneration" views even those in their 20s as outdated in their tech skills (Rosen 2010).

Technology may be creating greater generation gaps than in the past, but many Americans are enthusiastic about technological advances. There is a greater division of opinion over whether social networking sites or Internet blogs have been changes for the better (see *Table 16.2*). Overall, according to a recent national survey, 68 percent of Americans see technology companies as having a positive impact on the country, compared with large corporations and the federal government (25 percent each), Congress (24 percent), and banks and financial institutions (22 percent) (Allen 2010).

Should the average American be less upbeat about technology? Recently, a computer scientist noted that every technology has a "dark side" (Lohr 2011: A1). It's difficult to separate the costs from the benefits, but let's first look at some of technology's major benefits.

### Some Benefits

Technological advances have for some time offered benefits such as performing surgery, tracking people's medications, and monitoring children and older family members from afar using webcams. In other cases, robotic devices have performed repetitive, dirty, or dangerous tasks, including digging deep wells, testing military equipment, and locating bombs.

© Purdue University News Service/David Umberger

Junko Kimura/Getty Images

Japan leads the world in robotics, especially in developing intelligent robots that are useful in the real world. Waka-maru (pictured left), which is 3 feet tall and costs about $14,000, wakes up and follows its owner around, keeps her/him on a schedule, recites the day's news headlines, and alerts friends or relatives if the owner doesn't respond to a message. AIBO (pictured right), which sells for about $2,000, seems to be almost as effective as a live dog in relieving the loneliness of nursing home residents. When petted or talked to, AIBO responds by wagging its tail, barking, and blinking its lights (Banks et al. 2008).

DNA testing has provided millions of people with information about their genetic predispositions for diseases such as emphysema, cancer, and Huntington's disease (an incurable neurological disorder). Having such information has helped doctors and patients make more informed health care decisions (Harmon 2008).

Recent computer-vision systems can watch a hospital room and remind doctors and nurses to wash their hands (a significant source of hospital-acquired infections), and warn hospital staff of restless patients who are in danger of falling out of bed. These systems can also monitor shopping malls, schoolyards, subway platforms, and office complexes (Lohr 2011).

In the criminal justice system, technicians can monitor someone on probation or parole (see Chapter 7) required to wear ankle cuffs. If an alarm goes off, the monitoring computer can call the person to find out where she or he is, what they're doing, and notify police or probation officers about any suspicious activity that they should investigate (Vance 2011).

More than one-third of Americans age 65 or older falls each year. The economic cost of falls, including hip fractures and replacements, is about $75 billion a year. Low-cost wireless sensors in carpets, clothing, and rooms allow doctors to monitor an older person's walking and activity. As a result, physicians are reducing falls by devising exercise programs for specific muscles or changing medications to eliminate dizziness (Lohr 2009).

Some scientists also maintain that use of the Internet has enhanced human intelligence. With greater access to information, they argue, people become smarter and make better decisions. In effect, as the digital systems we rely on become faster and more sophisticated, so will our capabilities for solving problems such as growing population density, the spread of pandemics, and global pollution (Cascio 2009).

## Some Costs

Smart phones, the Internet, and other forms of telecommunication bring people together and provide quick access to a wealth of information. Some argue, however, that the Internet, especially, has made us more sedentary and lazy, and that the ease of online searching and browsing has limited our ability to concentrate, to read without distractions, and to think (Carr 2008). Others worry that the iGeneration expects an instant response from everyone they communicate with and has little patience for anything else (Rosen 2010).

Medical technology has improved the quality of many people's lives and increased their longevity (see Chapter 14). Some researchers, however, have found

### TABLE 16.2

## Opinions About Technological Changes

| TECHNOLOGICAL CHANGES | PERCENTAGE SAYING THAT THE CHANGE HAS BEEN "FOR THE BETTER" |
|---|---|
| Cell phones | 69 |
| Email | 68 |
| Internet | 65 |
| Increased surveillance/security | 58 |
| BlackBerrys/iPhones | 56 |
| Online shopping | 54 |
| Social networking sites | 35 |
| Internet blogs | 29 |

Source: Based on Kohut et al. 2009, p. 2.

that technology may increase unnecessary and expensive surgery. For example, after a Wisconsin hospital purchased robotic surgery technology for $2 million, the number of prostate removals doubled within three months compared with hospitals that hadn't acquired the technology. The latter recommended less intrusive and risky treatments such as radiation and "watchful waiting." The researchers were especially surprised by the number of robotic prostate removals because the incidence of men's prostate cancer has decreased since 2007 (Neuner et al. 2012).

Consumers may be devoted to their Apple computers, iPhones, and iPads, but what about Apple employees? In 2011, Apple stores sold $16 billion in merchandise worldwide, and the executives are millionaires. However, about 30,000 of the 43,000 Apple employees in the United States, earned only $25,000 a year. Most of the salespeople are young males who believe that working at Apple will enhance their résumés to get a better job. Most haven't been able to do so, however, because many similar technology companies have a huge number of new applicants every year (Segal 2012).

The most common concern is that almost every technological advance in telecommunications reduces privacy. Many "data mining" companies routinely collect information about people as the latter click from site to site on the Internet. Much of this web tracking is done anonymously, but a new crop of "snooper" sites is making it easier than ever before for anyone with Internet access to assemble and sell personal information, including your name, Social Security number, address, what you buy, where you go, whom you love, and which sites you've visited on almost any topic, product, or service.

The data come from a variety of sources, including public records on campaign contributions, property sales, and court cases; networking sites where people provide information about themselves, their jobs, relatives, and friends; and even Netflix, where marketers can track the movies that customers have watched and rated (Sarno 2009; Singer 2009, 2010; Stein 2011; Applegate 2011/2012).

Millions of people download apps (computer applications for mobile phones and tablets). Apps are useful, for example, for checking the weather, tracking flights, avoiding traffic jams, locating the closest ATMs, and getting sport scores. However, they intrude on privacy and encourage illegal behavior. For example, some apps track where people are (without their knowledge), and tip people off about drunken-driving checkpoints (Li 2011).

Many Americans are vulnerable because they're not very knowledgeable about privacy laws. For example, only 22 percent know that if a website has a privacy policy, the site can share information about you with other companies without your permission (Turow et al. 2009).

Another increasingly common intrusion on privacy is by health and life insurance companies, which pay other companies only about $15 per search for health "credit reports." They have accessed at least 200 million Americans' 5-year history of purchases of prescription drugs, their dosages and refills, and possible medical conditions. Using such reports, insurance companies can charge some customers higher premiums or exclude some medical conditions from policies. Even worse, according to some health experts, some insurance companies have misinterpreted the information and denied coverage (Nakashima 2008; Terhune 2008).

## 16-3c Some Ethical Issues

As you saw in *Chapter 12*, many Americans yearn to be young. As a result, RealAge (www.realage.com), which promises to shave years off your biological age, has become one of the most popular quizzes on the Internet. More than 27 million people have taken the test, which asks about 150 questions about lifestyle and family medical history and then recommends how to become younger.

Ralph Hagen/Cartoonstock

Throughout the test, RealAge prompts people to become free subscribers. Millions have done so, and their test results go into a marketing database, a clearinghouse for drug corporations. The corporations then suggest new medications that people don't need for problems that a doctor hasn't even diagnosed. The RealAge privacy policy states, in part, that it will share members' personal data with third parties, but few people read the policy statement (Clifford 2009). Thus, RealAge and similar sites profit by preying on people's anxieties and creating huge databases of private health information.

Biotechnology sparks intense ethical controversies. One ethical dilemma, for example, is that biotechnological advances, like other technological developments, are most readily available to the wealthy, enabling them to prolong life. For example, nearly 90,000 Americans are on lists for organ transplants, mostly kidneys and livers. Because of the scarcity of organs, many patients have turned to the Internet in search of donors. Proponents argue that the websites used by these people save lives because they motivate others to donate organs when they are touched by heart-wrenching personal stories. Opponents maintain that such pleas give an edge to those who are affluent, educated, and computer-literate (Stearns 2006; Morais 2007). Thus, technological advances are fraught with both promise and ethical pitfalls.

# REFERENCES

The references that are new to this edition are printed in red.

AAA Foundation for Traffic Safety. 2008. "Cell Phones and Driving: Research Update." Retrieved December 15, 2009 (www.AAAfoundation.org).

———. 2011. "Safety Culture: 'Do As I Say, Not as I Do' Rules Drivers' Decisions on Cell Phone Use and Texting." Accessed November 15, 2011 (www.AAAFoundation.org).

AAUW. 2011. "The Simple Truth About the Gender Pay Gap." Accessed December 20, 2011 (www.aauw.org).

Aberle, David. 1982. *The Peyote Religion among the Navaho*, 2nd edition. Chicago: Aldine.

"About Teen Pregnancy." 2011. Centers for Disease Control and Prevention, April 21. Accessed March 1, 2012 (www.cdc.gov).

Abrams, Simon. 2011. "Let's Stay Together." A.V. Club, January 19. Accessed December 6, 2011 (www.avclub.com).

Abu Dhabi Gallup Center. 2011. "Muslim Americans: Faith, Freedom, and the Future." August. Accessed June 9, 2012 (www.gallup.com).

Abudabbeh, Nuha. 1996. "Arab Families." Pp. 333–346 in *Ethnicity and Family Therapy*, 2nd edition, edited by Monica McGoldrick, Joe Giordano, and John K. Pearce. New York: Guilford Press.

Acierno, Ron, Melba Hernandez-Tejada, Wendy Muzzy, and Kenneth Steve. 2009. "National Elder Mistreatment Study." U.S. Department of Justice, March. Accessed March 5, 2012 (www.ncjrs.gov).

Acs, Gregory, and Austin Nichols. 2010. "Changes in the Economic Security of American Families." Urban Institute, February. Retrieved April 4, 2010 (www.urban.org).

ACT. 2007. *ACT National Curriculum Survey 2005–2006*. Retrieved July 29, 2007 (www.act.org).

Adams, Bert N., and R. A. Sydie. 2001. *Sociological Theory*. Thousand Oaks, CA: Pine Forge Press.

Adams, John S. 2012. "Federal Judge Forwards Racially Charged Email About Obama." *Great Falls Tribune*, March 1. Accessed March 21, 2012 (www.greatfallstribune.com).

Adamson, David M. 2010. "The Influence of Personal, Family, and School Factors on Early Adolescent Substance Use." Rand Health. Accessed August 20, 2011 (www.rand.org).

"Adding Up the Government's Total Bailout Tab." 2011. *New York Times*, July 24. Accessed September 29, 2011 (www.nytimes.com).

Adelman, Clifford. 2006. *The Toolbox Revisited: Paths to Degree Completion from High School Through College*. Washington, DC: U.S. Department of Education.

Adler, Jerry. 2005. "In Search of the Spiritual." *Newsweek*, August 20–September 5, 44–64.

———. 2007. "How to Design a Healthier Planet." *Newsweek*, March 6, 66–70.

Administration on Aging. "A Profile of Older Americans: 2011." Accessed March 6, 2012 (www.aoa.gov).

"After Much Debate, Veils Face Fines." 2011. *Time*, April 25, 15.

Agnvall, Elizabeth. 2012. "7 Medical Procedures You Don't Need." *AARP Bulletin*, May, 10–14.

Aizenman, N. C. 2005. "In Afghanistan, New Misgivings about an Old but Risky Practice." *Washington Post*, April 17, A16.

Akers, Ronald L. 1997. *Criminological Theories: Introduction and Evaluation*, 2nd edition. Los Angeles: Roxbury.

Akkad, Dania. 2009. "Mass Emigration Compounds Iraqi Refugee Crowding." *Christian Science Monitor*, September 18. Retrieved May 2, 2010 (www.csmonitor.com).

"Alcohol and Public Health." 2011. Centers for Disease Control and Prevention, October 28. Accessed April 24, 2012 (www.cdc.gov).

"Alcohol Use and Health." 2011. Centers for Disease Control and Prevention, October 28. Accessed April 24, 2012 (www.cdc.gov).

Alesina, Alberto, Edward Glaeser, and Bruce Sacerdote. 2005. "Work and Leisure in the U.S. and Europe: Why So Different?" National Bureau of Economic Research, April. Retrieved June 2, 2007 (www.nber.org).

Ali, Ayaan Hirsi. 2012. "Christophobia." *Newsweek*, February 13, 28–35.

Allegretto, Sylvia A. 2011. "Great Recession Exacerbated Existing Wealth Disparities in U.S." Economic Policy Institute, March 29. Accessed April 8, 2011 (www.epi.org).

Allegretto, Sylvia A., Ken Jacobs, and Laurel Lucia. 2011. "The Wrong Target: Public Sector Unions and State Budget Deficits." UC Berkeley Labor Center, October. Accessed November 15, 2011 (www.irle.berkeley.edu).

Allen, Jodie T. 2010. "United We Stand . . . on Technology." Pew Research Center, May 5. Retrieved May 6, 2010 (pewresearch.org).

Allen, Jodie T., and Richard Auxier. 2010. "Ask the Expert." Pew Research Center, December 29. Accessed December 30, 2010 (pewresearch.org).

Allen, Katherine R., and Ben K. Beitin. 2007. "Gender and Class in Culturally Diverse Families." Pp. 63–79 in *Cultural Diversity and Families: Expanding Perspectives*, edited by Bahira Sherif Trask and Raeann R. Hamon. Thousand Oaks, CA: Sage.

Alliance for Board Diversity. 2011. *Missing Pieces: Women and Minorities on Fortune 500 Boards*. Updated datasheet, July 21. Accessed August 13, 2011 (www.theabd.org).

Allport, Gordon W. 1954. *The Nature of Prejudice*. Reading, MA: Addison-Wesley.

Almeida, R. V., ed. 1994. *Expansions of Feminist Family Theory through Diversity*. New York: Haworth Press.

Almond, Douglas, Lena Edlund, and Kevin Milligan. 2009. "Son Preference and the Persistence of Culture: Evidence from Asian Immigrants to Canada." National Bureau of Economic Research, October. Retrieved May 1, 2010 (www.nber.org).

Almond, Lucinda. 2007. *The Abortion Controversy*. Farmington Hills, MI: Greenhaven Press.

Altonji, Joseph G., Sarah Cattan, and Iain Ware. 2010. "Identifying Sibling Influence on Teenage Substance Use." National Bureau of Economic Research, October. Accessed July 24, 2011 (www.nber.org).

Alvarez, Lizette. 2009. "Women at Arms: G.I. Jane Breaks the Combat Barrier." *New York Times*, August 16, A1.

American Academy of Pediatrics. 2001. "Children, Adolescents, and Television." *Pediatrics* 109 (November): 423–426.

American Association of Suicidology. 2009. "Elderly Suicide Fact Sheet." June 23. Retrieved December 17, 2009 (www.suicidology.org).

American Cancer Society. 2011. *Cancer Facts & Figures*. Atlanta, GA: American Cancer Society.

"American Indian and Alaska Native Heritage Month: November 2011." 2011. U.S. Census Bureau News, November 1. Accessed December 7, 2011 (www.census.gov).

American Rivers. 2005. "America's Most Endangered Rivers of 2005." Retrieved June 22, 2005 (www.americanrivers.org).

American Society for Aesthetic Plastic Surgery. 2011. "Demand for Plastic Surgery Rebounds by Almost 9%." April 4. Accessed July 20, 2011 (www.surgery.org).

American Sociological Association. 1999. *Code of Ethics and Policies and Procedures of the ASA Committee on Professional Ethics*. Retrieved January 10, 2010 (www.asanet.org).

Amnesty International. 2006. *Stonewalled: Police Abuse and Misconduct against Lesbian, Gay, Bisexual, and Transgender People in the U.S.* London: Amnesty International Publications.

Amusa, Malena. 2010. "'Precious' Pushes Past Controversy to Oscar Night." Women's eNews, March 5. Retrieved March 8, 2010 (www.womensenew.org).

Amy, N. K., A. Aalborg, P. Lyons, and L. Keranen. 2006. "Barriers to Routine Gynecological Cancer Screening for White and African-American Obese Women." *International Journal of Obesity* 30 (January): 147–155.

An, Ruopeng, and Roland Sturm. 2012. "School and Residential Neighborhood Food Environment and Diet Among California Youth." *American Journal of Preventive Medicine* 42 (February): 129–135.

Andersen, Margaret L., and Patricia Hill Collins. 2010. "Why Race, Class, and Gender Still Matter." Pp. 1–16 in *Race, Class, and Gender: An Anthology*," 7th edition, edited by Margaret L. Andersen and Patricia Hill Collins. Belmont, CA: Wadsworth.

Anderson, Craig A., et al. 2003. "The Influence of Media Violence on Youth." *Psychological Science in the Public Interest* 4 (December): 81–110.

Anderson, Elijah. 1999. *Code of the Street: Decency, Violence, and the Moral Life of the Inner City*. New York: W.W. Norton.

Anderson, James F., and Laronistine Dyson. 2002. *Criminological Theories: Understanding Crime in America*. Lanham, MD: University Press of America.

Anderson, John W. 2001. "Iran's Cultural Backlash." *Washington Post*, August 16, A1, A20.

Anderson, Sarah, Chuck Collins, Scott Klinger, and Sam Pizzigati. 2011. "Executive Excess 2011: The Massive CEO Rewards for Tax Dodging." Institute for Policy Studies, August 31. Accessed September 22, 2011 (www.ips-dc.org).

Anderson, Stephen A., and Ronald M. Sabatelli. 2007. *Family Interaction: A Multicultural Developmental Perspective*, 4th edition. Boston: Allyn & Bacon.

Applegate, Evan. 2011/2012. "Your Data, Their Score." *Bloomburg Businessweek*, December 26–January 8, 67.

Arendt, Hannah. 2004. *The Origins of Totalitarianism*. New York: Schocken.

Arias, Elizabeth. 2010. "United States Life Tables by Hispanic Origin." *Vital and Health Statistics* 2 (October): 1–15.

Ariès, Phillippe. 1962. *Centuries of Childhood*. New York: Vintage.

Armstrong, Elizabeth A., Laura Hamilton, and Paula England. 2010. "Is Hooking Bad for Young Women?" *Contexts* 9 (Summer): 22–27.

Arnoldy, Ben. 2011. "India Renews Effort to Save Females." *Christian Science Monitor*, June 13, 12.

Arum, Richard, and Josipa Roksa. 2011. *Academically Adrift: Limited Learning on College Campuses*. Chicago: University of Chicago Press.

Asch, Solomon. 1952. *Social Psychology*. Englewood Cliffs, NJ: Prentice-Hall.

Ashburn, Elyse. 2007. "A Race to Rescue Native Tongues." *Chronicle of Higher Education*, September 28, B15.

Ashford, Lori. S. 2006. "How HIV and AIDS Affect Populations." Population Reference Bureau. Retrieved February 1, 2008 (www.prb.org).

"Asian/Pacific American Heritage Month: May 2011." 2011. U.S. Census Bureau News, March 8. Accessed December 7, 2011 (www.census.gov).

"Asian/Pacific American Heritage Month: May 2012." 2012. U.S. Census Bureau News, March 21. Accessed April 7, 2012 (www.census.gov).

"Asthma in the U.S." 2011. Centers for Disease Control and Prevention, May. Accessed April 24, 2012 (www.cdc.gov).

Astin, Alexander W., and Helen S. Astin. 2005. "Spirituality in Higher Education." Higher Education Research Institute. Retrieved December 15, 2005 (www.gseis.ucla.edu).

Astin, Alexander W., Helen S. Astin, and Jennifer A. Lindholm. 2010. *Cultivating the Spirit: How College Can Enhance Students' Inner Lives*. New York: Jossey-Bass.

Atchley, Robert C., and Amanda S. Barusch. 2004. *Social Forces and Aging: An Introduction to Social Gerontology*, 10th edition. Belmont, CA: Wadsworth.

Atkinson, Anthony B., Thomas Piketty, and Emmanuel Saez. 2011. "Top Incomes in the Long Run of History." *Journal of Economic Literature* 49 (1): 3–71.

Attinasi, John J. 1994. "Racism, Language Variety, and Urban U.S. Minorities: Issues in Bilingualism and Bidialectalism." Pp. 319–347 in *Race*, edited by Steven Gregory and Roger Sanjek. New Brunswick, NJ: Rutgers University Press.

Aud, Susan, William Hussar, Grace Kena, Kevin Bianco, Lauren Frohlich, Jana Kemp, and Kim Tahan. *The Condition of Education 2011*. U.S. Department of Education, National Center for Education Statistics. Washington, DC: U.S. Government Printing Office.

Auguste, Byron, Paul Kihn, and Matt Miller. 2010. "Closing the Talent Gap: Attracting and Retaining Top-Third Graduates to Careers in Teaching." McKinsey & Company, September. Accessed March 20, 2012 (www.mckinsey.com).

Aumann, Kerstin, Ellen Galinsky, and Kenneth Matos. 2011. "The New Male Mystique." Families and Work Institute. Accessed October 25, 2011 (www.familiesandwork.org).

Aunola, Kaisa, and Jari-Erik Nurmi. 2005. "The Role of Parenting Styles in Children's Problem Behavior." *Child Development* 76 (November/December): 1144–1159.

Austen, Ian. 2002. "A Leg with a Mind of Its Own." *New York Times*, January 3, G1.

"Australian Passports Now Offer 'M' for Male, 'F' for Female or 'X'." 2011. *Baltimore Sun*, September 18, 24.

Auxier, Richard. 2010. "Congress in a Wordle." Pew Research Center, March 22. Retrieved March 26, 2010 (pewresearch.org).

Avellar, Sarah, and Pamela Smock. 2005. "The Economic Consequences of the Dissolution of Cohabiting Unions." *Journal of Marriage and Family* 67 (May): 315–327.

Avishai, Orit. 2010. "Women of God." *Contexts* 9 (Fall): 46–51.

Axtell, Roger E., Tami Briggs, Margaret Corcoran, and Mary Beth Lamb. 1997. *Do's and Taboos Around the World for Women in Business*. New York: John Wiley & Sons.

Babbie, Earl. 2002. *The Basics of Social Research*, 2nd edition. Belmont, CA: Wadsworth.

Babcock, Philip, and Mindy Marks. 2011. "The Falling Time Cost of College: Evidence from Half a Century of Time Use Data." *The Review of Economics and Statistics* 83 (May): 468–478.

Baenninger, MaryAnn. 2011. "For Women on Campuses, Access Doesn't Equal Success." *Chronicle of Higher Education*, October 7, A26.

Bainbridge, William S. 1997. *The Sociology of Religious Movements*. New York: Routledge.

Baio, Jon. 2012. "Prevalence of Autism Spectrum Disorders—Autism and Developmental Disabilities Monitoring Network, 14 Sites, United States, 2008." *MMWR* 61 (March 30): 1–19. Accessed April 24, 2012 (www.cdc.gov).

Bakan, Joel. 2004. *The Corporation: The Pathological Pursuit of Profit and Power*. New York: Free Press.

———. 2005. *The Corporation: The Pathological Pursuit of Profit and Power*. New York: Free Press.

Baldauf, Scott. 2010. "In Africa, New Heat on Gays." *Christian Science Monitor*, December 13, 12.

Bandura, Albert, and Richard H. Walters. 1963. *Social Learning and Personality Development*. New York: Holt, Rinehart & Winston.

Banerjee, Abhijit, Esther Duflo, Maitreesh Ghatak, and Jeanne Lafortune. 2009. "Marry for What? Caste and Mate Selection in Modern India." National Bureau of Economic Research, May. Retrieved March 2, 2010 (www.nber.org).

Banerjee, Neela. 2006. "Clergywomen Find Hard Path to Bigger Pulpit." *New York Times*, August 26, A1, A12.

Banerjee, Neela. 2011. "Obama Won't Develop Rules to Cut Smog." *Baltimore Sun*, September 3, 8.

Banfield, Edward C. 1974. *The Unheavenly City Revisited*. Boston: Little, Brown.

Banks, Duren, and Tracey Kyckelhahn. 2011. "Characteristics of Suspected Human Trafficking Incidents, 2008–2010." Office of Justice Programs, April. Accessed October 25, 2011 (www.ojp.gov).

Banks, Ingrid. 2000. *Hair matters: Beauty, Power, and Black Women's Consciousness*. New York: New York University Press.

Banks, Marian R., Lisa M. Willoughby, and William A. Banks. 2008. "Animal-Assisted Therapy and Loneliness in Nursing Homes: Use of Robotic versus Living Dogs." *Journal of the American Medical Directors Association* 9 (March): 173–177.

Barash, David P. 2002. "Evolution, Males, and Violence." *Chronicle of Higher Education*, May 24, B7–B9.

Barber, Benjamin. 2012. "What Democracy Looks Like." *Contexts* 11 (Spring): 14–16.

Barber, Bernard, and Lyle S. Lobel. 1952. "Fashion in Women's Clothes and the American Social System." *Social Forces* 31 (December): 124–131.

Bar-Cohen, Yoseph, and Cynthia Breazeal. 2003. *Biologically Inspired Intelligent Robots*. Bellingham, WA: Spie Press.

Barker, James B. 1993. "Tightening the Iron Cage: Concertive Control in Self-Managing Teams." *Administrative Science Quarterly* 38 (September): 408–437.

Barnard, Chester. 1938. *The Functions of the Executive*. Cambridge, MA: Harvard University Press.

Barnes, Cynthia. 2006. "China's 'Kingdom of Women'." *Slate*, November 17. Retrieved June 12, 2007 (www.slate.com).

Barnes, Robert. 2010. "Supreme Court Rules on Employer Monitoring of Cellphone, Computer Conversations." *Washington Post*, June 18, A1.

———. 2011. "Limits on Video Games Rejected." *Washington Post*, June 28, A1.

Barnett, Rosalind, and Caryl Rivers. 2004. *Same Difference: How Gender Myths Are Hurting Our Relationships, Our Children, and Our Jobs*. New York: Basic Books.

Barringer, Felicity. 2012. "As 'Yuck Factor' Subsides, Treated Wastewater Flows From Taps." *New York Times*, February 10, A1.

Barro, Robert J., and Rachel M. McCleary. 2003. "Religion and Economic Growth across Countries." *American Sociological Review* 68 (October): 760–781.

Barrows, Sydney B. 1989. *Mayflower Madam: The Secret Life of Sydney Biddle Barrows*. Westminster, MD: Arbor House.

Bart, Pauline B., and Eileen Geil Moran, eds. 1993. *Violence Against Women: The Bloody Footprints*. Thousand Oaks, CA: Sage.

Bartlett, Thomas. 2007. "'I Suffer Not a Woman to Teach'." *Chronicle of Higher Education*, April 13, A10–A12.

Bartlett, Tom. 2011. "Caffeine Is Definitely Good/Bad for You." *Chronicle of Higher Education*, January 20. Accessed January 25, 2011 (www.chronicle.com/blogs).

Barton, Allen H. 1980. "A Diagnosis of Bureaucratic Maladies." Pp. 27–36 in *Making Bureaucracies Work*, edited by Carol H. Weiss and Allen H. Barton. Beverly Hills, CA: Sage.

Basken, Paul. 2011a. "Obama Administration Tightens Rules on Financial Conflicts of Interests in Science." *Chronicle of Higher Education*, August 23. Accessed August 24, 2011 (www.chronicle.com).

———. 2011b. "U. of Pennsylvania Professor Accuses Colleagues of Slanted Research." *Chronicle of Higher Education*, July 12. Accessed July 24, 2011 (www.chronicle.com).

Batalova, Jeanne. 2011. "Asian Immigrants in the United States." Migration Information Source, May 24. Accessed December 7, 2011 (www.migrationinformation.org).

Baum, Katrina. 2006. "Identity Theft, 2004." Bureau of Justice Statistics Bulletin, April. Retrieved March 10, 2008 (wwwl.ojp.usdoj.gov).

Baum, Sandy, and Kathleen Payea. 2011. "Trends in Student Aid 2011." College Board Advocacy & Policy Center. Accessed April 12, 2012 (www.collegeboard.org).

Baumrind, Diana. 1968. "Authoritarian versus Authoritative Parental Control." *Adolescence* 3 (11): 255–272.

———. 1989. "Rearing Competent Children." Pp. 349–378 in *Child Development Today and Tomorrow*, edited by William Damon. San Francisco: Jossey-Bass.

Beadle, Amanda Peterson. 2012. "Minnesota Senator Thinks Abortion Pill Is Wrong, While Viagra Is a 'Wonderful Drug'." Alternet, May 3. Accessed May 24 (www.alternet.org).

Becker, Howard S. 1963. *Outsiders: Studies in the Sociology of Deviance*. New York: Free Press.

Beckles, Gloria L., and Benedict I. Truman. 2011. "Education and Income—United States, 2005 and 2009." Centers for Disease Control and Prevention. *MMWR* 60 (Suppl): 13–17.

Beede, David, Tiffany Julian, David Langdon, George McKittrick, Beethika Khan, and Mark Doms. 2011. "Women in STEM: A Gender Gap to Innovation." Economics and Statistics Administration, Issue Brief #04-11, August. Accessed October 25, 2011 (www.esa.gov).

Beeghley, Leonard. 2000. *The Structure of Social Stratification in the United States*, 3rd edition. Boston: Allyn & Bacon.

Begala, Paul. 2011. "Who You Calling Lazy?" Newsweek, December 15, 8.

Begley, Sharon. 2010. "Sins of the Grandfathers." *Newsweek*, November 8, 48–50.

Belkin, Lisa. 2008. "Smoother Transitions." *New York Times*, September 4, 2.

Belknap, Joanne. 2001. *The Invisible Woman: Gender, Crime, and Justice*, 2nd edition. Belmont, CA: Wadsworth.

———. 2007. *The Invisible Woman: Gender, Crime, and Justice*, 3rd edition. Belmont, CA: Wadsworth Press.

Bennett, Brian. 2011. "Feds Screening Thousands of Iraqi Refugees in U.S." *Baltimore Sun*, July 19, 7.

Bennett, Drake. 2010. "This Will Be on the Midterm. You Feel Me?" *Slate*, March 24. Accessed March 26, 2010 (www.slate.com).

———. 2011/2012. "Housing." Bloomberg Businessweek, December 26–January 8, 34.

Bennett, Jessica. 2009. "Tales of a Modern Diva." *Newsweek*, April 6, 42–43.

———. 2010. "The Beauty Advantage." *Newsweek*, July 26, 46–47.

Benokraitis, Nijole V. 1997. *Subtle Sexism: Current Practices and Prospects for Change*. Thousand Oaks, CA: Sage.

———. 2011. *Marriages & Families: Changes, Choices, and Constraints*, 7th edition. Upper Saddle River, NJ: Prentice Hall.

———, ed. 2000. *Feuds about Families: Conservative, Centrist, Liberal, and Feminist Perspectives*. Upper Saddle River, NJ: Prentice Hall.

Benson, Michael L. 2002. *Crime and the Life Course*. Los Angeles: Roxbury.

Benson, Michael L., and Greer Litton Fox. 2004. "When Violence Hits Home: How Economics and Neighborhood Play a Role." National Institute of Justice, September. Retrieved October 1, 2004 (www.ojp.usdoj.gov).

Benton, Thomas H. 2011. "A Perfect Storm in Undergraduate Education." *Chronicle of Higher Education*, February 25, A43, A45. Part 2 in April 8, A45–A46.

Bergen, Raquel K., Jeffrey L. Edleson, and Claire M. Renzetti, eds. 2005. *Violence against Women: Classic Papers*. Boston: Allyn & Bacon.

Berger, Dan. 2012. "How the Tax Code Is Skewed in the 1%'s Favor." AlterNet, January 2. Accessed January 3, 2012 (www.alternet.org).

Berger, Peter L., and Thomas Luckmann. 1966. *The Social Construction of Reality: A Treatise in the Sociology of Knowledge*. New York: Doubleday.

Berger, Rachel, et al. 2011. "Abusive Head Trauma During a Time of Increased Unemployment: A Multicenter Analysis." *Pediatrics* 128 (October): 637–643.

Bergquist, Amy. 2006. "Pharmacist Refusals: Dispensing (With) Religious Accommodation under Title VII." *Minnesota Law Review* 90 (April): 1073–1106.

Berk, Richard A. 1974. *Collective Behavior*. Dubuque, IA: Wm. C Brown Company.

Berkos, Kristen M., Terre H. Allen, Patricia Kearney, and Timothy G. Plax. 2001. "When Norms Are Violated: Imagined Interactions as Processing and Coping Mechanisms." *Communication Monographs* 68 (September): 289–300.

Berkowitz, Bill. 2011. "Why Is Jerry Falwell's Evangelical University Getting Filthy Rich Off Your Tax Money?" AlterNet, June 29. Accessed June 29, 2011 (www.alternet.org).

Berlin, Overton B., and Paul Kay. 1969. *Basic Color Terms*. Berkeley: University of California Press.

Berman, Greg, and Aubrey Fox. 2009. "Lessons from the Battle over D.A.R.E.: The Complicated Relationship between Research and Practice." U.S. Department of Justice, Bureau of Justice Assistance, Center for Court Innovation. Retrieved May 16, 2010 (www.ojp.usdoj.gov).

Bernhardt, Annette, et al. 2011. "The Good Jobs Deficit: A Closer Look at Recent Job Loss and Job Growth Trends Using Occupational Data." National Employment Law Project, July. Accessed July 30, 2011 (www.nelp.org).

Bernstein, David E. 2011. "Overt Vs. Covert." New York Times, May 22. Accessed May 23, 2011 (www.newyorktimes.com).

Berrey, Ellen. 2009. "Sociology Finds Discrimination in the Law." *Contexts* 8 (Spring): 28–32.

Bertoni, Steven. 2010. "On Board a Flying Mansion." *Forbes*, October 11, 194–195.

Bertrand, Marianne, and Sendhil Mullainathan. 2003. "Are Emily and Greg More Employable than Lakisha and Jamal? A Field Experiment on Labor Market Discrimination." Massachusetts Institute of Technology, Department of Economics. Retrieved May 3, 2004 (papers.ssrn.com).

Bertrand, Marianne, and Jessica Pan. 2011. "The Trouble with Boys: Social Influences and the Gender Gap in Disruptive Behavior." National Bureau of Economic Research, October. Accessed April 14, 2012 (www.nber.org).

Berube, Alan. 2011. "The State of Metropolitan America: Suburbs and the 2010 Census." Brookings Metropolitan Policy Program, July 14. Accessed June 22, 2012 (www.brookings.edu).

Berube, Alan, Audrey Sinter, Jill H. Wilson, and William H. Frey. 2006. "Finding Exurbia: America's Fast-Growing Communities at the Metropolitan Fringe." Brookings Institution. Retrieved February 2, 2008 (www.brookings.edu).

Best, Joel. 2006. *Flavor of the Month: Why Smart People Fall for Fads*. Berkeley: University of California Press.

Betcher, R. William, and William S. Pollack. 1993. *In a Time of Fallen Heroes: The Re-creation of Masculinity*. New York: Atheneum.

Bialik, Carl. 2011. "Irreconcilable Claim: Facebook Causes 1 in 5 Divorces." *Wall Street Journal*, March 12. Accessed November 8, 2011 (online.wsj.com).

Bianchi, Suzanne M., John P. Robinson, and Melissa A. Milkie. 2006. *Changing Rhythms of American Family Life*. New York: Russell Sage Foundation.

Billingsley, Andrew. 1992. *Climbing Jacob's Ladder: The Enduring Legacy of African-American Families*. New York: Simon & Schuster.

Binder, Amy. 1993. "Constructing Racial Rhetoric: Media Depictions of Harm in Heavy Metal and Rap Music." *American Sociological Review* 58 (December): 753–767.

Bingham, Shawn C., and Alexander A. Hernandez. 2009. "'Laughing Matters': The Comedian as Social Observer, Teacher, and Conduit of the Sociological Perspective." *Teaching Sociology* 37 (October): 335–352.

Bird, Chloe E., et al. 2010. "Neighborhood Socioeconomic Status and Biological 'Wear & Tear' in a Nationally Representative Sample of U.S. Adults." *Journal of Epidemiology and Community Health* 64 (October): 860–865.

Bivens, L. Josh. 2008. "Trade, Jobs, and Wages: Are the Public's Worries about Globalization Justified?" Economic Policy Institute, May 6. Retrieved July 4, 2008 (www.epi.org).

Black, M. C., and M. J. Breiding. 2008. "Adverse Health Conditions and Health Risk Behaviors Associated with Intimate Partner Violence—United States, 2005." *MMWR Weekly,* 57 (February 8): 113–117.

Black, M. C., et al. 2011. "National Intimate Partner and Sexual Violence Survey: 2010 Summary Report." Centers for Disease Control and Prevention, National Center for Injury Prevention and Control. Accessed March 3, 2012 (www.cdc.gov).

*Black-Owned Firms: 2002.* 2006. U.S. Census Bureau, SB02-00CS-BLK (RV). Retrieved May 1, 2007 (www.census.gov).

Blacksmith, Nikki, and Jim Harter. 2011. "Majority of American Workers Not Engaged in Their Jobs." Gallup, October 28. Accessed December 20, 2011 (www.gallup.com).

Blanchard, Dallas A. 1994. *The Anti-Abortion Movement and the Rise of the Religious Right: From Polite to Fiery Protest.* New York: Twayne Publishers.

Bland, Scott. 2010. "The Odds on State Gambling." *Christian Science Monitor,* July 19, 21.

Blau, Francine B., and Lawrence M. Kahn. 2006. "The Gender Pay Gap: Going, Going . . . but Not Gone." Pp. 37–66 in *The Declining Significance of Gender?* edited by Francine D. Blau, Mary C. Brinton, and David B. Grutsky. New York: Russell Sage Foundation.

Blau, Peter M. 1986. *Exchange and Power in Social Life,* revised edtion. New Brunswick, NJ: Transaction.

Blau, Peter M., and Marshall W. Meyer. 1987. *Bureaucracy in Modern Society,* 3rd edition. New York: Random House.

Blumberg, Stephen, et al. 2011. "Wireless Substitution: State-level Estimates From the National Health Interview Survey, January 2007–June 2010." *National Health Statistics Reports,* Number 39, April 20. Accessed January 17, 2011 (www.cdc.gov).

Blumenstyk, Goldie. 2009. "Company Says Research It Sponsored at Pitt and Hopkins Was Fraudulent." *Chronicle of Higher Education,* September 4. Retrieved September 6, 2009 (www.chronicle.com).

Blumer, Herbert. 1946. "Collective Behavior." Pp. 65–121 in *New Outline of the Principles of Sociology,* edited by Alfred M. Lee. New York: Barnes & Noble.

———. 1969. *Symbolic Interactionism: Perspective and Method.* Englewood Cliffs, NJ: Prentice Hall.

Blumgart, Jake. 2011. "4 Ways Government Policy Favors the Rich and Keeps the Rest of Us Poor." AlterNet, September 2. Accessed September 3, 2011 (www.alternet.org).

Boak, Joshua. 2009. "Layoffs Are Driving Change Among the Amish." *Los Angeles Times,* April 20. Retrieved April 22, 2009 (www.latimes.com).

Board of Trustees, Federal Old-Age and Survivors Insurance and Federal Disability Insurance Trust Funds. 2009. *The 2009 Annual Report of the Board of Trustees of the Federal Old-Age and Survivors Insurance and Disability Insurance Trust Funds.* Washington, DC: U.S. Government Printing Office.

Bobroff-Hajal, Anne. 2006. "Why Cousin Marriage Matters in Iraq." *Christian Science Monitor,* December 26, 9.

Bogle, Kathleen A. 2008. *Hooking Up: Sex, Dating, and Relationships on Campus.* New York: New York University Press.

Bohannon, Lisa. 2000. "Is Your Body Language on Your Side?" *Career World* 29 (November/December): 21–23.

Bohannon, Paul, ed. 1971. *Divorce and After.* New York: Doubleday.

Bollag, Burton. 2007. "Credential Creep." *Chronicle of Higher Education,* June 22, A10–A12.

Bonacich, Edna. 1972. "A Theory of Ethnic Antagonism: The Split Labor Market." *American Sociological Review* 37 (October): 547–559.

Bonner, Robert. 2007. "Research—Male Elementary Teachers: Myths and Realities." Men Teach. Retrieved June 12, 2008 (www.menteach.org).

Boonstra, Heather D., Rachel B. Gold, Cory L. Richards, and Lawrence B. Finer. 2006. "Abortion in Women's Lives." Guttmacher Institute. Retrieved December 21, 2006 (www.guttmacher.org).

Boraas, Stephanie, and William M. Rodgers III. 2003. "How Does Gender Play a Role in the Earnings Gap? An Update." *Monthly Labor Review* 126 (March): 9–15.

Borgerhoff Mulder, Monique. 2009. "Serial Monogamy as Polygyny or Polyandry?" *Human Nature* 20 (Summer): 130–150.

Borzekowski, Dina L., and Thomas N. Robinson. 2005. "The Remote, the Mouse, and the No. 2 Pencil." *Archives of Pediatrics & Adolescent Medicine* 159 (July): 607–613.

Bosk, Charles. 1979. *Forgive and Remember: Managing Medical Failure.* Chicago: University of Chicago Press.

"Boss Sells His Company, Shares Proceeds." 1999. *Baltimore Sun,* September 12, 24A.

Boston, Rob. 2011. "The Relentless Christian Crusade to Prevent Kids from Learning Science." AlterNet, July 11. Accessed July 12, 2011 (www.alternet.org).

Bourdieu, Pierre. 1984. *Distinction: A Social Critique of the Judgement of Taste.* Translated by Richard Nice. Cambridge, MA: Harvard University Press.

Bourke, Michael L., and Andres E. Hernandez. 2009. "The 'Butner Study' Redux: Report of the Incidence of Hands-on Child Victimization by Child Pornography Offenders." *Journal of Family Violence* 24 (April): 183–191.

Boushey, Heather, Jessica Arons, and Lauren Smith. 2010. "Families Can't Afford the Gender Wage Gap." Center for American Progress, April. Accessed April 30, 2010 (www.americanprogress.org).

Bowles, Samuel, and Herbert Gintis. 1977. *Schooling in Capitalist America: Educational Reform and the Contradictions of Economic Life.* New York: Basic Books.

Bradshaw, York W., and Michael Wallace. 1996. *Global Inequalities.* Thousand Oaks, CA: Pine Forge Press.

Braga, Anthony A. 2003. "Systematic Review of the Effects of Hot Spots Policing on Crime." Unpublished paper. Retrieved May 14, 2005 (www.campbellcollaboration.org/doc-pdf/hotspots.pdf).

Braitman, Keli A., Neil K. Chaudhary, and Anne T. McCart. 2011. "Effect of Passenger Presence on Older Drivers' Risk of Fatal Crash Involvement." Insurance Institute for Highway Safety, March. Accessed July 4, 2011 (www.iihs.org).

Bramlett, Mosher D., and William D. Mosher. 2002. *Cohabitation, Marriage, Divorce, and Remarriage in the United States.* Centers for Disease Control and Prevention, *Vital and Health Statistics* (Series 23, No. 25). Hyatsville, MD: National Center for Health Statistics.

Brea, Jorge A. 2003. "Population Dynamics in Latin America." *Population Bulletin* 58 (March): 1–36.

Breazeal, Cynthia L. 2002. *Designing Sociable Robots.* Cambridge, MA: MIT Press.

Bremner, Jason, Carl Haub, Marlene Lee, Mark Mather, and Eric Zuehlke. 2009. "World Population Highlights: Key Findings from PRB's 2009 World Population Data Sheet." *Population Bulletin* 64 (September): 1–12.

Brescoll, Victoria L., and Eric L. Uhlmann. 2008. "Can an Angry Woman Get Ahead? Status Conferral, Gender, and Expression of Emotion in the Workplace." *Psychological Science* 19 (March): 268–275.

Brewis, Alexandra A., Amber Wutich, Ashlan Falletta-Cowden, and Isa Rodriguez-Soto. 2011. "Body Norms and Fat Stigma in Global Perspective." *Current Anthropology* 52 (April): 269–276.

Briggs, Derek C. 2009. "Preparation for College Admission Exams." National Association for College Admission Counseling. Retrieved April 16, 2010 (www.nacacnet.org).

Brittingham, Angela, and G. Patricia de la Cruz. 2005. "We the People of Arab Ancestry in the United States." Census 2000 Special Reports, CENSR-21, March. Retrieved March 26, 2007 (www.census.gov).

Broder, John. 2007. "Industry Flexes Muscle, Weaker Energy Bill Passes." *New York Times,* December 14, 29.

Brody, Gene H., Shannon Dorsey, Rex Forehand, and Lisa Armistead. 2002. "Unique and Protective Contributions of Parenting and Classroom Processes to the Adjustment of African American Children Living in Single-Parent Families." *Child Development* 73 (January–February): 274–286.

Brooks, Arthur C. 2007. "I Love My Work." *The American* (online): September–October. Retrieved September 30, 2007 (www.theamericanmag.com).

Brooks, David. 2011. "It's Not About You." *New York Times,* May 31, A23.

Brown, David. 2005. "Polio Outbreak Occurs Among Amish Families in Minnesota." *Washington Post,* October 14, A3.

Brown, DeNeen L. 2009. "The High Cost of Poverty: Why the Poor Pay More." *Washington Post,* May 18, C1.

Brown, Matthew Hay. 2005. "Storm Victims Rely on Faith to Help Them Through Crisis." *Baltimore Sun*, September 12, 6A.

Brown, Meta, Andrew Haughwout, Donghoon Lee, Maricar Mebutas, and Wilbert van der Klaauw. 2012. "Grading Student Loans." Liberty Street Economics, March 5. Accessed March 16, 2012 (libertystreeteconomics.newyorkfed.org).

Brown, Pamela A., and Elizabeth McGann. 2011. "Medication Error Prevention: A Shared Responsibility." Medscape Medical News, June 14. Accessed April 24, 2012 (www.medscape.com).

Brown, Susan I. 2005. "How Cohabitation Is Reshaping American Families." *Contexts* 4 (Summer): 33–37.

Bruder, Jessica. 2011. "Tussle Over Taxes." Christian Science Monitor, October 17, 32–37.

Bruinius, Harry. 2010. "March of the Megacities." *Christian Science Monitor*, May 10, 26–30.

Brunvand, Jan H. 2001. *The Truth Never Stands in the Way of a Good Story*. Urbana and Chicago: University of Illinois Press.

Buechler, Steven M. 2000. *Social Movements in Advanced Capitalism: The Political Economy and Cultural Construction of Social Activism*. New York: Oxford University Press.

Buffet, Warren. 2011. "Stop Coddling the Super-Rich." *New York Times*, August 14, B21.

Bullock, Karen. 2005. "Grandfathers and the Impact of Raising Grandchildren." *Journal of Sociology and Social Welfare* 32 (March): 43–59.

Bumiller, Elisabeth. 2005. "In the Struggle over the Iraq War, Women Are On the Front Line." *New York Times*, August 29, 11.

Bureau of Justice Statistics. 2011a. "Direct Expenditures by Criminal Justice Function, 1982–2007. August 26. Accessed August 27, 2011 (bjs.ojp.usdoj.gov).

———. 2011b. "Violent Victimization Rates by Gender, 1973–2009." August 20. Accessed August 21, 2011 (bjs.ojp.usdoj.gov).

Bureau of Labor Statistics. 2005. "Time-Use Survey—First Results Announced by BLS." January 12. Accessed October 25, 2011 (www.bls.gov).

———. 2010. "Employment Situation Summary." April 2. Retrieved April 6, 2010 (www.bls.gov).

———. 2011. "American Time Use Survey—2010 Results." June 22. Accessed October 25, 2011 (www.bls.gov).

———. 2011. "Labor Force Statistics from the Current Population Survey." Accessed December 28, 2011 (www.bls.gov).

———. 2011. "Usual Weekly Earning of Wage and Salary Workers Second Quarter 2011." July 19. Accessed October 25, 2011 (www.bls.gov).

Bureau of Labor Statistics News Release. 2012. "The Employment Situation—December 2011." U.S. Department of Labor, January 6. Accessed January 10, 2012 (www.bls.gov).

———. 2012. "Physicians and Surgeons." *Occupational Outlook Handbook, 2012–13 Edition*. Accessed April 15, 2012 (www.bls.gov).

Bustillo, Miguel. 2006. "Farmers Say They've Got Fruit but No Labor." *Los Angeles Times*, April 17. Retrieved April 28, 2006 (www.latimes.com).

Butterfield, Fox. 2002. "Father Steals Best: Crime in an American Family." *New York Times* (August 15): 1A.

Cabrera, Natasha J., Jay Fagan, and Danielle Farrie. 2008. "Explaining the Long Reach of Fathers' Prenatal Involvement on Later Paternal Engagement." *Journal of Marriage and Family* 70 (December): 1094–1107.

Cadge, Wendy, and Courtney Bender. 2004. "Yoga and Rebirth in America: Asian Religions Are Here to Stay." *Contexts* 3 (Winter): 45–51.

Caesar, Stephen. 2012. "In States Across U.S., New Traffic Laws Gain Traction." *Baltimore Sun,* January 3, 11.

Cameron, Deborah. 2007. *The Myth of Mars and Venus*. New York: Oxford University Press.

Campbell, Duncan. 2008. "World Is Moving Towards Banning Death Penalty, Says Reprieve." *Guardian*, October 9. Accessed August 26, 2011 (www.guardian.co.uk).

Campion-Vincent, Véronique. 2005. "From Evil Others to Evil Elites: A Dominant Pattern in Conspiracy Theories Today." Pp. 103–122 in *Rumor Mills: The Social Impact of Rumor and Legend*, edited by Gary Alan Fine, Véronique Campion-Vincent, and Chip Heath. New Brunswick, NJ: Transaction Publishers.

Campo-Flores, Arian. 2005. "The Battle for Latino Souls." *Newsweek*, March 31, 50–51.

Caperton, Gaston. 2009. "Test Data Allow Better Decisions." *U.S. News & World Report*, September, 24.

Capgemini and Merrill Lynch Wealth Management. 2011. "World Wealth Report." Accessed September 22, 2011 (www.capgemini.com).

Caplan, Paula J., and Jeremy B. Caplan. 2009. *Thinking Critically About Research on Sex and Gender*, 3rd edition. Boston: Allyn & Bacon.

Cappiello, Dina. 2012. "EPA: Power Plant are Main Global Warming Culprits." *USA Today*, January 11. Accessed January 12, 2012 (www.usatoday.com).

Carasso, A., E. Steurle, G. Reynolds, T. Vericker, and J. Macomber. 2008. "Kids' Share 2008: How Children Fare in the Federal Budget." Urban Institute. Accessed September 27, 2009 (www.urban.org).

Card, David, and Laura Giuliano. 2011. "Peer Effects and Multiple Equilibria in the Risky Behavior of Friends." National Bureau of Economic Research, May. Accessed July 24, 2011 (www.nber.org).

Card, Josefina J., Marvin B. Eisen, James L. Peterson, and Bonnie Sherman-Williams. 1994. "Evaluating Teenage Pregnancy Prevention and Other Social Programs: Ten Stages of Program Assessment." *Family Planning Perspectives* 26 (May): 116–131.

Carey, Benedict. 2004. "Long After Kinsey, Only the Brave Study Sex." *New York Times*, November 9, F1.

———. 2010. "Revising Book on Disorders of the Mind." *New York Times,* February 10, 1.

Carey, Bill. 2011. "Airlines Stay Profitable Despite Escalating Fuel Costs." Aviation International News, August 1. Accessed August 13, 2011 (www.aionline.com).

Carl, Traci. 2002. "Amid Latte, Mocha Craze, Coffee Growers Go Hungry in Paradise." Retrieved September 19, 2002 (story.news.yahoo.com).

Carlson, Marcia, Sara McLanahan, Paula England, and Barbara Devaney. 2005. "What We Know about Unmarried Parents: Implications for Building Strong Families Programs." Mathematica Policy Research, Inc., January. Retrieved June 5, 2007 (www.mathematica-mpr.com).

Carmichael, Mary. 2007. "Troubled Waters." *Newsweek*, June 4, 52–56.

Carmon, Irin. 2011. "Why Does Everything Seem to Be Going Wrong for Women's Progress?" AlterNet, December 21. Accessed January 5, 2012 (www.alternet.org).

Carnagey, Nicholas L., and Craig A. Anderson. 2005. "The Effects of Reward and Punishment in Violent Video Games on Aggressive Affect, Cognition, and Behavior." *Psychological Science* 16 (November): 882–889.

Carnevale, Anthony P., Stephen J. Rose, and Ban Cheah. 2011a. "The College Payoff: Education, Occupations, Lifetime Earnings." Georgetown University Center on Education and the Workforce. Accessed April 10, 2012 (cew.georgetown.edu).

Carnevale, Anthony P., Nicole Smith, and Jeff Strohl. 2010. "Help Wanted: Projections of Jobs and Education Requirements Through 2018." Georgetown University Center on Education and the Workforce, June. Accessed April 10, 2012 (cew.georgetown.edu).

Carnevale, Anthony P., Jeff Strohl, and Michelle Melton. 2011b. "What's It Worth? The Economic Value of College Majors." Georgetown University Center on Education and the Workforce, May. Accessed April 10, 2012 (cew.georgetown.edu).

Carney, Ginny. 1997. "Native American Loanwords in American English." *Wicazo SA Review*, 12 (Spring): 189–203.

Carr, Nicholas. 2008. "Is Google Making Us Stupid?" *Atlantic Monthly*, July/August. Retrieved May 11, 2010 (www.theatlantic.com).

Carrell, Scott E., Mark Hoekstra, and James E. West. 2010. "Is Poor Fitness Contagious? Evidence from Randomly Assigned Friends." National Bureau of Economic Research, November. Accessed July 24, 2011 (www.nber.org).

Carroll, Jason S., Laura M. Padilla-Walker, Larry J. Nelson, Chad D. Olson, Carolyn McNamara Barry, and Stephanie D. Madsen. 2008. "Generation XXX: Pornography Acceptance and Use Among Emerging Adults." *Journal of Adolescent Research* 23 (January): 6–30.

Carroll, Jill. 2002. "Getting Good Teaching Evaluations Without Stand-Up Comedy." *Chronicle of Higher Education*, April 15. Retrieved April 15, 2002 (chronicle.com).

Carroll, John. 1956. "Introduction." *Language, Thought & Reality: Selected Writings of Benjamin Lee Whorf*. Cambridge, MA: MIT Press.

Carroll, Stephen J., and Emre Erkut. 2009. *The Benefits to Taxpayers from Increases in Students' Educational Attainment*. Santa Monica, CA: Rand.

Cascio, Jamais. 2009. "Get Smarter." *Atlantic Monthly*, July/August. Retrieved May 11, 2010 (www.theatlantic.com).

Case, Nancy Humphrey. 2010. "When Employees Rule." *Christian Science Monitor*. March 29, 22.

Caston, Richard J. 1998. *Life in a Business-Oriented Society: A Sociological Perspective*. Boston: Allyn & Bacon.

Catalano, Shannan. 2007. "Intimate Partner Violence in the United States." Bureau of Justice Statistics. Retrieved April 9, 2010 (www.ojp.usdoj.gov/bjs).

Cech, Erin, Brian Rubineau, Susan Silbey, and Caroll Seron. 2011. "Professional Role Confidence and Gendered Persistence in Engineering." *American Sociological Review* 70 (October): 641–660.

Cellini, Stephanie R., Signe-Mary McKernan, and Caroline Ratcliffe. 2008. "The Dynamics of Poverty in the United States: A Review of Data, Methods, and Findings." *Journal of Policy Analysis and Management* 27 (Summer): 577–605.

"Census Bureau Reports Minority Business Ownership Increasing at More Than Twice the National Rate." 2010. U.S. Census Bureau Newsroom, July 13. Accessed November 20, 2011 (www.census.gov).

Center for American Progress. 2006. "Public Recognizes Debt as a Fast Growing Problem in U.S." July 19. Retrieved June 3, 2007 (www.americanprogress.org).

Center for American Women and Politics. 2011. "Women in Elective Office 2011." Accessed October 25, 2011 (www.cawp.rutgers.edu).

Center for Army Leadership. 2011. "CASAL: Army Leaders' Perceptions of Army Leaders and Army Leadership Practices." Special Report 2011-1, June. Accessed August 13, 2011 (usacac.leavenworth.army.mil).

Center for Responsive Politics. 2011. "Average Wealth of Members of Congress." Accessed February 15, 2012 (www.opensecrets.org).

Center for the Study of Global Christianity. 2007. "Global Table 5: Status of Global Mission, Presence and Activities, AD 1800–2005." Gordon-Conwell Theological Seminary. Retrieved August 24, 2007 (www.gcts.edu).

Centers for Disease Control and Prevention. 2011. "Trends in Current Cigarette Smoking Among High School Students and Adults, United States, 1965–2010." November 3. Accessed April 24, 2012 (www.cdc.gov).

Centers for Medicare & Medicaid Services. 2010. "NHE Tables for Selected Calendar 1960–2010." April 11. Accessed April 24, 2012 (www.cms.gov).

———. 2012. "National Health Expenditure Projections 2010–2020." April 11. Accessed April 24, 2012 (www.cms.gov).

Chamberlain, Catherine J., Leonidas G. Koniaris, Albert W. Wu, and Timothy M. Pawlik. 2012. "Disclosure of 'Nonharmful' Medical Errors and Other Events." *Archives of Surgery* 147 (March): 282–286.

Chambliss, William J., and Robert B. Seidman. 1982. *Law, Order, and Power*, 2nd edition. Reading, MA: Addison-Wesley.

Chan, Sam. 1997. "Families with Asian Roots." Pp. 251–344 in *Developing Cross-Cultural Competence: A Guide for Working with Children and Their Families*, 2nd edition, edited by Eleanor W. Lynch and Marci J. Hanson. Baltimore, MD: Paul H. Brookes.

Chandler, Tertius, and Gerald Fox. 1974. *3000 Years of Urban Growth*. New York: Academic Press.

Chandra, A., G. A. Martinez, W. D. Mosher, J. C. Abma, and J. Jones. 2005. "Fertility, Family Planning, and Reproductive Health of U.S. Women: Data from the 2002 National Survey of Family Growth." *Vital and Health Statistics* (Series 23, No. 25). Hyattsville, MD: National Center for Health Statistics.

Chandra, Anjani, William D. Mosher, Casey Copen, and Catlainn Sionean. 2011. "Sexual Behavior, Sexual Attraction, and Sexual Identity in the United States: Data from the 2006–2008 National Survey of Family Growth." *National Health Statistics Reports*, no. 36. Hyattsville, MD: National Center for Health Statistics.

Chandy, Laurence, and Geoffrey Gertz. 2011. "Poverty in Numbers: The Changing State of Global Poverty from 2005 to 2015." Brookings Institution, Policy Brief, January. Accessed September 22, 2011 (www.brookings.edu).

Chapman, Paige. 2010. "To Ban or Not? Gossip Web Sites Still Pose Troubling Questions for Colleges." *Chronicle of Higher Education*, September 27. Accessed September 28, 2010 (chronicle.com).

Chapman, Tony, and Jenny Hockey, eds. 1999. *Ideal Homes? Social Change and Domestic Life*. New York: Routledge.

Charles, Maria. 2011. "What Gender Is Science?" *Contexts* 10 (Spring): 22–28.

Chasin, Barbara H. 2004. *Inequality & Violence in the United States: Casualties of Capitalism*, second edition. Amherst, NY: Humanity Books.

Chelala, César. 2002. "World Violence Against Women a Great Unspoken Pandemic." *Philadelphia Inquirer*, November 4. Retrieved November 7, 2002 (www.commondreams.org).

Chen, Michelle. 2012. "Health Care System Leaves Patients Frustrated—Nurses Work for a Solution." Alternet, June 4. Accessed June 5, 2012 (www.alternet.org).

Chesler, Phyllis. 2006. "The Failure of Feminism." *Chronicle of Higher Education*, February 26, B12.

Chesney-Lind, Meda, and Katherine Irwin. 2008. *Beyond Bad Girls: Gender, Violence, and Hype*. New York: Routledge.

Chesney-Lind, Meda, and Nikki Jones, eds. 2010. *Fighting for Girls: New Perspectives on Gender and Violence*. Albany: State University of New York Press.

Chesney-Lind, Meda, and Lisa Pasko. 2004. *The Female Offender: Girls, Women, and Crime*, 2nd edition. Thousand Oaks, CA: Sage.

Child Trends. 2010. "Parental Expectations for Children's Academic Attainment." Accessed July 27, 2011 (www.childtrendsdatabank.org).

Children's Defense Fund. 2008. *The State of America's Children 2008*. Retrieved July 12, 2009 (www.unicef.org).

"China's Population to Peak at 1.4 Billion Around 2026." 2012. U.S. Census Bureau Newsroom, March 22. Accessed May 15, 2012 (www.census.gov).

Choice, Pamela, and Leanne K. Lamke. 1997. "A Conceptual Approach to Understanding Abused Women's Stay/Leave Decisions." *Journal of Family Issues* 18 (May): 290–314.

Chou, Roger, et al. 2011. "Screening for Prostate Cancer: A Review of the Evidence for the U.S. Preventive Services Task Force." *Annals of Internal Medicine* 155 (April 3): 762–771.

Choudhury, Tufyal, Mohammed Aziz, Duaa Izzidien, Intissar Khreeji, and Dilwar Hussein. 2006. "Perceptions of Discrimination and Islamophobia: Voices from Members of Muslim Communities in the European Union." Retrieved September 1, 2007 (www.eumc.eu).

Christenson, Matthew, Thomas M. McDevitt, and Karen A. Stanecki. 2004. *Global Population Profile: 2002*. U.S. Census Bureau, International Population Reports WP/02. Washington, DC: U.S. Government Printing Office.

Christian, Steve. 2009. "Children of Incarcerated Parents." National Conference of State Legislatures, March. Accessed July 25, 2011 (www.cga.ct.gov).

Christina, Greta. 2011. "Wealthy, Handsome, Strong, Packing Endless Hard-Ons: The Impossible Ideals Men Are Expected to Meet." Independent Media Institute, June 20. Accessed June 21, 2011 (www.alternet.org).

*Chronicle of Higher Education*. 2010. "Student Demographics." *Almanac of Higher Education* 2010–11, 52 (August 27): 32.

Chu, Henry. 2007. "A Gift for India's Inter-Caste Couples." *Los Angeles Times*, November 4. Retrieved November 6, 2007 (www.latimes.com).

Chu, Kathy. 2010. "Fast-Food Chains in Asia Cater Menus to Customers." *USA Today*, September 7. Accessed September 8, 2010 (www.usatoday.com).

Churchill, Ward. 1997. *A Little Matter of Genocide: Holocaust and Denial in the Americas, 1492 to the Present*. San Francisco: City Lights Books.

*CIA World Factbook*. 2010. "Country Comparison: Distribution of Family Income, Gini Index." March 8. Accessed September 23, 2011 (www.cia.gov).

Ciabattari, Teresa. 2004. "Cohabitation and Housework: The Effects of Marital Intentions." *Journal of Marriage and Family* 66 (February): 118–125.

Cichocki, Mary K. 1981. "Women's Travel Patterns in a Suburban Development." Pp. 151–163 in *New Space for Women*, edited by Gerda R. Wekerle, Rebecca Peterson, and David Morley. Boulder, CO: Westview Press.

Citizens Against Government Waste. 2011. "2010 Pig Book Summary." Accessed February 15, 2012 (www.cagw.org).

Claxton, Gary, et al. 2011. "Employer Health Benefits, 2011 Annual Survey." Kaiser Family Foundation and Health Research & Educational Trust. Accessed April 15, 2012 (ehbs.kff.org).

Clayson, Dennis E., and Debra A. Haley. 2011. "Are Students Telling Us the Truth? A Critical Look at the Student Evaluation of Teaching." *Marketing Education Review* 21 (Summer): 101–112.

Clayton, Mark. 2010. "Earth's Growing Nitrogen Threat." *Christian Science Monitor*, January 10, 36–37.

Clement, Scott. 2011. "Workplace Harassment Drawing Wide Concern." *Washington Post*, November 16. Accessed February 25, 2012 (washingtonpost.com).

Clements, David D. 2012. *Corporations Are Not People: Why They Have More Rights Than You Do and What You Can Do About It*. San Francisco: Berrett-Koehler.

Clifford, Stephanie. 2009. "Online Age Quiz Is a Window for Drug Makers." *New York Times*, March 26, 1.

Clifford, Stephanie. 2012. "Unexpected Ally Helps Wal-Mart Cut Waste." *New York Times*, April 13, B1.

Clifford, Stephanie, and Andrew Martin. 2011. "In Time of Scrimping, Fun Stuff Is Still Selling." *New York Times*, September 24, A1.

Clifton, Donna, and Ashley Frost. 2011. "The World's Women and Girls 2011 Data Sheet." Population Reference Bureau. Accessed October 25, 2011 (www.prb.org).

Clifton, Jon. 2010. "Roughly 6.2 Million Mexicans Express Desire to Move to U.S." Gallup, June 7. Accessed November 15, 2011 (www.gallup.com).

Cloke, Kenneth, and Joan Goldsmith. 2002. *End of Management and the Rise of Organizational Democracy.* San Francisco: Jossey-Bass.

Cloud, John. 2010. "How to Recruit Better Teachers." *Time*, September 20, 47–52.

Cochrane, Megan. 2012. "Fewer Americans Report Healthy Eating Habits in 2011." Gallup, April 11. Accessed April 22, 2012 (www.gallup.com).

Cohan, Catherine I., and Stacey Kleinbaum. 2002. "Toward a Greater Understanding of the Cohabitation Effect: Premarital Cohabitation and Marital Communication." *Journal of Marriage and Family* 64 (February): 180–192.

Cohen, Jere. 2002. *Protestantism and Capitalism: The Mechanisms of Influence.* New York: Aldine de Gruyter.

Cohen, Naomi. 2011. "Define Gender Gap? Look Up Wikipedia's Contributor List." *New York Times*, January 31, A1.

Cohen, Philip N. 2012. "Recession and Divorce in the United States: Economic Conditions and the Odds of Divorce, 2008–2010." Maryland Population Research Center, Working Paper. Accessed May 24, 2012 (www.popcenter.umd.edu).

Cohn, D'Vera. 2010. "Cell Phone Challenge for the Census." Pew Research Center, December 17. Accessed July 6, 2011 (pewresearch.org).

———. 2011. "India Census Offers Three Gender Options." Pew Research Center, February 7. Accessed October 25, 2011 (www.pewresearch.org).

Cohn, D'Vera, Jeffrey S. Passel, Wendy Wang, and Gretchen Livingston. 2011. "Barely Half of U.S. Adults Are Married—A Record Low." Pew Research Center, December 14. Accessed March 1, 2012 (www.pewsocialtrends.org).

Colapinto, John. 1997. "The True Story of John/Joan." *Rolling Stone* (December 11): 54–73, 92–97.

———. 2001. *As Nature Made Him: The Boy Who Was Raised as a Girl.* New York: Harper Perennial.

———. 2004. "What Were the Real Reasons Behind David Reimer's Suicide?" *Slate*, June 3. Retrieved April 24, 2008 (www.slate.com).

Coley, Rebekah L. 2002. "What Mothers Teach, What Daughters Learn: Gender Mistrust and Self-Sufficiency Among Low-Income Women." Pp. 97–106 in *Just Living Together: Implications of Cohabitation on Families, Children, and Social Policy,*

edited by Alan Booth and Ann C. Crouter. Mahwah, NJ: Lawrence Erlbaum Associates.

Coll, Steve. 2008. *The Bin Ladens: An Arabian Family in the American Century.* New York: Penguin.

College Board. 2012. "The 8th Annual AP Report to the Nation." February 8. Accessed March 15, 2012 (apreport.collegeboard.org).

Collins, Patricia Hill. 1990. *Black Feminist Thought: Knowledge, Consciousness, and the Politics of Empowerment.* New York: Routledge.

Collins, Randall. 2002. "Credential Inflation and the Future of Universities." Pp. 23–46 in *The Future of the City of Intellect: The Changing American University,* edited by Steven Brint. Stanford, CA: Stanford University Press.

Conklin, John E. 2009. *Criminology,* 10th edition. Boston: Allyn & Bacon.

Conley, Dalton. 2004. *The Pecking Order: Which Siblings Succeed and Why.* New York: Pantheon.

Conlin, M. 2007. "The Kids Are All Right." *Business Week*, October 8, 18.

Connell, Dave. 2011. "30 Days of H20: Final Days." February 22. Accessed June 6, 2012 (blog.nature.org).

Connolly, Frank W. 2001. "My Students Don't Know What They're Missing." *Chronicle of Higher Education*, December 21, B5.

Conrad, Peter, and Kristin K. Barker. 2010. "The Social Construction of Illness: Key Insights and Policy Implications." *Journal of Health and Social Behavior* 51 (Suppl.): S67–S79.

Considine, Austin. 2011. "For Asian-American Stars, Many Web Fans." *New York Times*, July 31, ST6.

Conway, Kevin P., and Joan McCord. 2005. "Co-Offending and Patterns of Juvenile Crime." Washington, DC: National Institute of Justice. Retrieved October 12, 2006 (www.ojp.usdoj.gov/nij).

Cook, Bryan, and Young Kim. 2012. *The American College President 2012.* Washington, DC: American Council on Education.

Cook, Thomas D., and Donald T. Campbell. 1979. *Quasi-Experimentation: Design and Analysis Issues for Field Settings.* Chicago: Rand McNally.

Cooley, Charles Horton. 1909/1983. *Social Organization: A Study of the Larger Mind.* New Brunswick, NJ: Transaction Books.

Coontz, Stephanie. 2005. *Marriage: A History: How Love Conquered Marriage.* New York, Penguin.

Cooper, O. R., et al. 2010. "Increasing Springtime Ozone Mixing Ratios in the Free Troposphere over Western North America." *Nature* 463 (January 21): 344–348.

Cooperman, Alan. 2010. "Ask the Expert." Pew Research Center, December 29. Accessed December 30, 2010 (pewresearch.org).

Coser, Lewis. 1956. *The Functions of Social Conflict.* New York: Free Press.

Cowell, Alan, and James Kanter. 2011. "E. Coli Strain Was Previously Unknown, Official Says." *New York Times*, June 2, A11.

Crabtree, Steve. 2010. "Religiosity Highest in World's Poorest Nations." Gallup, June 3. Accessed August 31, 2012 (www.gallup.com).

Cressey, Donald R. 1953. *Other People's Money: A Study in the Social Psychology of Embezzlement.* Glencoe, IL: Free Press.

Crockett, Ariel. 2011. "Is the Black Television Series Dead?" Hello Beautiful, June 3. Accessed December 6, 2011 (hellobeautiful.com).

Crosby, Alex E., LaVonne Ortega, and Mark R. Stevens. 2011. "Suicides—United States, 1999–2007." Morbidity *and Mortality Weekly Report* 60 (January 14): 56–59. Accessed July 22, 2011 (www.cdc.gov).

Crothers, Charles. 1979. "On the Myth of Rural Tranquility: Comment on Webb and Collette." *American Journal of Sociology* 84 (May): 1441–1445.

Crowe, Edward. 2010. "Measuring What Matters: A Stronger Accountability Model for Teacher Education." Center for American Progress, July. Accessed March 11, 2012 (www.americanprogress.org).

Crowley, Michael. 2010. "What $120 Million Buys." Time, October 11, 35–37.

Cruikshank, Margaret. 2009. *Learning to Be Old: Gender, Culture, and Aging.* Lanham, MD: Rowman & Littlefield.

Cudd, Ann E., and Nancy Holstrom. 2010. Capitalism, For and Against: A Feminist Debate. New York: Cambridge University Press.

Culbert, Samuel A. 2011. "Why Your Boss Is Wrong About You." *New York Times*, March 3, A25.

Cullen, Lisa T., and Coco Masters. 2008. "We Just Clicked." *Time*, January 28, 86–89.

Cumberworth, Erin. 2010. "Homeboy Industries." *Pathways*, Summer, 22–23.

Currie, E. 1985. *Confronting Crime: An American Challenge.* New York: Pantheon.

Currie, Janet. 2011. "Inequality at Birth: Some Causes and Consequences." National Bureau of Economic Research, February. Accessed September 22, 2011 (www.nber.org).

Currie, Janet, and Erdal Tekin. 2006. "Does Child Abuse Cause Crime?" Cambridge, MA: National Bureau of Economic Research, Working Paper 12171. Retrieved June 12, 2006 (papers.nber.org).

Curtiss, Susan. 1977. *Genie.* New York: Academic Press.

Cutler, David M., and Mary Beth Landrum. 2011. "Dimensions of Health in the Elderly Population." National Bureau of Economic Research, June. Accessed April 24, 2012 (www.nber.org).

Cynkar, Peter, and Elizabeth Mendes. 2011. "More Than One in Six American Workers Also Act as Caregivers." Gallup, July 26. Accessed March 10, 2012 (www.gallup.com).

Dadigan, Marc. 2011. "What Makes a Tribe?" *Christian Science Monitor*, August 15 & 22, 30–31.

Dahl, Melissa. 2011. "Dangerous Beauty: FDA Discusses Contaminated Cosmetics." MSNBC News, November 30. Accessed May 10, 2012 (msnbc.msn.com).

Dahl, Robert A. 1961. *Who Governs? Democracy and Power in an American City.* New Haven. CT: Yale University Press.

Dalton, Madeline A., et al. 2005. "Use of Cigarettes and Alcohol by Preschoolers While Role-Playing as Adults." *Archives of Pediatrics & Adolescent Medicine* 159 (September): 854–859.

Damaske, Sarah. 2011. "A 'Major Career Woman'? How Women Develop Early Expectations about Work." *Gender & Society* 25 (August): 409–430.

Dandaneau, Steven P. 2001. *Taking It Big: Developing Sociological Consciousness in Postmodern Times*. Thousand Oaks, CA: Pine Forge Press.

Daniel, Lisa. 2011. "Panel Says Rescind Policy on Women in Combat." March 8. Accessed February 22, 2012 (www.army.mil).

Dante, Ed. 2010. "The Shadow Scholar." *The Chronicle Review*, November 19, B6–B9.

Danziger, S. 2008. "The Price of Independence: The Economics of Early Adulthood." *Family Focus* 53 (March): F7–F8.

Dao, James. 2011. "After Combat, the Unexpected Perils of Coming Home." *New York Times*, May 29, A1.

———. 2011. "In California, Indian Tribes with Casino Money Cast Off Members." *New York Times*, December 11, A20.

Databeast. 2012. "How American Singles Really Feel About Sex." *Newsweek*, February 13, 15.

Davey, Monica, and Susan Saulny. 2009. "Blagojevich Charged with 16 Corruption Felonies." *New York Times*, April 3, A1.

Davidson, Paul. 2011. "Season of Part-Time Jobs Kicks Off with Holidays." USA Today, November 25–27, A1–A2.

Davies, James B. 2008. "The World Distribution of Household Wealth." United Nations University, February. Accessed September 21, 2011 (www.wider.unu.edu).

Davies, James C. 1962. "Toward a Theory of Revolution." *American Sociology Review* 27 (February): 5–19.

———. 1979. "The J-Curve of Rising and Declining Satisfaction as a Cause of Revolution and Rebellion." Pp. 413–436 in *Violence in America: Historical and Comparative Perspectives*, edited by Hugh D. Graham and Ted R. Gurr. Beverly Hills, CA: Sage.

Davis, F. J. 1991. *Who is Black?* University Park: Pennsylvania State University Press.

Davis, Kenneth C. 2010. "America's True History of Religious Tolerance." *Smithsonian* magazine, October. Accessed September 28, 2010 (www.smithsonian.com).

Davis, Kingsley, and Wilbert E. Moore. 1945. "Some Principles of Stratification." *American Sociological Review* 10 (April): 242–249.

Day, Jennifer Cheeseman, and Kelly Holder. 2004. "Voting and Registration in the Election of November 2002." U.S. Census Bureau, Current Population Reports, P20-552. Retrieved August 2, 2004 (www.census.gov).

de la Cruz, G. Patricia, and Angela Brittingham. 2003. "The Arab Population: 2000." U.S. Census Bureau. Retrieved May 3, 2004 (www.census.gov).

de las Casas, Bartolome. 1992. *The Devastation of the Indies: A Brief Account*. Baltimore, MD: Johns Hopkins University Press.

De Long, J. Bradford, and Andrei Shleifer. 1992. "Princes and Merchants: European City Growth Before the Industrial Revolution." National Bureau of Economic Research, December. Retrieved January 10, 2008 (www.nper.org).

Death Penalty Information Center. 2009. "The Death Penalty in 2009: Year End Report." Retrieved February 25, 2010 (www.deathpenaltyinfo.org).

———. 2012. "Facts About the Death Penalty." August 19. Accessed August 20, 2011 (www.deathpenaltyinfo.org).

Deaton, Angus S. 2009. "Aging, Religion, and Health." National Bureau of Economic Research, August. Retrieved April 26, 2010 (www.nber.org).

Dee, Thomas S. 2006. "The Why Chromosome." *Education Next* No 4. (Fall): 69–75.

Deegan, Mary Jo. 1986. *Jane Addams and the Men of the Chicago School, 1892–1918*. New Brunswick, NJ: Transaction Books.

Degler, Carl. 1981. *At Odds: Women and the Family in America from the Revolution to the Present*. New York: Oxford University Press.

DeGraw, David. 2011. "Americans Don't Realize Just How Badly We're Getting Screwed by the Top 0.1 Percent Hoarding the Country's Wealth." AlterNet, August 14. Accessed August 15, 2011 (www.alternet.org).

DeKeseredy, Walter S. 2011. *Violence against Women: Myths, Facts, Controversies*. Toronto: University of Toronto Press.

DeLeire, Thomas, and Leonard M. Lopoo. 2010. "Family Structure and the Economic Mobility of Children." Economic Mobility Project, April. Accessed February 5, 2011 (www.economicmobility.org).

DeLisi, Matt, Anna Kosloski,, Molly Sween, Emily Hachmeister, Matt Moore, and Alan Drury. 2010. "Murder by Numbers: Monetary Costs Imposed by a Sample of Homicide Offenders." *Journal of Forensic Psychiatry & Psychology* 21 (August): 501–513.

Dell, Kristina. 2011. "I Owe U." *Time*, October 31, 40–44.

della Porta, Donatella, and Mario Diani. 1999. *Social Movements: An Introduction*. Malden, MA: Blackwell.

DeMaris, Alfred. 2001. "The Influence of Intimate Violence on Transitions Out of Cohabitation." *Journal of Marriage and Family* 63 (February): 235–246.

Demos, John. 1986. *Past, Present, and Personal: The Family and the Life Course in American History*. New York: Oxford University Press.

DeNavas-Walt, Carmen, Bernadette D. Proctor, and Jessica C. Smith. 2011. *Income, Poverty, and Health Insurance Coverage in the United States: 2010*. U.S. Census Bureau, Current Population Reports, P60-239. Washington, DC: U.S. Government Printing Office.

Denton-Borhaug, Kelly. 2004. "The Complex and Rich Landscape of Student Spirituality: Findings from the Goucher College Spirituality Survey." *Religion & Education* 31 (Fall): 41–61.

DeParle, Jason. 2012. "Harder for Americans to Rise From Lower Rungs." New York Times, January 5, A1.

Deresiewicz, William. 2009. "Faux Friendship." *The Chronicle Review*, December 11, B6–B10.

Desmond-Harris, Jenee. 2009. "Why Michelle's Hair Matters." *Time*, September 7, 55–57.

DeVega, Chauncey. 2012. "The 10 Most Racist Moments of the GOP Primary (So Far)." AlterNet, January 25. Accessed January 26, 2012 (www.alternet.org).

Deveny, Kathleen. 2008. "They're No Baby Einsteins." *Newsweek*, January 14, 61.

Dew, Jeffrey. 2009. "Bank on It: Thrifty Couples Are the Happiest." Pp. 23–30 in *The State of Our Union, Marriage in America 2009: Money & Marriage*, edited by W.

Bradford Wilcox and Elizabeth Marquardt. Charlottesville, VA: The National Marriage Project.

Dewan, Shaila, and Robert Gebeloff. 2012. "Among the Wealthiest One Percent, Many Variations." New York Times, January 15, A1.

Dey, Judy Goldberg, and Catherine Hill. 2007. "Behind the Pay Gap." American Association of University Women Educational Foundation. Retrieved March 1, 2008 (www.aauw.org).

Diamond, Marie. 2011. "Contrary to GOP Claims, U.S. Has Second Lowest Corporate Taxes in the Developed World." AlterNet, July 6. Accessed July 7, 2011 (www.alternet.org).

———. 2012. "Santorum Staffer Says Women Shouldn't Be President Because It's Against God's Will." Think Progress, January 17. Accessed April 10, 2012 (thinkprogress.org).

Diamond, Milton, and H. Keith Sigmundson. 1997. "Sex Reassignment at Birth: Long-Term Review and Clinical Implications." *Archives of Pediatrics & Adolescent Medicine* 15 (March): 298–304.

Dickert-Conlin, Stacy, and Ross Rubenstein. 2007. *Economic Inequality and Higher Education: Access Persistence and Success*. New York: Russell Sage Foundation.

Dilworth-Anderson, Peggye, Linda M. Burton, and William L. Turner. 1993. "The Importance of Values in the Study of Culturally Diverse Families." *Family Relations* 42 (July): 238–242.

Dines, Gail. 2010. *Pornland: How Porn Has Hijacked Our Sexuality*. Boston: Beacon Press.

DiSalvo, David. 2010. "Are Social Networks Messing with Your Head?" *Scientific American Mind* 20 (January/February): 48–55.

DiversityInc. 2008. "Fortune 500 Black, Latino, Asian CEOs." July 22. Retrieved September 11, 2008 (www.diversity.com).

Dixon, Robyn. 2009. "Africa's Bitter Cycle of Child Slavery." *Los Angeles Times*, July 12. Retrieved July 13, 2009 (www.latimes.com).

Do, Hien Duc. 1999. *The Vietnamese Americans*. Westport, CT: Greenwood Press.

Dodson, Lisa, and Wendy Luttrell. 2011. "Families Facing Untenable Choices." *Contexts* 10 (Winter): 38–42.

Dodson, Robin E., Marcia Nishioka, Laurel J. Standley, Laura J. Perovich, Julia Green Brody, and Ruthann A. Rudel. 2012. "Endocrine Disruptors and Asthma-Associated Chemicals in Consumer Products." *Environmental Health Perspectives*, March 8. Accessed May 10, 2012 (ehponline.org).

Doeringer, Peter B., and Michael J. Piore. 1971. *Internal Labor Markets and Manpower Analysis*. Lexington, MA: Heath-Lexington Books.

Dokoupil, Tony. 2011. "Mad as Hell." *Newsweek*, June 6, 6–7.

Dolnick, Sam. 2011. "Dance, Laugh, Drink, Save the Date: It's a Ghanaian Funeral." *New York Times*, April 12, A1.

Domhoff, G. William. 2006. *Who Rules America? Power, Politics, & Social Change*, 5th edition. Boston: McGraw Hill.

Domosh, Mona, and Joni Seager. 2001. *Feminist Geographers Make Sense of the World*. New York: Guilford Press.

Dorius, Cassandra J., Stephen J. Bahr, John P. Hoffman, and Elizabeth L. Harmon. 2004. "Parenting Practices as Moderators of the Relationship between Peers and Adolescent Marijuana Use." *Journal of Marriage and Family* 66 (February): 163–178.

Douglas, Susan J., and Meredith W. Michaels. 2004. *The Mommy Myth: The Idealization of Motherhood and How It Has Undermined Women.* New York: Free Press.

Dovidio, John F. 2009. "Racial Bias, Unspoken but Heard." *Science* 326 (December 18): 1641–1642.

Doyle, Francis X. 2011. "Bishops' Missed Opportunity." *Baltimore Sun*, November 14, 15.

Draffan, George. 2003. "Profile of Boeing." Retrieved September 5, 2004 (www.endgame.org).

Dreazen, Yochi J. 2009. "Muslim Population in the Military Raises Difficult Issues." *Wall Street Journal*, November 9. Retrieved November 12, 2009 (online.wsj.com).

Drogin, Bob. 2009. "Same-Sex Vote Draws U.S. Focus." *Baltimore Sun*, November 4, 10.

Du Bois, W. E. B. 1986. *The Souls of Black Folk.* New York: Library of America.

Dubowitz, Tamara, Melonie Heron, Ricardo Basurto-Davila, Chloe E. Bird, Nicole Lurie, and José J. Escarce. 2011. "Racial/Ethnic Differences in U.S. Health Behaviors: A Decomposition Analysis." *American Journal of Health Behavior* 35 (May): 290–304.

Duda, Jeremy. 2011. "Judge Upholds Major Portions of Alabama's SB1070-Style Law." *Arizona Capitol Times*, September 29. Accessed November 15, 2011 (azcapitoltimes.com).

Duhigg, Charles. 2009a. "Clean Water Laws Are Neglected, at a Cost of Suffering." *New York Times*, August 10, A1.

———. 2009b. "Millions in U.S. Drink Dirty Water, Records Show." *New York Times*, December 8, A1.

———. 2010. "Saving U.S. Water and Sewer Systems Would Be Costly." *New York Times*, March 14, A1.

Duncan, Greg J., Johanne Boisjoly, Michael Kremer, Dan M. Levy, and Jacque Eccles. 2005. "Peer Effects in Drug Use and Sex Among College Students." *Journal of Abnormal Child Psychology* 33 (June): 375–385.

Dunifon, Rachel, and Lori Kowaleski-Jones. 2007. "The Influence of Grandparents in Single-Mother Families." *Journal of Marriage and Family* 69 (May): 465–481.

Durden, Tyler. 2011. "Here Are the 29 Public Companies with More Cash Than the US Treasury." Zero Hedge, July 15. Accessed January 9, 2012 (www.zerohedge.com).

Durkheim, Emile. 1893/1964. *The Division of Labor in Society.* New York: Free Press.

———. 1897/1951. *Suicide: A Study in Sociology*, translated by John A. Spaulding and George Simpson, edited by George Simpson. New York: Free Press.

———. 1898/1956. *Education and Sociology*, translated by Sherwood D. Fox. Glencoe, IL: Free Press.

———. 1961. *The Elementary Forms of the Religious Life.* New York: Collier Books.

Durose, Matthew R., and Patrick A. Langan. 2004. "Felony Sentences in State Courts, 2002." Washington, DC: U.S. Department of Justice, Bureau of Justice Statistics.

Dush, Kamp, Catherine Cohan, and Paul Amato. 2003. "The Relationship between Cohabitation and Marital Quality and Stability: Change Across Cohorts?" *Journal of Marriage and Family* (August): 539–549.

Duster, Troy. 2005. "Race and Reification in Science." *Science* 307, February 18, 1050–1051.

Dye, Thomas R., and Harmon Ziegler. 2003. *The Irony of Democracy: An Uncommon Introduction to American Politics*, 12th edition. Belmont, CA: Wadsworth.

Eberstadt, Nicholas. 2009. "Poor Statistics." *Forbes*, March 2, 26.

Eckstein, Rick, Rebecca Schoenike, and Kevin Delaney. 1995. "The Voice of Sociology: Obstacles to Teaching and Learning the Sociological Imagination." *Teaching Sociology* 23 (October): 353–363.

"Economy Dominates Public's Agenda, Dims Hopes for the Future." 2011. Pew Research Center for the People & the Press, January 20. Accessed November 15, 2011 (www.people-press.org).

Edin, Kathryn, and Laura Lein. 1997. *Making Ends Meet: How Single Mothers Survive Welfare and Low-Wage Work.* New York: Russell Sage Foundation.

Egan, Timothy. 2006. "The Rise of Shrinking-Vacation Syndrome." *New York Times*, August 20, A18.

Egley, Arlen, Jr., and Christina E. O'Donnell. 2009. "Highlights of the 2007 National Youth Gang Survey." Office of Justice Programs, April. Retrieved February 22, 2010 (www.ojp.usdoj.gov).

Ehrenreich, Barbara. 2001. *Nickel and Dimed: On (Not) Getting by in America.* New York: Metropolitan Books.

Ehrlich, Paul. 1971. *The Population Bomb*, 2nd edition. San Francisco: Freeman.

Ehrlich, Paul R., and Anne H. Ehrlich. 2008. *The Dominant Animal: Human Evolution and the Environment.* Washington, DC: Island Press.

Eibner, Christine, and Carter C. Price. 2012. "The Effect of the Affordable Care Act on Enrollment and Premiums, With and Without the Individual Mandate." RAND Health. Accessed April 15, 2012 (www.rand.org).

Eisenberg, Abne M., and Ralph R. Smith, Jr. 1971. *Nonverbal Communication.* Indianapolis: Bobbs-Merrill.

Eisenberg, Nancy, et al., 2005. "Relations among Positive Parenting, Children's Effortful Control, and Externalizing Problems: A Three-Wave Longitudinal Study." *Child Development* 76 (September/October): 1055–1071.

Ekman, Paul. 1985. *Telling Lies: Clues to Deceit in the Marketplace, Politics, and Marriage.* New York: W.W. Norton & Company.

Ekman, Paul, and Wallace V. Friesen. 1984. *Unmasking the Face: A Guide to Recognizing Emotions from Facial Clues.* Palo Alto, CA: Consulting Psychologists Press.

El Nasser, Haya, and Paul Overberg. 2011. "In Many Neighborhoods, Kids Are Only a Memory." *USA Today*, June 3, 1A, 6A.

Elgin, Ben. 2008. "Green—Up to a Point." *Business Week*, March 3, 25–26.

Ellingwood, Ken. 2009. "Not Being on Time a High Art in Mexico." *Los Angeles Times*, September 12. Retrieved September 22, 2009 (www.latimes.com).

Ellis, Bill. 2005. "Legend/AntiLegend: Humor as an Integral part of the Contemporary Legend Process." Pp. 123–140 in *Rumor Mills: The Social Impact of Rumor and Legend*, edited by Gary Alan Fine, Véronique Campion-Vincent, and Chip Heath. New Brunswick, NJ: Transaction Publishers.

Ellison, Christopher G., Amy M. Burdette, and W. Bradford Wilcox. 2010. "The Couple That Prays Together: Race and Ethnicity, Religion, and Relationship Quality Among Working-Age Adults." *Journal of Marriage and Family* 72 (August): 963–975.

Ellison, Jesse. 2011. "The 2011 Global Women's Progress Report." *Newsweek*, September 26, 27–33.

Ellsberg, Michael. 2011. *The Education of Millionaires: It's Not What You Think and It's Not Too Late.* New York: Portfolio/Penguin.

Emanuel, Ezekiel J. 2011. "Billions Wasted on Billing." *New York Times*, November 12. Accessed April 24, 2012 (www.nytimes.com).

Engemann, Kristie M., and Michael T. Owyang. 2005. "So Much for That Merit Raise: The Link between Wages and Appearance." *Regional Economist*, April, 10–11.

England, Paula. 2010. "The Gender Revolution: Uneven and Stalled." *Gender & Society* 24 (April): 149–166.

England, Paula, Emily Fitzgibbons Shafer, and Alison C. K. Fogarty. 2007. "Hooking Up and Forming Romantic Relationships on Today's College Campuses." Pp. 531–547 in *The Gendered Society Reader*, 3rd edition, edited by Michael Kimmel. New York: Oxford University Press.

England, Paula, and Reuben J. Thomas. 2009. "The Decline of the Date and the Rise of the College Hook Up." Pp. 141–152 in *Family in Transition*, 15th edition, edited by Arlene S. Skolnick and Jerome H. Skolnick. Boston: Pearson Higher Education.

English, Cynthia. 2011. "Most Americans See College as Essential to Getting a Good Job." Gallup, August 18. Accessed April 3, 2012 (www.gallup.com).

Ennis, Sharon R., Merarys Rios-Vargas, and Nora G. Gilbert. 2011. "The Hispanic Population: 2010." U.S. Census Bureau Brief, May. Accessed December 7, 2011 (www.census.gov).

Environmental Integrity Project. 2007. "Paying Less to Pollute: Environmental Enforcement Under the Bush Administration." Retrieved February 12, 2008 (www.enrironmentalintegrity.org).

EPA. 2012. "Municipal Solid Waste." Environmental Protection Agency, April 3. Accessed June 22, 2012 (www.epa.gov).

Epstein, Joseph. 2011. *Gossip: The Untrivial Pursuit.* New York: Houghton Mifflin Harcourt.

Erikson, Kai T. 1966. *Wayward Puritans: A Study in the Sociology of Deviance.* New York: John Wiley & Sons.

Esipova, Neli, Julie Ray, and Rajesh Srinivasan. 2010. "Young, Less Educated Yearn to Migrate to the U.S." Gallup, April 30. Retrieved April 30, 2010 (www.gallup.com).

Esposito, John L., and Dalia Mogahed. 2007. *Who Speaks for Islam? What a Billion Muslims Really Think*. New York: Gallup Press.

Essed, Philomena, and David Theo Goldberg, eds. 2002. *Race Critical Theories: Text and Context*. Malden, MA: Blackwell Publishers.

Esty, Daniel C., M. A. Levy, C. H. Kim, A. de Sherbinin, T. Srebojnak, and V. Mara. 2008. *2008 Environmental Performance Index*. New Haven, CT: Yale Center for Environmental Law and Policy.

Evans, Harold, Gail Buckland, and David Lefer. 2006. *They Made America: From the Steam Engine to the Search Engine: Two Centuries of Innovators*. New York: Little, Brown.

Evans, M. D. R., Jonathan Kelley, Joanna Sikora, and Donald J. Treiman. 2010. "Family Scholarly Culture and Educational Success: Books and Schooling in 27 Nations." *Research in Social Stratification and Mobility*, in press. Accessed July 20, 2011 (doi:10.1016/j.rssm.2010.01.002).

Ewen, Lynda Ann. 1998. *Social Stratification and Power in America: A View from Below*. Six Hills, NY: General Hall.

Ewers, Justin. 2004. "Drowning in Applications." *U.S. News & World Report*, December 20, 64–65.

Expedia.com. 2008. "2008 International Vacation Deprivation Survey Results." Retrieved July 4, 2008 (www.expedia.com).

"Face Veil Banned in Universities." 2010. *Time*, August 2, 12.

Fagan, Brian. 2008. *The Great Warming: Climate Change and the Rise and Fall of Civilizations*. New York: Bloomsbury Press.

Fahrenthold, David A. 2009. "Environmentalists Seek to Wipe Out Plush Toilet Paper." *Washington Post*, September 24, A1.

Fakhraie, Fatemeh. 2009. "Feminists Don't Understand Muslim Women." Double X, May 20. Retrieved April 26, 2010 (www.doublex.com).

Falah, Ghazi-Wald, and Caroline Nagel, eds. 2005. *Geographies of Muslim Women: Gender, Religion, and Space*. New York: Guilford Press.

Fallis, David S., Scott Higham, and Kimberly Kindy. 2012. "Congressional Earmarks Sometimes Used to Fund Projects Near Lawmakers' Properties." Washington Post, February 6. Accessed February 7, 2012 (www.washingtonpost.com).

Faris, Robert, and Diane Felmlee. 2011. "Status Struggles: Network Centrality and Gender Segregation in Same- and Cross-Gender Aggression." *American Sociological Review* 76 (February): 48–73.

Farley, John E., and Gregory D. Squires. 2005. "Fences and Neighbors: Segregation in 21st-century America." *Contexts* 4 (Winter): 33–39.

Farley, Melissa. 2001. "Prostitution: The Business of Sexual Exploitation." Pp. 879–891 in *Encyclopedia of Women and Gender*, volume 2, edited by Judith Worrell. New York: Academic Press.

Farley, Melissa, et al. 2011. "Comparing Sex Buyers with Men Who Don't Buy Sex: 'You can have a good time with the servitude' vs. 'You're supporting a system of degradation'." Paper presented at

Psychologists for Social Responsibility Annual Meeting, July 15, 2011, Boston.

Farley, Sally D., Amie M. Ashcraft, Mark F. Stasson, and Rebecca L. Nusbaum. 2010. "Nonverbal Reactions to Conversational Interruption: A Test of Complementarity Theory and the Status/Gender Parallel." *Journal of Nonverbal Behavior* 34 (December): 193–206.

Farnam, T. W. 2012. "White House Visitor Logs Provide Window Into Lobbying Industry." *Washington Post*, May 20. Accessed May 21, 2012 (www.washingtonpost.com).

Farr, Kathryn. 2005. *Sex Trafficking: The Global Market in Women and Children*. New York: Worth.

Farrell, Michael B. 2010. "Trial Raises Stakes in Gay-Marriage Debate." *Christian Science Monitor*, January 24, 19–20.

Faussett, Richard. 2011. "Alabama Sets New Standard." *Baltimore Sun*, June 10, 6.

Feagin, Joe R. 2001. "Social Justice and Sociology: Agendas for the Twenty-First Century." *American Sociological Review* 66 (February): 1–20.

Feagin, Joe R., and Clairece Booher Feagin. 2008. *Racial and Ethnic Relations*, 8th edition. Upper Saddle River, NJ: Prentice Hall.

Feagin, Joe R., and Melvin P. Sikes. 1994. *Living with Racism: The Black Middle-Class Experience*. Boston: Beacon Press.

Feagin, Joe R., and Robert Parker. 1990. *Building American Cities: The Urban Real Estate Game*, 2nd edition. Englewood Cliffs, NJ: Prentice Hall.

Federal Bureau of Investigation. 2009. *Crime in the United States 2008*. Retrieved February 12, 2010 (www.fbi.gov).

———. 2009. "Family Child Abductions." Retrieved December 18, 2009 (www.fbi.gov).

———. 2011. *Crime in the United States, 2010*. Accessed August 21, 2011 (www.fbi .gov/ucr).

Federal Interagency Forum on Child and Family Statistics. 2011. *America's Children: Key National Indicators of Well-Being, 2011*. Washington, DC: U.S. Government Printing Office.

Feldman-Jacobs, Charlotte, and Donna Clifton. 2010. "Female Genital Mutilation/Cutting: Data and Trends, Update 2010." Population Reference Bureau, February. Accessed April 20, 2012 (www.prb.org).

Fenton, Justin. 2010. "Bad News Spreads Fast." *Baltimore Sun*, October 21, 1, 8.

Ferguson, Niall. 2012. "Rich America, Poor America." *Newsweek*, January 23, 42-47.

Ferree, Myra M. 2005. "It's Time to Mainstream Research on Gender." *Chronicle Review*, August 12, B10.

Ferrie, Joseph P., and Karen Rolf. 2011. "Socioeconomic Status in Childhood and Health After Age 70: A New Longitudinal Analysis for the U.S., 1895–2005. National Bureau of Economic Research, June. Accessed September 22, 2011 (www.nber.org).

Fields, Jason, and Lynne. M. Casper. 2001. "America's Families and Living Arrangements: 2000." U.S. Census Bureau, Current Population Reports, P20-537. Retrieved February 12, 2003 (www.census .gov).

File, Thom, and Sarah Crissey. 2010. "Voting and Registration in the Election of November

2008: Population Characteristics." U.S. Census Bureau, Current Population Reports, P20-572. Retrieved May 15, 2010 (www .census.gov).

Fine, Eve. 2010. "Is Biology Still Destiny? Recent Studies of Sex and Gender Differences." *Feminist Collections* 31 (Summer): 1–7.

Finer, Lawrence B., and Mia R. Zolna. 2011. "Unintended Pregnancy in the United States: Incidence and Disparities, 2006." *Contraception*, August 25, published online. Accessed October 25, 2011 (www.contraceptionjournal.org).

Finke, Roger, and Rodney Stark. 1992. *The Churching of America, 1776–1990: Winners and Losers in Our Religious Economy*. New Brunswick, NJ: Rutgers University Press.

Finkelhor, David, Heather Turner, Sherry Hamby, and Richard Ormrod. 2011. "Polyvictimization: Children's Exposure to Multiple Types of Violence, Crime, and Abuse." Office of Justice Programs, October. Accessed March 7, 2012 (www.ojp.usdoj .gov).

Finkelstein, Eric A., et al. 2012. "Obesity and Severe Obesity Forecasts Through 2030." *American Journal of Preventive Medicine* 42 (6). Accessed May 14, 2012 (ajponline .org).

Finn, Jeremy D. 2006. *The Adult Lives of At-Risk Students: The Roles of Attainment and Engagement in High School* (NCES 2006-328). Washington, DC: U.S. Department of Education.

"Firms Suggest Limits on Food Marketed to Kids." 2011. *Baltimore Sun*, July 15, 12.

Fischer, Emily, Rob Sargent, and Elizabeth Ridlington. 2012. "Building a Better America: Reducing Pollution and Saving Money with Efficiency." Environment America Research & Policy Center, March. Accessed June 22, 2012 (www .environmentamericacenter.org).

Fischman, Josh. 2009. "Global Warming Before Smokestacks." *Chronicle Review*, November 6, B11–B12.

Flaherty, Joan. 2011. "Ill-Mannered Students Can Wreck More Than Your Lecture." *Chronicle of Higher Education*, October 28, A31.

Flanagan, William G. 1990. *Urban Sociology: Images and Structure*. Boston: Allyn & Bacon.

Flavin, Jeanne. 2001. "Feminism for the Mainstream Criminologist: An Invitation." *Journal of Criminal Justice* 29 (July/August): 271–285.

Flegal, Katherine M., Margaret D. Carroll, Cynthia L. Ogden, and Lester R. Curtin. 2010. "Prevalence and Trends in Obesity among U.S. Adults, 1999–2008." *JAMA* 303 (January 20): 235–241.

Fleischman, Howard L., Paul J. Hopstock, Marisa P. Pelczar, and Brooke E. Shelley. 2010. *Highlights from PISA 2009: Performance of U.S. 15-Year-Old Students in Reading, Mathematics, and Science Literacy in an International Context*. U.S. Department of Education, National Center for Education Statistics. Washington, DC: U.S. Government Printing Office.

Fleishman, Jeffrey, and Amro Hassan. 2009. "Gadget to Help Women Feign Virginity Angers Many in Egypt." *Los Angeles Times*,

October 7. Retrieved October 28, 2009 (www.latimes.com).

Fletcher, Douglass Scott, and Ian M. Taplin. 2002. *Understanding Organizational Evolution: Its Impact on Management and Performance.* Westport, CT: Quorum Books.

Flower, Shawn M. 2010. "Gender-Responsive Strategies for Women Offenders." National Institute of Corrections, November. Accessed August 20, 2011 (www.nicic.gov).

Fogg, Piper. 2005. "Don't Stand So Close to Me." *Chronicle of Higher Education,* April 29, A10–A12.

Fong, Timothy P. 2002. *The Contemporary Asian American Experience: Beyond the Model Minority,* 2nd edition. Upper Saddle River, NJ: Prentice Hall.

Food and Agriculture Organization of the United Nations. 2009. "Hunger in the Face of Crisis." Policy Brief No. 6. Retrieved May 2, 2010 (ftp.fao.org).

Food and Agriculture Organization of the United Nations. 2011. "The State of the World's Land and Water Resources for Food and Agriculture, Managing Systems at Risk: Summary Report." Accessed June 20, 2012 (www.fao.org).

Food & Water Watch. 2007. "Take Back the Tap: Why Choosing Tap Water over Bottled Water Is Better for Your Health, Your Pocketbook, and the Environment." Retrieved May 1, 2010 (www.foodandwaterwatch.org).

"Food Companies Propose Cutting Back on Junk Food Marketing Aimed at Children." 2011. *Washington Post,* July 14. Accessed July 20, 2011 (washingtonpost.com).

Ford, Michael F. 2012. "Five Myths about the American Dream." *Washington Post,* January 6. Accessed January 7, 2012 (www.washingtonpost.com).

Ford, Peter. 2007. "Pollution Puts China Lake Off Limits." *Christian Science Monitor,* June 4, 7.

———. 2010. "China's Crib Conundrum." *Christian Science Monitor,* December 27, 23, 27.

———. 2011. "Why Japan Will Rebound." *Christian Science Monitor,* March 28, 16–29.

"Formaldehyde." 2004. Environmental Defense. Retrieved July 10, 2005 (www.scorecard.org).

Foroohar, Rana. 2011. "Whatever Happened to Upward Mobility?" *Time,* November 14, 27–34.

Fortuny, Karina, and Ajay Chaudry. 2011. "Children of Immigrants: Growing National and State Diversity." Urban Institute, October. Accessed November 5, 2011 (www.urban.org).

Foucault, Michel. 1975. *The Birth of the Clinic: An Archaeology of Medical Perception.* New York: Vintage Books.

Foust-Cummings, Heather, Laura Sabattini, and Nancy Carter. 2008. "Women in Technology: Maximizing Talent, Minimizing Barriers." *Catalyst.* Retrieved April 28, 2008 (www.catalyst.org).

Fox, James Alan, and Marianne W. Zawitz. 2004. "Homicide Trends in the United States: 2002 Update." Washington, DC: U.S. Department of Justice, Bureau of Justice Statistics.

Fox, Susannah. 2006. "Are 'Wired Seniors' Sitting Ducks?" Pew Internet & American Life Project, April. Retrieved September 20, 2007 (www.pewinternet.org).

Frank, T. A. 2006. "A Brief History of Wal-Mart." CorpWatch. Retrieved February 19, 2007 (www.corpwatch.org).

Franklin, Rachel S. 2003. "Migration of the Young, Single, and College Educated: 1995 to 2000." U.S. Census Bureau, Census 2000 Special Reports CENSR-12. Washington, DC: U.S. Government Printing Office.

Freese, Jeremy. 2008. "Genetics and the Social Science Explanation of Individual Outcomes." *American Journal of Sociology* 114 (Suppl.): S1–S35.

Freitas, Donna. 2008. *Sex & the Soul: Juggling Sexuality, Spirituality, Romance, and Religion on America's College Campuses.* New York: Oxford University Press.

Frey, William H. 2008. "Race, Immigration, and America's Changing Electorate." Brookings Institution, February 18. Retrieved June 23, 2008 (www.brookings.edu).

———. 2011. "Race & Ethnicity." *The State of Metropolitan America,* 132–143. Brookings Institution. Accessed June 24, 2012 (www.brookings.edu).

———. 2011. "The Uneven Aging and 'Younging' of America: State and Metropolitan Trends in the 2010 Census." Brookings Metropolitan Policy Program. Accessed June 22, 2012 (www.brookings.edu).

Friedan, Betty. 1963. *The Feminine Mystique.* New York: Norton.

Frieden, Thomas R. 2011. "Foreword." Centers for Disease Control and Prevention. *MMWR 60* (Suppl): 1–2.

Friedman, Jaclyn. 2011. *What You Really Really Want: The Smart Girl's Shame-Free Guide to Sex and Safety.* Berkeley, CA: Seal Press.

Friedrichs, David O. 2004. *Trusted Criminals: White Collar Crime in Contemporary Society,* 2nd edition. Belmont, CA: Wadsworth.

Frier, Sarah. 2012. "Insurers Profit From Health Law They Fought Against." Bloomberg, January 5. Accessed April 15, 2012 (www.bloomberg.com).

Frosch, Dan. 2007. "18 Air Force Cadets Exit Over Cheating." *New York Times,* May 2, 18.

"Frustration with Congress Could Hurt Republican Incumbents." 2011. Pew Research Center for the People & the Press, December 15. Accessed December 17, 2011 (www.people-press.org).

Fry, Richard. 2010. "Hispanics, High School Dropouts and the GED." Pew Hispanic Center, May 13. Accessed March 20, 2012 (www.pewhispanic.org).

———. 2011. "Hispanic College Enrollment Spikes, Narrowing Gaps with Other Groups." Pew Research Center, August 25. Accessed April 15, 2012 (www.pewhispanic.org).

Fry, Richard, and D'Vera Cohn. 2010. "Women, Men, and the New Economics of Marriage." Pew Research Center, January 19. Accessed January 25, 2010 (pewsocialtrends.org).

———. 2011. "Living Together: The Economics of Cohabitation." Pew Research Center, June 27. Accessed March 1, 2012 (www.pewsocialtrends.org).

Fryer, Roland G., Jr., Devah Pager, and Jörg L. Spenkuch. 2011. "Racial Disparities in Job Finding and Offered Wages." National Bureau of Economic Research, September. Accessed December 7, 2011 (www.nber.org).

Gabriel, Trip. 2010. "To Stop Cheats, Colleges Learn Their Trickery." *New York Times,* July 5, A1.

Gamson, William. 1990. *The Strategy of Social Protest,* 2nd edition. Belmont, CA: Wadsworth.

Gans, Herbert J. 1962. "Urbanism and Suburbanism as Ways of Life: A Reevaluation of Definitions." Pp. 625–648 in *Human Behavior and Social Processes: An Interactionist Approach,* edited by Arnold M. Rose. Boston: Houghton Mifflin.

———. 1971. "The Uses Of Poverty: The Poor Pay All." *Social Policy,* July/August, 78–81.

———. 2005. "Race as Class." *Contexts* 4 (Fall): 17–21.

———. 2005. "Wishes for the Discipline's Future." *Chronicle Review,* August 12, B9.

Gardner, Marilyn. 2008. "Happiness Is a Warm 'Thank You'." *Christian Science Monitor,* January 28, 13, 16.

Garfinkel, Harold. 1956. "Conditions of Successful Degradation Ceremonies." *American Journal of Sociology* 61 (March): 420–424.

———. 1967. *Studies in Ethnomethodology.* Englewood Cliffs, NJ: Prentice-Hall.

Garrett, Danielle. 2012. "Turning to Fairness: Insurance Discrimination Against Women Today and the Affordable Care Act." National Women's Law Center, March. Accessed April 15, 2012 (www.nwlc.org).

Garrison, Michelle M., and Dimitri A. Christakis. 2005. "A Teacher in the Living Room? Educational Media for Babies, Toddlers and Preschoolers." Henry J. Kaiser Family Foundation, December. Retrieved April 12, 2008 (www.kff.org).

Gates, Gary, L. M. V. Badgett, Jennifer E. Macomber, and Kate Chambers. 2007. "Adoption and Foster Care by Gay and Lesbian Parents in the United States." Urban Institute, March. Retrieved April 10, 2010 (www.urban.org).

Gates, Gary, and Jason Ost. 2004. *The Gay & Lesbian Atlas.* Washington, DC: Urban Institute Press.

Gaudin, Nicolas. 2011. "IARC Classified Radiofrequency Electromagnetic Fields as Possibly Carcinogenic to Humans." World Health Organization, International Agency for Research on Cancer, May 31. Accessed July 7, 2011 (www.iarc.fr).

Gavrilos, Dina. 2006. "U.S. News Magazine Coverage of Latinos: 2006 Report." National Association of Hispanic Journalists, June. Retrieved April 2, 2007 (www.nahj.org).

Gaylin, Willard. 1992. *The Male Ego.* New York: Viking.

Geertz, Clifford. 1966. "Religion as a Cultural System." Pp. 1–46 in *Anthropological Approaches to the Study of Religion,* edited by Michael Banton. London: Tavistock.

Gelbard, Alene, Carl Haub, and Mary M. Kent. 1999. "World Population Beyond Six Billion." *Population Bulletin* 54 (March): 1–44.

Gelles, Richard J. 1997. *Intimate Violence in Families,* 3rd edition. Thousand Oaks, CA: Sage.

"Gender Equality Universally Embraced, but Inequalities Acknowledged." 2010. Pew

Research Center, Global Attitudes Project, July 1. Accessed November 1, 2011 (www .pewglobal.org).

Gerth, H. H., and C. Wright Mills, eds. 1946. *Max Weber: Essays in Sociology*. New York: Oxford University Press.

Gewertz, Catherine. 2009. "Do Men Deserve a Break in College Admissions?" *Education Week* online, December 9. Retrieved December 13, 2010 (blogs.edweek.org).

Ghosh, Bobby. 2007. "Why They Hate Each Other." *Time*, March 5, 29–40.

Gibson, Campbell, and Kay Jung. 2006. "Historical Census Statistics on the Foreign-Born Population of the United States: 1850 to 2000." U.S. Census Bureau, Population Division, February. Accessed December 7, 2011 (www.census.gov).

Gibson, David. 2012. "Vatican Orders Crackdown on American Nuns." *USA Today*, April 18. Accessed April 20, 2012 (www.usatoday .com).

Gibson, Megan. 2012. "Celebrities Offering Scientific 'Facts'? Just Say No." *Time* Newsfeed, January 3. Accessed January 24, 2012 (newsfeed.time.com).

Gilbert, Dennis. 2008. *The American Class Structure in an Age of Growing Inequality*, 7th edition. Belmont, CA: Wadsworth Press.

Gilbert, Dennis, and Joseph A. Kahl. 1993. *The American Class Structure: A New Synthesis*, 4th edition. Homewood, IL: Dorsey Press.

Gillen, Andrew, Matthew Denhart, and Jonathan Robe. 2011. "Who Subsidizes Whom? An Analysis of Educational Costs and Revenues." Center for College Affordability and Productivity, March. Accessed March 15, 2012 (centerforcollegeafordability.org).

Gilligan, Carol. 1982. *In a Different Voice: Psychological Theory and Women's Development*. Cambridge, MA: Harvard University Press.

Giugni, Marco, Doug McAdam, and Charles Tilley, eds. 1999. *How Social Movements Matter*. Minneapolis: University of Minnesota Press.

Giuliano, Laura, David I. Levine, and Jonathan Leonard. 2009. "Manager Race and the Race of New Hires." *Journal of Labor Economics* 27 (October): 589–631.

Givhan, Robin. 2010. "Fashion: Michelle Obama Acknowledges Indian Fashion Industry During Trip." *Washington Post*, November 11, C2.

———. 2010. "First Lady Michelle Obama: 'Let's Move' and Work on Childhood Obesity Problem." *Washington Post*, February 10, C1.

Glassner, Barry. 2010. "Still Fearful After All These Years." *Chronicle Review*, January 22, B11–B12.

Glauber, Bill. 2011. "Do Monarchies Still Matter?" *Christian Science Monitor*, February 21, 27–31.

Glaze, Lauren E., Thomas P. Bonczar, and Fan Zhang. 2010. "Probation and Parole in the United States, 2009." Bureau of Justice Statistics Bulletin, December. Accessed August 20, 2011 (bjs.ojp.usdoj.gov).

Glenn, David. 2011. "For Business Majors, Easy Does It." *Chronicle of Higher Education*, April 22, A1, A3–A5.

Glenn, Norval D. 2001. "Social Science Findings and the 'Family Wars'." *Society* 38 (May/June): 13–19.

———. 2005. "With This Ring . . . : A National Survey on Marriage in America." National Fatherhood Initiative. Retrieved April 2, 2006 (www.fatherhood.org).

Glionna, John M. 2009. "Aceh's Morality Police on the Prowl for Violators." *Los Angeles Times*, November 8. Retrieved November 28, 2009 (www.latimes.com).

———. 2009. "South Korean Kids Get a Taste of Boot Camp." *Los Angeles Times*, August 22, A1.

"Global Christianity: A Report on the Size and Distribution of the World's Christian Population." 2011. Pew Research Forum, December. Pew Research Center, December 1. Accessed April 8, 2012 (pewresearch .org).

Glynn, Sarah Jane, and Audrey Powers. 2012. "10 Wildly Depressing Facts About the Gender Wage Gap." Alternet, April 16. Accessed April 17, 2012 (www.alternet.org).

Godofsky, Jessica, Cliff Zukin, and Carl Van Horn. 2011. "Unfulfilled Expectations: Recent College Graduates Struggle in a Troubled Economy." Work Trends, May. Accessed June 5, 2011 (www.heldrich.rutgers.edu).

Goffman, Erving. 1959. *The Presentation of Self in Everyday Life*. New York: Doubleday Anchor Books.

———. 1961. *Asylums: Essays on the Social Situation of Mental Patients and Other Inmates*. Garden City, NY: Anchor Books.

———. 1963. *Stigma: Notes on the Management of Spoiled Identity*. Englewood Cliffs, NJ: Prentice-Hall.

———. 1967. *Interaction Ritual: Essays on Face-to-Face Behavior*. New York: Anchor Books.

———. 1969. *Strategic Interaction*. Philadelphia: University of Pennsylvania Press.

Goldberg, Jonah. 2009. "America Through the Reality Lens." *Los Angeles Times*, December 15. Retrieved December 22, 2009 (www .latimes.com).

Golden, Daniel. 2006. *The Price of Admission: How America's Ruling Class Buys Its Way into Elite Colleges—and Who Gets Left Outside the Gates*. New York: Crown.

Goldin, Claudia, Lawrence F. Katz, and Ilyana Kuziemko. 2006. "The Homecoming of American College Women: The Reversal of the College Gender Gap." National Bureau of Economic Research. Retrieved July 25, 2007 (www.nber.org/papgers/w12139).

Goldin, Claudia, and Maria Shim. 2004. "Making a Name: Women's Surnames at Marriage and Beyond." *Journal of Economic Perspectives* 18 (Spring): 143–160.

Goodale, Gloria. 2009. "A Wake-Up Call on Water Use." *Christian Science Monitor*, June 10. Retrieved June 12, 2009 (www.csmonitor .com).

———. 2010. "Birthplace of the Drive-Thru Bans Them to Curb Obesity." *Christian Science Monitor*, August 2. Accessed August 3, 2011 (www.csmonitor.com).

———. 2010. "The 'Real' Costs of Air Travel." *Christian Science Monitor*, May 3, 21.

Gooding, Gretchen E., and Rose M. Kreider. 2010. "Women's Marital Naming Choices in a Nationally Representative Sample." *Journal of Family Issues* 31 (5): 681–701.

Goodman, Peter S. 2010. "Despite Signs of Recovery, Chronic Joblessness Rises." *New York Times*, February 20, 1.

Goodnough, Abby. 2010. "Making It Clear That a Clear Parking Space Isn't." *New York Times*, December 28, A10.

Goodstein, Laurie. 2010. "Lutherans Offer Warm Welcome to Gay Pastors." *New York Times*, July 25, A13.

———. 2011. "Bishops Say Rules on Gay Parents Limit Freedom of Religion." *New York Times*, December 29, A10.

Gopnik, Alison, Andrew N. Meltzoff, and Patricia K. Kuhl. 2001. *The Scientist in the Crib: What Early Learning Tells Us about the Mind*. New York: Perennial.

Gorman, Anna. 2010. "Immigrants Often See Peril in Reporting Domestic Abuse." *Los Angeles Times*, January 25. Retrieved January 26, 2010 (www.latimes.com).

Gorman, Bill. 2011. "'CSI: Crime Scene Investigation' Is the Most-Watch Drama Series in the World!" TV by the numbers, June 13. Accessed August 22, 2011 (tvbythenumbers.zap2it.com).

Gottdiener, Mark, and Ray Hutchison. 2000. *The New Urban Sociology*, 2nd edition. New York: McGraw-Hill.

Gottlieb, Mark. 2011. "Attention to Duty." *Christian Science Monitor*, June 6, 44.

Gottman, John M. 1994. *What Predicts Divorce? The Relationships between Marital Processes and Marital Outcome*. Hillsdale, NJ: Lawrence Erlbaum Associates.

Gould, Deborah B. 2012. "Occupy's Political Emotions." *Contexts* 11 (Spring): 20–21.

Gould, Elise. 2012. "A Decade of Declines in Employer-Sponsored Health Insurance Coverage." Economic Policy Institute, February 23. Accessed April 15, 2012 (www.epi.org).

Gouldner, Alvin W. 1962. "Anti-Minotaur: The Myth of a Value-Free Sociology." *Social Problems* 9 (Winter): 199–212.

"Government Resolves to Start Making Sense Under New Law Forbidding Federal Gibberish." 2011. *Washington Post*, May 20. Accessed May 21, 2011 (www .washingtonpost.com).

Grabe, S., L. M. Ward, and J. S. Hyde. 2008. "The Role of the Media in Body Image Concerns Among Women: A Meta-Analysis of Experimental and Correlational Studies." *Psychological Bulletin* 134 (May): 460–476.

Graif, Corina, and Robert J. Sampson. 2009. "Spatial Heterogeneity in the Effects of Immigration and Diversity on Neighborhood Homicide Rates." *Homicide Studies* 13 (August): 242-260.

Grall, Timothy. 2011. "Custodial Mothers and Fathers and Their Child Support: 2009." U.S. Census Bureau, Current Population Reports, P60-240, December. Accessed March 10, 2012 (www.census.gov).

Grant, Alexis. 2008. "In Cameroon, Polygamy Doesn't Pay." *Christian Science Monitor*, November 13, 20.

Grant, Gerald, and Christine E. Murray. 1999. *Teaching in America: The Slow Revolution*. Cambridge, MA: Harvard University Press.

Grauerholz, Liz, and Sharon Bouma-Holtrop. 2003. "Exploring Critical Sociological Thinking." *Teaching Sociology* 31 (October): 485–496.

Graves, Joseph L., Jr. 2001. *The Emperor's New Clothes: Biological Theories of Race at the Millennium*. New Brunswick, NJ: Rutgers University Press.

Gray, Paul S., John B. Williamson, David R. Karp, and John R. Dalphin. 2007. *The Research Imagination: An Introduction to Qualitative and Quantitative Methods.* New York: Cambridge University Press.

Grazian, David. 2008. *On the Make: The Hustle of Urban Nightlife.* Chicago: University of Chicago Press.

Greeley, Brendan. 2011. "The Union, Jacked." *Bloomberg Businessweek,* February 28–March 6, 8–9.

Green, Andrew, and Josh Bivens. 2011. "Lack of Jobs, Not Lack of Skills, Explains Underemployment Rate." Economic Policy Institute, June 15. Accessed January 13, 2012 (www.epi.org).

Green, Jeff. 2011. "The Silencing of Sexual Harassment." *Bloomberg Businessweek,* November 21–27, 27–28.

Greene, Bob. 2011. "Did Cell Phones Unleash Our Inner Rudeness?" CNN News, April 10. Accessed April 15, 2011 (www.cnn.com).

Grieco, Elizabeth M., and Edward N. Trevelyan. 2010. "Place of Birth of the Foreign-Born Population: 2009." U.S. Census Bureau, American Community Survey Briefs, October. Accessed December 7, 2011 (www.census.gov).

Griffis, Margaret, ed. 2012. "Casualties in Iraq." AntiWar.com, February 11. Accessed April 24, 2012 (antiwar.com).

Gross, Rita M. 1996. *Feminism and Religion: An Introduction.* Boston: Beacon Press.

Grossman, David. 2011. "Fliers Come to Fisticuffs Over Reclined Seat." *USA Today,* June 6. Accessed June 7, 2011 (usatoday.org).

Groves, Robert, and Frank Vitrano. 2011. "The U.S. Decennial Census and the American Community Survey: Looking Back and Looking Ahead." Population Reference Bureau, April. Accessed November 9, 2011 (www.prb.org).

"Growing Number of Americans Say Obama Is a Muslim." 2010. Pew Research Center, August 19. Accessed April 8, 2012 (pewresearch.org).

Gruber, Jonathan. 2005. "Religious Market Structure, Religious Participation, and Outcomes: Is Religion Good for You?" National Bureau of Economic Research, May. Retrieved August 28, 2007 (www.nber.org).

Guerino, Paul, Paige M. Harrison, and William J. Sabol. 2011. "Prisoners in 2010." Bureau of Justice Statistics, Office of Justice Programs, December. Accessed January 5, 2012 (www.bjs.gov).

Gurney, Joan M., and Kathleen T. Tierney. 1982. "Relative Deprivation and Social Movements: A Critical Look at Twenty Years of Theory and Research." *Sociological Quarterly* 23 (Winter): 33–47.

Gurwitt, Rob. 2005. "The Nose That Knows." *Mother Jones,* March/April, 24.

Guterman, Lila. 2005. "Lost Count." *Chronicle of Higher Education,* February 4, A10–A13.

Guterman, Stanley S. 1969. "In Defense of Wirth's 'Urbanism as a Way of Life'." *American Journal of Sociology* 74 (March): 492–499.

Guttmacher Institute. 2009. "A Real-Time Look at the Impact of the Recession on Women's Family Planning and Pregnancy Decisions." September. Retrieved April 12, 2010 (www z.guttmacher.org).

———. 2011. "Facts on Induced Abortion in the United States." August. Accessed October 25, 2011 (www.guttmacher.org).

Gwynne, Kristen. 2011. "Drug Company Profiteering, Pill Mills and Thousands of Addicts: How Oxycontin Has Spread Through America." Alternet, June 20. Accessed April 24, 2012 (www.alternet.org).

Hacker, Andrew, and Claudia Dreifus. 2011. *Higher Education?: How Colleges Are Wasting Our Money and Failing Our Kids—and What We Can Do About It.* New York: St. Martin's Griffin.

Hacker, Jacob S., and Paul Pierson. 2010. *Winner-Take-All Politics: How Washington Made the Rich Richer and Turned Its Back on the Middle Class.* New York: Simon and Schuster.

Hagan, Frank E. 2008. *Introduction to Criminology: Theories, Methods, and Criminal Behavior,* 6th edition. Thousand Oaks, CA: Sage.

Hagenbaugh, Barbara. 2006. "U.S. Manufacturers Getting Desperate for Skilled People." *USA Today,* December 5, 1.

Halfond, Jay A. 2004. "Grade Inflation Is Not a Victimless Crime." *Christian Science Monitor,* May 3, 9.

Hall, Edward T. 1959. *The Silent Language.* New York: Doubleday.

———. 1966. *The Hidden Dimension.* Garden City, NY: Doubleday.

Halsey, Ashley, III. 2010. "Study: Older People Are Driving More, Having Fewer Accidents." *Washington Post,* June 22, A10.

———. 2011. "Seat in Lap Leads to Scuffle in the Air." *Washington Post,* June 1, A1.

Hamburger, Tom, and Kim Geiger. 2010. "Obama Rebuked Over Science." *Baltimore Sun,* July 12, 1, 10.

Hamby, Sherry, and David Finkelhor. 2001. "Choosing and Using Child Victimization Questionnaires." *OJJDP Juvenile Justice Bulletin.* Washington, DC: U.S. Department of Justice.

Hamby, Sherry, David Finkelhor, Heather Turner, and Richard Ormrod. 2011. "Children's Exposure to Intimate Partner Violence and Other Family Violence." Office of Justice Programs, October. Accessed March 7, 2012 (www.ojp.usdoj.gov).

Hamermesh, Daniel S., and Amy W. Parker. 2003. "Beauty in the Classroom: Professors' Pulchritude and Putative Pedagogical Productivity." Working Paper 9853, July. Cambridge, MA: National Bureau of Economic Research.

Hamilton, Anita. 2011. "Droughtbusters." *Time,* October 13, B1–B7.

Hamilton, Brady E., and Stephanie J. Ventura. 2012. "Birth Rates for U.S. Teenagers Reach Historic Lows for All Age and Ethnic Groups." NCHS Data Brief, No. 89, April. Accessed April 25, 2012 (www.cdc.gov).

Hamilton, Malcolm B. 1995. *The Sociology of Religion: Theoretical and Comparative Perspectives.* New York: Routledge.

———. 2001. *The Sociology of Religion,* 2nd edition. New York: Routledge.

Hamm, Steve. 2007. "The Trouble with India." *Business Week,* March 19, 49–58.

Hampton, Keith N., Lauren Sessions, Eun Ja Her, and Lee Rainie. 2009. "Social Isolation and New Technology." Pew Internet & American Life Project, November. Retrieved February 5, 2010 (www.pewinternet.org).

Hampton, Keith N., Lauren Sessions Goulet, Lee Rainie, and Kristin Purcell. 2011. "Social Networking Sites and Our Lives." Pew Internet & American Life Project, July 16. Accessed July 20, 2011 (pewinternet.org).

Hamre, Bridget K., and Robert C. Pianta. 2001. "Early Teacher-Child Relationships and the Trajectory of Children's School Outcomes Through Eighth Grade." *Child Development* 72 (March/April): 625–638.

Handelsman, Jo, et al. 2005. "Careers in Science." *Science* 309 (August 19): 1190–1191.

Hanes, Stephanie. 2010. "Texting Bans Ineffective?" *Christian Science Monitor,* October 11, 20.

———. 2011. "Pretty in Pink?" *Christian Science Monitor,* September 26, 26–31.

Haney, Craig, Curtis Banks, and Philip Zimbardo. 1973. "Interpersonal Dynamics in a Simulated Prison." *International Journal of Criminology and Psychology* 1: 69–97.

Hanowski, Rich. 2009. "New Data from VTTI Provides Insight into Cell Phone Use and Driving Distraction." Virginia Tech Transportation Institute. Retrieved December 15, 2009 (www.vtti.vt.edu).

Hansen, Lawrence A. 2010. "Noxious Groupthink." *Chronicle of Higher Education,* November 12, B6–B8.

Hanushek, Eric A., and Ludger Woessman. 2005. "Does Educational Tracking Affect Performance and Inequality? Differences-in-Differences Evidence Across Countries." National Bureau of Economic Research, February. Retrieved July 25, 2007 (www.nber.org).

Haq, Husna. 2009. "Ethnic Malls Are Buzzing." *Christian Science Monitor,* August 30, 30–31.

———. 2010. "Why Prisons Are Less Full." *Christian Science Monitor,* July 12, 16–17.

Harder, Joshua, and Jon A. Krosnick. 2008. "Why Do People Vote? A Psychological Analysis of the Causes of Voter Turnout." *Journal of Social Issues,* 64 (September): 525–549.

Hargrove, Thomas. 2011. "Disappearing Jobs Illustrate U.S. Manufacturing Revolution." Knox News, November 13. Accessed January 9, 2012 (www.knoxnews.com).

Harlow, Caroline Wolf. 2003. "Education and Correctional Populations." U.S. Department of Justice, Bureau of Justice Statistics. Retrieved January 16, 2003 (www.ojp.usdoj.gov/bjs).

Harlow, Harry E., and Margaret K. Harlow. 1962. "Social Deprivation in Monkeys." *Scientific American,* 206 (November): 137–146.

Harman, Danna. 2007. "Qatar Reformed by a Modern Marriage." *Christian Science Monitor,* March 6, 20.

Harmon, Amy. 2008. "The DNA Age—Insurance Fears Lead Many to Shun DNA Tests." *New York Times,* February 24, 1.

Harms, Roger W. 2011. "Can You Get Pregnant from Pre-Ejaculation Fluid?" Mayo Clinic, September 17. Accessed October 25, 2011 (www.mayoclinic.com).

Harrell, Erika. 2007. "Black Victims of Violent Crime." Washington, DC: U.S. Department of Justice, Bureau of Justice Statistics.

Harrell, Margaret C., and Nancy Berglass. 2011. "Losing the Battle: The Challenge of Military Suicide." Center for a New American Security, October. Accessed April 24, 2012 (www.cnas.org).

Harris, Chauncey D., and Edward L. Ullman. 1945. "The Nature of Cities." *Annals* 242: 7–17.

Harris, Gardiner. 2011. "When the Nurse Wants to Be Called 'Doctor'." *New York Times*, June 6, A23.

Hart, Timothy C., and Callie Rennison. 2003. "Reporting Crime to the Police, 1992–2000." Washington, DC: U.S. Department of Justice, Bureau of Justice Statistics.

Harter, Jim. 2011. "Engaged Workers Report Twice as Much Job Creation." Gallup, August 9. Accessed August 21, 2011 (www.gallup.com).

Harter, Jim, and Sangeeta Agrawal. 2011. "Actively Disengaged Workers and Jobless in Equally Poor Health." Gallup, April 20. Accessed April 21, 2011 (www.gallup.com).

Hartocollis, Anemona. 2012. "To Gulp or to Sip? Debating a Crackdown on Big Sugary Drinks." *New York Times*, June 1, A22.

Haskins, Ron. 2008a. "Education and Economic Mobility." Pp. 91-104 in *Getting Ahead or Losing Ground: Economic Mobility in America,* edited by Julia B. Isaacs, Isabel V. Sawhill, and Ron Haskins. Washington DC: Brookings Institution.

———. 2008b. "Immigration: Wages, Education, and Mobility." Pp. 81-90 in *Getting Ahead or Losing Ground: Economic Mobility in America,* edited by Julia B. Isaacs, Isabel V. Sawhill, and Ron Haskins. Washington DC: Brookings Institution.

———. 2008c. "Wealth and Economic Mobility." Pp. 47–60 in *Getting Ahead or Losing Ground: Economic Mobility in America*, edited by Julia B. Isaacs, Isabel V. Sawhill, and Ron Haskins. Washington, DC: Brookings Institution.

Haub, Carl. 2011. "The Caste Census: A Feudal Classification of Society in India?" Population Reference Bureau, September 7. Accessed September 12, 2011 (prbblog.org).

Haub, Carl, and James Gribble. 2011. "The World at 7 Billion." *Population Bulletin* 66 (July): 1–12.

Haub, Carl, and Toshiko Kaneda. 2011. "2011 World Population Data Sheet." Population Reference Bureau, wall poster.

Hausmann, Ricardo, Laura D. Tyson, and Saadia Zahidi. 2011. *The Global Gender Gap 2011*. Geneva, Switzerland, World Economic Forum. Accessed January 25, 2012 (www.weforum.org).

Hayden, Dolores. 2002. *Redesigning the American Dream: Gender, Housing, and Family Life*. New York: W.W. Norton.

Haygood, Wil. 2010. "One Family's Plunge from the Middle Class into Poverty." *Washington Post*, November 19, A1.

"Hazardous Air Pollution—A National Overview." 2005. Environmental Defense. Retrieved July 10, 2005 (www.scorecard.org).

He, Wan, and Mark N. Muenchrath. 2011. "90+ in the United States: 2006–2008." U.S. Census Bureau, ACS-17, November. Accessed March 10, 2012 (www.census.gov).

Health Resources and Services Administration. 2010. "Assuring Access to Essential Health Care." Accessed July 15, 2011 (www.plainlanguage.gov).

Heath, Chip. 2005. "Introduction." Pp. 81–85 in *Rumor Mills: The Social Impact of Rumor and Legend*, edited by Gary Alan Fine, Véronique Campion-Vincent, and Chip Heath. New Brunswick, NJ: Transaction Publishers.

Heath, Jennifer. 2008. *The Veil: Women Writers on Its History, Lore, and Politics*. Berkeley: University of California Press.

Hechter, Michael, and Karl-Dieter Opp. 2001. "Introduction." Pp. xi–xx in *Social Norms*, edited by Michael Hechter and Karl-Dieter Opp. New York: Russell Sage Foundation.

Hedges, Chris. 2011. "Our Public Schools Are Churning Out Drones for the Corporate State." AlterNet, April 11. Accessed April 14, 2011 (www.alternet.org).

Hegewisch, Ariane, Claudia Williams, and Amber Henderson. 2011. "The Gender Wage Gap by Occupation." Institute for Women's Policy Research, April. Accessed May 2, 2011 (www.iwpr.org).

Heider, Elanor R., and Donald C. Olivier. 1972. "The Structure of the Color Space in Naming and Memory for Two Languages." *Cognitive Psychology*, 3: 337–354.

Heilbroner, R. L., and L. C. Thurow. 1998. *Economics Explained: Everything You Need to Know about How the Economy Works and Where It's Going*. New York: Touchstone.

Heimer, Karen, and Candace Kruttschnitt. 2006. "Introduction: New Insights into the Gendered Nature of Crime and Victimization." Pp. 1–14 in *Gender and Crime: Patterns of Victimization and Offending*, edited by Karen Heimer and Candace Kruttschnitt. New York: New York University Press.

Heimer, Karen, Stacy Wittrock, and Halime Ünal. 2006. "The Crimes of Poverty: Economic Marginalization and the Gender Gap in Crime." Pp. 115–126 in *Gender and Crime: Patterns of Victimization and Offending*, edited by Karen Heimer and Candace Kruttschnitt. New York: New York University Press.

Helgesen, Sally. 2008. "Female Leadership: Changing Business for the Better." *Christian Science Monitor*, January 17, 9.

Hendrix, Steve. 2009. "In D.C., Tattoos Are Largely Taboo from 9 to 5." *Washington Post*, December 10, A1.

Herbert, Bob. 2009. "A Culture Soaked in Blood." *New York Times*, April 25, 19.

Hernandez, Donald J. 2011. "Double Jeopardy: How Third-Grade Reading Skills and Poverty Influence High School Education." Annie E. Casey Foundation. Accessed April 2, 2012 (www.aecf.org).

Hertsgaard, Mark. 2012. "Emissions Impossible." *Mother Jones*, May/June, 37–46.

Herzberg, David. 2009. *Happy Pills in America: From Miltown to Prozac*. Baltimore, MD: Johns Hopkins University Press.

Hesse, Monica. 2010. "Coffee, Tea, or Flee? JetBlue Attendant's Exit Strategy Serves Crummy Job Right." *Washington Post*, August 11, C1.

Hetherington, E. Mavis, Ross D. Parke, and Virginia Otis Locke. 2006. *Child*

*Psychology: A Contemporary Viewpoint*, 6th edition. Boston: McGraw-Hill.

Higham, Scott, Kimberly Kindy, and David S. Fallis. 2012. "Capitol Assets: Some Legislators Send Millions to Groups Connected to Their Relatives." *Washington Post,* February 6. Accessed February 7, 2012 (www.washingtonpost.com).

Higher Education Research Institute. 2011. "The American Freshman: National Norms Fall 2010." HERI Research Brief, January. Accessed June 8, 2012 (www.heri.ucla.edu).

"Highlights of Women's Earnings in 2010." 2011. Bureau of Labor Statistics, July. Accessed October 25, 2011 (www.bls.org).

Hilbert, Richard A. 1992. *The Classical Roots of Ethnomethodology: Durkheim, Weber, and Garfinkel*. Chapel Hill: University of North Carolina Press.

Hill, Catherine, Christianne Corbett, and Andresse St. Rose. 2010. "Why So Few? Women in Science, Technology, Engineering, and Mathematics." American Association of University Women. Accessed October 25, 2011 (www.aauw.org).

Hillaker, B. D., H. E. Brophy-Herb, F. A. Villarruel, and B. E. Hass. 2008. "The Contributions of Parenting to Social Competencies and Positive Values in Middle School Youth: Positive Family Communication, Maintaining Standards, and Supportive Family Relationships." *Family Relations* 57 (December): 591–601.

Hilzenrath, David S. 2011. "Justice Department, SEC Investigations Often Rely on Companies' Internal Probes." *Washington Post*, May 22, A1.

Himmelstein, Kathryn E. W., and Hannah Brückner. 2011. "Criminal-Justice and School Sanctions Against Nonheterosexual Youth: A National Longitudinal Study." *Pediatrics* 127 (January): 49–57.

Hing, Julianne. 2011. "5 Ways Alabama's New Anti-Immigrant Law Is Even Worse Than Arizona's SB 1070." AlterNet, June 24. Accessed July 2, 2011 (www.alternet.org).

Hinkle, Stephen, and John Schopler. 1986. "Bias in the Evaluation of In-Group and Out-Group Performance." Pp. 196–212 in *Psychology of Everyday Intergroup Relations*, 2nd edition, edited by Stephen Worchel and William G. Austin. Chicago: Nelson-Hall.

Hira, Ron. 2008. "An Overview of the Offshoring of U.S. Jobs." *Population Bulletin* 63 (June): 14–15.

"Hispanic Heritage Month 2011: Sept. 15–Oct. 15." U.S. Census Bureau News, August 26. Accessed December 7, 2011 (www.census z.gov).

*Hispanic-Owned Firms: 2002*. 2006. U.S. Census Bureau, SB02-00CS-HISP (RV). Retrieved May 1, 2007 (www.census.gov).

"HIV in the United States: At a Glance." 2012. Centers for Disease Control and Prevention, March. Accessed April 24, 2012 (www.cdc.gov).

Hochschild, Arlie Russell. 1983. *The Managed Heart: Commercialization of Human Feeling*. Berkeley: University of California Press.

———. 1989. *The Second Shift: Working Parents and the Revolution at Home*. New York: Penguin.

Hoecker-Drysdale, Susan. 1992. *Harriet Martineau: First Woman Sociologist*. Providence, RI: Berg.

Hoefer, Michael, Nancy Rytina, and Bryan C. Baker. 2011. "Estimates of the Unauthorized Immigrant Population Residing in the United States: January 2010." Office of Immigration Statistics, Homeland Security, Population Estimates, February. Accessed November 25, 2011 (www.dhs.gov).

Holder, Kelly. 2006. "Voting and Registration in the Election of November 2004." U.S. Census Bureau, Current Population Reports, P20-556. Retrieved May 15, 2007 (www.census.gov).

Holland, Joshua. 2011. "Anthony Weiner's Uncensored Penis Picture Plus 10 Other Images That Are Even More Obscene." AlterNet, June 9. Accessed June 10, 2011 (www.alternet.org).

———. 2011. "9 Countries That Do It Better: Why Does Europe Take Better Care of Its People Than America?" Alternet, June 15. Accessed April 24, 2012 (www.alternet .org).

Holmes, Oliver Wendell. 2011. "Over-Prescribed: How Taking Too Many Pills Is Hurting America." Alternet, June 22. Accessed April 24, 2012 (www.alternet.org).

Holzer, Harry. 2007. "Better Workers for Better Jobs: Improving Worker Advancement in the Low-Wage Labor Market." Urban Institute, December 12. Retrieved May 18, 2008 (www.urban.org).

Homans, George. 1974. Social Behavior: Its Elementary Forms, revised edition. New York: Harcourt Brace Jovanovich.

Hondagneu-Sotelo, Pierrette. 2001. Doméstica: Immigrant Workers Cleaning and Caring in the Shadows of Affluence. Berkeley: University of California Press.

hooks, bell. 2000. Feminism Is for Everybody: Passionate Politics. Cambridge, MA: South End Press.

Houser, Ari. 2007. "Women & Long-Term Care." AARP Public Policy Institute. Retrieved April 10, 2010 (assets.aarp.org).

Howard, Jeff. 2003. "Still at Risk: The Causes and Costs of Failure to Educate Poor and Minority Children for the Twenty-First Century." Pp. 81–97 in A Nation Reformed? American Education 20 Years after A Nation at Risk, edited by David T. Gordon and Patricia A. Graham. Cambridge, MA: Harvard Education Press.

Howard, Judith A., and Jocelyn A. Hollander. 1997. Gendered Situations, Gendered Selves: A Gender Lens on Social Psychology. Thousand Oaks, CA: Sage.

Hoyt, Homer. 1939. The Structure and Growth of Residential Neighborhoods in American Cities. Washington, DC: Federal Housing Administration.

Hsieh, Paul. 2011. "Beware the Next Health Reform: 'Accountable' Care." Christian Science Monitor, June 6, 33.

Hubbard, Ruth. 1990. The Politics of Women's Biology. New Brunswick, NJ: Rutgers University Press.

Huffington Post. 2011. "New American Bible Changes Some Words, Including 'Holocaust'." March 3. Accessed July 16, 2011 (www.huffingtonpost.com).

Hughes, Everett C. 1945. "Dilemmas and Contradictions of Status." American Journal of Sociology 50: 353–359.

Human Rights Now. 2011. "The World's Worst Places to Be a Woman." Women's Rights Group, June 17. Accessed October 25, 2011 (blog.amnestyusa.org).

Humes, Karen R., Nicholas A. Jones, and Roberto R. Ramirez. 2011. "Overview of Race and Hispanic Origin: 2010." U.S. Census Brief, March. Accessed December 7, 2011 (www .census.gov).

Hunsinger, Dana. 2011. "Long-Term Unemployed Face Stigmas in Job Search." USA Today, January 23. Accessed January 24, 2011 (www.usatoday.com).

Hunt, Jennifer. 2010. "Why Do Women Leave Science and Engineering?" National Bureau of Economic Research, March. Retrieved April 20, 2010 (www.nber.org).

Hunter, James Davison, and Joshua Yates. 2002. "In the Vanguard of Globalization: The World of American Globalizers." Pp. 323–357 in Many Globalizations: Cultural Diversity in the Contemporary World, edited by Peter L. Berger and Samuel P. Huntington. New York: Oxford University Press.

Hurh, Won Moo. 1998. The Korean Americans. Westport, CT: Greenwood Press.

Hurwitz, Michael. 2011. "The Impact of Legacy Status on Undergraduate Admissions at Elite Colleges and Universities." Economics of Education Review 30 (June): 480–492.

Hutter, Mark. 2007. Experiencing Cities. Boston: Allyn & Bacon.

Hvistendahl, Mara. 2012. Unnatural Selection: Choosing Boys Over Girls, and the Consequences of a World Full of Men. New York: Public Affairs.

Hyde, Janet S. 2005. "The Gender Similarities Hypothesis." American Psychologist 60 (September): 581–592.

Hyman, Mark. 2011. "The Dangers of the Medical Industrial Complex." April 29. Accessed April 15, 2012 (drhyman.com).

"Impact of Alabama's Extreme HB 56 Law, The." 2011. Huffington Post, November 4. Accessed December 4, 2011 (www. huffingtonpost.com).

Independent Sector. 2012. "Independent Sector's Value of Volunteer Time." Accessed May 4, 2012 (www.independentsector.org).

Ingraham, Nathan. 2012. "Hybrid Vehicle Ownership Has Been on Decline, Study Finds." Washington Post, April 9. Accessed June 21, 2012 (www.washingtonpost.com).

"Injuries, Illnesses, and Fatalities." 2011. Bureau of Labor Statistics. Accessed April 24, 2012 (www.bls.gov).

Institute for Women's Leadership. 2011. "Women Heads of State." Women's Leadership Fact Sheet, May. Accessed February 17, 2012 (www.iwl.rutgers.edu).

Institute of Medicine. 2011. Climate Change, the Indoor Environment, and Health. Washington, DC: National Academies Press.

Insurance Institute for Highway Safety. 2010. "Q&A: Older Drivers." Accessed October 28, 2011 (www.iihs.org).

———. 2010. "Status Report." 45 (September 28). Accessed August 18, 2011 (www.iihs.org).

Intergovernmental Panel on Climate Change. 2011. "The IPCC Special Report on Managing the Risks of Extreme Events and Disasters to Advance Climate Change Adaptation." Accessed June 22, 2012 (www .ipcc.ch).

International Federation of Health Plans. 2011. "2011 Comparative Price Report: Medical and Hospital Fees by Country." Accessed April 15, 2012 (www.ifhp.com).

International Institute for Democracy and Electoral Assistance. 2007. "Turnout over Time: Advances and Retreats in Electoral Participation." Retrieved May 24, 2007 (www.idea.int).

Internet Crime Complaint Center. 2011. "2010 Internet Crime Report." Accessed August 22, 2011 (www.nw3c.org).

Inter-Parliamentary Union. 2011. "Women in Parliaments: World Classification." Accessed February 16, 2012 (www.ipu.org).

Ioannidis, John. 2005. "Contradicted and Initially Stronger Effects in Highly Cited Clinical Research." Journal of the American Medical Association 294 (July 13): 218–228.

Isaacs, Julia B. 2008a. "Economic Mobility of Black and White Families." Pp. 71-80 in Getting Ahead or Losing Ground: Economic Mobility in America, edited by Julia B. Isaacs, Isabel V. Sawhill, and Ron Haskins. Washington DC: Brookings Institution.

———. 2008b. "Economic Mobility of Families Across Generations." Pp. 15–26 in Getting Ahead or Losing Ground: Economic Mobility in America, edited by Julia B. Isaacs, Isabel V. Sawhill, and Ron Haskins. Washington, DC: Brookings Institution.

———. 2008c. "Economic Mobility of Men and Women." Pp. 61–70 in Getting Ahead or Losing Ground: Economic Mobility in America, edited by Julia B. Isaacs, Isabel V. Sawhill, and Ron Haskins. Washington, DC: Brookings Institution.

———. 2008d. "International Comparisons of Economic Mobility." Pp. 37–46 in Getting Ahead or Losing Ground: Economic Mobility in America, edited by Julia B. Isaacs, Isabel V. Sawhill, and Ron Haskins. Washington, DC: Brookings Institution.

Jackson, C. Kirabo. 2012. "Do College-Prep Programs Improve Long-Term Outcomes?" National Bureau of Economic Research, February. Accessed April 14, 2012 (www .nber.org).

Jackson, D. D. 1998. "'This Hole in Our Heart': Urban Indian Identity and the Power of Silence." American Indian Culture and Research Journal 22 (4): 227–254.

Jackson, Henry C. 2010. "Ag Secretary Pushes School Nutrition Plan." Washington Post, February 8. Retrieved February 15, 2010 (www.washingtonpost.com).

Jackson, Shelly L., and Thomas L. Hafemeister. 2011. "Financial Abuse of Elderly People vs. Other Forms of Elder Abuse: Assessing Their Dynamics, Risk Factors, and Society's Response." U.S. Department of Justice, February. Accessed March 5, 2012 (www. ncjrs.gov).

Jacobe, Dennis. 2012. "Americans Still Prioritize Economic Growth Over Environment." Gallup, March 29. Accessed June 22, 2012 (www.gallup.com).

———. 2012. "Health Costs, Gov't Regulations Curb Small Business Hiring." Gallup, February 15. Accessed April 22, 2012 (www.gallup.com).

Jacobe, Dennis, and Jeffrey M. Jones. 2009. "Lack of Money/Wages Top Family Financial Problem in the U.S." Gallup, November 19. Retrieved April 6, 2010 (www.gallup.com).

Jacobs, Jerry A. 2005. "Multiple Methods in *ASR*." *Footnotes* 33 (December): 1, 4.

Jacobs, Ken, Dave Graham-Squire, and Stephanie Luce. 2011. "Living Wage Policies and Big-Box Retail: How a Higher Wage Standard Would Impact Walmart Workers and Shoppers." Center for Labor Research and Education, April. Accessed May 5, 2011 (laborcenter.berkeley.edu).

Jacobsen, Linda A., Mary Kent, Marlene Lee, and Mark Mather. 2011. "America's Aging Population." *Population Bulletin* 66 (February): 1–16.

Jagger, Alison M., and Paula S. Rothenberg, eds. 1984. *Feminist Frameworks*, 2nd edition. New York: McGraw-Hill.

Jamison, Tyler B., and Lawrence Ganong. 2011. "'We're Not Living Together': Stayover Relationships Among College-Educated Emerging Adults." *Journal of Social and Personal Relationships* 28 (June): 536–557.

Jandt, Fred E. 2001. *Intercultural Communication: An Introduction*. Thousand Oaks, CA: Sage.

Janis, Irving L. 1972. *Victims of Groupthink: A Psychological Study of Foreign-Policy Decisions and Fiascoes*. Boston: Houghton Mifflin.

Janofsky, Michael. 2003. "Young Brides Stir New Outcry on Utah Polygamy." *New York Times*, February 28, 1.

Jansen, Jim. 2011. "The Civic and Community Engagement of Religiously Active Americans." Pew Research Center, November 23. Accessed April 8, 2012 (pewresearch.org).

Jäntti, Markus, et al. 2006. "American Exceptionalism in a New Light: A Comparison of Intergenerational Earnings Mobility in the Nordic Countries, the United Kingdom and the United States." Institute for the Study of Labor, January. Accessed February 4, 2012 (ffp.iza.org).

Jeffery, Clara. 2006. "Poor Losers." *Mother Jones*, July/August, 20–21.

Jeffreys, Sheila. 2011. *Man's Dominion: The Rise of Religion and the Eclipse of Women's Rights*. New York: Routledge.

Jemal, A., R. Siegel, E. Ward, Y. Hao, J. Xu, T. Murray, and M. J. Thun. 2008. "Cancer Statistics, 2008." *CA: Cancer Journal for Clinicians* 58 (March–April): 71–96.

Jenkins, J. Craig. 1983. "Resource Mobilization Theory and the Study of Social Movements." *Annual Review of Sociology* 9 (August): 527–553.

Jenness, Valerie, and Ryken Grattet. 2001. *Making Hate a Crime: From Social Movement to Law Enforcement*. New York: Russell Sage Foundation.

Jensen, Peter. 2010. "It's a Crying Shame." *Baltimore Sun*, December 18, 13.

Jernigan, David. 2010. "Alcohol Marketing and Youth: Why It's a Problem and What You Can Do." Center for Alcohol Marketing and Youth, December 14. Accessed April 12, 2012 (www.camy.org).

Jeune, Bernard, and James W. Vaupel, eds. 1995. *Exceptional Longevity: From Prehistory to the Present*. Denmark: Odense University Press.

Jha, Prabhat, et al. 2011. "Trends in Selective Abortion of Girls in India: Analysis of Nationally Representative Birth Histories from 1990 to 2005 and Census Data from 1991 to 2011." *Lancet* 377 (June 4): 1921–1928.

Jilani, Zaid. 2011. "Profits at Largest 500 Corporations Grew by 81 Percent in 2010." Think Progress, May 5. Accessed May 7, 2011 (www.alternet.org).

Jiménez, Tomás R. 2007. "The Next Americans." *Los Angeles Times*, May 27, M1.

Johnson, Allan G. 2008. "Our House Is on Fire." Paper presented March 17. Retrieved June 17, 2008 (uhavax.hartford.edu/agjohnson/kellogg.htm).

Johnson, Carolyn Y. 2010. "Author on Leave After Harvard Inquiry." *Boston Globe*, August 10. Accessed August 12, 2010 (www.bostonglobe.com).

Johnson, Hank, and Bert Klandermans, eds. 1995. *Social Movements and Culture*. Minneapolis: University of Minnesota Press.

Johnson, Jean, and Jon Rochkind. 2009. *With Their Whole Lives Ahead of Them: Myths and Realities about Why So Many Students Fail to Finish College*. Public Agenda. Retrieved April 20, 2010 (www.publicagenda.org).

Johnson, Leslie, and Justine Lloyd. 2004. *Sentenced to Everyday Life: Feminism and the Housewife*. New York: Berg.

Johnson, Rachel, James Nunns, Jeffrey Rohaly, Eric Toder, and Roberton Williams. 2011. "Why Some Tax Units Pay No Income Tax." Tax Policy Center, July. Accessed September 22, 2011 (www.taxpolicycenter.org).

Johnston, Pamela. 2005. "Dressing the Part." *Chronicle of Higher Education*, August 10. Retrieved Augusts 11, 2005 (chronicle.com/jobs).

Jones, Jeffrey M. 2007. "Public: Family of Four Needs to Earn Average of $52,000 to Get By." Gallup News Service, February 9. Retrieved June 2, 2007 (www.galluppoll.com).

———. 2008. "Fewer Americans Favor Cutting Back Immigration." Gallup, July 10. Retrieved September 9, 2008 (www.gallup.com).

———. 2010. "Americans Still Perceive Crime as on the Rise." Gallup, November 18. Accessed August 22, 2011 (www.gallup.com).

———. 2011. "Americans Most Confident in Military, Least in Congress." Gallup, June 23. Accessed August 15, 2011 (www.gallup.com).

———. 2011. "Approval of Labor Unions Holds Near Its Low, at 52%." Gallup, August 31. Accessed December 20, 2011 (www.gallup.com).

———. 2011. "In U.S., 3 in 10 Say They Take the Bible Literally." Gallup, July 8. Accessed April 8, 2012 (www.gallup.com).

———. 2011. "Jobs, Economy Remain Dominant Concerns for Americans." Gallup, June 8. Accessed November 14, 2011 (www.gallup.com).

———. 2011. "Record-High 86% Approve of Black-White Marriages." Gallup, September 11. Accessed December 7, 2011 (www.gallup.com).

———. 2012. "Expected Retirement Age in U.S. Up to 67." Gallup, April 27. Accessed June 6, 2012 (www.gallup.com).

———. 2012. "Record-High 40% of Americans Identify as Independents in '11." Gallup, January 9. Accessed January 20, 2012 (www.gallup.com).

———. 2012. "Texas Widens Gap Over Other States in Percentage Uninsured." Gallup, March 2. Accessed April 22, 2012 (www.gallup.com).

Jones, Rachel K., and Joerg Dreweke. 2011. "Countering Conventional Wisdom: New Evidence on Religion and Contraceptive Use." Guttmacher Institute, April. Accessed April 24, 2012 (www.guttmacher.org).

Jones, Rachel K., and Megan L. Kavanaugh. 2011. "Changes in Abortion Rates Between 2000 and 2008 and Lifetime Incidence of Abortion." *Obstetrics & Gynecology* 117 (June): 1358–1366.

Jones, Susan S., and Hye-Won Hong. 2001. "Onset of Voluntary Communication: Smiling Looks to Mother." *Infancy* 2 (3): 353–370.

Jonsson, Jan O., David B. Grusky, Reinhard Pollak, and Matthew Di Carlo. 2009. "Recent Trends in Social Mobility in the United States: A New Approach to Modeling Trend in Big Class, Gradational, and Microclass Reproduction." Stanford Center for the Study of Poverty and Inequality, September. Accessed September 22, 2011 (inequality.com).

Jonsson, Patrik. 2007. "In US Justice, How Much Bias?" *Christian Science Monitor*, September 21, 1, 10.

———. 2010. "'Tea Party' Activists: Who They Are, What They Want." *Christian Science Monitor*, January 31, 21.

———. 2012. "A Showdown Over Voter ID." *Christian Science Monitor*, January 16, 16–17.

Josephson Institute. 2011. "What Would Honest Abe Say?" February 10. Accessed April 3, 2012 (www.josephsoninstitute.org).

Joyce, Amy. 2006. "Vacation Deprivation." *Washington Post*, June 25, D4.

"Judge Tosses Lawsuit from Tribes on Fighting Sioux Nickname." 2012. Fox News, May 3. Accessed June 6, 2012 (www.foxnews.com).

Juergensmeyer, Mark. 2003. "Thinking Globally about Religion." Pp. 3–13 in *Global Religions: An Introduction*, edited by Mark Juergensmeyer. New York: Oxford University Press.

Julian, Tiffany, and Robert Kominski. 2011. "Education and Synthetic Work-Life Earnings Estimates." U.S. Census Bureau, American Community Survey Reports, September. Accessed March 12, 2012 (www.census.gov).

Kagan, Jerome. 2011. "Want Better Students? Teach Their Parents." *Christian Science Monitor*, February 21, 34.

Kahan, Dan M., Hank Jenkins-Smith, and Donald Braman. 2011. "Cultural Cognition of Scientific Consensus." *Journal of Risk Research* 14 (2): 147–174.

Kahlenberg, Richard D. 2010. "10 Myths About Legacy Preferences in College Admissions." *Chronicle of Higher Education*, October 1, A23, A25.

Kaiser Family Foundation. 2011. "Health Care Spending in the United States and Selected OECD Countries." April. Accessed April 15, 2012 (www.kff.org).

Kalev, Alexandra, Frank Dobbin, and Erin Kelly. 2006. "Best Practices or Best Guesses?

Assessing the Efficacy of Corporate Affirmative Action and Diversity Policies." *American Sociological Review* 71 (August): 589–617.

Kalil, Ariel. 2002. "Cohabitation and Child Development." Pp. 153–160 in *Just Living Together: Implications of Cohabitation on Families, Children, and Social Policy*, edited by Alan Booth, Ann C. Crouter, and Nancy S. Landale. Mahwah, NJ: Lawrence Erlbaum Associates.

Kamber, Richard, and Mary Biggs. 2002. "Grade Conflation: A Question of Credibility." *Chronicle of Higher Education*, April 12, B14.

Kaminski, Deborah, and Cheryl Geisler. 2012. "Survival Analysis of Faculty Retention in Science and Engineering by Gender." *Science* 335 (February 17): 864–866.

Kantrowitz, Mark. 2011. "The Distribution of Grants and Scholarships by Race." September 2. Accessed March 20, 2012 (www.finaid.org).

Kapferer, Jean-Noel. 1992. "How Rumors Are Born." *Society* 29 (July/August): 53–60.

Kaplan, David A. 2010. "The Best Company to Work For." *Fortune*, February 8, 57–72.

Kaplan, Don. 2010. "JetBlue Loony Gets to Slide." *New York Post*, October 20. Accessed August 18, 2011 (www.nypost.com).

Karnow, Stanley. 2004. "Keep Your Tired, Poor Stereotypes About Immigrants." *Baltimore Sun*, March 5, 13A.

Karukstis, Kerry K. 2009. "Women in Science, Beyond the Research University: Overlooked and Undervalued." *Chronicle of Higher Education*, July 10, A23.

Katz, Jackson. 2006. *The Macho Paradox: Why Some Men Hurt Women and How All Men Can Help*. Naperville, IL: Sourcebooks.

Katzenbach, Jon. 2003. *Why Pride Matters More Than Money: The Power of the World's Greatest Motivational Force*. New York: Crown Business.

Keehan, Sean P., et al. 2011. "National Health Spending Projections Through 2020: Economic Recovery and Reform Drive Faster Spending Growth." *Health Affairs* 30 (July). Accessed April 15, 2012 (www.healthaffairs.org).

Keen, Judy. 2011. "Blacks Exodus Reshapes Cities." *USA Today*, May 20. Accessed June 21, 2012 (www.usatoday.com).

Keenan, Nora L., and Kimberly A. Rosendorf. 2011. "Prevalence of Hypertension and Controlled Hypertension—United States, 2005–2008." Centers for Disease Control and Prevention. *MMWR* 60 (Suppl): 62–66.

Keeter, Scott. 2009. "New Tricks for Old—and New—Dogs: Challenges and Opportunities Facing Communications Research." Pew Research Center Publications, March 3. Retrieved December 31, 2009 (pewresearch.org).

———. 2010. "Ask the Expert." Pew Research Center, December 29. Accessed December 30, 2010 (pewresearch.org).

———. 2011. "Ask the Expert." Pew Research Center, June 30. Accessed July 5, 2011 (pewresearch.org).

Keeter, Scott, et al. 2010. "A Year after Obama's Election: Blacks Upbeat about Black Progress, Prospects." Pew Research Center, January 12. Retrieved March 13, 2010 (www.pewresearchcenter.org).

Kelley, Lauren. 2011. "5 Companies That Did Something Good for the World This Year." AlterNet, December 25. Accessed June 23, 2012 (www.alternet.org).

Kemp, Alice Abel. 1994. *Women's Work: Degraded and Devalued*. Englewood Cliffs, NJ: Prentice Hall.

Kempner, Joanna, Clifford S. Perlis, and Jon F. Merz. 2005. "Ethics: Forbidden Knowledge." *Science* 307 (February 11): 854.

Kendall, Diana. 2002. *The Power of Good Deeds: Privileged Women and the Social Reproduction of the Upper Class*. Lanham, MD: Rowman & Littlefield.

Kennedy, Allison, Katherine LaVail, Glen Nowak, Michelle Basket, and Sarah Landry. 2011. "Confidence About Vaccines in the United States: Understanding Parents' Perceptions." *Health Affairs* 30 (June): 1151–1159.

Kennedy, Tracy L. M., Aaron Smith, Amy Tracy Wells, and Barry Wellman. 2008. "Networked Families." Pew Internet & American Life Project, October 19. Retrieved September 20, 2008 (www.pewinternet.org).

Kent, Mary Mederios. 2007. "Immigration and America's Black Population." *Population Bulletin* 62 (December): 1–16.

———. 2010. "In U.S., Who Is at Greatest Risk for Suicides?" Population Reference Bureau, November. Accessed May 20, 2011 (www.prb.org).

———. 2011. "U.S. Women Delay Marriage and Children for College." Population Reference Bureau, January. Accessed March 1, 2012 (www.prb.org).

Kestin, Sally, and Peter Franceschina. 2007. "Federal Watchdog Examines Seminoles' Gambling Profits." *South Florida Sun-Sentinel*, December 8. Retrieved December 10, 2007 (www.sun-sentinel.com).

Keyes, Charles. 2011. "Army Still Grappling with Soldier Suicides." CNN, November 18. Accessed May 10, 2012 (articles.cnn.com).

Kilar, Steve, and Justin Fenton. 2012. "New Federal Definition of Rape is Now Formally Approved." *Baltimore Sun*, January 7, 3.

Kilbourne, Jean. 1999. *Deadly Persuasion: Why Women and Girls Must Fight the Addictive Power of Advertising*. New York: Free Press.

Kim, Young M. 2011. "Minorities in Higher Education." American Council on Education. Accessed October 25, 2011 (www.acenet.edu).

King, C. Wendell. 1956. *Social Movements in the United States*. New York: Random House.

King, Michael, and Annie Bartlett. 2006. "What Same Sex Civil Partnerships May Mean for Health." *Journal of Epidemiology and Community Health* 60 (March): 188–191.

Kinnard, Christophe, Christian M. Zdanowicz, David A. Fisher, Elisabeth Isaksson, Anne de Vernal, and Lonnie G. Thompson. 2011. "Reconstructed Changes in Arctic Sea Ice Over the Past 1,450 Years." *Nature* 479 (November): 509–512.

Kinsella, Kevin, and David R. Phillips. 2005. "Global Aging: The Challenge of Success." *Population Bulletin* 60 (March): 1–40.

Kirkpatrick, David D. 2010. *The Facebook Effect: The Inside Story of the Company That Is Connecting the World*. New York: Simon & Schuster.

Kivel, Paul. 2004. *You Call This a Democracy? Who Benefits, Who Pays and Who Really Decides*. New York: Apex Press.

Kivisto, Peter, and Dan Pittman. 2001. "Goffman's Dramaturgical Sociology: Personal Sales and Service in a Commodified World." Pp. 311–334 in *Illuminating Social Life: Classical and Contemporary Theory Revisited*, 2nd edition. Thousand Oaks, CA: Pine Forge Press.

Klandermans, Bert. 1984. "Mobilization and Participation: Social Psychological Explanations of Resource Mobilization Theory." *American Sociological Review* 49 (October): 583–600.

Klaus, Marshall H., John H. Kennell, and Phyllis H. Klaus. 1995. *Bonding: Building the Foundations of Secure Attachment and Independence*. Reading, MA: Addison-Wesley.

Klaus, Patsy. 2006. "Crime and the Nation's Households, 2004." Washington, DC: U.S. Department of Justice, Bureau of Justice Statistics.

Klein, Allison, and Josh White. 2011. "Technology, Police Tactics Are Reining in Car Thefts." *Washington Post*, July 24, A1.

Klein, Ezra. 2011. "Do We Still Need Unions? Yes: Why They're Worth Fighting For." *Newsweek*, March 7, 18.

Klein, Joe. 2011. "As Goes Wisconsin . . . So Goes the Nation." *Time*, March 7, 36–39.

Klinenberg, Eric. 2012. "The Solo Economy." *Fortune*, February 6, 129–133.

Klopfenstein, Kristin, and M. Kathleen Thomas. 2005. "The Advance Placement Performance Advantage: Fact or Fiction?" Retrieved February 10, 2005 (www.utdallas.edu).

Kneebone, Elizabeth. 2009. "Job Sprawl Revisited: The Changing Geography of Metropolitan Employment." Brookings Institute, Metropolitan Policy Program, April. Retrieved May 1, 2010 (www.brookings.edu).

Kneebone, Elizabeth, and Emily Garr. 2011. "Income & Poverty." *The State of Metropolitan America*, 132–143. Brookings Institution. Accessed June 24, 2012 (www.brookings.edu).

Kneebone, Elizabeth, Carey Nadeau, and Alan Berube. 2011. "The Re-emergence of Concentrated Poverty: Metropolitan Trends in the 2000s." Brookings, November. Accessed December 20, 2011 (www.brookings.edu).

Knickerbocker, Brad. 2005. "Fallout of Marijuana Verdict." *Christian Science Monitor*, June 8, 3.

Koch, Wendy. 2012. "Workplaces Ban Not Only Smoking, but Smokers Themselves." *USA Today*, January 3. Accessed January 4, 2011 (www.usatoday.com).

Kochhar, Rakesh. 2007. "1995–2005: Foreign-Born Latinos Make Progress on Wages." Pew Hispanic Center, August 21. Retrieved October 10, 2007 (www.pewhispanic.org).

———. 2011. "In Two Years of Economic Recovery, Women Lost Jobs, Men Found Them." Pew Research Center, July 6. Accessed September 22, 2011 (www.pewsocialtrends.org).

Kochhar, Rakesh, Richard Fry, and Paul Taylor. 2011. "Twenty-to-One: Wealth Gaps Rise to Record Highs Between Whites, Blacks and

Hispanics." Pew Research Center, July 26. Accessed September 22, 2011 (www.pewsocialtrends.org).

———. 2011. "Wealth Gaps Rise to Record Highs Between Whites, Blacks and Hispanics." Pew Research Center, July 26.

Kocieniewski, David. 2011. "G.E.'s Strategies Let It Avoid Taxes Altogether." *New York Times*, March 24, A1.

Kohut, Andrew, Carroll Doherty, Michael Dimock, and Scott Keeter. 2009. "Current Decade Rated as Worst in 50 Years." Pew Research Center for the People & the Press, December 21. Retrieved May 6, 2010 (www.people-press.org).

———. 2011. "Beyond Red vs. Blue Political Typology." Pew Research Center for the People & the Press, May 4. Accessed July 9, 2011 (www.people-press.org).

Kohut, Andrew, et al. 2007. "Blacks See Growing Values Gap between Poor and Middle Class." Pew Research Center, November 13. Retrieved December 10, 2007 (www.pewsocialtrends.org).

Kolet, Ilan, and Shobhana Chandra. 2012. *Bloomberg Businessweek*, February 6–12, 20–21.

Kopczuk, Wojciech, Emmanuel Saez, and Jae Song. 2009. "Earnings Inequality and Mobility in the United States: Evidence from Social Security Data since 1937." February 3. Accessed September 25, 2011 (elsa.berkeley.edu).

Kornhauser, William. 1959. *The Politics of Mass Society*. New York: Free Press.

Kort-Butler, Lisa A., and Kelley J. Sittner Hartshorn. 2011. "Watching the Detectives: Crime Programming, Fear of Crime, and Attitudes about the Criminal Justice System." *Sociological Quarterly* 52 (Winter): 36-55.

Kosmin, Barry A., and Ariela Keysar. 2009. *American Religious Identification Survey [ARIS 2008]*. Summary Report, March. Retrieved April 25, 2010 (www.americanreligionsurvey-aris.org).

Kosova, Weston, and Pat Wingert. 2009. "Crazy Talk." *Newsweek*, June 8, 54–62.

Kost, Kathryn, and Stanley Henshaw. 2012. "U.S. Teenage Pregnancies, Births and Abortions, 2008: National Trends by Age, Race and Ethnicity." Guttmacher Institute, February. Accessed March 1, 2012 (www.guttmacher.org).

Kowitt, Beth, and Rupali Arora. 2011. "The 50 Most Powerful Women." *Fortune*, October 17, 125–131.

Kozol, Jonathan. 2005. *The Shame of the Nation: The Restoration of Apartheid Schooling in America*. New York: Crown.

Krache, D. 2008. "How to Ground a 'Helicopter Parent'." CNN, August 19. Retrieved July 10, 2009 (www.cnn.com).

Kramer, Laura. 2005. *The Sociology of Gender: A Brief Introduction*, 2nd edition. Los Angeles: Roxbury.

Kramer, Mattea. 2011. "Many Egg Producers Still Not Complying with Food-Sanitation Rules." *Washington Post*, October 1. Accessed April 24, 2012 (www.washingtonpost.com).

Kraska, Peter B. 2004. *Theorizing Criminal Justice: Eight Essential Orientations*. Long Grove, IL: Waveland Press.

Kreager, Derek A., and Jeremy Staff. 2009. "The Sexual Double Standard and Adolescent Peer Acceptance." *Social Psychology Quarterly* 72 (June): 143–164.

Kreider, Rose M., and Diana B. Elliott. 2009. "America's Families and Living Arrangements: 2007." U.S. Census Bureau, Current Population Reports, P20-561, September. Accessed March 10, 2012 (www.census.gov).

Kreider, Rose M., and Renee Ellis. 2011. "Number, Timing, and Duration of Marriages and Divorces: 2009." U.S. Census Bureau, Current Population Reports, P70-125, May. Accessed March 10, 2012 (www.census.gov).

Kristof, Nicholas D. 2012. "A Veteran's Death, the Nation's Shame." *New York Times*, April 14, SR1.

Kroll, Luisa. 2011. "World's Richest Self-Made Women." *Forbes*, May 23, 22.

Krugman, Paul. 2002. "Crony Capitalism, U.S.A." *New York Times*, January 15, 21.

Kulczycki, Andrei, and Arun P. Lobo. 2002. "Patterns, Determinants, and Implications of Intermarriage among Arab Americans." *Journal of Marriage and the Family* 64 (February): 202–210.

Kulkami, Sandeep C., Alison Levin-Rector, Majid Ezzati, and Christopher J. L. Murray. 2011. "Falling Behind: Life Expectancy in U.S. Counties from 2000 to 2007 in an International Context." *Population Health Metrics* 9 (16): 1–17.

Kurdek, Lawrence A. 2004. "Are Gay and Lesbian Cohabiting Couples *Really* Different from Heterosexual Married Couples?" *Journal of Marriage and Family* 66 (November): 880–900.

Kurlantzick, Joshua. 2012. "Freedom's Fizzle." *Bloomberg Businessweek,* January 23–29, 6–7.

Kusenbach, Margarethe. 2009. "Salvaging Decency: Mobile Home Residents' Strategies of Managing the Stigma of 'Trailer' Living." *Qualitative Sociology* 32 (December): 399–428.

Kutner, Lawrence, and Cheryl Olson. 2008. *Grand Theft Childhood*. New York: Simon & Schuster.

Kwok, Alvin C., et al. 2011. "The Intensity and Variation of Surgical Care at the End of Life: A Retrospective Cohort Study." *The Lancet*, October 6. Accessed April 15, 2012 (www.thelancet.com).

"Labor Unions Seen as Good for Workers, Not U.S. Competitiveness." 2011. Pew Research Center, February 17. Accessed February 19, 2011 (www.people-press.org).

LaFraniere, Sharon. 2007. "African Crucible: Cast as Witches. Then Cast Out." *New York Times*, November 15, 1.

Lakoff, Robin. T. 1990. *Talking Power: The Politics of Language*. New York: Basic Books.

Lalasz, Robert. 2006. "Americans Flocking to Outer Suburbs in Record Numbers." Population Reference Bureau. Retrieved January 24, 2008 (www.prg.org).

Lamb, Gregory M. 2009. "National Parks Face New Threats." *Christian Science Monitor*, September 27, 36–37.

———. 2009. "Rise of the Public You." *Christian Science Monitor*, August 30, 13–16.

Lamb, Sharon, and Lyn Mikel Brown. 2007. *Packaging Girlhood: Rescuing Our Daughters from Marketers' Schemes*. New York: St. Martin's Press.

Lammers, Joris, Jenka I. Stoker, Jennifer Jordan, Monique Pollmann, and Diederik A. Stapel. 2011. "Power Increases Infidelity Among Men and Women." *Psychological Science* 22 (September): 1191–1197.

Lampman, Jane. 2007. "New Fight, Old Foe: Slavery." *Christian Science Monitor*, February 21, 13–14.

Landsberg, Mitchell. 2007. "They're More Interested in Money Than Mao." *Los Angeles Times*, June 26, A1.

Landsberger, Henry A. 1958. *Hawthorne Revisited*. Ithaca, NY: Cornell University Press.

Laneri, Raquel. 2011. "Slumdog Millions." *Forbes*, May 9, 100–101.

Lang, Kurt, and Gladys Engel Lang. 1961. *Collective Dynamics*. New York: Thomas Y. Crowell.

Langton, Lynn. 2011. "Identity Theft Reported by Households, 2005–2010." Bureau of Justice Statistics, November. Accessed March 12, 2012 (www.bjs.gov).

Lannutti, Pamela J., Melanie Laliker, and Jerold L. Hale. 2001. "Violations of Expectations and Social-Sexual Communication in Student/Professor Interactions." *Communication Education* 50 (January): 69–82.

Lantz, Paula M., Richard L. Lichtenstein, and Harold A. Pollack. 2007. Health Policy Approaches to Population Health: The Limits of Medicalization." *Health Affairs* 26 (5): 1253–1257.

Laraña, Enrique, Hank Johnston, and Joseph R. Gusfield, eds. 1994. *New Social Movements: From Ideology to Identity*. Philadelphia: Temple University Press.

LaRossa, Ralph, and Donald C. Reitzes. 1993. "Symbolic Interactionism and Family Studies." Pp. 135–163 in *Sourcebook of Family Theories and Methods: A Contextual Approach*, edited by Pauline G. Boss, William J. Doherty, Ralph LaRossa, Walter R. Schumm, and Suzanne K. Steinmetz. New York: Plenum Press.

Larson, Jeffry, and Rachel Hickman. 2004. "Are College Marriage Textbooks Teaching Students the Premarital Predictors of Marital Quality?" *Family Relations* 53 (July): 385–392.

Lasswell, Harold. 1936. *Politics: Who Gets What, When and How*. New York: McGraw-Hill.

Lathrop, Richard G., and John Hasse. 2007. "Tracing New Jersey's Dynamic Landscape: A Municipal Report Card on Urban Grown and Open Space Loss." Retrieved January 28, 2008 (www.crssa.rutgers.edu).

Lavelle, Marianne. 2007. "Water Woes." *U.S. News & World Report*, June 4, 37–46.

Lawless, Jennifer L., and Richard L. Fox. 2005. *It Takes a Candidate: Why Women Don't Run for Office*. New York: Cambridge University Press.

Layton, Lyndsey, and Dan Eggen. 2011. "Industries Lobby Against Voluntary Nutrition Guidelines for Food Marketed to Kids." *Washington Post*, July 9. Accessed July 10, 2011 (www.washingtonpost.com).

Lazonick, William. 2011. "How the SEC Let the Wolves into the Stock Market Chicken Coop." New Deal 2.0, July 22. Accessed January 14, 2012 (www.newdeal20.org).

Leaper, Campbell, and Melanie M. Ayres. 2007. "A Meta-Analytic Review of Gender Variations in Adults' Language Use: Talkativeness, Affiliative Speech, and Assertive Speech." Personality and Social Psychology Review 11 (November): 328–363.

Ledger, Kate. 2009. "Sociology and the Gene." Contexts 8 (Summer): 16–20.

Lee, Don. 2007. "In China, Income Disparity Takes a Great Leap." Los Angeles Times, June 10, C1.

Lee, J., W. Grigg, and G. Dion. 2007. The Nation's Report Card: Mathematics 2007 (NCES 2007-494). Washington, DC: National Center for Education Statistics, Institute of Education Sciences, U.S. Department of Education.

Lee, Marlene. 2009. "Aging, Family Structure, and Health." Population Reference Bureau, October. Retrieved April 10, 2010 (www.prb.org).

LegiStorm. 2011. "Former Lobbyists Working for Congress Outnumber Elected Officials." September 13. Accessed February 15, 2012 (www.legistorm.com).

Lemert, Edwin M. 1951. Social Pathology: A Systematic Approach to the Theory of Sociopathic Behavior. New York: McGraw-Hill.

———. 1967. Human Deviance, Social Problems and Social Control. Englewood Cliffs, NJ: Prentice Hall.

Lencioni, Patrick. 2002. The Five Dysfunctions of a Team: A Leadership Fable. San Francisco: Jossey-Bass.

Lengermann, Patricia Madoo, and Jill Niebrugge-Brantley. 1992. "Contemporary Feminist Theory." Pp. 308–357 in Contemporary Sociological Theory, 3rd edition, edited by George Ritzer. New York: McGraw-Hill.

Lenhart, Amanda, Mary Madden, and Aaron Smith. 2011. "Teens, Kindness and Cruelty on Social Network Sites." Pew Research Center, November 9. Accessed November 10, 2011 (pewinternet.org).

Leonardsen, Dag. 2004. Japan as a Low-Crime Nation. New York: Palgrove Macmillan.

Leopold, Les. 2011. "Galleon Hedge Fund Billionaire Guilty on All Counts: 7 Ways Hedge Funds Lie, Cheat and Steal. AlterNet, May 11. Accessed June 10, 2011 (www.alternet.org).

Lerner, Susan, and Deanna Bitetti. 2011. "Deep Drilling, Deep Pockets: Expenditures of the Natural Gas Industry in New York to Influence Public Policy." Common Cause, April. Accessed June 20, 2012 (www.commoncause.org).

Levi, Jeffrey, Laura M. Segal, and Rebecca St. Laurent. 2011. "Investing in America's Health: A State-by-State Look at Public Health Funding and Key Health Facts." Trust for America's Health, March. Accessed April 15, 2012 (www.healthyamericans.org).

Levine, Arthur. 2006. "Educating School Teachers." The Education Schools Project, September. Retrieved March 15, 2010 (www.edschools.org).

Levine, Marc V. 1994. "A Nation of Hamburger Flippers?" Baltimore Sun, July 31, 1E, 4E.

Lewin, Tamar. 2009. "No Einstein in Your Crib? Get a Refund." New York Times, October 24, 1.

Lewis, David Levering. 1993. W. E. B. Du Bois: Biography of a Race, 1868–1919. New York: Henry Holt.

Lewis, Oscar. 1966. "The Culture of Poverty." Scientific American 115 (October): 19–25.

Li, Shan. 2011. "Smartphones Have Dark Side." Baltimore Sun, July 22, 18.

Li, Wei. 2009. Ethnoburb: The New Ethnic Community in Urban America. Honolulu: University of Hawai'i Press.

Lichtblau, Eric, David Johnson, and Ron Nixon. 2008. "F.B.I. Struggles to Handles Financial Fraud Cases." New York Times, October 19, A1.

Lichter, Daniel T., Zhenchao Qian, and Leanna M. Mellott. 2006. "Marriage or Dissolution? Union Transitions among Poor Cohabiting Women." Demography 43 (May): 223–240.

"Life Size Barbie Shows Young Girls the Dangers of Unrealistic Body Expectations." 2011. Diets in Review, May 3. Accessed October 25, 2011 (www.dietsinreview.com).

Lilly, J. Robert, Francis T. Cullen, and Richard A. Ball. 1995. Criminological Theory: Context and Consequences, 2nd edition. Thousand Oaks, CA: Sage.

Lindsey, Linda L. 2005. Gender Roles: A Sociological Perspective, 4th edition. Upper Saddle River, NJ: Prentice Hall.

Lino, Mark. 2011. Expenditures on Children by Families, 2010. U.S. Department of Agriculture, Center for Nutrition Policy and Promotion. Accessed July 25, 2011 (www.cnpp.usda.gov).

Linton, Ralph. 1936. The Study of Man. New York: Appleton-Century-Crofts.

———. 1964. The Study of Man: An Introduction. New York: Appleton-Century-Crofts.

Lipman, Becca. 2011. "Here Are the 12 Female CEOs of Fortune 500 Companies." Business Insider, July 9. Accessed August 13, 2011 (www.businessinsider.com).

Liu, Runjuan, and Donald Trefler. 2008. "Much Ado about Nothing: American Jobs and the Rise of Service Outsourcing to China and India." National Bureau of Economic Research, Working Paper 14061, June. Retrieved April 6, 2010 (www.nber.org).

Livingston, Gretchen, and Kim Parker. 2011. "A Tale of Two Fathers: More Are Active, but More Are Absent." Pew Research Center, June 15. Accessed March 1, 2012 (www.pewsocialtrends.org).

Livingston, Gretchen, Kim Parker, and Susannah Fox. 2009. "Latinos Online, 2006–2008: Narrowing the Gap." Pew Hispanic Center, December 22. Retrieved February 5, 2010 (www.pewhispanic.org).

Llana, Sara Miller. 2005. "Can a $103 Fine Stop Students from Swearing?" Christian Science Monitor, December 7, 1–2.

"Lobbying for the Faithful: Religious Advocacy Groups in Washington, D.C." 2011. Pew Research Center, November. Accessed April 8, 2012 (pewresearch.org).

Locher, David A. 2002. Collective Behavior. Upper Saddle River, NJ: Prentice Hall.

Lofland, John. 1993. "Collective Behavior: The Elementary Forms." Pp. 70–75 in Collective Behavior and Social Movements, edited by Russell L. Curtis, Jr. and Benigno E. Aguirre. Boston: Allyn & Bacon.

———. 1996. Social Movement Organizations: Guide to Research on Insurgent Realities. New York: Aldine De Gruyter.

Lofquist, Daphne. 2011. "Same-Sex Couple Households." U.S. Census Bureau, ACSBR/10-03, September. Accessed March 10, 2012 (www.census.gov).

Lofquist, Daphne, Terry Lugaila, Martin O'Connell, and Sarah Feliz. 2012. "Households and Families: 2010." 2010 Census Briefs, C2010BR-14, April. Accessed May 15, 2012 (www.census.gov).

Logan, John, and Harvey Molotch. 1987. Urban Fortunes: The Political Economy of Place. Berkeley: University of California Press.

Logan, John R., and Wenquan Zhang. 2011. "Global Neighborhoods: New Evidence from Census 2010." US2010 Project, November. Accessed June 23, 2012 (www.s4.brown.edu).

Lohr, Steve. 2009. "Watch the Walk and Prevent a Fall." New York Times, November 8, 4.

———. 2011. "Computers That See You and Keep Watch Over You." New York Times, January 1, A1.

Lopez, Mark Hugo. 2009. "Latinos and Education: Explaining the Attainment Gap." Pew Hispanic Center, October 7. Retrieved April 2, 2010 (www.pewhispanic.org).

Lopez, Mark Hugo, and Gabriel Velasco. 2011a. "Childhood Poverty Among Hispanics Sets Record, Leads Nation." Pew Hispanic Center, September 28. Accessed November 15, 2011 (www.pewhispanic.org).

———. 2011b. "A Demographic Portrait of Puerto Ricans, 2009." Pew Hispanic Center, June 13. Accessed November 15, 2011 (www.pewhispanic.org).

———. 2011c. "The Toll of the Great Recession: Childhood Poverty Among Hispanics Sets Record, Leads Nation." Pew Hispanic Center, September 28. Accessed March 1, 2012 (www.pewhispanic.org).

Lorber, Judith. 2005. Gender Inequality: Feminist Theories and Politics, 3rd edition. Los Angeles: Roxbury.

Lorber, Judith, and Lisa Jean Moore. 2007. Gendered Bodies: Feminist Perspectives. Los Angeles: Roxbury.

Lord, Susan M., Michelle M. Camacho, Richard A. Layton, Russell A. Long, Matthew W. Ohland, and Mara H. Wasburn. 2009. "Who's Persisting in Engineering? A Comparative Analysis of Female and Male Asian, Black, Hispanic, Native American, and White Students." Journal of Women and Minorities in Science and Engineering 15 (September): 167–190.

Loucks, Eric B., Michal Abrahamowicz, Yongling Xiao, and John W. Lynch. 2011. "Associations of Education with 30 Year Life Course Blood Pressure Trajectories: Framingham Offspring Study." BMC Public Health 11 (February): 139–149.

Lowenberger, Amanda, Joanna Mauer, Andrew deLaski, Marianne DiMascio, Jennifer

Amann, and Steven Nadel. 2012. "The Efficiency Boom: Cashing in on the Savings from Appliance Standards." American Council for an Energy-Efficient Economy, and Appliance Standards Awareness Project, March. Accessed June 22, 2012 (www.aceee.org).

Lowenkamp, Christopher T., and Edward J. Latessa. 2005. "Developing Successful Reentry Programs: Lessons Learned from the `What Works' Research." *Corrections Today* 67 (April): 72–77.

Lublin, Joann S. 2010. "CEO Pay in 2010 Jumped 11%. *Wall Street Journal*, May 6. Accessed September 27, 2011 (online.wsj.com).

Luckey, John R. 2008. "CRS Report for Congress: The United States Flag: Federal Law Relating to Display and Associated Questions." Congressional Research Service, April 14. Accessed July 20, 2011 (www.senate.gov).

Lugaila, Terry. A. 1998. "Marital Status and Living Arrangements: March 1998 (Update)." U.S. Census Bureau, Current Population Reports, P20-514. Retrieved August 8, 2000 (www.census.gov).

Lundberg, Shelly, and Robert A. Pollak. 2007. "The American Family and Family Economics." *Journal of Economic Perspectives* 21 (Spring): 3–26.

Luo, Michael, and Mike McIntire. 2012. "Donors Gave as Santorum Won Earmarks." *New York Times*, January 10, A1.

Lutfey, Karen E., Kevin W. Eva, Eric Gerstenberger, Carol L. Link, and John B. McKinlay. 2010. "Physician Cognitive Processing as a Source of Diagnostic and Treatment Disparities in Coronary Heart Disease: Results of a Factorial Priming Experiment." *Journal of Health and Social Behavior* 51 (March): 16–29.

Lutz, William. 1989. *Doublespeak*. New York: Harper & Row.

Lynch, David J. 2012. "Take Two Years and Call Me in the Morning." *Bloomberg Businessweek*, Spring, B2–B18.

Lynch, Eleanor W., and Marci J. Hanson, eds. 1999. *Developing Cross-Cultural Competence: A Guide for Working with Children and Their Families*, 2nd edition. Baltimore, MD: Paul H. Brookes.

Lynn, David B. 1969. *Parental and Sex Role Identification: A Theoretical Formulation*. Berkeley, CA: McCutchen.

Lynn, M., and M. Todoroff. 1995. "Women's Work and Family Lives." Pp. 244–71 in *Feminist Issues: Race, Class, and Sexuality*, edited by Nancy Mandell. Scarborough, Ontario: Prentice Hall Canada.

Lyons, Linda. 2005. "Tracking U.S. Religious Preferences Over the Decades." Gallup Poll, May 24. Retrieved August 25, 2007 (www.galluppoll.com).

Maccoby, Eleanor E., and John A. Martin. 1983. "Socialization in the Context of the Family: Parent-Child Interaction." Pp. 1–101 in *Socialization, Personality, and Social Development: Vol. 4. Handbook of Child Psychology*, edited by E. Mavis Hetherington. New York: Wiley.

Macionis, John J., and Vincent N. Parrillo. 2007. *Cities and Urban Life*, 4th edition. Upper Saddle River, NJ: Prentice Hall.

Mack, Jessica. 2011. "Boy or Girl? Why Do More Americans Prefer Male Children?" AlterNet, July 11. Accessed October 25, 2011 (www.alternet.org).

MacKenzie, Donald, and Judy Wajcman, eds. 1999. *The Social Shaping of Technology*, 2nd edition. Philadelphia: Open University Press.

Madigan, Nick. 2003. "Suspect's Wife Is Said to Cite Polygamy Plan." *New York Times*, March 15, 1.

———. 2005. "Sheehan's Ad Says Bush Lied About Iraq War." *Baltimore Sun*, August 27, 1D, 4D–5D.

———. 2009. "Seniors Increasingly Targeted by Con Artists; Police 'Struggling to Keep Up'." *Baltimore Sun*, January 14, 3.

Magnier, Mark. 2009. "In Northern India, Village Elders Order 'Honor Killings'." *Los Angeles Times*, September 26. Retrieved September 28, 2009 (www.latimes.com).

Maines, David R. 2001. *The Faultline of Consciousness: A View of Interactionism in Sociology*. New York: Aldine De Gruyter.

Major, Brenda, Mark Appelbaum, Linda Beckman, Mary Ann Dutton, Nancy Felipe Russo, and Carolyn West. 2008. "Report of the APA Task Force on Mental Health and Abortion." American Psychological Association, August 13. Retrieved September 5, 2008 (www.apa.org).

Major, William. 2011. "Thoreau's Cellphone Experiment." *Chronicle of Higher Education*, January 21, A33, A35.

"Major Religions of the World Ranked by Number of Adherents." 2007. Retrieved August 24, 2007 (www.adherents.com).

Makino, Catherine. 2009. "Rape Victim Presses Case of Police Abuse in Japan." Women's eNews, May 5. Retrieved May 7, 2009 (www.womensenews.org).

Malkan, Stacy. 2007. *Not Just a Pretty Face: The Ugly Side of the Beauty Industry*. Gabriola Island, Canada: New Society Publishers.

Malthus, Thomas Robert. 1798/1965. *An Essay on Population*. New York: Augustus Kelley.

———. 1872/1991. *An Essay on the Principle of Population*, 7th edition. London: Reeves & Turner.

Mananzan, Mary John. 2002. "Theological Reflections on Violence against Women (A Catholic Perspective)." Pp. 205–212 in *Gendering the Spirit: Women, Religion & the Post-Colonial Response*, edited by Durre S. Ahmed. New York: Zed Books.

Mandell, Nina. 2012. "Birth Control Panel Photo—Made Up Completely of Men—Goes Viral." *New York Daily News*, February 16. Accessed April 22, 2012 (www.nydailynews.com).

Marcotte, Amanda. 2011. "10 States Where Abortion Is Virtually Illegal for Some Women." AlterNet, June 12. Accessed October 25, 2011 (www.alternet.org).

Markoff, John. 2009. "Scientists Worry Machines May Outsmart Man." *New York Times*, July 26, 1.

Marks, Joseph L. 2007. *Fact Book on Higher Education 2007*. Atlanta, GA: Southern Regional Education Board.

Marlar, Jenny. 2011. "World's Women Less Likely to Have Good Jobs." Gallup, June 23. Accessed October 25, 2011 (www.gallup.com).

Marquand, Robert. 2011. "Europe Rejects Multiculturalism." *Christian Science Monitor*, February 21, 14.

Marsh, John. 2011. "Why Education Is Not an Economic Panacea." *Chronicle of Higher Education*, September 2, B10–B13.

Martin, Joyce A., et al. 2011. "Births: Final Data for 2009." *National Vital Statistics Reports*, 60 (1), November 3. Accessed March 3, 2012 (www.cdc.gov).

Martin, Karin A. 2009. "Normalizing Heterosexuality: Mothers' Assumptions, Talk, and Strategies with Young Children." *American Sociological Review* 74 (April): 190–207.

Martin, Molly A. 2008. "The Intergenerational Correlation in Weight: How Genetic Resemblance Reveals the Social Role of Families." *American Journal of Sociology* 114 (Suppl.): S67–S105.

Martin, Philip, and Elizabeth Midgley. 2010. "Immigration in America 2010." Population Bulletin Update, June, pp. 1–6. Accessed November 15, 2011 (www.prb.org).

Martin, Philip, and Gottfried Zürcher. 2008. "Managing Migration: The Global Challenge." *Population Bulletin* 63 (March): 1–20.

Martin, Suzanne. 2006. "Advertising to Youth: What Youth Want and What Advertisers Need to Know." *Trends and Tudes* 5 (August): 1–5.

Martinez, G. M., A. Chandra, J. C. Abma, J. Jones, and W. D. Mosher. 2006. "Fertility, Contraception, and Fatherhood: Data on Men and Women from Cycle 6 (2002) of the National Survey of Family Growth. *Vital Health Statistics* 23 (26). Retrieved June 3, 2007 (www.cdc.gov/nchs).

Marx, Gary T., and Douglas McAdam. 1994. *Collective Behavior and Social Movements: Process and Structure*. Upper Saddle River, NJ: Prentice Hall.

Marx, Karl. 1844/1964. *Economic and Philosophic Manuscripts of 1844*. New York: International Publishers.

———. 1845/1972. "The German Ideology." Pp. 110–164 in *The Marx-Engels Reader*, edited by Robert C. Tucker. New York: W. W. Norton.

———. 1867/1967. *Capital*, edited by Friedrich Engels. New York: International Publishers.

———. 1934. *The Class Struggles in France*. New York: International Publishers.

———. 1964. *Karl Marx: Selected Writings in Sociology and Social Philosophy*, translated by T. B. Bottomore. New York: McGraw-Hill.

Masis, Julie. 2010. "Leashed for Safekeeping." *Christian Science Monitor*, July 26, 7.

Massey, Douglas S. 2007. *Categorically Unequal: The American Stratification System*. New York: Russell Sage Foundation.

Masters, Jonathan. 2011. "Militant Extremists in the United States." Council on Foreign Relations, February 7. Accessed July 19, 2011 (www.cfr.org).

Mather, Mark. 2008. "Population Losses Mount in U.S. Rural Areas." Population Reference Bureau, March. Retrieved July 29, 2008 (www.prb.org).

———. 2009. *Children in Immigrant Families Chart New Path*. Washington, DC: Population Reference Bureau.

Mather, Mark, Keven Pollard, and Linda A. Jacobsen. 2011. *First Results from the 2010 Census. Washington*, DC: Population Reference Bureau.

Matlin, Chadwick. 2010. "Airlines Haven't Been This Profitable Since 1978." CNNMoney.com, October 28. Accessed August 13, 2011 (money.cnn.com).

Mauss, Armand. 1975. *Social Problems as Social Movements*. Philadelphia: Lippincott.

Maynard, Micheline. 2009. "Even as Fares Creep Up, Airlines Tack on Fees, Too." *New York Times*, October 16, 3.

Mayo, Elton. 1945. *The Problems of an Industrial Civilization*. Cambridge, MA: Harvard University Press.

McAdam, Doug, and Ronnelle Paulsen. 1994. "Specifying the Relationship between Social Ties and Activism." *American Journal of Sociology* 99 (November): 640–667.

McCabe, Donald L., Kenneth D. Butterfield, and Linda K. Treviño. 2006. "Academic Dishonesty in Graduate Business Programs: Prevalence, Causes, and Proposed Action." *Academy of Management Learning Education* 5 (September): 294–305.

McCanna, Shaun. 2009. "New Afghan Crisis: No Room in Schools." *Christian Science Monitor*, December 6, 11.

McCarthy, Ellen. 2004. "Md. Professor Archives History of Dot-Com Bombs." *Washington Post*, October 28, E1.

McCarthy, John, and Mayer Zald. 1977. "Resource Mobilization and Social Movements: A Partial Theory." *American Journal of Sociology* 82 (May): 1212–1241.

McCarthy, Michael. 2012. "Asian Stereotypes in Coverage of Knicks' Jeremy Lin." *USA Today*, February 16. Accessed February 20, 2012 (www.usatoday.com).

McCauley, Mary, and Paul Chenowith. 2011. "Measles—United States, January–May 20, 2011." *Morbidity and Mortality Weekly*, May 27. Accessed November 8, 2011 (www.cdc.gov).

McCauley, Mary Beth. 2012. "The Faith Factor." *Christian Science Monitor*, April 2, 26–31.

McCormack, Kate. 2011. "The Feminization of the College Degree?" Women's Media Center, May 31. Accessed June 10, 2011 (www.womensmediacenter.com).

McCoy, J. Kelly, Gene H. Brody, and Zolinda Stoneman. 2002. "Temperament and the Quality of Best Friendships: Effect of Same-Sex Sibling Relationships." *Family Relations* 51 (July): 248–255.

McDonald, Renee, Ernest N. Jouriles, Suhasini Ramisetty-Mikler, Raul Caetano, and Charles E. Green. 2006. "Estimating the Number of American Children Living in Partner-Violent Families." *Journal of Family Psychology* 20 (March): 137–142.

McEwan, Melissa. 2012. "Unbelievable: Man Beats Wife, Judge Orders Him to Take Her Out to Red Lobster and the Bowling Alley." AlterNet, February 9. Accessed February 10, 2012 (www.alternet.org).

McFadden, Robert D., and Angela Macropoulos. 2008. "Wal-Mart Employee Trampled to Death." *New York Times*, November 29, A16.

McFalls, Joseph A., Jr. 2007. "Population: A Lively Introduction, 5th edition." *Population Bulletin* 62 (March): 1–31.

McGloin, Jean Marie, and Alex R. Piquero. 2010." On the Relationship between Co-Offending Network Redundancy and Offending Versatility." *Journal of Research in Crime and Delinquency* 47 (February): 163-90.

McGregor, Jena. 2011. "Anne Mulcahy on Women in the Boardroom." *Washington Post*, October 6. Accessed October 25, 2011 (www.washintonpost.com).

McGregor, Jena, and Steve Hamm. 2008. "Managing the Workforce." *Business Week*, January 28, 34–43.

McHale, Susan M. 2001. "Free-Time Activities in Middle Childhood: Links with Adjustment in Early Adolescence." *Child Development* 76 (November/December): 1764–1778.

McIntosh, Peggy. 1995. "White Privilege and Male Privilege: A Personal Account of Coming to See Correspondences through Work in Women's Studies." Pp. 76–87 in *Race, Class, and Gender: An Anthology*, 2nd edition, edited by Margaret L. Andersen and Patricia Hill Collins. Belmont, CA: Wadsworth.

McKinnon, Mark. 2011. "Do We Still Need Unions? No: Let's End a Privileged Class." *Newsweek*, March 7, 19.

McManus, Mike. 2012. "Let's Revive Marriage in America." *Baltimore Sun*, January 3, 13.

McNamee, Stephen J., and Robert K. Miller, Jr. 1998. "Inheritance and Stratification." Pp. 193–213 in *Inheritance and Wealth in America*, edited by Robert K. Miller, Jr., and Stephen J. McNamee. New York: Plenum Press.

McNeil, David G. 2010. "U.S. Apologizes for Syphilis Program in Guatemala." *New York Times*, October 1, A1.

McPhail, Clark, and Ronald T. Wohlstein. 1983. "Individual and Collective Behavior within Gatherings, Demonstrations, and Riots." *Annual Review of Sociology*, 9 (August): 579–600.

McRae, Susan. 1999. "Cohabitation or Marriage?" Pp. 172–190 in *The Sociology of the Family*, edited by Graham Allan. Malden, MA: Blackwell Publishers.

Mead, George Herbert. 1934. *Mind, Self, and Society*. Chicago: University of Chicago Press.

———. 1964. *On Social Psychology*. Chicago: University of Chicago Press.

Mead, Margaret. 1935. *Sex and Temperament in Three Primitive Societies*. New York: Morrow.

"Media Coverage of Hispanics." 2009. Pew Hispanic Center Project for Excellence in Journalism, December 7. Retrieved March 12, 2010 (pewhispanic.org).

Mehl, Matthias R., Simine Vazire, Nairán Ramírez-Esparza, Richard B. Slatcher, and James W. Pennebaker. 2007. "Are Women Really More Talkative Than Men?" *Science* 317 (July): 82.

Melucci, Alberto. 1995. "The New Social Movements Revisited: Reflections on a Sociological Misunderstanding." Pp. 107–119 in *Social Movements and Social Classes: The Future of Collective Action*, edited by Louis Maheu. Thousand Oaks, CA: Sage.

Menchik, Daniel A., and Xiaoli Tian. 2008. "Putting Social Context into Text: The Semiotics of E-mail Interaction." *American Journal of Sociology* 114 (September): 332–370.

Mendelsohn, Oliver, and Maria Vicziany. 1998. *The Untouchables, Subordination, Poverty and the State in Modern India*. New York: Cambridge University Press.

Mendes, Elizabeth. 2010. "In U.S., Health Disparities Across Incomes Are Wide-Ranging." Gallup, October 18. Accessed October 21, 2010 (www.gallup.com).

———. 2011. "Americans Don't Want Biases in Hiring Smokers, the Overweight." Gallup, July 22. Accessed April 22, 2012 (www.gallup.com).

———. 2012. "Fewer Americans Have Employer-Based Health Insurance." Gallup, February 14. Accessed April 22, 2012 (www.gallup.com).

———. 2012. "Lack of Retirement Funds Is Americans' Biggest Financial Worry." Gallup, March 6. Accessed March 10, 2012 (www.gallup.com).

———. 2012. "More Americans Uninsured in 2011." Gallup, January 24. Accessed April 22, 2012 (www.gallup.com).

———. 2012. "Most Caregivers Look After Elderly Parent; Invest a Lot of Time." Gallup, March 6. Accessed March 10, 2012 (www.gallup.com).

Mendes, Elizabeth, and Jenny Marlar. 2011. "Underemployed Americans' Wellbeing Continues to Suffer." Gallup, June 8. Accessed December 20, 2011 (www.gallup.com).

Mendillo, Michael. 2012. "Stop Letting High-School Courses Count for College Credit." *Chronicle of Higher Education*, January 6, A28.

Meng, Liu. 2011. "Chinese College Drops Plan to Discourage Kissing and Other 'Uncivilized Behavior' on Campus." *Global Times*, April 22. Accessed October 25, 2011 (china.globaltimes.cn).

Menissi, Fatima. 1991. *The Veil and the Male Elite: A Feminist Interpretation of Women's Rights in Islam*. Reading, MA: Addison-Wesley.

———. 1996. *Women's Rebellion and Islamic Memory*. Atlantic Highlands, NJ: Zed Books.

Mercer. 2011. "Latest Survey Finds Health Benefit Cost Growth for 2012 Likely to Be the Lowest in 15 Years." September 21. Accessed April 15, 2012 (www.mercer.com).

Mertens, Richard. 2009. "Indiana's Amish Return to Their Plows." *Christian Science Monitor*, May 24, 23.

Merton, Robert K. 1938. "Social Structure and Anomie." *American Sociological Review* 3 (December): 672–682.

———. 1948/1996. "The Self-Fulfilling Prophecy." Pp. 183–201 in *Robert K. Merton: On Social Structure and Science*, edited by Piotr Sztompka. Chicago: University of Chicago Press.

———. 1949. "Discrimination and the American Creed." Pp. 99–126 in *Discrimination and National Welfare*, edited by Robert M. MacIver. New York: Harper.

———. 1968. *Social Theory and Social Structure*. New York: Free Press.

Merton, Robert K., and Alice K. Rossi. 1950. "Contributions to the Theory of Reference Group Behavior." Pp. 40–105 in *Continuities in Social Research*, edited by Robert K.

Merton and Paul L. Lazarsfeld. New York: Free Press.

Meyer, Cheryl L., and Michelle Oberman. 2001. *Mothers Who Kill Their Children: Understanding the Acts of Moms from Susan Smith to the "Prom Mom."* New York: New York University Press.

Meyerson, Harold. 2011. "Corporate America, Paving a Downward Economic Slide." *Washington Post*, January 5. Accessed January 6, 2011 (www.washingtonpost.com).

Michels, Robert. 1911/1949. *Political Parties.* Glencoe, IL: Free Press.

Mikkelson, Barbara, and David Mikkelson. 2005. "Super Bull Sunday." Retrieved August 14, 2009 (www.snopes.com).

Milgram, Stanley. 1963. "Behavioral Study of Obedience." *Journal of Abnormal and Social Psychology* 67 (4): 371–378.

———. 1965. "Some Conditions of Obedience and Disobedience to Authority." *Human Relations* 18 (February): 57–76.

Milkman, Ruth. 2012. "Revolt of the College-Educated Millennials." *Contexts* 11 (Spring): 13–14.

Miller, D. W. 2001. "DARE Reinvents Itself—With Help From Its Social-Scientist Critics." *Chronicle of Higher Education*, October 16, A12-A-14.

Miller, Lisa. 2010. "A Woman's Place Is in the Church." *Newsweek*, April 12, 34–41.

———. 2012. "Feminism's Final Frontier? Religion." *Washington Post*, March 8. Accessed March 9, 2012 (www.washingtonpost.com).

Miller, Matthew. 2005. "The (Porn) Player." *Forbes*, July 4, 124, 126, 128.

Miller, S. M. 2001. "My Meritocratic Rise." *Tikkun* (March/April): 1–3. Retrieved January 20, 2003 (www.tikkun.org).

Mills, C. Wright. 1956. *The Power Elite.* New York: Oxford University Press.

———. 1959. *The Sociological Imagination.* New York: Oxford University Press.

Min, Pyong Gap. 2002. "Introduction." Pp. 1–14 in *Religions in Asian America: Building Faith Communities*, edited by Pyong Gap Min and Jung Ha Kim. Walnut Creek, CA: AltaMira Press.

Mincy, Ronald, ed. 2006. *Black Males Left Behind.* Washington, DC: Urban Institute Press.

Miners, Zach. 2009. "A New Look at Why Girls Don't Get In." *U.S. News Weekly*, November 13, 8.

Mischel, Walter. 1966. "A Social Learning View of Sex Differences." Pp. 57–81 in *The Development of Sex Differences*, edited by Eleanor E. Maccoby. Stanford, CA: Stanford University Press.

Mishel, Lawrence. 2011. "Huge Disparity in Share of Total Wealth Gain Since 1983." Economic Policy Institute, September 15. Accessed September 18, 2011 (www.epi.org).

Mitchell, Josh, and Maya Jackson-Randall. 2012. "Student-Loan Debt Tops $1 Trillion." *Wall Street Journal*, April 22, 5A.

Mocan, Naci H., and Duha Tore Altindag. 2012. "Education, Cognition, Health Knowledge, and Health Behavior." National Bureau of Economic Research, March. Accessed April 24, 2012 (www.nber.org).

"Modest Rise in Number Saying There Is 'Solid Evidence' of Global Warming." 2011. Pew Research Center, December 1. Accessed December 19, 2011 (www.people-press.org).

Mokhiber, Russell. 2007. "Twenty Things You Should Know about Corporate Crime." *Corporate Crime Reporter* 21 (June 12). Retrieved February 25, 2010 (corporatecrimereporter.com).

Money, John, and Anke A. Ehrhardt. 1972. *Man & Woman, Boy & Girl: The Differentiation and Dimorphism of Gender Identity from Conception to Maturity.* Baltimore, MD: Johns Hopkins University Press.

Montenegro, Xenia P. 2004. "The Divorce Experience: A Study of Divorce at Midlife and Beyond." *AARP the Magazine* (May): 40–79.

Montgomery, Lori. 2011. "Ever-Increasing Tax Breaks for U.S. Families Eclipse Benefits for Special Interests." *Washington Post*, September 18, A1, A14.

Moore, Elizabeth S. 2006. "It's Child's Play: Advergaming and the Online Marketing of Food to Children." Henry J. Kaiser Family Foundation, July. Retrieved April 12, 2008 (www.kff.org).

Moore, Kathleen. 2008. "Low-Income Homes Green—and Affordable." *Daily Gazette*, July 1. Retrieved September 21, 2008 (www.dailygazette.com).

Moore, Martha T. 2006. "Cows Power Plan for Alternative Fuel." *USA Today*, December 3, 11A.

Morais, Richard C. 2007. "Desperate Arrangements." *Forbes*, January 29, 72–79.

Morales, Lymari. 2011. "Americans Grow More Negative About Their Personal Finances." Gallup, October 19. Accessed December 20, 2011 (www.gallup.com).

———. 2011. "Fewer Americans See U.S. Divided into 'Haves,' 'Have Nots'." Gallup, December 15. Accessed June 1, 2012 (www.gallup.com).

———. 2011. "Near Record-Low Confidence in U.S. Public Schools." Gallup, July 29. Accessed April 3, 2012 (www.gallup.com).

———. 2011. "U.S. Adults Estimate that 25% of Americans Are Gay or Lesbian." Gallup, Mary 27. Accessed October 25, 2011 (www.gallup.com).

"More U.S. Soldiers Killed Themselves Than Died in Combat in 2010." 2011. Good Evening, January 27. Accessed May 10, 2012 (www.good.is).

Morgenson, Gretchen. 2004. "No Wonder C.E.O.'s Love Those Mergers." *New York Times*, July 18, C1.

Morin, Richard. 2009. "What Divides America?" Pew Research Center, September 24. Retrieved March 2, 2010 (pewresearch.org).

———. 2011. "The Public Renders a Split Verdict on Changes in Family Structure." Pew Research Center, February 16. Accessed March 1, 2012 (www.pewsocialtrends.org).

———. 2012. "Rising Share of Americans See Conflict Between Rich and Poor." Pew Research Center, January 11. Accessed January 15, 2012 (www.pewsocialtrends.org).

Morin, Richard, and D'Vera Cohn. 2008. "Women Call the Shots at Home; Public Mixed on Gender Roles in Jobs." Pew Research Center,

September 28. Retrieved November 20, 2008 (pewresearch.org).

"Mormons in America: Certain in Their Beliefs, Uncertain of Their Place in Society." 2012. Pew Research Center, January 12. Accessed April 8, 2012 (pewresearch.org).

Morris, Desmond. 1994. *Bodytalk: The Meaning of Human Gestures.* New York: Crown Trade Paperbacks.

Morrison, Denton E. 1971. "Some Notes Toward Theory on Relative Deprivation, Social Movements, and Social Change." *American Behavioral Scientist* 14 (May–June): 675–690.

Mortenson, Tom. 2011. "Economic Change Effects on Men and Implications for the Education of Boys." *EdWeek*, May 14. Accessed June 3, 2012 (www.edweek.org).

Motivans, Mark. 2004. "Intellectual Property Theft, 2002." Washington, DC: U.S. Department of Justice, Bureau of Justice Statistics.

Moyer, Imogene L. 2001. *Criminological Theories: Traditional and Nontraditional Voices and Themes.* Thousand Oaks, CA: Sage.

———. 2003. "Jane Addams: Pioneer in Criminology." *Women & Criminal Justice* 14 (3/4): 1–14.

Moynihan, Colin. 2011. "Protestors Find Wall Street Off Limits." *New York Times*, September 18, A22.

Muehlenhard, Charlene L., and Zoe D. Peterson. 2011. "Distinguishing Between *Sex* and *Gender*: History, Current Conceptualizations, and Implications." *Sex Roles* 64 (June): 791–803.

Mui, Ylan Q. 2009. "Bottled Water Boom Appears Tapped Out." *Washington Post*, August 13, A10.

Mukhopadhyay, Samhita. 2011. *Outdated: Why Dating Is Ruining Your Life.* Berkeley, CA: Seal Press.

Mulac, Anthony. 1998. "The Gender-Linked Language Effect: Do Language Differences Really Make a Difference?" Pp. 127–153 in *Sex Differences and Similarities in Communication: Critical Essays and Empirical Investigations of Sex and Gender in Interaction*, edited by Daniel J. Canary and Kathryn Dindia. Mahwah, NJ: Lawrence Erlbaum Associates.

Mullaney, Tim. 2012. "More Job Openings, but Job Hunters Don't Have Skills Needed." *USA Today*, February 7. Accessed March 3, 2012 (www.usatoday.com).

Mulligan, Thomas S. 2004. "Beanie Babies Rode Own Bubble." *Baltimore Sun*, August 31, 1C, 3C.

Mumford, Lewis. 1961. *The City in History: Its Origins, Transformations, and Its Prospects.* New York: Harcourt, Brace.

Mundy, Liza. 2012. *The Richer Sex: How the New Majority of Female Breadwinners Is Transforming Sex, Love, and Family.* New York: Simon & Schuster.

Munk-Olsen, Trine, Thomas Munk Laursen, Carsten B. Pedersen, Øjvind Lidegaard, and Preben Bo Mortensen. 2011. "Induced First-Trimester Abortion and Risk of Mental Disorder." *New England Journal of Medicine* 364 (4): 332–339.

Murdock, George P. 1940. "The Cross-Cultural Survey." *American Sociological Review*, 5: 361–370.

———. 1945. "The Common Denominator of Cultures." Pp. 123–142 in *The Science of Man in the World Crisis*, edited by Ralph Linton. New York: Columbia University Press.

———. 1967. "Ethnographic Atlas: A Summary." *Ethnology* 6, 109–236.

Murphey, David, Bonnie Mackintosh, and Marci McCoy-Roth. 2011. "Early Childhood Policy Focus: Healthy Eating and Physical Activity." Child Trends, July 25. Accessed April 24, 2012 (www.childtrends.org).

Murphy, Cait. 2005. "Fast-Forward to the Future." *Fortune*, September 19, 271.

Murphy, Caryle. 2009. "Behind the Veil." *Christian Science Monitor*, December 13, 12–17.

Murphy, Evelyn, and E. J. Graff. 2005. *Getting Even: Why Women Don't Get Paid Like Men—And What to Do About It*. New York: Simon & Schuster.

Murphy, John. 2004. "S. Africa's New Goal: Economic Equality." *Baltimore Sun*, April 27, 1A, 4A.

Murphy, Sherry L., Jiaquan Xu, and Kenneth D. Kochanek. 2012. "Deaths: Preliminary Data for 2010." *National Vital Statistics Reports* 60 (January 11): 1–69.

Murray, Charles. 2008. "College Daze." *Forbes*, September 1, 32.

———. 2012. *Coming Apart: The State of White America, 1960–2010*. New York: Crown Forum.

Mutzabaugh, Ben. 2011. "Baggy Pants Lead to Arrest of US Airways Passenger." *USA Today*, June 16. Accessed June 18, 2011 (www.usatoday.com).

———. 2011. "Reports: Thai Airline Recruits 'Third-Sex' Attendants." *USA Today*, January 29. Accessed October 25, 2011 (www.usatoday.com).

Myers, Candice A., Tim Slack, and Joachim Singelmann. 2010. "Understanding the Aftermath of Hurricanes Katrina and Rita." Population Reference Bureau, February. Retrieved April 30, 2010 (www.prb.org).

Myers, John P. 2007. *Dominant-Minority Relations in America: Convergence in the New World*, 2nd edition. Boston: Allyn & Bacon.

Myerson, J. A. 2011. "Obama Administration Has Many Ties to Big Banks—And Protestors Occupying Wall Street Know It." AlterNet, October 9. Accessed October 10, 2011 (www.alternet.org).

Mykyta, Laryssa, and Suzanne Macartney. 2011. "The Effects of Recession on Household Composition: 'Doubling Up' and Economic Well-Being." U.S. Census Bureau, SEHSD Working Paper Number 2011-4. Accessed January 19, 2012 (www.census.gov).

Naili, Hajer. 2011. "Study Details Sex-Traffic in Post-Saddam Iraq." Women's e-News, November 9. Accessed February 25, 2012 (www.womensenews.org).

Naisbitt, John, Nana Naisbitt, and Douglas Philips. 1999. *High Tech/High Touch: Technology and Our Search for Meaning*. New York: Broadway Books.

Nakao, Keiko, and Judith Treas. 1992. "The 1989 Socioeconomic Index of Occupations: Construction from the 1989 Occupational Prestige Scores," GSS Methodological Report No. 74. Chicago: National Opinion Research Center.

Nakashima, Ellen. 2008. "Prescription Data Used to Assess Consumers." *Washington Post*, August 4, A1.

Nanda, Rupashree. 2010. "It's Official: 37 PC Live Below Poverty Line." IBN Live, April 18. Accessed September 15, 2011 (ibnlive.in.com).

National Academy of Sciences. 2007. *Treatment of PTSD: An Assessment of the Evidence*. Washington, DC: National Academies Press.

———. 2008. *Science, Evolution, and Creationism*. Washington, DC: National Academy of Sciences.

National Academy of Sciences. 2012. "Water Reuse: Potential for Expanding the Nation's Water Supply through Reuse of Municipal Wastewater." Report in Brief. Accessed June 23, 2012 (www.nasonline.org).

National Center for Health Statistics. 2011. *Health, United States, 2010: With Special Feature on Death and Dying*. Hyattsville, MD.

———. 2012. *Health, United States, 2011: With Special Feature on Socioeconomic Status and Health*. Hyattsville, MD.

National Center on Addiction and Substance Abuse. 2008. "National Survey of American Attitudes on Substance Abuse XIII: Teens and Parents." Retrieved January 10, 2010 (www.casacolumbia.org).

National Center on Elder Abuse. 2005. "Elder Abuse Prevalence and Incidence." Retrieved June 23, 2007 (www.elderabusecenter.org).

National Coalition for the Homeless. 2009. "How Many People Experience Homelessness?" July. Retrieved March 4, 2010 (www.nationalhomeless.org).

National Council of Teachers of English Committee on Public Doublespeak. 2005. Retrieved February 12, 2006 (www.ncte.org).

National Drug Intelligence Center. 2011. *The Economic Impact of Illicit Drug Use on American Society*. Washington D.C.: United States Department of Justice.

National Healthcare Disparities Report. 2003. U.S. Department of Health and Human Services. Retrieved April 20, 2004 (www.qualitytools.ahrq.gov).

National Institute on Drug Abuse. 2009. "Principle of Drug Addiction Treatment: A Research Based Guide." Accessed August 25, 2011 (www.nida.nih.gov).

National Institutes of Health. 2006. "Stem Cell Information." Retrieved August 6, 2008 (www.stemcells.nih.gov/info).

———. 2006. *Women of Color Health Data Book: Adolescents to Seniors*. Accessed May 25, 2012 (orwh.od.nih.gov).

National Latino Alliance for the Elimination of Domestic Violence. 2005. "Domestic Violence Affects Families of All Racial, Ethnic, and Economic Backgrounds." Retrieved June 20, 2007 (www.dvalianza.org).

National Nanotechnology Coordination Office. 2008. "Nanotechnology: Big Things from a Tiny World." Accessed June 20, 2012 (www.nano.gov).

National Public Radio. 2010. "Black Male Privilege?" Interview transcript, March 4. Retrieved March 10, 2010 (www.npr.org).

National Science Board. 2012. *Science and Engineering Indicators 2012*. Arlington, VA: National Science Foundation.

National Survey of Student Engagement. 2011. *Fostering Student Engagement Campuswide—Annual Results 2011*. Bloomington, IN: Indiana University for Postsecondary Research.

National Women's Law Center. 2011. "Poverty Among Women and Families, 2000–2010." September. Accessed September 25, 2011 (www.nwlc.org).

———. 2011. "Second Anniversary of the Recovery Shows No Job Growth for Women." July. Accessed July 25, 2011 (www.nclc.org).

Natural Resources Defense Council. 2007. "Beach Pollution." Retrieved February 8, 2008 (www.ndrc.org).

Neelakantan, Shailaja. 2006. "In India, Conservatives Want Women Under Wraps." *Chronicle of Higher Education*, May 26, A47–A48.

———. 2011. "In India, Caste Discrimination Still Plagues University Campuses." Chronicle of Higher Education, December 16, A12–A15.

Neider, Linda L., and Chester A. Schriesheim, eds. 2005. *Understanding Teams*. Greenwich, CT: Information Age.

Nestle, Marion. 2012. "Genetically Modified Food Myths and Truths—A Critical Review of the Science." AlterNet, June 21. Accessed June 26, 2012 (www.alternet.org).

Netherlands Environmental Assessment Agency. 2007. "China Now No. 1 in CO2 Emissions; USA in Second Position." Retrieved February 12, 2008 (www.mnp.nl).

Neuner, Joan M., William A. See, Liliana E. Pezzin, Sergey Tarima, and Ann B. Nattinger. 2012. "The Association of Robotic Surgical Technology and Hospital Prostatectomy Volumes." *Cancer* 118 (January 15): 371–377.

Newman, David M. 2005. *Identities and Inequalities: Exploring the Intersections of Race, Class, Gender, and Sexuality*. New York: McGraw-Hill.

Newport, Frank. 2001. "Americans See Women as Emotional and Affectionate, Men as More Aggressive." *Gallup Poll Monthly* No. 425 (February): 34–38.

———. 2006. "Who Believes in God and Who Doesn't?" Gallup Poll, June 23. Retrieved June 24, 2006 (www.galluppoll.com).

———. 2010. "Four in 10 U.S. Workers Say Their Company Is Understaffed." Gallup, October 8. Accessed December 20, 2011 (www.gallup.com).

———. 2010. "In U.S., 54% Support Death Penalty in Cases of Murder." Gallup, November 8. Accessed August 22, 2011 (www.gallup.com).

———. 2010. "Near-Record High See Religion Losing Influence in America." Gallup, December 29. Accessed April 8, 2012 (www.gallup.com).

———. 2010. "Tea Party Supporters Overlap Republican Base." Gallup, July 2. Accessed January 20, 2012 (www.gallup.com).

———. 2011. "Americans Prefer Boys to Girls, Just As They Did in 1941." Gallup, June 23. Accessed October 25, 2011 (www.gallup.com).

———. 2011. "Americans Still Prefer Male Bosses; Many Have No Preference." Gallup, September 8. Accessed October 25, 2011 (www.gallup.com).

———. 2011. "Christianity Remains Dominant Religion in the United States." Gallup, December 23. Accessed April 8, 2012 (www.gallup.com).

———. 2011. "For First Time, Majority in U.S. Supports Public Smoking Ban." Gallup, July 15. Accessed August 22, 2011 (www.gallup.com).

———. 2011. "For First Time, Majority of Americans Favor Legal Gay Marriage." Gallup, May 20. Accessed October 25, 2011 (www.gallup.com).

———. 2011. "More Than 9 in 10 Americans Continue to Believe in God." Gallup, June 3. Accessed April 8, 2012 (www.gallup.com).

———. 2011. "Record-High 50% of Americans Favor Legalizing Marijuana Use." Gallup, October 17. Accessed April 22, 2012 (www.gallup.com).

———. 2012. "Americans' Economic Worries: Jobs, Debt, and Politicians." Gallup, January 12. Accessed January 20, 2012 (www.gallup.com).

———. 2012. "Americans Like Having a Rich Class, as They Did 22 Years Ago." Gallup, May 11. Accessed June 1, 2012 (www.gallup.com).

———. 2012. "Americans' Worries About Global Warming Up Slightly." Gallup, March 30. Accessed June 22, 2012 (www.gallup.com).

———. 2012. "Congress' Job Approval at New Low of 10%." Gallup, February 8. Accessed February 20, 2012 (www.gallup.com).

———. 2012. "Half of Americans Support Legal Gay Marriage." Gallup, May 8. Accessed May 9, 2012 (www.gallup.com).

Newport, Frank, Sangeeta Agrawal, and Dan Witters. 2010. "Very Religious Americans Report Less Depression, Worry." Gallup, December 1. Accessed April 8, 2012 (www.gallup.com).

Newport, Frank, Dan Witters, and Sangeeta Agrawal. 2012. "In U.S., Very Religious Americans Have Higher Wellbeing Across All Faiths." Gallup, December 1. Accessed April 8, 2012 (www.gallup.com).

Nichols, Austin, and Melissa Favreault. 2009. "A Detailed Picture of Intergenerational Transmission of Human Capital." Urban Institute, May 22. Retrieved March 6, 2010 (www.urban.org).

Niebuhr, H. Richard. 1929. *The Social Sources of Denominationalism.* New York: Meridian.

Nielsen Company. 2008. "College Spring Break Study." February 27. Retrieved April 10, 2008 (www.alcoholstats.com).

Niewyk, Donald L., and Francis R. Nicosia. 2000. *The Columbia Guide to the Holocaust.* New York: Columbia University Press.

Nisbett, Richard E. 2009. *Intelligence and How to Get It: Why Schools and Cultures Count.* New York: W. W. Norton.

Nitkin, David. 2008. "Hopkins' Carson to Get Medal of Freedom." *Baltimore Sun*, June 12, 7B.

"No Action on Greenhouse Gases." 2008. *Baltimore Sun*, July 12, 2A.

"No Consensus About Whether Nation Is Divided Into 'Haves' and 'Have-Nots'." 2011. Pew Research Center for People & the Press, September 29. Accessed September 29 (people-press.org).

NOAA National Climatic Data Center. 2011. *State of the Climate: Global Analysis for Annual 2011.* National Oceanic and Atmospheric Administration. Accessed June 22, 2012 (www.ncdc.noaa.gov).

Nocera, Joe. 2011. "You Call That Tough?" *New York Times*, May 7, A21.

Nordin, Kendra. 2011. "Little Farms in the City." *Christian Science Monitor*, September 19, 36–37.

Norris, Floyd. 2012. "The Number of Those Working Past 65 is at a Record High." *New York Times*, May 18, B3.

Norris, Tina, Paula L. Vines, and Elizabeth M. Hoeffel. 2012. "The American Indian and Alaska Native Population: 2010." U.S. Census Bureau. 2010 Census Briefs, January. Accessed March 20, 2012 (www.census.gov).

Norton, Michael I., and Dan Ariely. 2011. "Building a Better America—One Wealth Quintile at a Time." *Perspectives on Psychological Science* 6 (1): 9–12.

Norton, Michael I., and Samuel R. Sommers. 2011. "Whites See Racism as a Zero-Sum Game That They Are Now Losing." *Perspectives on Psychological Science* 6 (3): 215–218.

Nyseth, Hollie, Sarah Shannon, Kia Heise, and Suzy Maves McElrath. 2011. "Embedded Sociologists." *Contexts* 10 (Spring): 44–50.

Oakes, Jeannie. 1985. *Keeping Track: How Schools Structure Inequality.* New Haven, CT: Yale University Press.

Obach, Brian K. 2004. *Labor and the Environmental Movement: The Quest for Common Ground.* Cambridge, MA: MIT Press.

Oberschall, Anthony. 1973. *Social Conflict and Social Movements.* Englewood Cliffs, NJ: Prentice Hall.

———. 1995. *Social Movements: Ideologies, Interests, and Identities.* New Brunswick, NJ: Transaction.

O'Brien, Jodi, and Peter Kollock. 2001. *The Production of Reality: Essays and Readings on Social Interaction*, 3rd edition. Thousand Oaks, CA: Pine Forge Press.

O'Donnell, Victoria. 2007. *Television Criticism.* Thousand Oaks, CA: Sage.

OECD. 2011. *Education at a Glance 2011: OECD Indicators.* OECD Publishing. Accessed March 30, 2012 (dx.doi.org).

———. 2011. "Growing Income Inequality in OECD Countries: What Drives It and How Can Policy Tackle It?" May 2. Accessed September 20, 2011 (www.oecd.org/els/social/inequality).

Office of Justice Programs. 2011. "Adult Crime Solutions." Accessed August 25, 2011 (www.crimesolutions.gov).

Office of Juvenile Justice and Delinquency Prevention. 2010. "Best Practices to Address Community Gang Problems: OJJDP's Comprehensive Gang Model." Second edition, October. Accessed August 21, 2011 (www.ncjrs.gov).

Office of the Deputy Chief of Staff for Intelligence. 2006. "Arab Cultural Awareness: 58 Factsheets." U.S. Army Training and Doctrine Command, Ft. Leavenworth, Kansas, January. Retrieved February 15, 2006 (www.fas.org).

Ogburn, William F. 1922. *Social Change with Respect to Culture and Original Nature.* New York: Dell.

Ogden, Cynthia L., and Margaret D. Carroll. 2010. "Prevalence of Obesity Among Children and Adolescents: United States, Trends 1963–1965 Through 2007–2008." National Center for Health Statistics, June. Accessed July 20, 2011 (www.cdc.gov).

———. 2010. "Prevalence of Overweight, Obesity, and Extreme Obesity Among Adults: United States, Trends 1960–1962 Through 2007–2008." National Center for Health Statistics, June. Accessed March 11, 2012 (www.cdc.gov/nchs).

Ogden, Cynthia L., Margaret D. Carroll, Brian K. Kit, and Katherine M. Flegal. 2012. "Prevalence of Obesity in the United States, 2009–2010." NCHS Data Brief No. 82, January. Accessed April 24 (www.nchs.gov).

Ogden, Cynthia L., Molly M. Lamb, Margaret D. Carroll, and Katherine M. Flegal. 2010. "Obesity and Socioeconomic Status in Adults: United States, 2005–2008." NCHS Data Brief No. 50, December. Accessed April 24 (www.nchs.gov).

O'Hara, John M. 2010. *A New American Tea Party: The Counterrevolution against Bailouts, Handouts, Reckless Spending, and More Taxes.* Hoboken, NJ: John Wiley & Sons.

O'Hare, William P. 2002. "Tracking the Trends in Low-Income Working Families." *Population Today* 30 (August/September): 1–3.

O'Harrow, Jr., Robert. 2008. "Earmark Spending Makes a Comeback." *Washington Post*, June 13, A1.

Ohlemacher, Stephen. 2011. "Social Security Makes $6.5B in Overpayments." *USA Today*, June 14. Accessed June 15, 2011 (www.usatoday.com).

Ojito, Mirta. 2009. "Doctors in Cuba Start Over in the U.S." *New York Times*, August 4, D1.

O'Keeffe, Gwenn S., and Kathleen Clarke-Pearson. 2011. "The Impact of Social Media on Children, Adolescents, and Families." *Pediatrics* 127 (April 1): 800–804.

Olver, Mark E. 2010. "Sexuality, Sexual Deviance, and Sexual Offending." *Sex Roles* 63 (December): 900–903.

Oppel, Richard A. Jr. 2011. "Steady Decline in Major Crime Baffles Experts." *New York Times*, May 4, A17.

Orenstein, Peggy. 2011. *Cinderella Ate My Daughter: Dispatches from the Front Lines of the New Girlie-Girl Culture.* New York: HarperCollins.

Ornstein, Allan C. 2003. *Pushing the Envelope: Critical Issues in Education.* Upper Saddle River, NJ: Prentice Hall.

Ornstein, Charles, and Tracy Weber. 2011. "Medical Schools Plug Holes in Conflict-of-Interest Policies." ProPublica, May 19. Accessed on July 8, 2011 (www.propublica.org).

Orrenius, Pia M., and Madeline Zavodny. 2009. "Do Immigrants Work in Riskier Jobs?" *Demography* 46 (August): 535–551.

Orum, Anthony M. 2001. *Introduction to Political Sociology*, 4th edition. Upper Saddle River, NJ: Prentice Hall.

Packard, Vance. 1959. *The Status Seekers.* New York: David McKay.

Padgett, Tim. 2010. "Robes for Women." *Time*, September 27, 53–55.

Page, Susan, and Naomi Jagoda. 2010. "What Is the Tea Party? A Growing State of Mind." *USA Today*, July 1. Accessed July 5, 2010 (www.usatoday.com).

Palmeri, Christopher, and Peter S. Green. 2010. "Former Executives Bomb at the Ballot Box." *Bloomberg Businessweek*, November 8–14, 35–36.

Pan, L., et al. 2009. "Differences in Prevalence of Obesity among Black, White, and Hispanic Adults—United States, 2006–2008." *MMWR Weekly* 58 (July 17): 740–744.

Parents Television Council. 2010. "Best and Worst TV Shows of the Week." Retrieved January 24, 2010 (www.parentstv.org).

Park, Robert, and Ernest Burgess. 1921. *Introduction to the Science of Sociology*. Chicago: University of Chicago Press.

Parker, Kim. 2009. "The Harried Life of the Working Mother." Pew Research Center, October 1. Retrieved March 11, 2010 (pewresearch.org).

Parker, Kim, Richard Fry, D'Vera Cohn, and Wendy Wang. 2011. "Is College Worth It? College Presidents, Public Assess Value, Quality and Mission of Higher Education." Pew Research Center, May 16. Accessed April 15, 2012 (www.pewsocialtrends.org).

Parker-Pope, Tara. 2011. "Web of Popularity, Achieved by Bullying." *New York Times*, February 14, A1.

Parkinson, C. Northcote. 1962. *Parkinson's Law*, 2nd edition. Boston: Houghton Mifflin.

Parsons, Talcott. 1951. *The Social System*. Glencoe, IL: Free Press.

———. 1954. *Essays in Sociological Theory*, revised edition. New York: Free Press.

———. 1959. "The School Class as a Social System: Some of Its Functions in American Society." *Harvard Educational Review* 29 (Fall): 297–313.

———. 1960. *Structure and Process in Modern Societies*. New York: Free Press.

Parsons, Talcott, and Robert F. Bales, eds. 1955. *Family, Socialization and Interaction Process*. New York: Free Press.

Partnership for Public Service. 2011. "Scores by Effective Leadership." Accessed August 13, 2011 (www.bestplacestowork.org).

Pascoe, C. J. 2007. *Dude, You're a Fag: Masculinity and Sexuality in High School*. Berkeley: University of California Press.

Passel, Jeffrey S., Gretchen Livingston, and D'Vera Cohn. 2012. "Explaining Why Minority Births Now Outnumber White Births." Pew Research Center, May 17. Accessed May 18, 2012 (www.pewsocialtrends.org).

Passel, Jeffrey S., and D'Vera Cohn. 2008. "U.S. Population Projections: 2005–2050." Pew Research Center, February 11. Accessed November 15, 2011 (www.pewresearch.org).

———. 2011. "Unauthorized Immigrant Population: National and State Trends, 2010." Pew Hispanic Center, February 1. Accessed November 15, 2011 (www.pewhispanic.org).

Passel, Jeffrey S., D'Vera Cohn, and Ana Gonzalez-Barrera. 2012. "Net Migration from Mexico Falls to Zero—and Perhaps Less." Pew Hispanic Center, April 23. Accessed May 13, 2012 (www.pewhispanic.org).

Passel, Jeffrey S., D'Vera Cohn, and Mark Hugo Lopez. 2011. "Hispanics Account for More Than Half of Nation's Growth in Past Decade." Pew Research Center, March 24. Accessed December 7, 2011 (www.pewhispanic.org).

Passel, Jeffrey S., Wendy Wang, and Paul Taylor. 2010. "Marrying Out: One-in-Seven New U.S. Marriages Is Interracial or Interethnic." Pew Research Center, June 15. Accessed December 7, 2011 (www.pewresearchcenter.org).

Pastor, Manuel, Justin Scoggins, Jennifer Tran, and Rhonda Ortiz. 2010. "The Economic Benefits of Immigrant Authorization in California." Center for the Study of Immigrant Integration, January. Retrieved March 14, 2010 (csii.usc.edu).

"Pathways to Prosperity: Meeting the Challenge of Preparing Young Americans for the 21st Century." Harvard Graduate School of Education, February. Accessed April 2, 2012 (www.gse.harvard.edu).

Paul, Annie Murphy. 2010. *Origins: How the Nine Months Before Birth Shape the Rest of Our Lives*. New York: Free Press.

Paul, Noel. C. 2002. "The Birth of a Would-Be-Fad." *Christian Science Monitor*, September 23, 11, 14–16.

Paul, Pamela. 2005. *Pornified: How Pornography Is Transforming Our Lives, Our Relationships, and Our Families*. New York: Henry Holt.

Paul, Richard, and Linda Elder. 2007. *The Miniature Guide to Critical Thinking: Concepts and Tools*. Dillon Beach, CA: Foundation for Critical Thinking.

Paulson, Amanda. 2012. "School Lunches Get First Overhaul in 15 Years—but Pizza Still a Vegetable." *Christian Science Monitor*, January 25. Accessed January 27, 2012 (*www.csmonitor.com*).

Pay Equity Commission. 2012. "The Gender Wage Gap." February 7. Accessed June 1, 2012 (www.payequity.gov.on.ca).

Payne, K. K., and L. Gibbs. 2011. "Marital Duration at Divorce, 2010." NCFMR, FP-11-13. Accessed March 6, 2012 (ncfmr.bgsu.edu).

Pearce, Diana. 1978. "The Feminization of Poverty: Women, Work, and Welfare." *Urban and Social Change Review* 11, 28–36.

Pearlstein, Steven. 2011. "Why They're Winning on CEO Pay." *Washington Post*, June 24. Accessed June 25, 2011 (www.washingtonpost.com).

Pearsall, Beth. 2011. "Health Care: The High Cost of Being a Woman." *Outlook*, 105 (Winter): 8–11.

Pedersen, Paul. 1995. *The Five Stages of Culture Shock: Critical Incidents around the World*. Westport, CT: Greenwood Press.

Pell, Nicholas. 2011. "Beyond Occupy Wall Street: 11 American Uprisings You've Never Heart of That Changed the World." AlterNet, October 21. Accessed October 22, 2011 (www.alternet.org).

Pell Institute. 2011. "Developing 20/20 Vision on the 2020 Degree Attainment Goal: The Threat of Income-Based Inequality in Education." May. Accessed April 12, 2010 (www.pellinstitute.org).

Perlmutter, David D. 2001. "Students Are Blithely Ignorant; Professors Are Bitter." *Chronicle of Higher Education*, July 27, B20.

———. 2011. "Why Politicians Should Be More Like Professors." *Chronicle of Higher Education*, February 25, A27.

Pescosolido, Bernice A., Brea L. Perry, J. Scott Long, Jack K. Martin, John I. Nurnberger, Jr., and Victor Hesselbrock. 2008. "Under the Influence of Genetics: How Transdisciplinarity Leads Us to Rethink Social Pathways to Illness." *American Journal of Sociology* 114 (Suppl.): S171–S201.

Peter, Lawrence J., and Raymond Hull. 1969. *The Peter Principle*. New York: Morrow.

Peterman, Amber, Tia Palermo, and Caryn Bredenkamp. 2011. "Estimates and Determinants of Sexual Violence Against Women in the Democratic Republic of Congo." *American Journal of Public Health* 101 (June): 1060–1067.

Peterson, Scott. 2008. "In Iran, Barbie Seen as Cultural Invader." *Christian Science Monitor*, September 15, 4.

Petrosino, Anthony, Carolyn Turpin-Petrosino, and John Buehler. 2003. "Scared Straight and Other Juvenile Awareness Programs for Preventing Juvenile Delinquency: A Systematic Review of the Randomized Experimental Evidence." *Annals of the American Academy of Political and Social Science* 589 (September): 41–62.

Pew Center on the States. 2008. "One in 100: Behind Bars in America 2008." February. Retrieved May 11, 2008 (www.pewcenteronthestates.org).

———. 2009. *One in 31: The Long Reach of American Corrections*. Washington, DC: The Pew Charitable Trusts.

———. 2010. "Prison Count 2010." April. Accessed August 24, 2011 (www.pewcenteronthestates.org).

———. 2011. "State of Recidivism: The Revolving Door of America's Prisons." April. Accessed August 24, 2011 (www.pewcenteronthestates.org).

Pew Forum on Religion & Public Life. 2008. "U.S. Religious Landscape Survey: Religious Beliefs and Practices: Diverse and Politically Relevant." June. Retrieved June 26, 2008 (religions.pewforum.org).

———. 2009a. "Faith in Flux: Changes in Religious Affiliation in the U.S." April. Retrieved April 25, 2010 (www.pewforum.org).

———. 2009b. "Faith-Based Programs Still Popular, Less Visible." November 16. Retrieved April 25, 2010 (www.pewforum.org).

Pew Global Attitudes Project. 2006. "The Great Divide: How Westerners and Muslims View Each Other." June 22. Retrieved August 15, 2007 (www.pewglobal.org).

———. 2011. "The American-Western European Values Gap." Pew Research Center, November 17. Accessed June 9, 2012 (pewglobal.org).

Pew Initiative on Food and Biotechnology. 2004. "Genetically Modified Crops in the United States." August. Retrieved September 30, 2005 (www.pewagbiotech.org).

Pew Internet & American Life Project. 2011. "Demographics of Internet Users." Accessed August 4, 2011 (pewinternet.org).

Pew Research Center. 2008. "Inside the Middle Class: Bad Times Hit the Good Life." April 9. Retrieved April 10, 2008 (pewresearch.org).

Pew Research Center for the People & the Press. 2010. "83%—Support Christmas Displays in Public." Retrieved April 26, 2010 (pewresearch.org).

Pewewardy, Cornel. 1998. "Fluff and Feathers: Treatment of American Indians in the Literature and the Classroom." *Equity & Excellence in Education* 31 (April): 69–76.

Pflaumer, Alicia. 2011. "Texting Bride Video Goes Viral on the Web." *Christian Science Monitor,* November 1. Accessed November 3, 2011 (www.csmonitor.com).

Phelan, Jo C., Bruce G. Link, and Parisa Tehranifar. 2010. "Social Conditions as Fundamental Causes of Health Inequalities: Theory, Evidence, and Policy Implications." *Journal of Health and Social Behavior* 41 (Suppl.): S28–S40.

Pianta, Robert C., Jay Belsky, Renate Houts, Fred Morrison, and The National Institute of Child Health and Human Development (NICHD) Early Child Care Research Network. 2007. "Teaching: Opportunities to Learn in America's Elementary Classrooms." *Science* 315 (March 30): 1795–1796.

Pipher, Mary. 1994. *Reviving Ophelia: Saving the Selves of Adolescent Girls.* New York: Putnam.

Pittz, Will. 2005. "Closing the Gap: Solutions to Race-Based Health Disparities." Applied Research Center & Northwest Federation of Community Organizations, June. Retrieved April 20, 2007 (www.arc.org).

Planty, M., W. Hussar, T. Snyder, S. Provasnik, G. Kena, R. Dinkes, A. KewalRamani, and J. Kemp. 2008. *The Condition of Education 2008* (NCES 2008-031). Washington, DC: National Center for Education Statistics, Institute of Education Sciences, U.S. Department of Education.

Planty, M., S. Provasnik, W. Hussar, T. Snyder, G. Kena, G. Hampden-Thompson, R. Dinkes, and S. Choy. 2007. *The Condition of Education 2007* (NCES 2007-064). Washington, DC: National Center for Education Statistics, Institute of Education Sciences, U.S. Department of Education.

Plateris, Alexander A. 1973. *100 Years of Marriage and Divorce Statistics: 1867–1967.* Rockville, MD: National Center for Health Statistics.

Pollard, Kelvin. 2011. "The Gender Gap in College Enrollment and Graduation." *Population Reference Bureau,* April. Accessed March 10, 2012 (www.prb.org).

Polsby, Nelson W. 1959. "Three Problems in the Analysis of Community Power." *American Sociological Review* 24 (December): 796–803.

Pond, Allison, Gregory Smith, and Scott Clement. 2010. "Religion Among the Millennials: Less Religiously Active than Older Americans, but Fairly Traditional in Other Ways." Pew Research Center. Retrieved April 25, 2010 (www.pewforum.org).

Pong, Suet-ling, Lingxin Hao, and Erica Gardner. 2005. "The Roles of Parenting Styles and Social Capital in the School Performance of Immigrant Asian and Hispanic Adolescents." *Social Science Quarterly* 86 (December): 928–950.

Popenoe, David, and Barbara D. Whitehead. 2002. *Should We Live Together? What Young Adults Need to Know About Cohabitation Before Marriage—A Comprehensive Review of Recent Research,* 2nd edition. New

Brunswick, NJ: The National Marriage Project, Rutgers University. Retrieved July 12, 2003 (marriage.rutgers.edu).

———. 2006. *The State of Our Unions: 2006.* The National Marriage Project, Rutgers University. Retrieved July 28, 2006 (marriage.rutgers.edu).

Porter, Sheri. 2011. "AAFP Board Revises Retail Health Clinic Policy." February 24. Accessed August 12, 2011 (www.aafp.org).

Postel, Sandra. 2012. "Humanity's Growing Impact on the World's Freshwater." AlterNet, February 23. Accessed June 23, 2012 (www.alternet.org).

Powell, Lisa M., Glen Szczypka, Frank J. Chaloupka, and Carol L. Braunschweig. 2007. "Nutritional Content of Television Food Advertisements Seen by Children and Adolescents in the United States." *Pediatrics* 120 (September): 576–583.

Powers, Charles H. 2004. *Making Sense of Social Theory: A Practical Introduction.* Lanham, MD: Rowman & Littlefield.

Price, Barbara Raffel, and Natalie J. Sokoloff, eds. 2004. *The Criminal Justice System and Women: Offenders, Prisoners, Victims, & Workers,* 3rd edition. New York: McGraw-Hill.

Prior, Markus. 2009. "The Immensely Inflated News Audience: Assessing Bias in Self-Reported News Exposure." *Public Opinion Quarterly* 73 (Spring): 130–143.

Prins, Nomi. 2011. "Guess How Much More Wall St. Spends on Bonuses Than on Penalties for Torpedoing the Economy?" AlterNet, June 27. Accessed June 28, 2012 (www.alternet.org).

"A Profile of the Working Poor, 2007." 2009. U.S. Department of Labor, U.S. Bureau of Labor Statistics, Report 1012, March. Retrieved August 5, 2009 (www.dol.gov).

Project on Government Oversight. 2010. "Letter to NIH on Ghostwriting Academics." November 29. Accessed July 3, 2011 (www.pogo.org).

Prothrow-Stith, Deborah, and Howard R. Spivak. 2005. *Sugar and Spice and No Longer Nice: How We Can Stop Girls' Violence.* San Francisco: Jossey-Bass.

Pryor, John H. 2011. "The Changing First-Year Student: Challenges for 2011." Higher Education Research Institute at UCLA, January 27. Accessed April 10, 2012 (www.heri.ucla.edu).

Pryor, John H., Sylvia Hurtado, Victor B. Saenz, José Louis Santos, and William S. Korn. 2007. *The American Freshman: Forty Year Trends.* Los Angeles: Higher Education Research Institute, UCLA.

"Public Favors Tougher Border Controls and Path to Citizenship." Pew Research Center for the People & the Press, February 24. Accessed November 15, 2011 (www.people-press.org).

"Public Views of the Divide between Religion and Politics." 2012. Pew Research Center, February 27. Accessed April 9, 2012 (www.people-press.org).

Puddington, Arch. 2012. "Freedom in the World 2012: The Arab Uprisings and Their Global Repercussions." Freedom House. Accessed February 15, 2012 (www.freedomhouse.org).

Purcell, Kristen. 2011. "Search and Email Still Top the List of Most Popular Online Activities."

Pew Research Center, August 9. Accessed August 10, 2011 (pewinternet.org).

Putnam, Frank W. 2006. "The Impact of Trauma on Child Development." *Juvenile and Family Court Journal* 57 (Winter): 1–11.

Putnam, Robert D., and David E. Campbell. 2012. *American Grace: How Religion Divides and Unites Us.* New York: Simon & Schuster.

Qian, Zhenchao. 2005. "Breaking the Last Taboo: Interracial Marriage in America." *Contexts* 4 (Fall): 33–37.

Qian, Zhenchao, and Daniel T. Lichter. "Changing Patterns of Interracial Marriage in a Multiracial Society." *Journal of Marriage and Family* 73 (October): 1065–1084.

"Quality Counts 2007: From Cradle to Career." 2007. *Education Week,* January 4. Retrieved July 23, 2007 (www.edweek.org).

Quick, Becky. 2010. "No Perp Walks. No Jail Time. Why Prosecutors Are Going Easy on Wall Street." *Fortune,* July 5, 50.

Quinney, Richard. 1980. *Class, State, and Crime.* Boston: Little, Brown.

Quizon, Derek. 2011. "Old Sports Icons Never Die. Not These Ones, Anyhow." *Chronicle of Higher Education,* January 28. Accessed June 1, 2012 (www.chronicle.com).

Radwin, David. 2009. "High Response Rates Don't Ensure Survey Accuracy." *Chronicle Review,* October 9, B8–B9.

Rainie, Lee. 2011. "Asian-Americans and Technology." Pew Internet Project, January 6. Accessed August 4, 2011 (pewinternet.org).

Rainie, Lee, Kristen Purcell, and Aaron Smith. 2011. "The Social Side of the Internet." Pew Internet & American Life Project, January 1. Accessed August 10, 2011 (pewinternet.org).

Raintree Nutrition, Inc. 2008. "Rainforest Facts." Retrieved February 12, 2008 (www.rain-tree.com).

Rampell, Catherine. 2010. "In Law Schools, Grades Go Up, Just Like That." *New York Times,* June 21, A1.

———. 2011. "Companies Spend on Equipment, Not Workers." *New York Times,* June 10, A1.

Rand, Michael R. 2009. "Criminal Victimization, 2008." Bureau of Justice Statistics Bulletin, September. Retrieved February 25, 2010 (bjs.ojp.usdoj.gov).

Rank, Mark R. 2011. "Rethinking American Poverty." *Contexts* 10 (Spring): 16–21.

Rano, Jason, and Jane Houlihan. 2012. "Myths on Cosmetic Safety." Skin Deep Cosmetics Database. Accessed May 10, 2012 (www.ewg.org).

Raphael, Steven, and Michael A. Stoll. 2010. "Job Sprawl and the Suburbanization of Poverty." Brookings Institute, Metropolitan Policy Program, March. Retrieved May 1, 2010 (www.brookings.edu).

Rashad, Marwa. 2012. "Saudi Women Take Over Lingerie Shops." *Baltimore Sun,* January 12, 12.

Rattray, Sharon. 2010. "2010 Toy Sales." June 23. Accessed July 20, 2011 (www.suite101.com).

Ray, Rebecca, and John Schmitt. 2007. "No-Vacation Nation." Center for Economic and Policy Research, May. Retrieved July 4, 2008 (www.cepr.net).

Reaney, Patricia. 2012. "Average Cost of U.S. Weddings Hits $27,021." Reuters, May 23. Accessed June 2, 2012 (www.reuters.com).

Reardon, Sean F. 2011. "The Widening Academic Achievement Gap Between the Rich and the Poor: New Evidence and Possible Explanations." Pp. 91–116 in *Whither Opportunity? Rising Inequality and the Uncertain Life Chances of Low-Income Children*, edited by R. Murnane and G. Duncan. New York: Russell Sage Foundation Press.

Reardon, Sean F., and Kendra Bischoff. 2011. "More Unequal and More Separate: Growth in the Residential Segregation of Families by Income, 1970–2009." US2010 Project, November. Accessed June 23, 2012 (www.s4.brown.edu).

Redd, Zakia, Tahilin Sanchez Karver, David Murphey, Kristin Anderson Moore, and Dylan Knewstub. 2011. "Two Generations in Poverty: Status and Trends among Parents and Children in the United States, 2000–2010." Child Trends Research Brief, November. Accessed January 12, 2012 (www.childtrends.org).

Reich, Robert B. 2011. *Aftershock: The Next Economy and America's Future*. New York: Vintage Books.

Reiman, Jeffrey, and Paul Leighton. 2010. *The Rich Get Richer and the Poor Get Prison: Ideology, Class, and Criminal Justice*, 9th edition. Upper Saddle River, NJ: Prentice Hall.

Reimer, Susan. 2012. "In Birth Control Debate, Where Are the Women's Voices?" *Baltimore Sun*, February 16, 1, 3.

Rein, Lisa. 2011. "Federal Workers Tell us What Should be Cut From the Budget." *Washington Post*, April 14, B4.

Reiss, Steven. 2004. "The Sixteen Strivings for God." *Zygon* 39 (June): 303–320.

"Religions of the World: Number of Adherents, Names of Houses of Worship . . .". 2007. Religious Tolerance. Retrieved August 24, 2007 (www.religioustolerance.org).

Rennison, Callie M. 2003. *Intimate Partner Violence, 1993–2001*. Washington, DC: U.S. Department of Justice.

Rheault, Magali, and Kyley McGeeney. 2011. "Education Is a Key Predictor of Emotional Health After 65." Gallup, August 19. Accessed April 3, 2012 (www.gallup.com).

———. 2011. "Standard of Living, Health Key to Emotional Wellbeing After 65." Gallup, September 1. Accessed April 22, 2012 (www.gallup.com).

Rhoads, Steven E. 2004. *Taking Sex Differences Seriously*. San Francisco: Encounter Books.

Ricciardelli, Rosemary, Kimberley A. Clow, and Philip White. 2010. "Investigating Hegemonic Masculinity: Portrayals of Masculinity in Men's Lifestyle Magazines." *Sex Roles* 63 (March): 64–78.

Rice, Doyle. 2012. "2012 Is USA's Warmest Year on Record, So Far." *USA Today*, May 8. Accessed June 21, 2012 (www.usatoday.com).

Richburg, Keith B. 2009. "States Seek Less Costly Substitutes for Prison." *Washington Post*, July 13, A1.

Riche, Martha Farnsworth. 2000. "America's Diversity and Growth: Signposts for the 21st Century." *Population Bulletin* 55 (June): 1–41.

Richtel, Matt, and Alexei Barrionuevo. 2005. "Wendy's Restaurants." *New York Times*, April 22, A9.

Rideout, Victoria. 2007. "Parents, Children & Media." Henry J. Kaiser Family Foundation, June. Retrieved April 12, 2008 (www.kff.org).

Rideout, Victoria, Ulla G. Foehr, and Donald F. Roberts. 2010. *Generation M2: Media in the Lives of 8- to 18-Year-Olds*. Kaiser Family Foundation, January. Retrieved January 30, 2010 (www.kff.org).

Ridge, Mian. 2009. "A Peaceful Train for Women." *Christian Science Monitor*, November 1, 4.

Riesman, David. 1953. *The Lonely Crowd*. New York: Doubleday.

Riordan, Cornelius. 1997. *Equality and Achievement: An Introduction to the Sociology of Education*. New York: Addison-Wesley Longman.

Ripley, Amanda. 2011. "Meet Your Government Workers." *Time*, March 7, 41–44.

"Rising Restrictions on Religion." 2011. Pew Research Center, August. Accessed April 8, 2012 (pewresearch.org).

Ritzer, George. 1992. *Contemporary Sociological Theory*, 3rd edition. New York: McGraw-Hill.

———. 1996. *The McDonaldization of Society: An Investigation into the Changing Character of Contemporary Social Life*. Thousand Oaks, CA: Pine Forge Press.

———. 2008. *The McDonaldization of Society*, 5th edition. Los Angeles: Pine Forge Press.

Rivers, Caryl, and Rosalind C. Barnett. 2011. "'Mancession' Focus Masks Women's Real Losses." Women's e-News, May 4. Accessed May 6, 2011 (www.womensenews.org).

Rivlin, Gary. 2011. "The Billion-Dollar Bank Heist." *Newsweek*, July 18, 9–11.

Roan, Shari. 2008. "Cut in Paid Sick Days Leave Unhealthy Employees Stuck in the Workplace." *Los Angeles Times*, July 7. Retrieved July 7, 2008 (www.latimes.com).

Robert Wood Johnson Foundation. 2010. "California Nurse Ratio Law Saves Lives, Improves Nurse Morale, Study Finds." May 26. Accessed June 6, 2012 (www.rwjf.org).

Roberts, Keith A. 2004. *Religion in Sociological Perspective*, 4th edition. Belmont, CA: Wadsworth.

Roberts, Sam, and Peter Baker. 2010. "Asked to Declare His Race, Obama Checks 'Black'." *New York Times*, April 2, 9.

Robertson, Ruth, and Sara R. Collins. 2011. "Realizing Health Reform's Potential." Commonwealth Fund, May. Accessed April 24, 2012 (www.commonwealthfund.org).

Robey, Elizabeth B., Daniel J. Canary, and Cynthia S. Burggraf. 1998. "Conversational Maintenance Behaviors of Husbands and Wives: An Observational Analysis." Pp. 373–392 in *Sex Differences and Similarities in Communication: Critical Essays and Empirical Investigations of Sex and Gender in Interaction*, edited by Daniel J. Canary and Kathryn Dindia. Mahwah, NJ: Lawrence Erlbaum Associates.

Robinson, B. A. 2011a. "Conflicts Regarding Evolution, Intelligent Design & Creationism in U.S. Public Schools." Religious Tolerance, February 12. Accessed April 10, 2012 (www.religioustolerance.org).

———. 2011b. "Religiously-Based Civil Unrest and Warfare." Religious Tolerance, October 1. Accessed April 10, 2012 (www.religioustolerance.org).

Robinson, Laurie O., and Jeff Slowikowski. 2011. "Scary—and Ineffective." *Baltimore Sun*, February 1, 11.

Rochman, Bonnie. 2011. "Ingredient Anxiety: Hyping What's Not There." *Time*, November 14, 69.

———. 2012. "The End of an Epidemic?" *Time*, February 6, 16.

Roethlisberger, F. J., and William J. Dickson. 1939/1942. *Management and the Worker: An Account of a Research Program Conducted at the Western Electric Company, Hawthorne Works, Chicago*. Cambridge, MA: Harvard University Press.

Rojstaczer, Stuart. 2009. "Grade Inflation Gone Wild." *Christian Science Monitor*, March 24, 9.

"Role of STD Prevention and Treatment in HIV Prevention, The." 2010. Centers for Disease Control and Prevention. Accessed April 24, 2012 (www.cdc.gov).

Roller, Dough. 2010. "8 Immediate Cost Benefits of Health Care Reform." U.S. News, September 24. Accessed April 15, 2012 (money.usnews.com).

Romer, Dan. 2011. "After 11 Years of Setting the Record Straight, Stories about Holiday Suicides Still Outnumber Those Debunking the Myth." Annenberg Public Policy Center, December 13. Accessed December 14, 2011 (www.annenbergpublicpolicycenter.org).

Romero, Simon, and Marc Lacey. 2010. "Looting Flares Where Authority Breaks Down." *New York Times*, January 17, A1.

Ronfeldt, Matthew, Hamilton Lankford, Susanna Loeb, and James Wyckoff. 2011. "How Teacher Turnover Harms Student Achievement." National Bureau of Economic Research, June. Accessed April 14, 2012 (www.nber.org).

Roscigno, Vincent J. 2010. "Ageism in the American Workplace." *Contexts* 9 (Winter): 16–21.

Rose, Fred. 1997. "Toward a Class-Cultural Theory of Social Movements: Reinterpreting New Social Movements." *Sociological Forum* 12 (September): 461–494.

Rose, Peter I. 1997. *They and We: Racial and Ethnic Relations in the United States*, 5th edition. New York: McGraw-Hill.

Rose, Stephen J., and Scott Winship. 2009. "Ups and Downs: Does the American Economy Still Promote Upward Mobility?" Economic Mobility Project. Accessed September 20, 2011 (www.economicmobility.org).

Rosen, Anne Farris. 2010. "A Brief History of Religion and the U.S. Census." January 26. Retrieved April 26, 2010 (pewresearch.org).

Rosen, Jacob, and Blake Hannaford. 2006. "Doc at a Distance." *IEEE Spectrum* (October): 34–39.

Rosen, Larry D. 2010. *Rewired: Understanding the iGeneration and the Way They Learn*. New York: Palgrave Macmillan.

Rosenberg, Janice. 1993. "Just the Two of Us." Pp. 301–307 in *Reinventing Love: Six Women Talk About Lust, Sex, and Romance*, edited by Laurie Abraham, Laura Green, Magda Krance, Janice Rosenberg, Janice Somerville, and Carroll Stoner. New York: Plume.

Rosenberg, Martha. "Some Asthma Drugs Kill More People Than Asthma: Why Is Big Pharma Allowed to Hawk Deadly Pills?"

Alternet, November 3. Accessed April 24, 2012 (www.alternet.org).

Rosenfeld, Jake. 2010. "Little Labor: How Union Decline Is Changing the American Landscape." *Pathways*, Summer, 3–6.

Rosin, Hanna. 2010. "The End of Men." *The Atlantic*, July/August. Accessed October 25, 2011 (www.theatlantic.com).

Ross, Jeffrey Ian, and Stephen C. Richards. 2002. *Behind Bars: Surviving Prison*. Indianapolis, IN: Alpha Books.

Rothenberg, Paula S. 2008. *White Privilege: Essential Reading on the Other Side of Racism*, 3rd edition. New York: Worth.

Rothlin, Phillippe, and Peter R. Werder. 2008. *Boreout! Overcoming Workplace Demotivation*. Philadelphia: Kogan Page.

Rothman, Sheila M. 1978. *Women's Proper Place: A History of Changing Ideals and Practices, 1870 to the Present*. New York: Basic Books.

Rowe, Meredith L., and Susan Goldin-Meadow. 2009. "Differences in Early Gesture Explain SES Disparities in Child Vocabulary Size at School Entry." *Science* 323 (February 13): 951–953.

Rowe-Finkbeiner, Kristin. 2004. *The F-Word: Feminism in Jeopardy, Women, Politics, and the Future*. Emeryville, CA: Seal Press.

Royster, Deirdre A. 2003. *Race and the Invisible Hand: How White Networks Exclude Black Men from Blue-Collar Jobs*. Berkeley: University of California Press.

Rubin, Kenneth, William Bukowski, and Jeffrey G. Parker. 1998. "Peer Interactions, Relationships, and Groups." Pp. 619–700 in *Handbook of Child Psychology: Vol. 3. Social, Emotional, and Personality Development*, edited by William Damon and Nancy Eisenberg. New York: Wiley.

Rubin, Rita. 2004. "'Smart Pills' Make Headway." *USA Today*, July 7, 1D.

Rugh, Jacob S., and Douglas S. Massey. 2010. "Racial Segregation and the American Foreclosure Crisis." *American Sociological Review* 75 (October): 629–651.

Rushe, Dominic. 2011. "Explosive New Report: CEO Pay Skyrocketed 27% Last Year, Top 10 Earners Pocket More Than $770 Million Between Them." Alternet, December 14. Accessed April 24, 2012 (www.alternet.org).

Rutherford, Markella B. 2009. "Children's Autonomy and Responsibility: An Analysis of Childrearing Advice." *Qualitative Sociology* 32 (December): 337–353.

Ryan, Missy. 2008. "In US, Record Numbers Seeking Food Stamps." *Christian Science Monitor*, May 8, 4.

Ryerson, William. 2004. "Responses to: Demographic 'Bomb' May Only Go 'Pop!'" *Pop!ulation Press* 10 (Fall): 21.

Saad, Lydia. 2001. "Majority Considers Sex Before Marriage Morally Okay." *Gallup Poll Monthly*, No. 428 (May): 46–48.

———. 2006. "Families of Drug and Alcohol Abusers Pay an Emotional Toll." Gallup News Service, August 25. Retrieved August 27, 2006 (www.galluppoll.com).

———. 2008. "Cultural Tolerance for Divorce Grows to 70%." Gallup News Service, May 19. Retrieved July 8, 2008 (www.gallup.com).

———. 2009. "Honesty and Ethics Poll Finds Congress' Image Tarnished." Gallup, December 9. Retrieved March 26, 2010 (www.gallup.com).

———. 2009. "Two in Three Americans Worry About Identity Theft." Gallup, October 16. Accessed August 22, 2011 (www.gallup.com).

———. 2010. "Four Moral Issues Sharply Divide Americans." Gallup, May 26. Accessed July 15, 2011 (www.gallup.com).

———. 2010. "Nearly 4 in 10 Americans Still Fear Walking Alone at Night." Gallup, November 5. Accessed August 22, 2011 (www.gallup.com).

———. 2011. "Common State Abortion Restrictions Spark Mixed Reviews." Gallup, July 25. Accessed October 25, 2011 (www.gallup.com).

———. 2011. "One in Four Americans Have Less Respect for Smokers." Gallup, August 5. Accessed April 22, 2012 (www.gallup.com).

———. 2012. "Conservatives Remain the Largest Ideological Group in U.S." Gallup, January 12. Accessed January 20, 2012 (www.gallup.com).

———. 2012. "U.S. Economy Most Toxic of 24 Issues." Gallup, January 23. Accessed January 24, 2012 (www.gallup.com).

Sabattini, Laura, Nancy M. Carter, Jeanine Prime, and David Megathlin. 2007. "The Double-Bind Dilemma for Women in Leadership: Damned If You Do, Doomed If You Don't." Catalyst. Retrieved April 28, 2008 (www.catalyst.org).

Sachs, Jeffrey S. 2005. "Confusion over Population: Growth or Dearth?" *Pop!ulation Press* 11 (Winter/Spring): 17.

Sacks, Peter. 2007. *Tearing Down the Gates: Confronting the Class Divide in American Education*. Berkeley: University of California Press.

Saegert, Susan, and Gary Winkel. 1981. "The Home: A Critical Problem for Changing Sex Roles." Pp. 41–63 in *New Space for Women*, edited by Gerda R. Wekerle, Rebecca Peterson, and David Morley. Boulder, CO: Westview Press.

Saenz, Rogelio. 2010. "Latinos in the United States 2010." Population Reference Bureau, December. Accessed December 7, 2011 (www.prb.org).

Sagarin, Edward. 1975. *Deviants and Deviance*. New York: Praeger.

Sanburn, Josh. 2010. "Brief History: Secret Medical Testing." *Time*, October 18, 30.

———. 2011. "This Time, Men Are Finding Jobs Faster Than Women." *Time*, July 8. Accessed September 27, 2011 (www.time.com).

Sandstrom, Kent L., Daniel D. Martin, and Gary Alan Fine. 2006. *Symbols, Selves, and Social Reality: A Symbolic Interactionist Approach to Social Psychology and Sociology*, 2nd edition. Los Angeles: Roxbury.

Sang-Hun, Choe. 2009. "Group Resists Korean Stigma for Unwed Mothers." *New York Times*, October 8, 6.

Sapir, Edward. 1929. "The Status of Linguistics as a Science." *Language* 5 (4): 207–214.

Sarno, David. 2009. "Online, Your Private Life Is Searchable." *Los Angeles Times*, August 16. Retrieved August 18, 2009 (www.latimes.com).

"Saudi Arabia and Its Women." 2011. *New York Times*, September 27. Accessed October 25, 2011 (www.nytimes.com).

"Saudi King Announces New Rights for Women." 2011. *Baltimore Sun*, September 26, 8.

Saulny, Susan. 2011. "Counting by Race Can Throw Off Some Numbers." *New York Times*, February 10, A1.

Sawchuck, Stephen. 2012. "Teacher Quality, Status Entwined Among Top-Performing Nations." *Education Week* 31 (16): 12–16.

Sawhill, Isabel V. 2008. "Trends in Intergenerational Mobility." Pp. 27–36 in *Getting Ahead or Losing Ground: Economic Mobility in America*, edited by Julia B. Isaacs, Isabel V. Sawhill, and Ron Haskins. Washington, DC: Brookings Institution.

Scarlett, W. George, Sophie Naudeau, Dorothy Salonius-Pasternak, and Iris Ponte. 2005. *Children's Play*. Thousand Oaks, CA: Sage.

Scelfo, Julie. 2007. "Come Back, Mr. Chips." *Newsweek*, September 17, 44.

Schachter, Jason P. 2004. "Geographical Mobility: 2002 to 2003." U.S. Census Bureau, Current Population Reports, P20-549. Retrieved May 15, 2007 (www.census.gov).

Schaeffer, Robert K. 2003. *Understanding Globalization: The Social Consequences of Political, Economic, and Environmental Change*, 2nd edition. Lanham, MD: Rowman & Littlefield.

Schehr, Robert C. 1997. *Dynamic Utopia: Establishing Intentional Communities as a New Social Movement*. Westport, CT: Bergin & Garvey.

Schieman, Scott. 2010. "Socioeconomic Status and Beliefs about God's Influence in Everyday Life." *Sociology of Religion* 71 (Spring): 25–51.

Schiesel, Seth. 2011. "Supreme Court Has Ruled; Now Games Have a Duty." *New York Times*, July 29, C1.

Schirber, Michael. 2007. "Why Desalination Doesn't Work (Yet)." *Live Science*, June 25. Retrieved February 7, 2008 (www.livescience.com).

Schlesinger, Izchak M. 1991. "The Wax and Wane of Whorfian Views." Pp. 7–44 in *Influence of Language on Culture & Thought*, edited by Robert Cooper and Bernard Spolsky. New York: Mounton de Gruyter.

Schlesinger, Robert. 2011. "Two Takes: Collective Bargaining Rights for Public Sector Unions?" *U.S. News Weekly*, February 25, 15–16.

Schmall, Emily. 2007. "The Cult of Chick-fil-A." *Forbes*, July 23, 80, 83.

Schmidt, Steffen W., Mack C. Shelley, and Barbara A. Bardes. 2001. *American Government and Politics Today*. Belmont, CA: Wadsworth.

Schmitz, Joelle. 2010. "Women in Politics? The U.S. Is Sliding." *USA Today*, October 13. Accessed October 14, 2010 (www.usatoday.com).

Schnall, Marianne. 2012. "Letting Girls Be Girls—A Global Campaign." Women's Media Center, January 25. Accessed January 26, 2012 (womensmediacenter.com).

Schneider, Nathan. 2011. "From Occupy Wall Street to Occupy Everywhere." *The Nation*, October 31. Accessed November 1, 2011 (www.thenation.com).

Schnittker, Jason. 2008. "Happiness and Success: Genes, Families, and the Psychological Effects of Socioeconomic Position and Social Support." *American Journal of Sociology* 114 (Suppl.): S233–S259.

———. 2009. "Mirage of Health in the Era of Biomedicalization: Evaluating Change in the Threshold of Illness, 1972–1996." *Social Forces* 87 (June): 2155–2182.

Schoen, Robert, and Vladimir Canudas-Romo. 2006. "Timing Effects on Divorce: 20th Century Experience in the United States." *Journal of Marriage and Family*, 68 (August): 749–758.

Schoenman, Julie A., and Nancy Chockley. 2011. "Understanding U.S. Health Care Spending." NIHCM Foundation Data Brief, July. Accessed April 15, 2012 (www.nihcm.org).

Schultz, Ellen E. 2011. *Retirement Heist: How Companies Plunder and Profit From the Nest Eggs of American Workers.* New York: Penguin.

Schur, Edwin M. 1968. *Law and Society: A Sociological View.* New York: Random House.

Schurman-Kauflin, Deborah. 2000. *The New Predator: Women Who Kill.* New York: Algora.

Schutz, Alfred. 1967. *The Phenomenology of the Social World.* Evanston, IL: Northwestern University Press.

Schwalbe, Michael. 2001. *The Sociologically Examined Life: Pieces of the Conversation,* 2nd edition. Mountain View, CA: Mayfield.

Schwartz, Marlene B., Lenny R. Vartanian, Brian A. Nosek, and Kelly D. Brownwell. 2006. "The Influence of One's Own Body Weight on Implicit and Explicit Anti-fat Bias." *Obesity* 14 (July): 440–447.

Schwartz, Nelson D. 2011. "Bank Closings Tilt Toward Poor Areas." *New York Times,* February 23, B1.

Schweizer, Peter. 2011. *Throw Them All Out: How Politicians and Their Friends Get Rich Off Insider Stock Tips, Land Deals, and Cronyism That Would Send the Rest of Us to Prison.* New York: Houghton Mifflin.

Schwyzer, Hugo. 2011. "Why We Must Change the Myth That Men Can't Change." AlterNet, April 19. Accessed October 25, 2011 (www.alternet.org).

Scommegna, Paola. 2005. "Clean Water's Historic Effect on U.S. Mortality Rates Provides Hope for Developing Countries." Population Reference Bureau. Retrieved June 25, 2005 (www.prb.org).

———. 2011. "Least Segregated U.S. Metros Concentrated in Fast-Growing South and West." Population Reference Bureau, September 11. Accessed June 23, 2012 (www.prb.org).

———. 2011a. "More of Us on Track to Reach Age 100: Genes, Habits, Baboons Examined for Longevity Clues." Population Reference Bureau, July. Accessed March 6, 2012 (www.prb.org).

———. 2011b. "U.S. Parents Who Have Children With More Than One Partner." Population Reference Bureau, June. Accessed March 6, 2012 (www.prb.org).

Scott, Katherine V. 1995. *Gender and Development: Rethinking Modernization and Dependency Theory.* Boulder, CO: Lynne Rienner.

Scott, Robert E. 2010. "Unfair China Trade Costs Local Jobs." Economic Policy Institute, March 23. Retrieved April 4, 2010 (www.epi.org).

Segal, David. 2012. "Apple's Retail Army, Long on Loyalty, but Short on Pay." *New York Times,* June 24, A1.

Seager, Joni. 2009. *The Penguin Atlas of Women in the World,* 4th edition. New York: Penguin Books.

"Second Anniversary of the Recovery Shows No Job Growth for Women." 2011. National Women's Law Center, July. Accessed September 22, 2011 (www.nwlc.org).

"Second National Report on Human Exposure to Environmental Chemicals." 2003. Centers for Disease Control and Prevention, January. Retrieved June 25, 2005 (www.cdc.gov).

Segura, Denise. A. 1994. "Working at Motherhood: Chicana and Mexican Immigrant Mothers and Employment." Pp. 211–233 in *Mothering: Ideology, Experience, and Agency,* edited by Evelyn N. Glenn, Grace Chang, and Linda R. Forcey. New York: Routledge.

Seife, Charles. 2010. *Proofiness: The Dark Arts of Mathematical Deception.* New York: Viking.

Seltzer, Judith A. 2004. "Cohabitation and Family Change." Pp. 57–78 in *Handbook of Contemporary Families: Considering the Past, Contemplating the Future,* edited by Marilyn Coleman and Lawrence H. Ganong. Thousand Oaks, CA: Sage.

Semuels, Alana. 2008. "Gay Marriage May Be a Gift to California's Economy." *Los Angeles Times,* June 2. Retrieved June 9, 2008 (www .latimes.com).

Semuels, Alana. 2010. "Can a Prison Save a Town?" *Los Angeles Times,* May 3. Accessed May 4, 2010 (latimes.com).

Sequist, Thomas D., Garrett M. Fitzmaurice, Richard Marshall, Shimon Shaykevich, Dena Gelb Safran, and John Z. Ayanian. 2008. "Physical Performance and Racial Disparities in Diabetes Mellitus Care." *Archives of Internal Medicine* 168 (June 9): 1145–1151.

Settersten, Richard A. Jr., and Barbara Ray. 2010. "What's Going on with Young People Today? The Long and Twisting Path to Adulthood." *The Future of Children* 20 (Spring): 19–36.

"7 People Charged in Fatal Mob Attack." 2002. *Baltimore Sun,* August 4, 2A.

"75%—A Nation of Flag Wavers." 2011. Pew Research Center. Accessed July 15, 2011 (pewresearch.org/databank).

Severson, Kim. 2011. "Trying to Hold Down Blue Language on a Red-Letter Day." *New York Times,* February 13, A18.

Shaefer, H. Luke, and Kathryn Edin. 2012. "Extreme Poverty in the United States, 1996 to 2011." National Poverty Center, February. Accessed March5, 2012 (www.npc.umich.edu).

Shaer, Matthew. 2010. "Why Facebook Falters but Still Flies." *Christian Science Monitor,* August 16 &23, 40–41.

Shamoo, Adil E., and Bonnie Bricker. 2011. "The Threat of Bad Science." *Baltimore Sun,* January 11, 13.

Shanahan, Michael J., Shawn Bauldry, and Jason Freeman. 2010. "Beyond Mendel's Ghost." *Contexts* 3 (Fall): 34–39.

Shapiro, Danielle. 2009. "Children Targeted as Witches in the Congo." Women's eNews, November 22. Retrieved November 23, 2009 (www.womensenews.org).

Sharifzadeh, Virginia-Shirin. 1997. "Families with Middle Eastern Roots." Pp. 441–482 in *Developing Cross-Cultural Competence: A Guide for Working with Children and Families,* edited by Eleanor W. Lynch and Marci J. Hanson. Baltimore, MD: Paul H. Brookes.

Shaw-Taylor, Yoku, and Nijole V. Benokraitis. 1995. "The Presentation of Minorities in Marriage and Family Textbooks." *Teaching Sociology* 23 (April): 122–135.

Shelton, Deborah, and Julie Deardorff. 2012. "Doctor Organizations List Overused Tests, Procedures." *Baltimore Sun,* April 4, 8.

Shepherd, Julianne Escobedo. 2011. "Alabama Latinos Flee Immigration Law, Leaving Insufficient Workforce for Tuscaloosa Tornado Cleanup." AlterNet, June 30. Accessed July 2, 2011 (www.alternet.org).

———. 2011. "Hard-Partying Rich Boy Kills Two in Hit-and-Run, Buys Himself Out of a Prison Sentence." AlterNet, June 6. Accessed June 10, 2011 (www.alternet.org).

———. 2011. "Louisiana Man Gets Life Sentence . . . For Weed." AlterNet, May 9. Accessed May 10, 2011 (www.alternet.org).

Sheppard, Kate. 2011. "The Hackers and the Hockey Stick." *Mother Jones,* May/June, 33–45.

Sherwood, Jessica Holden. 2010. *Wealth, Whiteness, and the Matrix of Privilege: The View from the Country Club.* Lanham, MD: Lexington Books.

Shibutani, Tamotsu. 1986. *Social Process: An Introduction to Sociology.* Berkeley: University of California Press.

Shierholz, Heidi. 2009. "Nine Years of Job Growth Wiped Out." Economic Policy Institute, July 2. Retrieved August 9, 2009 (www.epi.org).

———. 2010. "The Effects of Citizenship on Family Income and Poverty." Economic Policy Institute, February 24. Retrieved March 15, 2010 (www.epi.org).

———. 2011. "Ten Facts About the Recovery." Economic Policy Institute, July 6. Accessed January 13, 2012 (www.epi.org).

———. 2012a. "Nearly Three Years of a Job-Seekers Ratio Above 4-to-1." Economic Policy Institute, January 12. Accessed January 13, 2012 (www.epi.org).

———. 2012b. "A Solid Step in the Right Direction for the Labor Market." Economic Policy Institute, January 6. Accessed January 13, 2012 (www.epi.org).

Shilo, Guy, and Riki Savaya. 2011. "Effects of Family and Friend Support on LGB Youths' Mental Health and Sexual Orientation Milestones." *Family Relations* 60 (July): 318–330.

Shin, Hyon B., and Robert A. Kominski. 2010. "Language Use in the United States: 2007." U.S. Census Bureau, American Community Survey Reports, April. Accessed December 7, 2011 (www.census.gov).

Shore, Rima, and Barbara Shore. 2009. "Reducing Infant Mortality." Annie E. Casey Foundation, July. Accessed April 24, 2012 (www.aecf.org).

Shorto, Russell. 2008. "No Babies?" *New York Times,* June 29, 34.

Siegel, Andrea F. 2009. "Dog Killer Gets 3 Years." *Baltimore Sun,* August 21, 10.

Siegman, Aron W., and Stanley Feldstein, eds. 1987. *Nonverbal Behavior and Communication.* Hillsdale, NJ: Lawrence Erlbaum Associates.

Sifferlin, Alexandra. 2012. "Skinny Jeans and High Heels: What Health Dangers Lurk in

Your Closet?" *Time*, February 23. Accessed February 25, 2012 (healthland.time.com).

Silverman, Rachel E. 2003. "Provisions Boost Rights of Couples Living Together." *Wall Street Journal*, March 5, D1.

Silverstein, Ken. 2002. "Unjust Rewards." *Mother Jones*, May/June, 69–86.

Simon, Robin W., and Leda E. Nath. 2004. "Gender and Emotion in the United States: Do Men and Women Differ in Self-Reports of Feelings and Expressive Behavior?" *American Journal of Sociology* 109 (March): 1137–1176.

Simpson, Ian, and Lisa Baertlein. 2012. "Law Puts More Vegetables, Fruits on Schoolkids' Plates." *Baltimore Sun*, January 26, 3.

Simpson, Sally S., and Carole Gibbs. 2006. "Making Sense of Intersections." Pp. 269–302 in *Gender and Crime: Patterns of Victimization and Offending*, edited by Karen Heimer and Candace Kruttschnitt. New York: New York University Press.

Singer, Natasha. 2009. "When 2+2 Equals a Privacy Question." *New York Times*, October 18, 4.

———. 2010. "Shoppers Who Can't Have Secrets." *New York Times*, April 30, 5.

Sjoberg, Gideon. 1960. *The Preindustrial City: Past and Present*. Glencoe, IL: Free Press.

Slackman, Michael. 2007. "In Arab Hub, Poor Are Left in Dire Straits." *New York Times*, March 1, A1, A4.

Slevin, Peter. 2005. "In Heartland, Stem Cell Research Meets Fierce Opposition." *Washington Post*, August 10, A1.

"Slowing Giant: U.S. Loses Some of Its Lead." 2011. *Forbes*, March 28, 108–109.

Smallwood, Scott, and Alex Richards. 2011. "How Educated Are State Legislators?" *Chronicle of Higher Education*, June 17, A1, A3–A4.

Smedley, Audrey. 2007. *Race in North America: Origin and Evolution of a Worldview*, 3rd edition. Boulder, CO: Westview Press.

Smelser, Neil J. 1962. *Theory of Collective Behavior*. New York: Free Press.

———. 1988. "Social Structure." Pp. 103–129 in *Handbook of Sociology*, edited by Neil J. Smelser. Newbury Park, CA: Sage.

Smith, Aaron. 2010. "Neighbors Online." Pew Internet & American Life Project, June 9. Accessed June 12, 2010 (pewinternet.org).

———. 2011. "35% of American Adults Own a Smartphone." Pew Research Center, August 11. Accessed August 13, 2011 (pewinternet .org).

Smith, Adam. 1776/1937. *An Inquiry into the Nature and Causes of the Wealth of Nations*. New York: Modern Library.

Smith, Dorothy E. 1987. *The Everyday World as Problematic: A Feminist Sociology*. Toronto: University of Toronto Press.

Smith, Jane I. 1994. "Women in Islam." Pp. 303–325 in *Today's Woman in World Religions*, edited by Arvind Sharma. Albany: State University of New York Press.

Smith, Robert B. 2011. "Saving the 'Lost Boys' of Higher Education." *Chronicle of Higher Education*, October 7, A26–A27.

Smith, S. E. 2011. "$260 Million After Death? How Rich Executives Make Money From Beyond the Grave." AlterNet, September 19. Accessed September 21, 2011 (www.alternet.org).

Smith, Stacy L., and Marc Choueiti. 2011. "Gender Inequality in Cinematic Content?

A Look at Females on Screen & Behind-the-Camera in Top-Grossing 2008 Films." Annenberg School for Communication & Journalism, University of Southern California." April 22. Accessed October 25, 2011 (annenberg.usc.edu).

"Smoking and Tobacco Use." 2012. Centers for Disease Control and Prevention, January 24. Accessed April 24, 2012 (www.cdc.gov).

Snell, Marilyn B. 2007. "The Talking Way." *Mother Jones*, January/February, 30–35.

Snipp, C. Matthew. 1996. "A Demographic Comeback for American Indians." *Population Today* 24 (November): 4–5.

Snow, Kate, John Gever, and Dan Childs. 2009. "Experts Debunk Health Care Reform Bill's 'Death Panel' Rule." ABC News, August 11. Retrieved May 9, 2010 (abcnews.go.com).

Snyder, Howard N., and Melissa Sickmund. 2006. *Juvenile Offenders and Victims: 2006 National Report*. Washington, DC: U.S. Department of Justice, Office of Justice Programs, Office of Juvenile Justice and Delinquency Prevention.

Snyder, Thomas D., and Sally A. Dillow. 2010. *Digest of Education Statistics 2009*. Washington, DC: National Center for Education Statistics, Institute of Education Sciences, U.S. Department of Education.

———. 2011. *Digest of Education Statistics 2010*. National Center for Education Statistics, April. Accessed October 25, 2011 (nces .ed.gov).

Soguel, Dominique. 2009. "Wage Gap Study Arrives in Time for Equal Pay Day." Women's eNews, April 28. Retrieved April 28, 2009 (www.womensnews.org).

Southern Poverty Law Center. 2011. "Active U.S. Hat Groups." Accessed July 20, 2011 (http:// www.splcenter.org).

Spalter-Roth, Roberta, and Nicole Van Vooren. 2008. "What Are They Doing with a Bachelor's Degree in Sociology? Data Brief on Current Jobs." American Sociological Association, Department of Research and Development, January. Retrieved March 27, 2008 (www.asanet.org).

Sparks, Sarah D. 2011. "Report Points to Widening Gap in Boys' Educational Attainment." *Education Week*, May 17. Accessed October 25, 2011 (blogs.edweek .org).

Spector, Malcolm, and John Kitsuse. 1977. *Constructing Social Problems*. Menlo Park, CA: Cummings.

Spectorsky, A. C. 1955. *The Exurbanites*. Philadelphia: J. B. Lippincott.

Spencer, Herbert. 1862/1901. *First Principles*. New York: P. F. Collier & Son.

Spiess, Michele. 2003. "Juveniles and Drugs." Rockville, MD: ONDCP Drug Policy Information Clearinghouse.

"Splicing Nanofiber, Stem Cells." 2012. *Baltimore Sun*, June 24, 13.

Spotts, Peter N. 2004. "Blowing in the Wind: Transatlantic Pollution." *Christian Science Monitor*, August 5, 14, 17.

———. 2009. "New Climate Change Signal: Oceans Turning Acidic." *Christian Science Monitor*, December 9. Retrieved December 12, 2010 (www.csmonitor.com).

Spradley, J. P., and M. Phillips. 1972. "Culture and Stress: A Quantitative Analysis." *American Anthropologist* 74 (3): 518–529.

"States Enact Record Number of Abortion Restrictions in First Half of 2011." 2011. Guttmacher Institute, July 13. Accessed October 25, 2011 (www.guttmacher .org).

Stearns, Matt. 2006. "Organ Transplants Called Biased for the Well-to-Do." *Philadelphia Inquirer*, June 7, A10.

Steffensmeier, Darrell, Jennifer Schwartz, Hua Zhong, and Jeff Ackerman. 2005. "An Assessment of Recent Trends in Girls' Violence Using Diverse Longitudinal Sources: Is the Gender Gap Closing?" *Criminology* 43 (May): 355–405.

Stein, Joel. 2011. "Your Data, Yourself." *Time*, March 21, 40–46.

Steinhauer, Jennifer. 2011. "Millionaires on Food Stamps and Jobless Pay? G.O.P. Is On It." *New York Times*, December 13, A18.

Stephenson, John B. 2007. "Environmental Protection: EPA-State Enforcement Partnership Has Improved, but EPA's Oversight Needs Further Enhancement." United States Government Accountability Office, July, GAO 07-883. Retrieved February 12, 2008 (www.gao.gov).

Stepney, Chloe. 2011. "Waste Water: Time to Drink It?" *Christian Science Monitor*, August 29, 21.

Sternberg, Robert. 2010. *College Admissions for the 21st Century*. Cambridge, MA: Harvard University Press.

Sternbergh, Adam. 2008. "Why White People Like 'Stuff White People Like'." *New Republic*, March 17. Retrieved November 6, 2008 (www.tnr.com).

Sternheimer, Karen. 2007. "Do Video Games Kill?" *Contexts* 6 (Winter): 13–17.

Stevenson, Betsey, and Justin Wolfers. 2007. "Marriage and Divorce: Changes and Their Driving Forces." *Journal of Economic Perspectives* 21 (Spring): 27–52.

Stewart, Pearl, and Katia Paz Goldfarb. 2007. "Historical Trends in the Study of Diverse Families." Pp. 3–19 in *Cultural Diversity and Families: Expanding Perspectives*, edited by Bahira Sherif Trask and Raeann R. Hamon. Thousand Oaks, CA: Sage.

Stich, Sally. 2010. "How Connected Are We?" *Women's Day*, November 1, 18–24.

Stillars, Alan L. 1991. "Behavioral Observation." Pp. 197–218 in *Studying Interpersonal Interaction*, edited by B. M. Montgomery and S. Duck. New York: Guilford Press.

Stolley, Giordano, and Somchai Taphaneeyapan. 2002. "Stomaching Bugs in Thailand." *Baltimore Sun*, June 20, 2A.

Stone, Pamela. 2007. *Opting Out? Why Women Really Quit Careers and Head Home*. Berkeley: University of California Press.

Strasburger, Victor C., et al. 2006. "Children, Adolescents, and Advertising." *Pediatrics* 118 (December 6): 2563–2569.

Strauss, Valerie. 2011. "Probe: Widespread Cheating on Tests Detailed in Atlanta." *Washington Post*, July 5. Accessed July 6, 2011 (www.washingtonpost.com).

"Streaking." 2005. Wikipedia Encyclopedia. Retrieved September 8, 2005 (www .en.wikipedia.org/wiki).

Strean, William B. 2009. "Remembering Instructors: Play, Pain, and Pedagogy." *Qualitative Research in Sport and Exercise* 1 (November): 210–220.

Streib, Jessi. 2011. "Class Reproduction by Four Year Olds." *Qualitative Sociology* 34 (June): 337–352.

Stroebe, Margaret, Maarten van Son, Wolfgang Stroebe, Rolf Kleber, Henk Schut, and Jan van den Bout. 2000. "On the Classification and Diagnosis of Pathological Grief." *Clinical Psychology Review* 20 (January): 57–75.

Substance Abuse and Mental Health Services Administration. 2012. "Mental Health, United States, 2010." Accessed April 24, 2012 (www.samhsa.gov).

———. 2012. *Results from the 2010 National Survey on Drug Use and Health: Summary of National Findings*, NSDUH Series H-41, HHS Publication No. (SMA) 11-4658. Rockville, MD: Substance Abuse and Mental Health Services Administration.

Sullivan, Andrew. 2011. "Why Gay Marriage Is Good for Straight America." *Newsweek*, July 25, 12–14.

———. 2012. "How State Beat Church." *Newsweek*, February 20, 41–45.

Sullivan, Evelin. 2001. *The Concise Book of Lying.* New York: Farrar, Straus and Giroux.

Sullivan, Will. 2007 "Road Warriors." *US News & World Report*, May 7, 40–49.

Sulzberger, A. G. 2011. "Hispanics Reviving Faded Towns on the Plains." *New York Times*, November 13, A1.

Summers, Nick. 2010. "Do Fines Ever Make Corporations Change?" *Newsweek*, November 13, 56.

Sumner, William G. 1906. *Folkways.* New York: Ginn.

Sunstein, Cass R. 2009. *On Rumors: How Falsehoods Spread, Why We Believe Them, What Can Be Done.* New York: Farrar, Straus and Giroux.

Sutherland, Edwin H. 1949. *White Collar Crime.* New York: Holt, Rinehart, and Winston.

Sutherland, Edwin H., and D. R. Cressey. 1970. *Criminology*, 8th edition. Philadelphia: Lippincott.

Sutton, Philip W. 2000. *Explaining Environmentalism: In Search of a New Social Movement.* Burlington, VT: Ashgate.

Switzer, Jacqueline V. 2003. *Disabled Rights: American Disability Policy and the Fight for Equality.* Washington, DC: Georgetown University Press.

Szarota, Piotr. 2010. "The Mystery of the European Smile: A Comparison Based on Individual Photographs Provided by Internet Users." *Journal of Nonverbal Behavior* 34 (December): 249–256.

Tabachnick, Rachel. 2011. "The 'Christian' Dogma Pushed by Religious Schools That Are Supported by Your Tax Dollars." AlterNet, May 23. Accessed July 12, 2011 (www.alternet.org).

Tabarrok, Alex. 2012. "Tuning in to Dropping Out." *Chronicle of Higher Education*, March 9, B4–B5.

Tach, Laura, and Sarah Halpern-Meekin. 2009. "How Does Premarital Cohabitation Affect Trajectories of Marital Quality?" *Journal of Marriage and Family* 71 (May): 298–317.

Tajfel, Henri. 1982. "Social Psychology of Intergroup Relations." *Annual Review of Psychology*, 1–39.

Tannen, Deborah. 1990. *You Just Don't Understand: Women and Men in Conversation.* New York: Ballantine.

Tannenbaum, Frank. 1938. *Crime and the Community.* New York: Columbia University Press.

Tanur, Judith M. 1994. "The Trustworthiness of Survey Research." *Chronicle of Higher Education*, May 25, B1–B3.

Tavernise, Sabrina. 2011. "Married Couples Are No Longer a Majority, Census Finds." *New York Times*, May 26, A22.

———. 2011. "Soaring Poverty Casts Spotlight on 'Lost Decade'." *New York Times*, September 14, A1.

Tavernise, Sabrina, and Robert Gebeloff. 2010. "Immigrants Make Paths to Suburbia, Not Cities." *New York Times*, December 14, A15.

Tavernise, Sabrina, and Jeff Zeleny. 2010. "South and West See Large Gains in Latest Census." *New York Times*, December 21, A1.

Taylor, Frederick W. 1911/1967. *The Principles of Scientific Management.* New York: W. W. Norton & Company.

Taylor, Jay. 1993. *The Rise and Fall of Totalitarianism in the Twentieth Century.* New York: Paragon House.

Taylor, Jonathan B., and Joseph P. Kalt. 2005. *American Indians on Reservations: A Databook of Socioeconomic Change between the 1990 and 2000 Censuses.* Harvard Project on American Indian Economic Development, January. Retrieved April 20, 2007 (www.ksg .harvard.edu).

Taylor, Paul. 2010. "The Decline of Marriage and Rise of New Families." Pew Research Center, November 18. Accessed March 1, 2012 (www.pewsocialtrends.org).

Taylor, Paul, Cary Funk, and April Clark. 2007. "As Marriage and Parenthood Drift Apart, Public Is Concerned about Social Impact." Pew Research Center, July 1. Retrieved June 15, 2009 (www.pewresearch.org).

Taylor, Paul, and Mark Hugo Lopez. 2011. "The Mexican-American Boom: Births Overtake Immigration." Pew Hispanic Center, July 14. Accessed November 15, 2011 (www .pewhispanic.org).

Taylor, Paul, Rich Morin, Kim Parker, and D'Vera Cohn. 2009. "Growing Old in America: Expectations vs. Reality." Pew Research Center, June 29. Retrieved April 30, 2010 (pewsocialtrends.org).

Taylor, Susan C. 2003. *Brown Skin: Dr. Susan Taylor's Prescription for Flawless Skin, Hair, and Nails.* New York: HarperCollins.

Teicher, Martin. 2000. "Wounds That Time Won't Heal: The Neurobiology of Child Abuse." *Cerebrum* 2 (Fall). Retrieved July 7, 2002 (www.dana.org).

Tejada-Vera, B., and P. D. Sutton. 2009. "Births, Marriages, Divorces, and Deaths: Provisional Data for 2008." *National Vital Statistics Reports* 57 (July 29): 1–6.

Telles, Edward E. 2010. "Mexican Americans and Immigrant Incorporation." *Contexts* 9 (Winter): 28–33.

Teranishi, Robert. 2011. "Asian Americans and Pacific Islanders: Facts, not Fiction—Setting the Record Straight." CARE and College Board. Accessed December 7, 2011 (www .nyu.edu/projects/care).

Terhune, Chad. 2008. "They Know What's in Your Medicine Cabinet." *Business Week*, August 4, 48–52.

Tétreault, Mary Ann. 2001. "A State of Two Minds: State Cultures, Women, and Politics in Kuwait." *International Journal of Middle East Studies* 33 (May): 203–220.

Thibaut, John W., and Harold H. Kelley. 1959. *The Social Psychology of Groups.* New York: Wiley.

Thomas, Anita Jones, Jason Daniel Hacker, and Denada Hoxha. 2011. "Gendered Racial Identity of Black Young Women." *Sex Roles* 64 (April): 530–542.

Thomas, W. I., and Dorothy Swaine Thomas. 1928. *The Child in America.* New York: Alfred A. Knopf.

Thompson, Arienne. 2010. "16, Pregnant . . . and Famous: Teen Moms Are Newest Stars." *USA Today*, November 23. Accessed November 24, 2011 (www.usatoday.com).

Thompson, Michael. 2010. "Voter Turnout: Other Nations Overshadow U.S." Associated Content. Retrieved March 29, 2010 (www .associatedcontent.com).

Thompson, Robert S., et al. 2006. "Intimate Partner Violence: Prevalence, Types, and Chronicity in Adult Women." *American Journal of Preventive Medicine* 30 (June): 447–457.

Thornton, Michael. 2011. "11 Reasons Why the Unemployment Crisis Is Even Worse Than You Think." AlterNet, September 14. Accessed September 15, 2011 (www.alternet.org).

Thumma, Scott, Dave Travis, and Warren Bird. 2005. "Megachurches Today 2005." Hartford Institute for Religion Research. Retrieved August 14, 2007 (hirr.hartsem .edu).

"Ties That Bind, The." 2000. *Public Perspective* 11 (May–June): 10.

Tilly, Charles. 1978. *From Mobilization to Revolution.* Reading, MA: Addison-Wesley.

*Time/Money* Poll. 2011. Time, October 10, 29.

Tizon, Tomas. 2008. "An Alaskan Village Prepares to Move." *Christian Science Monitor*, January 10, 17.

Tobar, Hector. 2009. "Language as a Bridge and an Identity." *Los Angeles Times*, September 22. Retrieved September 24, 2009 (www .latimes.com).

Toder, Eric J. 2005. "What Will Happen to Poverty Rates Among Older Americans in the Future and Why?" The Urban Institute, November. Retrieved August 16, 2002 (www.urban .org).

Tormey, Simon. 1995. *Making Sense of Tyranny: Interpretations of Totalitarianism.* New York: Manchester University Press.

Tornatzky, Louis G., Richard Cutler, and Jongho Lee. 2002. "College Knowledge: What Latino Parents Need to Know and Why They Don't Know It." The Tomás Rivera Policy Institute. Retrieved January 2, 2005 (www.trpi.org).

Toth, Emily. 2011. "No Girls Aloud." *Chronicle of Higher Education*, April 1, A26.

Touraine, Alain. 1981. *The Voice and the Eye: An Analysis of Social Movements.* Cambridge, UK: Cambridge University Press.

———. 2002. "The Importance of Social Movements." *Social Movement Studies* 1 (April): 89–96.

Tresniowski, Alex, Liz McNeil, Kathy Ehrich Dowd, Diane Herbst, Mary S. Park, and Nicole Weisensee Egan. 2011. "A Charmed Life, a Tragic Death." *People*, January 20, 58–61.

Truman, Jennifer L., and Michael R. Rand. 2010. "Criminal Victimization, 2009." Bureau of Justice Statistics Bulletin, October. Accessed August 20, 2011 (bjs.ojp.usdoj.gov).

Trumbull, Mark. 2011. "Navigating the Jobless Economy." *Christian Science Monitor*, September 5, 26–31.

Tu, Ha T. 2011. "Surprising Decline in Consumers Seeking Health Information." Health System Change, November. Accessed April 24 (www.hschange.org).

Tucker, Cynthia. 2007. "Lingering Sexism Impedes Women's Path to Highest Level of Power." *Baltimore Sun,* January 8, A9.

Tucker, Eric. 2011. "Former Chairman Gets 30 Years for $3 Billion Mortgage Fraud." *USA Today*, June 30. Accessed July 1, 2011 (www.usatoday.com).

Tumin, Melvin M. 1953. "Some Principles of Stratification: A Critical Analysis." *American Sociological Review* 18 (August): 387–393.

Tumulty, Karen, and Kate Pickert with Alice Park. 2010. "America, the Doctor Will See You Now." *Time*, April 5, 24–31.

Turk, Austin T. 1969. *Criminality and the Legal Order*. Chicago: Rand-McNally.

———. 1976. "Law as a Weapon in Social Conflict." *Social Problems* 23 (February): 276–291.

Turner, Margery Austin, and Zachary Dade. 2011. "Neighborhood Diversity: Immigration Brings Big Changes to Urban Neighborhoods." Urban Institute, Metro Trends. Accessed June 24, 2012 (www.metrotrends.org).

Turner, Ralph H., and Lewis M. Killian. 1987. *Collective Behavior*, 3rd edition. Englewood Cliffs, NJ: Prentice Hall.

Turow, Joseph, Jennifer King, Chris Jay Hoofnagle, Amy Bleakley, and Michael Hennessy. 2009. "Americans Reject Tailored Advertising." Social Science Research Network, September. Retrieved May 11, 2010 (papers.ssrn.com).

Twitchell, Geoffrey R., Gregory L. Hanna, Edwin H. Cook, Scott F. Stoltenberg, Hiram E. Fitzgerald, and Robert A. Zucker. 2001. "Serotonin Transporter Promoter Polymorphism Genotype Is Associated With Behavioral Disinhibition and Negative Affect in Children of Alcoholics." *Alcoholism: Clinical and Experimental Research* 25 (July): 953–959.

United Conference of Mayors. 2010. "Hunger and Homelessness Survey: A Status Report on Hunger and Homelessness in America's Cities: A 27-City Survey." December. Accessed September 25, 2011 (www.usmayors.org).

United Human Rights Council. 2004. "History of Genocide." Retrieved April 29, 2004 (www.unitedhumanrights.org).

United Nations Department of Economic and Social Affairs. 2012. "World Urbanization Prospects: The 2011 Revision, Highlights." March. Accessed June 23, 2012 (esa.un.org).

United Nations Development Fund for Women. 2007. "Violence against Women—Facts and Figures." March. Retrieved March 11, 2007 (www.unifem.org).

United Nations Development Programme. 2006. *Human Development Report 2006: Beyond Scarcity: Power, Poverty, and the Global Water Crisis*. New York: United Nations Development Programme.

———. 2010. *Human Development Report 2010*. Accessed September 22, 2011 (hdr.undp.org).

United Nations World Water Development Report 2. 2006. "Water: A Shared Responsibility." Retrieved February 3, 2008 (www.unesco.org).

U.S. Bureau of Labor Statistics. 2008. "Table 39: Median Weekly Earning of Full-Time Wage and Salary Workers by Detailed Occupation and Sex." Retrieved July 2, 2008 (www.bls.gov).

U.S. Bureau of Labor Statistics. 2012. "Employment Status." Table 2. Accessed May 22, 2012 (www.bls.gov).

———. 2010. *Statistical Abstract of the United States: 2010*, 129th edition. Washington, DC: U.S. Government Printing Office.

———. 2010. "Voting and Registration in the Election of November 2008—Detailed Tables." Retrieved March 26, 2010 (www.ccensus.gov).

———. 2011. *Statistical Abstract of the United States: 2011*, 130th ed. Washington, DC: U.S. Government Printing Office.

———. 2012. "Educational Attainment in the United States, 2011—Detailed Tables." Accessed March 1, 2012 (www.census.gov).

———. 2012. *Statistical Abstract of the United States: 2012,* 131st edition. Washington, DC: Government Printing Office.

U.S. Census Bureau FactFinder. 2011. "Marital Status." S1201. Accessed March 1, 2012 (www.census.gov).

U.S. Census Bureau News. 2008. "An Older and More Diverse Nation by Midcentury." August 14. Retrieved August 18, 2008 (www.census.gov).

———. 2011. "Father's Day: June 19, 2011." April 20. Accessed March 10, 2012 (www.census.gov).

———. 2011. "Unmarried and Single Americans Week Sept. 18–24, 2011." August 26. Accessed March 10, 2012 (www.census.gov).

———. 2012. "Father's Day: June 17, 2012." May 2. Accessed May 21, 2012 (www.census.gov).

———. 2012. "Valentine's Day 2012: Feb. 14." January 4. Accessed March 10, 2012 (www.census.gov).

U.S. Census Bureau Population Division. 2008. "2008 National Population Projections Tables and Charts." August 14. Retrieved September 9, 2008 (www.census.gov/population/www/projections/tablesandcharts.html).

———. 2012. "Table 3. Annual Estimates of the Resident Population by Sex, Race, and Hispanic Origin for the United States: April 1, 2010 to July 1, 2011 (NC-EST2011-03)." Accessed May 15, 2012 (www.census.gov).

U.S. Chamber of Commerce, Center for American Progress, and Frederick M. Hess. 2009. *Leaders and Laggards: A State-by-State Report Card on Educational Innovation*. November. Retrieved April 22, 2010 (www.americanprogress.org).

U.S. Congress Joint Economic Committee. 2012. "Mother's Day Report: Paycheck Fairness Helps Families, Not Just Women." May 9. Accessed June 1, 2012 (www.jec.sen.gov).

U.S. Department of Health and Human Services. 1999. *Mental Health: A Report of the Surgeon General*. Rockville, MD: U.S. Department of Health and Human Services, National Institutes of Health, National Institute of Mental Health.

———. 2011. "Child Maltreatment 2010." Administration for Children and Families, Children's Bureau. Accessed March 5, 2012 (www.acf.hhs.gov).

———. 2012. *Preventing Tobacco Use Among Youth and Young Adults: A Report of the Surgeon General*. Atlanta, GA: National Center for Chronic Disease Prevention and Health promotion, Office on Smoking and Health.

U.S. Department of Justice. 1996. "Policing Drug Hot Spots." National Institute of Justice, January. Retrieved May 3, 2005 (www.ncjrs.org).

———. 2011. "U.S. Attorney Announces Drug Endangered Children Task Force." Press Notice, May 31. Accessed July 25, 2011 (www.usdoj.gov).

U.S. Department of Labor. 2002. *Working in the 21st Century*. Washington, DC: U.S. Government Printing Office.

U.S. Department of Labor Statistics. 2011. "Household Data Annual Averages." Table 39. Accessed January 3, 2012 (www.bls.gov).

U.S. Department of Labor Wage and Hour Division. 2012. "Minimum Wage Laws in the States." January 1. Accessed January 11, 2012 (www.dol.gov).

U.S. Department of State. 2006. *Trafficking in Persons Report*. June. Retrieved November 1, 2006 (www.state.gov).

U.S. Environmental Protection Agency. 2009. "EPA's Endangerment Finding." December 7. Retrieved May 1, 2010 (www.epa.gov).

U.S. Equal Employment Opportunity Commission. 2011a. "Pregnancy Discrimination Charges, EEOC & FEPAs Combined: FY 1997–FY 2011." Accessed January 25, 2012 (www.eeoc.gov).

———. 2011b. "Sexual Harassment Charges, EEOC & FEPAs Combined: FY 1997–FY 2011." Accessed January 25, 2012 (www.eeoc.gov).

U.S. General Accounting Office. 2004. "Defense of Marriage Act: Update to Prior Report." www.gao.gov (accessed July 5, 2007).

U.S. Senate Special Committee on Aging, American Association of Retired Persons, Federal Council on the Aging, and U.S. Administration on Aging. 1991. *Aging America: Trends and projections, 1991*. Washington, DC: Department of Health and Human Services.

United States Senate. 2004. *Report on the U.S. Intelligence Community's Prewar Intelligence Assessments on Iraq*. Retrieved September 13, 2006 (www.gpoaccess.gov).

U.S. Senate Special Committee on Aging, American Association of Retired Persons, Federal Council on the Aging, and U.S. Administration on Aging. 1991. *Aging America: Trends and Projections, 1991*. Washington, DC: U.S. Department of Health and Human Services.

Urbina, Ian. 2006. "In Online Mourning, Don't Speak Ill of the Dead." *New York Times*, November 5, 1.

———. 2007. "Court Rejects Law Limiting Pornography on Internet." *New York Times*, March 23, A11.

———. 2009. "For Runaways, Sex Buys Survival." *New York Times*, October 27, A1.

"Usual Weekly Earnings of Wage and Salary Workers Fourth Quarter 2011." 2012. Bureau of Labor Statistics, January 24. Accessed March 15, 2012 (www.bls.gov).

Uwimana, Solange. 2012. "Whites More Likely to Be Drug Addicts Than Blacks: So Why Do Racial Drug Stereotypes Persist?" AlterNet, February 24. Accessed February 25, 2012 (www.alternet.org).

Valenti, Jessica. 2007. "How the Web Became a Sexists' Paradise." *The Guardian*, April 6. Retrieved April 9, 2007 (www.guardian .co.uk).

van Ginneken, Jaap. 2003. *Collective Behavior and Public Opinion: Rapid Shifts in Opinion and Communication.* Mahwah, NJ: Lawrence Erlbaum Associates.

Van Hook, Jennifer, and Claire E. Altman. 2012. "Competitive Food Sales in Schools and Childhood Obesity: A Longitudinal Study." *Sociology of Education* 85 (January): 23–39.

Van Vooren, Nicole, and Roberta Spalter-Roth. 2010. "Tracking Master's Students through Programs and into Careers." *Footnotes* (September/October): 10–11.

Vance, Ashlee. 2011. "The Data Knows." *Bloomburg Businessweek*, September 12–September 18, 71–74.

Vandewater, Elizabeth A., Victoria J. Rideout, Ellen A. Wartella, Xuan Huang, June H. Lee, and Mi-suk Shim. 2007. "Digital Childhood: Electronic Media and Technology Use Among Infants, Toddlers, and Preschoolers." *Pediatrics* 119 (May): e1006–e1015.

Vartabedian, Ralph, and Ken Bensinger. 2010. "Toyota Faces $16.4 Million Penalty. *Baltimore Sun*, April 6, 6.

Veblen, Thorstein. 1899/1953. *The Theory of the Leisure Class.* New York: New American Library.

Vedantam, Shankar. 2002. "Negative View? It May Be Brain 'Knob'." *Seattle Times*, February 12. Retrieved February 13, 2002 (seattletimes.nwsource.com).

Velasquez-Manoff, Moises. 2009. "Pressure Builds over Bottled Water." *Christian Science Monitor*, October 18, 36–37.

Venkatesh, Sudhir. 2008. *Gang Leader for a Day: A Rogue Sociologist Takes to the Streets.* New York: Penguin Press.

Vigdor, Jacob L. 2008. "Measuring Immigrant Assimilation in the United States." Center for Civic Innovation, May. Retrieved June 6, 2008 (www.manhattan-institute.org).

Vold, George B. 1958. *Theoretical Criminology.* New York: Oxford University Press.

Waldref, J. 2008. "Women at Work Find Reinforced Glass Ceilings." Women's eNews, August. Retrieved September 2, 2008 (www .womensenews.org).

Waldron, Travis. 2011. "Pastor at Kentucky Church That Banned Interracial Couples Calls for Vote to Reverse Decision." AlterNet, December 3. Accessed December 7, 2011 (www.alternet.org).

Walker, Lenore E. 2000. *The Battered Woman Syndrome,* 2nd edition. New York: Springer.

Wallace, Bruce. 2006. "Japanese Schools to Teach Patriotism." *Los Angeles Times*, December 16. Retrieved December 18, 2006 (www .latimes.com).

Wallis, Cara. 2011. "Performing Gender: A Content Analysis of Gender Display in Music Videos." *Sex Roles* 64 (February): 160–172.

Walls, Mark. 2011. "High Court Strikes Down Calif. Law on Violent Video Games." *Education Week*, June 27. Accessed June 28, 2011 (blogs.edweek.org).

Walsh, Bryan. 2011. "Parched Earth." *Time*, August 22, 41–45.

Walters, Joanna. 2011. "Occupy America: Protests Against Wall Street and Inequality Hit 70 Cities." *Guardian*, October 8. Accessed October 27, 2011 (www.guardian.co.uk).

Wan, William. "Chinese Dog Eaters and Dog Lovers Spar Over Animal Rights." *Washington Post*, May 28, A14.

Wang, Jing, Tonja R. Nansel, and Ronald J. Ianotti. 2010. "Cyber and Traditional Bullying: Differential Association with Depression." *Journal of Adolescent Health* 48 (April): 415–417.

Wang, Wendy. 2012. "The Rise of Intermarriage: Rates, Characteristics Vary by Race and Gender." Pew Research Center, February 16. Accessed March 1, 2012 (www .pewsocialtrends.org).

Wang, Wendy, and Kim Parker. 2011. "Women See Value and Benefits of College; Men Lag on Both Fronts, Survey Finds." Pew Research Center, August 17. Accessed April 15, 2012 (www.pewsocialtrends.org).

Warner, Margaret, Li Hui Chen, Diane M. Makuc, Robert N. Anderson, and Arialdi M. Miniño. 2011. "Drug Poisoning Deaths in the United States, 1980–2008." NCHS Data Brief No. 81, December. Accessed April 15, 2012 (www.nchs.org).

Warren, Jocelyn T., S. Marie Harvey, and Jillian T. Henderson. 2010. "Do Depression and Low Self-Esteem Follow Abortion Among Adolescents? Evidence from a National Study." *Perspectives on Sexual and Reproductive Health* 42 (December): 230–235.

Wasley, Paula. 2007. "46 Students Are Disciplined for Cheating at Indiana University's Dental School." *Chronicle of Higher Education*, May 9. Retrieved May 10, 2007 (www .chronicle.com).

———. 2008. "The Syllabus Becomes a Repository of Legalese." *Chronicle of Higher Education* 54, March 14, A1, A8–A11.

Water Quality & Health Council. 2005. "Facts About Chlorine and Drinking Water." Retrieved July 3, 2005 (www. waterandhealth.org).

Waters, Rob. 2011. "Are Psychiatrists Inventing Mental Illness to Feed Americans More Pills?" Alternet, December 27. Accessed April 24, 2012 (www.alternet.org).

Weber, Lynn, Tina Hancock, and Elizabeth Higginbotham. 1997. "Women, Power, and Mental Health." Pp. 380–396 in *Women's Health: Complexities and Differences,* edited by Sheryl B. Ruzek, Virginia L. Olesen, and Adele E. Clarke. Columbus: Ohio State University Press.

Weber, Max. 1920/1958. *The Protestant Ethic and the Spirit of Capitalism* (1904–1905), translated by Talcott Parsons. New York: Charles Scribner's Sons.

———. 1925/1947. *The Theory of Social and Economic Organization.* New York: Free Press.

———. 1925/1978. *Economy and Society*, edited by Guenther Roth and Claus Wittich. Berkeley: University of California Press.

———. 1946. *From Max Weber: Essays in Sociology*, translated and edited by H. H. Gerth and C. Wright Mills. Berkeley: University of California Press.

Weisbuch, Max, Kristin Pauker, and Nalini Ambady. 2009. "The Subtle Transmission of Race Bias via Televised Nonverbal Behavior." *Science* 326 (December 18): 1711–1714.

Weiser, Wendy R., and Lawrence Norden. 2011. "Voting Law Changes in 2012." Brennan Center for Justice, October 3. Accessed January 20, 2012 (www.brennancenter.org).

Weiss, Carol H. 1998. *Evaluation: Methods for Studying Programs and Policies*, 2nd edition. Upper Saddle River, NJ: Prentice Hall.

Weiss, Rick. 2004. "Nanomedicine's Promise Is Anything but Tiny." *Washington Post*, January 31, A8.

———. 2005. "The Power to Divide." *National Geographic* 208 (July): 3–27.

Weitz, Rose. 2013. *The Sociology of Health, Illness, and Health Care: A Critical Approach*, 6th edition. Boston: Wadsworth.

Welch, David. 2011. "For the UAW, a Bargaining Dilemma." *Bloomberg Businessweek*, September 19–25, 23–24.

Welch, Susan, John Gruhl, John Comer, and Susan Rigdon. 2004. *Understanding American Government*, 7th edition. Belmont, CA: Wadsworth.

Wellins, Richard S., William C. Byham, and Jeanne M. Wilson. 1991. *Empowered Teams: Creating Self-Directed Work Groups That Improve Quality, Productivity, and Participation.* San Francisco: Jossey-Bass.

Wennberg, J. E., E. S. Fisher, D. C. Goodman, and J. S. Skinner. 2008. *Tracing the Care of Patients with Severe Chronic Illness.* Dartmouth Institute for Health Policy and Clinical Practice. Retrieved April 9, 2010 (www.dartmouthatlas.org).

Wenneras, Christine, and Agnes Wold. 1997. "Nepotism and Sexism in Peer Review." *Nature* 387 (May 22): 341–343.

West, Candace, and Don H. Zimmerman. 1987. "Doing Gender." *Gender and Society* 1 (June): 125–51.

———. 2009. "Accounting for Doing Gender." *Gender & Society* 23 (February): 112–122.

Whelan, Christine B. 2009. "A Feminist-Friendly Recession?" Pp. 55–62 in *The State of Our Unions, Marriage in America 2009: Money & Marriage*, edited by W. Bradford Wilcox. The National Marriage Project and the Institute for American Values. Retrieved December 20, 2009 (www.stateofourunions .org).

Whelan, David. 2007. "The Sentencing Game." *Forbes*, February 12, 40.

———. 2011. "In One Pocket, Out the Other." *Forbes*, June 6, 30, 32.

Whisnant, Rebecca, and Christine Stark, eds. 2004. *Not for Sale: Feminists Resisting Prostitution and Pornography.* North Melbourne, Australia: Spinifex Press.

White, James W. 2005. *Advancing Family Theories*. Thousand Oaks, CA: Sage.

White House Council on Women and Girls. 2011. *Women in America: Indicators of Social and Economic Well-Being*. U.S. Department of Commerce and Bureau of Justice Statistics, March. Accessed June 1, 2011 (www.whitehouse.gov).

Whitehead, Jaye Cee. 2011. "The Wrong Reasons for Same-Sex Marriage." *New York Times*, May 16, A21.

Whitelaw, Kevin. 2000. "But What to Call It?" *U.S. News & World Report*, October 16, 42.

Whiteman, Shawn, D., Susan M. McHale, and Anna Soli. 2011. "Theoretical Perspectives on Sibling Relationships." *Journal of Family Theory & Review* 3 (June): 124-139.

Whitman, Steven, Jennifer Orsi, and Marc Hurlbert. 2012. "The Racial Disparity in Breast Cancer Mortality in the 25 Largest Cities in the United States." *Cancer Epidemiology* 36 (April): e147–e141.

Whorf, Benjamin Lee. 1956. *Language, Thought, and Reality*. Cambridge, MA: MIT Press.

Whoriskey, Peter. 2011. "Congressional Net Worth More Than Doubles Since 1984." *Washington Post*, December 26. Accessed December 27, 2011 (www.washingtonpost.com).

"Who Votes, Who Doesn't, and Why: Regular Voters, Intermittent Voters, and Those Who Don't." 2006. Pew Research Center for the People & the Press, October 18. Retrieved May 3, 2007 (www.peoplepress.org).

Wickenden, Dorothy. 2006. "Top of the Class." *New Yorker*, October 2. Retrieved July 26, 2007 (www.newyorker.com).

Wicker, Christine. 2009. "How Spiritual Are We?" *Parade*, October 4, 4–5.

Wiehe, Vernon R. 1997. *Sibling Abuse: Hidden Physical, Emotional, and Sexual Trauma*, 2nd edition. Thousand Oaks, CA: Sage.

Wilcox, W. Bradford, and Elizabeth Marquardt. 2011. "The State of Our Unions: Marriage in America 2011." University of Virginia, National Marriage Project. Accessed March 4, 2012 (www.virginia.edu/marriageproject).

Wildsmith, Elizabeth, Nicole R. Steward-Streng, and Jennifer Manlove. 2011. "Childbearing Outside Marriage: Estimates and Trends in the United States." Child Trends, November. Accessed March 4, 2012 (www.childtrends.org).

Wilkinson, Charles. 2006. *Blood Struggle: The Rise of Modern Indian Nations*. New York: W.W. Norton & Company.

Wilkinson, Richard, and Kate Pickett. 2009. *The Spirit Level: Why Greater Equality Makes Societies Stronger*. New York: Bloomsbury Press.

Williams, David R., and Michelle Sternthal. 2010. "Understanding Racial-Ethnic Disparities in Health: Sociological Contributions." *Journal of Health and Social Behavior* 41 (Suppl.): S15–S27.

Williams, Frank P., III, and Marilyn D. McShane. 2004. *Criminological Theory*, 4th edition. Upper Saddle River, NJ: Prentice Hall.

Williams, Robin M., Jr. 1970. *American Society: A Sociological Interpretation*, 3rd edition. New York: Knopf.

Williams, Timothy. 2011. "Tackling Infant Mortality Rates Among Blacks." *New York Times*, October 14, A10.

Williams, Wendy H., and Stephen J. Ceci. 2012. "When Scientists Choose Motherhood." *American Scientist* 100 (March/April). Accessed February 20, 2012 (www.amricanscientist.org).

Willie, Charles Vert, and Richard J. Reddick. 2003. *A New Look at Black Families*, 5th edition. Walnut Creek, CA: AltaMira Press.

Wilson, Thomas C. 1993. "Urbanism and Kinship Bonds: A Test of Four Generalizations." *Social Forces* 71 (March): 703–712.

Wilson, William Julius. 1996. *When Work Disappears: The World of the New Urban Poor*. New York: Knopf.

Wiltenburg, Mary. 2002. "Minority." *Christian Science Monitor*, January 31, 14.

Wingfield, Adia Harvey. 2010. "Are Some Emotions Marked 'Whites Only'? Racialized Feeling Rules in Professional Workplaces." *Social Problems* 57 (May): 258–268.

Winograd, Morley, and Michael D. Hais. 2011. "Millennial Generation Redefines Faith in America." *Christian Science Monitor*, September 26, 34.

Winseman, Albert L. 2004. "Women in the Clergy: Perception and Reality." Gallup Organization, March 30. Retrieved March 30, 2004 (www.gallup.com).

Winship, Scott. 2011. "Mobility Impaired." *National Review*, November 7. Accessed February 4, 2012 (www.nationalreview.com).

Winter, Michael. 2011. "Reports: King Revokes Lashing of Saudi Woman Who Drove." *USA Today*, September 28. Accessed October 25, 2011 (www.usatoday.com).

———. 2011. "13 More Charged in SAT Cheating Scandal in N.Y." *USA Today*, November 23. Accessed November 24, 2011 (www.usatoday.com).

Wirth, Louis. 1938. "Urbanism as a Way of Life." *American Journal of Sociology* 44 (July): 1–24.

Wiseman, Paul. 2010. "When the Textile Mill Goes, So Does a Way of Life." *USA Today*, March 10, 1A.

Witters, Dan. 2010. "The Flu Season That Wasn't." Gallup, May 6. Retrieved May 6, 2010 (www.gallup.com).

———. 2010. "Large Metro Areas Top Small Towns, Rural Areas in Wellbeing." Gallup, May 17. Retrieved May 20, 2010 (www.gallup.com).

———. 2012. "Caregiving Costs U.S. Economy $25.2 Billion in Lost Productivity." Gallup, March 6. Accessed March 10, 2012 (www.gallup.com).

———. 2012. "Key Chronic Diseases Decline in U.S." Gallup, January 20. Accessed April 22, 2012 (www.gallup.com).

Witters, Dan, and Sangeeta Agrawal. 2011. "Unhealthy U.S. Workers' Absenteeism Costs $153 Billion." Gallup, October 17. Accessed April 22, 2012 (www.gallup.com).

Wolfe, Alan. 2008. "Pew in the Pews." *Chronicle of Higher Education*, March 21, B5–B6.

Wolff, Edward N. 2010. "Recent Trends in Household Wealth in the United States: Rising Debt and the Middle-Class Squeeze—an Update to 2007." Levy Economics Institute, March. Accessed September 15, 2011 (www.levyinstitute.org).

Wolfson, Mark. 2001. *The Fight Against Big Tobacco: The Movement, the State, and the Public's Health*. New York: Aldine de Gruyter.

Wolgin, Philip E., and Angela Maria Kelley. 2011. "Your State Can't Afford It: The Fiscal Impact of States' Anti-Immigrant Legislation." Center for American Progress, July. Accessed November 20, 2011 (www.americanprogress.org).

*Women in the Labor Force: A Databook*. 2011. December 2010, Report 1026. Accessed October 25, 2011 (www.bls.gov).

Wood, Julia T. 2011. *Gendered Lives: Communication, Gender, & Culture*, 9th edition. Belmont, CA: Wadsworth.

Woolf, Alex. 2004. *Fundamentalism*. Chicago: Raintree.

World Bank. 2008. *World Development Indicators 2008*. Washington, DC: The International Bank.

———. 2011. "Gross National Income 2010, Atlas Method." July 1. Accessed September 23, 2011 (www.worldbank.org).

———. 2012. *Gender Equality and Development*. Washington, DC. Accessed September 22, 2011 (www.worldbank.org).

World Health Organization. 2005. *WHO Multi-country Study on Women's Health and Domestic Violence against Women: Summary Report of Initial Results on Prevalence, Health Outcomes and Women's Responses*. Geneva: World Health Organization.

———. 2012. "Good Health Adds Life to Years: Global Brief for World Health Day." Accessed April 24, 2012 (www.who.int).

World Water Assessment Programme. 2009. *The United Nations World Water Development Report 3: Water in a Changing World*. Paris: UNESCO and London: Earthscan.

Wright, Wynne, and Elizabeth Ransom. 2005. "Stratification on the Menu: Using Restaurant Menus to Examine Social Class." *Teaching Sociology* 33 (July): 310–316.

Wulfhorst, Ellen. 2006. "US Mothers Deserve $134,121 In Salary." May 3. Retrieved May 10, 2006 (today.reuters.com).

Wyatt, Edward. 2009. "More Than Ever, You Can Say That on Television." *New York Times*, November 14, A1.

"X Factor, The." 2011. *Fortune*, October 17, S1–S3.

Yamazaki, Tomoko, and Komaki Ito. 2010. "Japan: Boosting Growth with Day Care." *Bloomburg Businessweek,* December 28, 2009 and January 4, 2010, 96–97.

Yan, Sophia. 2009. "Anonymous Gossip Sites." *Time*, December 7, 97–98.

Yardley, Jim. 2009. "Indian Women Find New Peace in Rail Commute." *New York Times*, September 16, A1.

———. 2010. "Soaring Above India's Poverty, a 27-Story Home." *New York Times*, October 28, A1.

Yarrow, Andrew L. 2009. "State of Mind." *Education Week*, October 21. Retrieved April 21, 2010 (www.edweek.org).

Yokota, Fumise, and Kimberly M. Thompson. 2000. "Violence in G-rated Animated Films." *JAMA: Journal of the American Medical Association* 283 (May 24/31): 2716–2720.

York, Emily Bryson. 2012. "Form Follows Content at McD's." *Baltimore Sun*, February 2, 14.

Young, Jeffrey R. 2011. "Programmed for Love." *Chronicle of Higher Education*, January 21, B6–B11.

Yu, Roger. 2007. "Indian-Americans Book Years of Success." *USA Today*, April 18, 1B–2B.

Zahedi, Ashraf. 2008. "Concealing and Revealing Female Hair: Veiling Dynamics in Contemporary Iran." Pp. 250–265 in *The Veil: Women Writers on Its History, Lore, and Politics*, edited by Jennifer Heath. Berkeley: University of California Press.

Zajonc, Robert B., and Gregory B. Markus. 1975. "Birth Order and Intellectual Development." *Psychological Review* 82 (January): 74–88.

Zezima, Katie. 2011. "For Many, 'Washroom' Seems to Be Just a Name." *New York Times*, September 13, A14.

Zhang, Xiao, Ana P. Martinez-Donate, Daphne Kuo, Nathan R. Jones, and Karen A. Palmersheim. 2011. "Trends in Home Smoking Bans in USA, 1995–2007: Prevalence, Discrepancies, and Disparities." *Tobacco Control* 20 (August): 330–338.

Zhu, Wei Xing, Li Lu, and Therese Hesketh. 2009. "China's Excess Males, Sex Selective Abortion, and One Child Polity: Analysis of Data from 2005 National Intercensus Survey." *British Medical Journal* 338. Accessed May 12, 2012 (www.bmj.com).

Zimbardo, Philip G., Christina Maslach, and Craig Haney. 2000. "Reflections on the Stanford Prison Experiment: Genesis, Transformations, Consequences." Pp. 193–237 in *Obedience to Authority: Current Perspectives on the Milgram Paradigm*, edited by Thomas Blass. Mahwah, NJ: Lawrence Erlbaum Associates.

Zimbardo, Philip. G. 1975. "Transforming Experimental Research into Advocacy for Social Change." Pp. 33–66 in *Applying Social Psychology: Implications for Research, Practice, and Training*, edited by Morton Deutsch and Harvey A. Hornstein. Hillsdale, NJ: Erlbaum.

Zimmer, Carl. 2011. "Rise of the Superbacteria." *Newsweek*, June 13 & 20, 11–12.

Zimmerman, Eilene. 2012. "Modern Romance: From Hookup to Stayover, Millennials' Route to Marriage Is Redefined." *Christian Science Monitor*, February 13, 26–31.

Zimmerman, F. J., D. A. Christakis, and A. N. Meltzoff. 2007. "Associations Between Media Viewing and Language Development in Children under Age 2 Years." *Journal of Pediatrics* 151 (October): 364–368.

Zirulnick, Ariel. 2011. "The Five Most Dangerous Countries for Women." *Christian Science Monitor*, June 16. Accessed October 25, 2011 (www.csmonitor.com).

———. 2011. "How Educating Girls Can Save the World." *Christian Science Monitor*, October 31, 13.

Zoroya, Gregg. 2010. "Civilian Soldiers' Suicide Rate Alarming." *USA Today*, November 25. Accessed November 26, 2010 (www.usatoday.com).

Zuckerman, Phil. 2011. "Taking Leave of Religion." *Chronicle Review*, November 25, B10–B12.

Zuehlke, Eric. 2009. "Immigrants Work in Riskier and More Dangerous Jobs in the United States." Population Reference Bureau, November. Retrieved March 15, 2010 (www.prb.org).

Zurbriggen, Eileen L., and Aurora M. Sherman. 2010. "Race and Gender in the 2008 U.S. Presidential Election: A Content Analysis of Editorial Cartoons." *Analyses of Social Issues and Public Policy* 10 (December): 223–247.

Zweigenhaft, Richard L., and G. William Domhoff. 1998. *Diversity in the Power Elite: Have Women and Minorities Reached the Top?* New Haven, CT: Yale University Press.

———. 2006. *Diversity in the Power Elite: How It Happened, Why It Matters.* Lanham, MD: Rowman & Littlefield.

# USE THE TOOLS.

• Rip out the Review Cards in the back of your book to study.

**Or Visit CourseMate to:**

• Read, search, highlight, and take notes in the Interactive eBook

• Review Flashcards (Print or Online) to master key terms

• Test yourself with Auto-Graded Quizzes

• Bring concepts to life with Games, Videos, and Animations!

Go to CourseMate for **SOC3** to begin using these tools.

Access at **www.cengagebrain.com**

Complete the Speak Up
survey in CourseMate at
**www.cengagebrain.com**

Follow us at
**www.facebook.com/4ltrpress**

# NAME INDEX

# C

Cabrera, Natasha J., 232
Cadge, Wendy, 321
Cameron, Deborah, 90
Campbell, David E., 269
Campbell, Donald T., 33
Campbell, Duncan, 135
Campion-Vincent, Véronique, 315
Campo-Flores, Arian, 264
Canudas-Romo, Vladimir, 229
Caperton, Gaston, 251
Caplan, Jeremy B., 157, 158
Caplan, Paula J., 157, 158
Carasso, A., 243
Card, David, 73
Card, Josefina J., 34
Carey, Benedict, 37, 131
Carey, Bill, 112
Carl, Traci, 151
Carlson, Marcia, 233
Carmichael, Mary, 308
Carmon, Irin, 159
Carnagey, Nicholas L., 74
Carnevale, Anthony P., 249, 252, 253
Carney, Ginny, 44
Carr, Nicholas, 329
Carrell, Scott E., 73
Carroll, Jason S., 168
Carroll, Jill, 259
Carroll, John, 42
Carroll, Margaret D., 53, 115
Carroll, Stephen J., 248
Carson, Benjamin S., 149
Cascio, Jamais, 329
Case, Nancy Humphrey, 110
Casper, Lynne M., 197
Caston, Richard J., 202
Castro, Fidel, 215
Catalano, Shannan, 237
Ceasar, 124
Cech, Erin, 176
Ceci, Stephen J., 254
Cellini, Stephanie R., 146
Chamberlain, Catherine J., 275
Chambliss, William J., 126
Chan, Sam, 236
Chandler, Tertius, 300
Chandra, A. G. A., 234
Chandra, Anjani, 164, 165
Chandra, Shobhana, 288
Chandy, Laurence, 150
Chapman, Paige, 97
Charles, Maria, 170
Chasin, Barbara H., 194
Chávez, César, 215
Chelala, César, 64
Chenowith, Paul, 23
Chesler, Phyllis, 17
Chesney-Lind, Meda, 64, 129
Chockley, Nancy, 282
Choice, Pamela, 245
Chou, Roger, 287
Choudhury, Tufyal, 270
Choueiti, Marc, 165
Christakis, Dimitri A., 74
Christenson, Matthew, 296
Christian, Steve, 76
Christina, Greta, 74
Chu, Henry, 226
Chu, Kathy, 54
Churchill, Ward, 184
Ciabattari, Teresa, 232
Cichocki, Mary K., 305
Clarke-Pearson, Kathleen, 97
Claxton, Gary, 282
Clayson, Dennis E., 28
Clayton, Mark, 308
Clement, Scott, 163
Clements, David D., 209
Clifford, Stephanie, 154, 311, 330

Clifton, Donna, 170, 174, 175
Clifton, Jon, 181
Clinton, Hillary Rodham, 223
Cloke, Kenneth, 110
Cloud, John, 258
Cochrane, Megan, 280
Cohan, Catherine I., 232
Cohen, Jere, 267
Cohen, Philip N., 230
Cohn, D'Vera, 28, 159, 171, 174, 179, 181, 206, 231, 232, 311
Colapinto, John, 63
Colbert, Stephen, 112
Coley, Rebekah L., 232
Coll, Steve, 228
Collins, Patricia Hill, 17, 175
Collins, Randall, 252
Collins, Sara R., 288
Comte, Auguste, 7, 13
Conley, Dalton, 67
Conlin, M., 76
Connell, Dave, 307
Connolly, Frank W., 96
Conrad, Peter, 291
Considine, Austin, 96
Conway, Kevin P., 130
Cook, Bryan, 253
Cook, Thomas D., 33
Cooley, Charles Horton, 67–68, 100
Coontz, Stephanie, 226
Cooper, Martin, 56
Cooper, O. R., 308
Cooperman, Alan, 28
Coser, Lewis, 101
Cowell, Alan, 275
Crabtree, Steve, 266
Cressey, Donald R., 126, 130
Crissey, Sarah, 217, 218
Crockett, Ariel, 196
Crooks, Hulda, 240
Crothers, Charles, 305
Crowe, Edward, 258
Crowley, Michael, 222
Cruikshank, Margaret, 243
Cudd, Ann E., 210
Culbert, Samuel A., 112
Cullen, Lisa T., 228
Cumberworth, Erin, 135
Currie, E., 132
Currie, Janet, 144, 239
Curtiss, Susan, 62
Cutler, David M., 276
Cynkar, Peter, 228, 241

# D

Dade, Zachary, 301
Dadigan, Marc, 192
Dahl, Melissa, 287
Dahl, Robert A., 219
Dais, Kenneth C., 267
Dalton, Madeline A., 66
Damaske, Sarah, 175
Dandaneau, Steven P., 3
Daniel, Lisa, 177
Dante, Ed, 260
Danziger, S., 78
Dao, James, 79
Davey, Monica, 221
Davidson, Paul, 205
Davies, James B., 150
Davies, James C., 323
Davis, F. J., 322
Davis, James A., 141
Davis, Kingsley, 151–152, 153
Day, Jennifer Cheeseman, 217
Deardorff, Julie, 287
Deaton, Angus S., 263
De Bois, W. E. B., 12
Dee, Thomas S., 161
Deegan, Mary Jo, 11
de Gaulle, Charles, 213

Degler, Carl, 76
DeGraw, David, 155
DeKeseredy, Walter S., 128
de la Cruz, G. Patricia, 193, 235
de las Casas, Bartolome, 184
DeLeire, Thomas, 149
De Lisi, Matt, 133
Dell, Kristina, 250
della Porta, Donatella, 326
De Long, J. Bradford, 300
DeMaris, Alfred, 232
de Mestral, George, 56
Demos, John, 76
DeNavas-Walt, Carmen, 3, 77, 145, 146, 189, 205, 227, 235, 281
Denton-Borhaug, Kelly, 260
Deparle, Jason, 148
Deresiewicz, William, 96
Desmond-Harris, Jenee, 317
DeVega, Chauncey, 185
Dew, Jeffrey, 230
Dewan, Shaila, 143
Dey, Judy Goldberg, 162
Diamond, Marie, 201, 270
Diamond, Milton, 63
Diani, Mario, 326
Dickert-Conlin, Stacy, 250
Dickson, William J., 108
Diggs, Taye, 197
Dillow, Sally A., 161, 248
Dilworth-Anderson, Peggye, 236
Dines, Gail, 168
DiSalvo, David, 96
Dixon, Robyn, 138
Do, Hien Duc, 245
Dodson, Lisa, 235, 245
Dodson, Robin E., 275
Doeringer, Peter B., 194
Dolnick, Sam, 51
Domhoff, G. William, 140, 142, 221
Domosh, Mona, 304
Dorius, Cassandra J., 71
Douglas, Susan J., 85
Dovidio, John F., 196
Doyle, Francis X., 268
Draffan, George, 202
Dreazen, Yochi J., 269
Dreifus, Claudia, 250
Drew, Charles R., 190
Dreweke, Joerg, 289
Drogin, Bob, 168
Du Bois, W. E. B., 14
Dubowitz, Tamara, 277
Dugard, Jaycee, 128
Duhigg, Charles, 306, 307, 308
Dunifon, Rachel, 26
Dunlap, Riley E., 311
Dunwoody, Ann E., 223
Durden, Tyler, 201
Durkheim, Émile, 8–9, 13, 124, 248, 266
Durose, Matthew R., 134
Dush, Kamp, 233
Duster, Troy, 180
Dye, Thomas R., 219
Dyson, Laronistine, 125, 132

# E

Eberstadt, Nicholas, 145
Eckstein, Rick, 5
Edin, Kathryn, 235
Egan, Timothy, 206
Eggen, Dan, 280
Egley, Arlen, 123
Ehrenreich, Barbara, 30
Ehrhardt, Anke A., 63
Ehrlich, Anne H., 298
Ehrlich, Paul R., 298
Eibner, Christine, 284
Eisenberg, Abne M., 92
Eisenberg, Nancy, 71
Ekman, Paul, 92

# SUBJECT INDEX

Economy. *See also* Capitalism; Economic
  inequality
  of China, 152
  conflict theory on, 209–210
  corporations and, 201–202
  crime and, 124
  divorce rates and, 230
  feminist theories on, 209, 210–211
  functionalist perspective on, 208–209
  gender and, 170
  global economic systems, 199–201
  of India, 152
  mixed economies, 200–201
  social institutions and, 114
  social mobility and, 148
  symbolic interactionist perspective on, 209,
    211
Ecosystem, 306
Edge cities, 302
Education. *See also* College education; Schools
  Advanced Placement (AP) courses, 251
  African Americans and, 190
  American Indians and, 192
  Asian Americans and, 191
  cheating, 259–260
  conflict theory on, 249, 250–253
  credentialism, 252
  defined, 247
  Democrats/Republicans on, 216
  divorce rates and, 230
  dropping out, 258–259
  feminist theories on, 249, 253–254
  functionalist perspective on, 247–250
  gender and, 160–161, 162
  grade inflation, 259
  health and, 277
  income and, 248, 249
  labeling in, 255
  latent functions of, 248–249
  manifest functions of, 247–248
  Middle Eastern Americans and, 193
  prestige and, 140
  privilege and, 252
  quality and quantity of, 256–257
  race and ethnicity, dropout rates by, 258
  SES (socioeconomic status) and, 250–251
  social class and, 250–251
  social control and, 251–252
  social institutions and, 114
  social mobility and, 149
  student engagement, assessment of, 255–256
  symbolic interactionist perspective on, 249,
    254–256
  teacher effectiveness and, 257–258
  tracking, 255
EEOC (Equal Employment Opportunity
  Commission)
  on pregnancy discrimination, 163
  on sexual harassment, 163
Efficiency
  McDonaldization of society and, 109
  as value, 45
Egalitarian family system, 228
Egypt, Cairo, 300
Elderly persons. *See* Older persons
Elderly support ratio, 241–242
Elective offices, women in, 163
Electronic surveys, 27
Email, miscommunication and, 96–97
Emergence of social movements, 324
Emergency medical technicians, earnings of, 288
Emigrants, 295
Emoticons, 97
Emotional health, abortion and, 167–168
Emotional labor, 89
Emotional support
  and family, 226
  religion providing, 266
Empirical study of society, 7
Employer and employee. *See also* Bureaucracies;
  Occupations and work

disengaged workers, 108
  guest workers, 182
  health insurance, employer-based, 282
Empty nest, 77–78
Endogamy, 226
English, languages and, 44
Enron scandal, 123
Environmental issues, 305–311
  air pollution, 308
  disease adn, 274–275
  global warming, 309–310
  sustainable development, 310–311
  toilet paper, 310
  water, 306–308
EPA (Environmental Protection Agency), 308
  sustainable development and, 310
Epidemiology, 274
Equal Employment Opportunity Commission
  (EEOC). *See* EEOC (Equal Employment
  Opportunity Commission)
Equality/inequality. *See also* Economic inequality;
  Gender inequality; Social inequality
  democracy and, 212
  global inequality, 150–151
  sexism, 159–160
  as value, 45
Equal Rights Party, 163
ESL (English as a second language), 184
Ethics
  social research and, 34–36
  technology and, 330
Ethnicity. *See* Race and ethnicity
Ethnoburbs, 303
Ethnocentrism, 49–50, 186
Ethnologies, 30
Ethnomethodology, 88
Euphemisms, 86
European Americans, 187–188
Evaluation research, 34
Evian, 307
Exchange theory, 244–245
Exogamy, 226
Experimental group, 32
Experiments, 32–34
Explanatory understanding, 10–11
Expressive crowds, 320
Expressive social movements, 321–322
Extended family, 227
  race and ethnicity and, 235–236
External pressures and culture, 56
Exurbs, 302
Eye contact, 92

# F

Facebook, 22, 94, 96, 97
Facial expressions, 92
Facts
  and common sense, 2
  nonmaterial facts, 8
Fads, 318
False consciousness, religion as, 268
Familism, 235–236
Family, 224–245. *See also* Divorce; Domestic violence; Extended family; Parenting; Same-sex
  marriage/families
  authority in, 227–228
  conflict theory on, 242, 243
  defined, 225
  Democrats/Republicans on, 216
  differences in, 226–229
  diversity in, 234–236
  economic security and, 226
  feminist theories on, 242, 243–244
  fictive kin, 236
  functionalist perspective on, 241–243
  functions of, 225–226
  gay and lesbian families, 236
  gender and, 160, 243–244
  multigenerational families, 240–241

nuclear family, 227
  online interaction and, 95–96
  personal space and, 94
  poverty and, 146, 235
  power in, 227–228
  as primary group, 100
  race and ethnicity and, 235–236
  residence patterns, 227
  scholarly culture, 77
  social class and, 234–235, 243
  as social institution, 114
  socialization and, 70–72, 226
  social mobility and, 149
  stepfamilies, 231
  symbolic interactionist perspective on, 242,
    244–245
Farming. *See* Agriculture
Fashion
  collective behavior and, 317–318
  and health, 287
Fast-food restaurants
  culture and, 54
  McDonaldization of society, 109
Fathers, stay-at-home, 234
FBI (Federal Bureau of Investigation)
  subcultures and, 51
  Uniform Crime Report (UCR), 119–120
FDA (Food and Drug Administration), 286
Fear and panic, 316–317
Feelings and facial expressions, 92
Felling rules, 89
Female genital mutilation/cutting (FGM/C), 175
Female infanticide, 297
*The Feminine Mystique* (Friedan), 324
Feminist theories, 15–17, 172, 174–176
  on crime, 123, 128–130
  culture and, 57, 58–59
  on economy and work, 209, 210–211
  on education, 249, 253–254
  evaluation of, 17
  on family and aging, 242, 243–244
  of health care and medicine, 285,
    287–289
  of politics, 220, 222–223
  on racial-ethnic inequality, 194, 195
  on religion, 267, 269–270
  on sexuality, 172, 174–176
  on social groups, 113
  social interaction and, 87, 89–90
  on social stratification, 152, 154–155
  on urbanization, 304–305
Feminist theories on, 123, 128–130
Feminization of poverty, 146, 154
Fertility, population and, 294
Fetal origins, 64
Fictive kin, 236
Field research, 29–30
Flag of United States, 41
Flexible roles, 84
Folk religions, 262
Folkways, 46–47
Foodborne illnesses, 275
Forced secularization, 265
Formal behaviors, 84
Formal deviance, 118
Formal organizations, 104–110
  bureaucracies, 105–106
  characteristics of, 104
  confidence of Americans in, 113
  voluntary associations, 105
Formal social controls an crime, 122
Fossil fuels, 308
Foster families, 225
Fragmentation of social movement, 325
Freedom as value, 45
Freeganism, 327
Friends, 72–73
  friends with benefits (FWB), 166
  gender and, 8–9
  peer groups, 72–73
Front stage behaviors, 88

Votes and voting (*continued*)
    gender and, 164
    marital status and, 217
    religion and, 218
    situational and structural factors, 218–219
    social class and, 217–218
    suffrage, 212

## W

Wage gap and gender, 162, 207–208, 210
Wages. *See* Salaries and wages
Walmart, 114, 311
Walt Disney Corporation, 200
Washington Redskins, 185
WASPs (white Anglo-Saxon Protestants), 14–15, 187–188

Water, 306–308
    contamination of, 306
    mismanagement of, 308
    pollution, 307
    privatization of, 307–308
Wealth, 139–140
    disparity in, 155
    gender gaps in, 154
White-collar crime, 126, 127
WHO (World Health Organization), 23
Widowhood, 79
*The Wire*, 4
Witch camps in Ghana, 119
Women. *See* Gender
Work. *See* Occupations and work
Working class, 143–144
Working poor, 144
Work teams, 110

World Bank, 150
World-system theory, 151
World Trade Center attacks. *See* 9/11 attacks

## Y

YouTube, 94, 96
Y2K rumor, 315
Yugoslavia, genocide in, 184

## Z

Zero population growth (ZPG), 299

# WHAT'S NEW IN SOC3?

*SOC3* differs from previous editions in several ways:

1. There is a new chapter on health and medicine (Chapter 14);
2. Four previous chapters have been combined into two: the economy and politics (Chapter 11), and education and religion (Chapter 13);
3. Each chapter has more material on cross-cultural variations;
4. A new feature, *Sociology in Your Life,* asks students to apply and think critically about sociological concepts, compare themselves with national attitudinal and behavioral surveys, and/or test their knowledge about topics such as immigration, poverty, aging, and gender roles.

I've revised the faculty "Prep Cards" and the student "In Review" cards. The latter now also include several short-essay questions to help students think critically, compare sociological theories, and prepare for tests.

Throughout *SOC3*, I've thoroughly updated all of the statistics, discussions of controversial issues (e.g., whether marijuana and same-sex marriage should be legalized in all states), the theoretical and empirical literature, and new examples to which students can relate (e.g., fashions, social networking, and worrying about finding a good job after graduation). There are almost 1,000 new references, and most are from 2011 or later.

**Unless noted as "new," other revisions, summarized below, note much of the updated and expanded coverage for each chapter.**

## Chapter 1: Thinking Like a Sociologist

- U.S. suicide rates by sex and age, the importance of symbols
- New: figure and discussion on women's and men's rights worldwide

## Chapter 2: Examining Our Social World

- Recent data on school bullying, new research on sexual orientation, and ethical issues
- New: emerging online research issues, and a new box on whether Facebook causes divorces

# WHAT'S NEW IN **SOC3**:

## Chapter 3: Culture

- How laws don't keep up with attitudes on issues such as marijuana and gambling; countercultures
- New: a box on translating federal gibberish into plain English; how technology has changed worldwide attitudes about slimness; Ghanaian funerals in the United States

## Chapter 4: Socialization

- Academic achievement and social class; boomerang children
- New: emerging field of fetal origins; Supreme Court's approving ultra-violent video games for adolescents; resocialization after deployment to Iraq and Afghanistan

## Chapter 5: Social Interaction and Social Structure

- Increased Internet and cell phone usage (variations by sex, age, race, ethnicity, and social class); social networking; cyberbullying and gossip; Facebook and privacy; degradation ceremonies
- New: airplane rage; racialized feeling rules in the workplace; "family scholarly culture" and its effects on children

## Chapter 6: Social Groups, Organizations, and Social Institutions

- Internet participation in groups; volunteering; disengaged and underemployed workers; employee–owned companies; electronic surveillance of workers; organizations' incompetence, waste, and corruption
- New: dramatic growth of corporate power; Americans' lack of confidence in organizations; recent efforts to curb obesity

## Chapter 7: Deviance, Crime, and the Criminal Justice System

- Crime (rates, offenders, and victims); corporate crimes and cybercrimes; incarceration and death penalties ; feminist explanations of deviance and crime; imprisonment rates by race, ethnicity, and sex
- New: FBI's revised definition of rape; employer stigmas against long-term unemployed; upper-class crime; labeling by American Psychiatric Association; economic costs of major crimes

## Chapter 8: Social Stratification: United States and Global

- Greater U.S. and global income and wealth inequality; how social class affects us; castes in India; how capitalism benefits the rich; recent effects of patriarchy; corporate welfare; symbolic interaction explanations of stratification
- New: a box on U.S. economic inequality; box on five ways the poor pay more; effect of Great Recession on poverty, homelessness, sex, age, race/ethnicity, and social class; recent trends in social mobility

## Chapter 9: Gender and Sexuality

- Cross-cultural variations in gender roles, sexual orientation, and sexuality; inequality in the workplace, childcare/ housework, education, and politics; the gender pay gap; gender stereotypes; sexism; homophobia; legality of same-sex marriages

# WHAT'S NEW IN **SOC3**:

- New: a box on gender and sexuality; the recent "sexy babes" trend and sexual scripts; figure and discussion of the gender pay gap; recent state efforts to limit or abolish abortion; new table and discussion of women in legislatures; "staying over"

## Chapter 10: Race and Ethnicity

- Americans' reactions to recent immigrants; recent racism and racial stereotypes; variations in the diversity and constraints of U.S. racial/ethnic groups; gendered racism; changes in interracial/interethnic relationships

- New: a box on U.S. ethnic/racial groups; recent U.S. Supreme Court ruling of some states' strict laws against undocumented immigrants; racial stereotyping on TV shows

## Chapter 11: The Economy and Politics

- Demise of unions and growth of offshoring; rise of low-wage jobs, unemployment, part-time work, and underemployment; increase of pork spending; growth of political inequality, globalization, and deindustrialization; political freedom; how Republicans and Democrats differ

- New: effect of Great Recession on workers and CEOs; increasing U.S. economic hardship; many Americans' disillusionment with politics; growth of the elite's power to shape political institutions; decrease of American women's influence on politics

## Chapter 12: Families and Aging

- Divorce rates, singlehood, postponing marriage, cohabitation, and nonmarital childbearing; growth of two-income families; stay-at-home dads; family violence and maltreatment; U.S. and global aging population and life expectancy

- New: social class and racial/ethnic family variations; discussion and figure of age dependency ratio; discussion and figure of experiencing intimate partner violence; caring for our graying population; the costs of paying for the elderly

## Chapter 13: Education and Religion

- U.S. educational attainment (by sex, race, ethnicity, and social class); gender inequality in STEM fields; student engagement, cheating, plagiarism, and grade inflation; low U.S. standardized test scores; religious beliefs and participation; secularization

- New: a box and material on who studies the least in college; the recent controversy on the value of a college degree; student debt; the "feminization of higher education;" gender gaps in education; relationship between politics and religion

## Chapter 14: Health and Medicine (New)

- The major topics are U.S. health and illness, comparisons of health care in the U.S. and around the world, and the four sociological perspectives on health and medicine. Some subtopics include epidemiology; reasons for contemporary illness and early death by age, sex, race, ethnicity, and social class; the sick role and physicians as gatekeepers; social inequality in, control of, and gender stratification in medicine and health care; and the medicalization of illness.

# 4

# WHAT'S NEW IN **SOC3**:

## Chapter 15: Population, Urbanization, and the Environment

- Discussion, figures, and tables on world urbanization trends, megacities, and the consequences of U.S. urbanization and suburbanization; the availability and consumption of water; the sources and effects of air pollution, global warming, climate change, and the greenhouse effect; sustainable development

- New: table on the world's largest countries; material on residential racial segregation and "ethnoburbs;" "global neighborhoods;" table and discussion of how much water it takes to make products, such as blue jeans; table on percentage of Americans who favor economic growth over protecting the environment

## Chapter 16: Social Change: Collective Behavior, Social Movements, and Technology

- Material on fads, fashion, riots, and recent innovations in technology

- New: discussion of recent social movements (e.g., Tea Party and Occupy Wall Street)

# Thinking Like a Sociologist

## CHAPTER 1 TOPICS

### 1 What Is Sociology?

*Sociology* is the systematic study of social interaction at a variety of levels. Sociologists use scientific research to discover patterns and create theories about who we are, how we interact with others, and why we do what we do. Sociology goes beyond common sense and conventional wisdom in understanding our social world, including small groups (e.g., families and friends), large organizations and institutions (e.g., your college), and entire societies (e.g., the United States).

### 2 What Is a Sociological Imagination?

The *sociological imagination*, which emphasizes the intersection between individual lives and external social influences, relies on both micro-level and macro-level approaches in examining the social world. *Microsociology* concentrates on the relationships between individuals, whereas *macrosociology* examines social dynamics across the breadth of a society. Macro-level systems shape society, often limiting our personal options on the micro level.

### 3 Why Study Sociology?

Regardless of your major, this course will help you (1) make more informed decisions, (2) understand diversity, (3) increase your input in shaping social policies and practices, (4) think critically, and (5) expand your career opportunities.

### 4 Some Origins of Sociological Theory

Sociologists use *theories* to explain why a phenomenon occurs among people, institutions, and societies. Theories produce knowledge, but can also offer solutions to everyday social problems. Some of the most influential theorists have included Auguste Comte, Harriet Martineau, Émile Durkheim, Karl Marx, Max Weber, Jane Addams, and W. E. B. Du Bois. Each brought to sociology a new level of understanding about our world.

### 5 Contemporary Sociological Theories

Sociologists typically use more than one theory to explain human behavior. The fullest understanding of society comes from using all four of these theories:

- *Functionalism explains society as interconnected social systems.* Critics contend that functionalism ignores social inequality and social conflict.

- *Conflict theory sees disagreement and the resulting changes in society as natural, inevitable, and even desirable.* Critics argue that conflict theory ignores the importance of harmony and cooperation.

- *Feminist theories, which build on conflict theory, maintain that sex inequality is central to all conflict.* Critics claim that these theories are too narrowly focused.

- *Symbolic interaction focuses on the meanings of micro-level interactions.* Critics maintain that this theoretical perspective overlooks the impact of macro-level factors on our everyday behavior.

**Example**: Critical Thinking versus Common Sense and Conventional Wisdom

When thinking critically, it's important to differentiate between common sense myths and facts. Here are a few examples:

**Myth**: Older people make up the largest group of those who are poor.
**Fact**: Children younger than 6, not older people, make up the largest group of those who are poor.
**Myth**: Divorce rates are higher today than ever before.
**Fact**: Divorce rates are lower today than they were between 1980 and 1990.

Now, based on the material you read in this chapter, construct your own "myth" and "fact."

## KEY TERMS

**sociology** the systematic study of social interaction at a variety of levels.

**sociological imagination** the ability to see the relationship between individual experiences and larger social influences.

**microsociology** a sociological approach that examines the patterns of individuals' social interaction in specific settings.

**macrosociology** the study of large-scale patterns and processes that characterize society as a whole.

**theory** a set of statements that explains why a phenomenon occurs.

**empirical** information that is based on observations, experiments, or other data collection rather than on ideology, religion, intuition, or conventional wisdom.

**social facts** aspects of social life, external to the individual, that can be measured.

**social solidarity** social cohesiveness and harmony.

**division of labor** an interdependence of different tasks and occupations, characteristic of industrialized societies, that produce social unity and facilitate change.

**capitalism** an economic system in which the ownership of the means of production—such as land, factories, large sums of money, and machines—is in private hands.

**alienation** the feeling of separation from one's group or society.

**value free** separating one's personal values, opinions, ideology, and beliefs from scientific research.

**functionalism (structural functionalism)** an approach that maintains that society is a complex system of interdependent parts that work together to ensure a society's survival.

**dysfunctions** social patterns that have a negative impact on a group or society.

**manifest functions** purposes and activities that are intended and recognized; they are present and clearly evident.

**latent functions** purposes and activities that are unintended and unrecognized; they are present but not immediately obvious.

**conflict theory** an approach that examines how and why groups disagree, struggle over power, and compete for scarce resources (such as property, wealth, and prestige).

**feminist theories** approaches that examine and seek to explain the social, economic, and political inequality of women in society.

**symbolic interactionism (interactionism)** a micro-level perspective that examines individuals' everyday behavior through the communication of knowledge, ideas, beliefs, and attitudes.

**social interaction** a process in which people take each other into account in their own behavior.

# TEST YOUR LEARNING

1. ____ looks at the relationship between individual characteristics; ____ examines the relationships between institutional characteristics.
   a. Microsociology; macrosociology
   b. Macrosociology; microsociology
   c. Metasociology; macrosociology
   d. Metasociology; microsociology

2. Which social class, as identified by Karl Marx, includes the ruling elite who own the means of production?
   a. Capitalists
   b. Communists
   c. Power elite
   d. Proletariat

3. James sees Julie laughing in the hallway with a friend and assumes that Julie is feeling happy. James is using Weber's
   a. explanatory understanding.
   b. surveillance understanding.
   c. common understanding.
   d. direct observational understanding.

4. Jeremy views society as a system of interrelated parts, but Thomas sees society as composed of groups competing for scarce resources. Jeremy would be considered a ____ theorist, and Thomas would be seen as a ____ theorist.
   a. symbolic interactionist; functionalist
   b. conflict; functionalist
   c. functionalist; symbolic interactionist
   d. functionalist; conflict

5. Many people buy designer clothes that they can't afford. The clothes are an example of a status symbol that reflects a
   a. latent function.
   b. manifest function.
   c. dysfunction.
   d. social system.

6. *True or False* In the definition of sociology, "systematic" means behavior that is built into the larger social structure of society.

7. *True or False* Émile Durkheim saw sociology as the scientific study of two aspects of society: social statics and social dynamics.

8. *True or False* Jane Addams was an early sociologist who published extensively on topics such as social disorganization, immigration, and urban neighborhoods.

9. *True or False* Much of contemporary functionalism grew out of the work of Auguste Comte and Émile Durkheim.

10. *True or False* Conflict theorists see society as cooperative and harmonious.

11. What does C. Wright Mills mean when he says that there's a connection between personal troubles and structural issues? Use an example, but not unemployment, to explain and illustrate the sociological imagination concept.

12. Suppose one of your friends is considering getting an abortion and seeks your advice. Would functionalist, conflict, feminist, and interactionists perspectives be useful in your response? Explain why or why not.

1. a  2. a  3. d  4. d  5. a  6. False  7. False  8. True  9. True  10. False

## TABLE 1.2

# Leading Contemporary Perspectives in Sociology

| THEORETICAL PERSPECTIVE | FUNCTIONALIST | CONFLICT | FEMINIST | SYMBOLIC INTERACTIONIST |
|---|---|---|---|---|
| Level of Analysis | Macro | Macro | Macro and Micro | Micro |
| Key Points | • Society is composed of interrelated, mutually dependent parts.<br>• Structures and functions maintain a society's or group's stability, cohesion, and continuity.<br>• Dysfunctional activities that threaten a society's or group's survival are controlled or eliminated. | • Life is a continuous struggle between the "haves" and the "have-nots."<br>• People compete for limited resources that are controlled by a small number of powerful groups.<br>• Society is based on inequality in terms of ethnicity, race, social class, and sex. | • Women experience widespread inequality in society because, as a group, they have little power.<br>• Sex, ethnicity, race, age, sexual orientation, and social class—rather than a person's intelligence and ability—explain many of our social interactions and lack of access to resources.<br>• Social change is possible only if we change our institutional structures and our day-to-day interactions. | • People act on the basis of the meaning they attribute to others. Meaning grows out of the social interaction that we have with others.<br>• People continuously reinterpret and reevaluate their knowledge and information in their everyday encounters. |

For full table, see Table 1.1 on page 19.

*For practice tests, printable flash cards, and more, visit 4ltrpress.cengage.com/soc.*

# Examining Our Social World

## CHAPTER 2 TOPICS

### 1 Doing Sociology: What Is Social Research?

*Social research* requires curiosity and imagination, but also an understanding of the rules and procedures that govern careful scientific study. The process involves choosing a socially relevant topic, asking a research question, developing and testing a hypothesis, and analyzing the findings. In contrast, many opinions in self-help publications often ignore the scientific method.

### 2 Why Is Sociological Research Important in Our Everyday Lives?

Much of our knowledge is based on tradition and authority. In contrast, sociological research creates new knowledge that helps us understand social life, exposes myths, affects social policies, sharpens our critical thinking skills, and helps us make informed decisions about our everyday lives.

### 3 The Scientific Method

The *scientific method* incorporates careful data collection, exact measurement, accurate recording and analysis of findings, thoughtful interpretation of results, and, when appropriate, a generalization of the findings to a larger group. A research question or a *hypothesis* examines the association between an *independent variable* and a *dependent variable*. Sociologists use both *qualitative* and *quantitative* approaches, and are always concerned about the *reliability* and *validity* of their measures.

### 4 Some Major Data Collection Methods

Six data collection methods are especially common in sociology (see Table 2.2 on the next page). In designing their studies, sociologists weigh the advantages and limitations of each data collection method. Because sociologists don't conduct research in a cultural vacuum, many groups use the findings to change current policies and practices.

**Example**: The Uneasy Relationship between Research and Practice

Many supporters of the DARE (Drug Abuse Resistance Education) program were unhappy when more than 30 research studies showed that DARE had negligible long-term effects in reducing teen drug use. Nonetheless, about 75 percent of U.S. school districts used the findings to revise the DARE curriculum. Many communities altered but continued DARE because they believed that the program built a positive relationship between police, students, parents, and educators (Berman and Fox 2009).

### 5 Ethics, Politics, and Sociological Research

Sociological research demands a strict code of ethics to avoid mistreating participants. For example, participants must give informed consent and must not be harmed, humiliated, abused, or coerced; researchers must honor their guarantees of privacy, confidentiality, and/or anonymity. Still, sociologists often encounter pressure from policy makers and others to limit their research to topics that won't stir controversy on sensitive issues.

## KEY TERMS

**social research** systematic study of human behavior.

**scientific method** a research process that includes careful data collection, exact measurement, accurate recording and analysis of the findings, thoughtful interpretation of results, and, when appropriate, a generalization of the findings to a larger group.

**concept** an abstract idea, mental image, or general notion that represents some aspect of our social life.

**variable** a characteristic that can change in value or magnitude under different conditions.

**independent variable** a characteristic that has an effect on the dependent variable.

**dependent variable** the outcome, which may be affected by the independent variable.

**control variable** a characteristic that is constant and unchanged during the research process.

**hypothesis** a statement of the expected relationship between two or more variables.

**reliability** the consistency with which the same measure produces similar results time after time.

**validity** the degree to which a measure is accurate and really measures what it claims to measure.

**deductive reasoning** an inquiry process that begins with a theory, prediction, or general principle that is then tested through data collection.

**inductive reasoning** an inquiry process that begins with a specific observation, followed by data collection, a conclusion about patterns or regularities, and the formulation of hypotheses that can lead to theory construction.

**population** any well-defined group of people (or things) about whom researchers want to know something.

**sample** a group of people (or things) that is representative of the population that researchers wish to study.

**probability sample** a sample in which each person has an equal chance of being selected because the selection is random.

**nonprobability sample** a sample for which there is little or no attempt to get a representative cross section of the population.

**qualitative research** research that examines nonnumerical material and interprets it.

**quantitative research** research that focuses on a numerical analysis of people's responses or specific characteristics.

**surveys** a systematic method for collecting data from respondents, including questionnaires, face-to-face or telephone interviews, or a combination.

**secondary analysis** examination of data that have been collected by someone else.

**field research** data collected by systematically observing people in their natural surroundings.

**content analysis** a data collection method that systematically examines some form of communication.

**experiment** a carefully controlled artificial situation that allows researchers to manipulate variables and measure the effects.

**experimental group** the group of participants in an experiment who are exposed to the independent variable.

**control group** the group of participants in an experiment who are not exposed to the independent variable.

**evaluation** research data collection method that uses all of the standard data collection methods to assess the effectiveness of social programs in both the public and private sectors.

# TEST YOUR LEARNING

1. Sociological research is important in our daily lives for a number of reasons. Which of the following is NOT one of those reasons?
   a. It exposes myths.
   b. It challenges the findings of psychology researchers.
   c. It affects social policy.
   d. It sharpens our critical thinking skills.

2. What is a hypothesis?
   a. A theory based on one's opinions.
   b. A measure of how accurate a study is.
   c. A statement of a relationship between two or more variables.
   d. An analysis of multiple studies that researchers use to make generalizations.

3. Alexandra notices that she performs best on tests that are given in the afternoon. She then begins to collect data of her test grades and asks close friends and classmates to do the same. Alexandra is using
   a. intuition.
   b. inferences.
   c. deductive reasoning.
   d. inductive reasoning.

4. Which of the following data collection methods is the most likely to suggest a cause-and-effect relationship?
   a. Experiment
   b. Survey
   c. Field research
   d. Content analysis

5. There are three golden rules in sociological research. Which of the following is NOT one of them?
   a. Do no harm.
   b. Be fully transparent with the subject.
   c. Get the subject's informed consent.
   d. Protect the subject's confidentiality.

6. **True or False** Social researchers continuously challenge the quality of existing studies.

7. **True or False** If researchers use a nonprobability sample, they can generalize the results to a larger population.

8. **True or False** Validity is the consistency that a measure produces similar results over time.

9. **True or False** Observing second graders interact with one another in their classroom is an example of field research.

10. **True or False** Correlation equals causation.

11. State a hypothesis about your GPA, the dependent variable. What three independent variables might you use? Explain why.

12. Suppose you want to examine the effects of student employment on class attendance. Which data collection method(s) would you use? Why?

1. b   2. c   3. d   4. a   5. b   6. True   7. False   8. False   9. True   10. False

## TABLE 2.2

# Some Data Collection Methods in Sociological Research

| METHOD | ADVANTAGES | DISADVANTAGES |
|---|---|---|
| Surveys | Questionnaires are fairly inexpensive and simple to administer; interviews have high response rates; findings are often generalizable | Mailed questionnaires may have low response rates; respondents tend to be self-selected; interviews are usually expensive |
| Secondary analysis | Usually accessible, convenient and inexpensive; often longitudinal and historical | Information may be incomplete; some documents may be inaccessible; some data can't be collected over time |
| Field research | Flexible; offers deeper understanding of social behavior; usually inexpensive | Difficult to quantify and to maintain observer/subject boundaries; the observer may be biased or judgmental; findings are not generalizable |
| Content analysis | Usually inexpensive; can recode errors easily; unobtrusive; permits comparisons over time | Can be labor intensive; coding is often subjective (and may be distorted); may reflect social class bias |
| Experiments | Usually inexpensive; plentiful supply of subjects; can be replicated | Subjects aren't representative of a larger population; the laboratory setting is artificial; findings can't be generalized |
| Evaluation research | Usually inexpensive; valuable in real-life applications | Often political; findings might be rejected |

For full table, see Table 2.2 on page 35.

For practice tests, printable flash cards, and more, visit 4ltrpress.cengage.com/soc.

# Culture

## CHAPTER 3 TOPICS

### 1 Culture and Society

*Culture* is learned, transmitted from one generation to another, adaptive, and always changing. A *society* shares a culture and sees itself as a social unit. People construct a *material culture* (such as buildings) and *nonmaterial culture* (such as rules for behavior) that influence each other (such as forbidding smoking in public buildings).

### 2 The Building Blocks of Culture

The following are some of the fundamental building blocks of culture:

- *Symbols* take many forms, can change over time, can unify or divide a society, and can affect cross-cultural views.

- *Language* can change over time; it can affect perceptions about sex, race, class, and ethnicity.

- *Values* provide general guidelines for behavior; they are usually emotion laden, vary across cultures, and change over time.

- *Norms*—whether folkways, mores, or laws—regulate our behavior; they vary across cultures and are subject to sanctions ranging from mild to severe.

**Example**: Sanctions for Violating the Dead

Sanctions are more severe for violating laws than folkways. Legacy.com, which carries a death notice or obituary for virtually all of the roughly 2.4 million Americans who die each year, dedicates at least 30 percent of its budget to weeding out comments (a relatively mild punishment) that "diss the dead" (Urbina 2006). In contrast, in many states, vandalizing a tombstone, a property crime, can result in a fine of up to $1,000, a jail sentence of up to a year, or both.

### 3 Some Cultural Similarities

Although many cultural characteristics vary across countries, *cultural universals* are common to all societies, such as some form of food taboo.

People who encounter an unfamiliar way of life or environment may experience *culture shock*.

### 4 Some Cultural Variations

*Subcultures* and *countercultures* account for some of the complexity within a society. The former differ from the larger society in some ways, whereas the latter oppose or reject some of the dominant culture's basic beliefs, values, and norms. *Ethnocentrism* has its benefits but can also lead to conflict and discrimination. *Cultural relativism*, the opposite of ethnocentrism, encourages cross-cultural understanding and respect. In *multiculturalism*, many cultures coexist without trying to dominate one another.

### 5 Popular Culture

*Popular culture*, which is widely shared among a population, includes television, music, radio, advertising, sports, hobbies, fads, fashions, and movies, as well as the food we eat, the people with whom we spend time, the gossip we share, and the jokes we pass along. Popular culture is typically spread through *mass media*, including television and the Internet, and has enormous power in shaping our perceptions and opinions.

### 6 Cultural Change and Technology

Some societies are relatively stable because of *cultural integration*, but all societies change over time because of diffusion, innovation and invention, discovery, external pressures, and changes in the physical environment. A *cultural lag* occurs when a culture's material side changes more rapidly than its nonmaterial side.

### 7 Sociological Perspectives on Culture

See Table 3.3 on the next page.

## KEY TERMS

**culture** the learned and shared behaviors, beliefs, attitudes, values, and material objects that characterize a particular group or society.

**society** a group of people who have lived and worked together long enough to become an organized population and to think of themselves as a social unit.

**material culture** the tangible objects that members of a society make, use, and share.

**nonmaterial culture** the shared set of meanings that people in a society use to interpret and understand the world.

**symbol** anything that stands for something else and has a particular meaning for people who share a culture.

**language** a system of shared symbols that enables people to communicate with one another.

**values** the standards by which members of a particular culture define what is good or bad, moral or immoral, proper or improper, desirable or undesirable, beautiful or ugly.

**norms** a society's specific rules concerning right and wrong behavior.

**folkways** norms that members of a society (or a group within a society) see as not being critical for a society's survival and that, consequently, are not severely punished when violated.

**mores** norms that members of a society consider very important because they maintain moral and ethical behavior.

**taboos** strong prohibitions of any act that is considered to be extremely offensive and forbidden because of social customs, religious or moral beliefs, or laws.

**laws** formal rules for behavior that are defined by a political authority that has the power to punish violators.

**sanctions** rewards for good or appropriate behavior and/or penalties for bad or inappropriate behavior.

**cultural universals** customs and practices that are common to all societies.

**ideal culture** the beliefs, values, and norms that people in a society say they hold or follow.

**real culture** the actual everyday behavior of people in a society.

**ethnocentrism** the belief that one's culture and way of life are superior to those of other groups.

**cultural relativism** the belief that no culture is better than another and that a culture should be judged by its own standards.

**subculture** a group or category of people whose distinctive ways of thinking, feeling, and acting differ somewhat from those of the larger society.

**counterculture** a group of people who deliberately oppose and consciously reject some of the basic beliefs, values, and norms of the dominant culture.

**multiculturalism (cultural pluralism)** the coexistence of several cultures in the same geographic area, without one culture dominating another.

**culture shock** a sense of confusion, uncertainty, disorientation, or anxiety that accompanies exposure to an unfamiliar way of life or environment.

**popular culture** the beliefs, practices, activities, and products that are widely shared among a population in everyday life.

**mass media** forms of communication designed to reach large numbers of people.

**cultural imperialism** the cultural values and products of one society influencing or dominating those of another.

**cultural integration** the consistency of various aspects of society that promotes order and stability.

**cultural lag** the gap when nonmaterial culture changes more slowly than material culture.

# TEST YOUR LEARNING

1. Society and culture are mutually
   a. dependent.
   b. exclusive.
   c. destructive.
   d. diversified.

2. _____ stand for something else and have a particular meaning for people who share a culture.
   a. Mores
   b. Symbols
   c. Values
   d. Norms

3. At a party, George was eating nachos and salsa. He took a bite of his chip and then dipped what was left of his chip back into the large bowl of salsa. His friend Brett gave him a disgusted look. Which of the following did George break?
   a. Mores
   b. A taboo
   c. A folkway
   d. A law

4. Which of the following is an example of a subculture?
   a. A group of adolescent Goths
   b. Roman Catholics
   c. A 40-and-over women's bridge club
   d. All of the above

5. Anaz is an 8-year-old Iranian girl who loves the female fashions she sees in American films and television shows. Lately she's been questioning why her mother wears a burka. Anaz's questioning of her family's traditions could best be explained by
   a. cultural relativism.
   b. multiculturalism.
   c. cultural imperialism.
   d. ethnocentrism.

6. **True or False** Culture is adapting and always changing.

7. **True or False** Language helps communicate ideas, but it's not considered a symbol.

8. **True or False** In most societies, the real culture matches the ideal culture.

9. **True or False** The legal controversy over file sharing and downloading music off the Web is an example of cultural lag.

10. **True or False** Symbolic interactionists rely on a macro approach in examining culture.

11. List five example of material culture and five of nonmaterial culture that affect you. Explain (a) how they intersect and (a) how they shape your attitudes and behavior.

12. Provide an example of each type of norm (folkways, mores, and laws) that you or someone you know has violated. Describe the negative sanctions for each violation.

1.a  2.b  3.c  4.d  5.c  6.True  7.False  8.False  9.True  10.False

## TABLE 3.3

# Sociological Explanations of Culture

| THEORETICAL PERSPECTIVE | FUNCTIONALIST | CONFLICT | FEMINIST | SYMBOLIC INTERACTIONIST |
|---|---|---|---|---|
| **Level of Analysis** | Macro | Macro | Macro and Micro | Micro |
| **Key Points** | • Similar beliefs bind people together and create stability.<br><br>• Sharing core values unifies a society and promotes cultural solidarity. | • Culture benefits some groups at the expense of others.<br><br>• As powerful economic monopolies increase worldwide, the rich get richer and the rest of us get poorer. | • Women and men often experience culture differently.<br><br>• Cultural values and norms can increase inequality because of sex, race/ethnicity, and social class. | • Cultural symbols forge identities (that change over time).<br><br>• Culture (such as norms and values) helps people merge into a society despite their differences. |

For full table, see Table 3.3 on page 57.

*For practice tests, printable flash cards, and more, visit 4ltrpress.cengage.com/soc.*

# Socialization

## CHAPTER 4 TOPICS

### 1 Socialization: Its Purpose and Importance

*Socialization*, a lifelong process, teaches us to be human. Socialization fulfills four key purposes: It establishes our social identity, teaches us role taking, controls our behavior (through *internalization*), and transmits culture to the next generation.

### 2 Nature and Nurture

Biologists tend to focus on the role of heredity (or genetics) in human development. In contrast, most social scientists, including sociologists, underscore the role of learning, socialization, and culture. This difference of opinion is often called the *nature-nurture debate*. *Sociobiologists* argue that genetics (nature) can explain much of our behavior, whereas most sociologists maintain that socialization and culture (nurture) shape even biological inputs.

### 3 Sociological Explanations of Socialization

Sociologists have offered many explanations of socialization, but two of the most influential, both at the micro level, have been social learning and symbolic interaction theories (see Table 4.2 on the next page).

### 4 Primary Socialization Agents

Parents are the first and most important *agents of socialization*, but siblings, grandparents, and other family members also play important roles. Other important socialization agents include play and peer groups, teachers and schools, popular culture, and the media. Advertising is an especially powerful force in socialization.

**Example**: Are Parents Realistic in Socializing Their Children?

Many parents tell their offspring that they are very intelligent. In fact, only about 5 percent of American kids can be considered "gifted" (endowed with significantly higher than average intellectual or other abilities), even though many are enrolled in gifted classes. Some educators argue that telling average—or even above average—children that they're superior does them a disservice: It gives them false expectations on how the world will treat them, encourages being self-centered, and increases anger and unhappiness when they don't succeed in college or the workplace (Deveny 2008).

### 5 Socialization Throughout Life

As we progress through the life course, we learn culturally approved norms, values, and roles. Infants are born with an enormous capacity for learning that parents and other caregivers can enrich and shape. In adolescence, these and other socialization agents teach children how to form relationships on their own, to get along with others, and to develop their social identity through play and peer groups. In adulthood, people must learn new roles that include singlehood, marriage, parenthood, divorce, work, and experiencing the death of a loved one. Socialization continues in later life when many people learn still new roles such as grandparents, retirees, older workers, and being widowed.

### 6 Resocialization

Much of *resocialization*, which can be voluntary or involuntary, takes place in *total institutions*, where people are isolated from the rest of society, stripped of their former identities, and required to conform to new rules and behavior.

## KEY TERMS

**socialization**  the lifelong process of social interaction in which the individual acquires a social identity and ways of thinking, feeling, and acting that are essential for effective participation in a society.

**internalization**  the process of learning cultural behaviors and expectations so deeply that we accept them without question.

**sociobiology**  a theoretical perspective that applies biological principles to explain the behavior of animals, including human beings.

**social learning theories**  maintain that people learn new attitudes, beliefs, and behaviors through social interaction, especially during childhood.

**self**  an awareness of one's social identity.

**looking-glass self**  a self-image based on how we think others see us.

**role taking**  learning to take the perspective of others.

**significant others**  the people who are important in one's life, such as parents (or other primary caregivers), siblings, and grandparents.

**anticipatory socialization**  the process of learning how to perform a role one doesn't yet occupy.

**generalized other**  people who don't have close ties to a child but who influence her or his internalization of society's norms and values.

**impression management**  the process of providing information and cues to others to present oneself in a favorable light while downplaying or concealing one's less appealing qualities.

**reference groups**  groups of people who shape an individual's self-image, behavior, values, and attitudes in different contexts.

**agents of socialization**  the individuals, groups, or institutions that teach us what we need to know to participate effectively in society.

**peer group**  people who are similar in age, social status, and interests.

**resocialization**  the process of unlearning old ways of doing things and adopting new attitudes, values, norms, and behavior.

**total institutions**  places where people are isolated from the rest of society, stripped of their former identities, and required to conform to new rules and behavior.

## TEST YOUR LEARNING

1. Emma learned from her mother that girls can do anything that boys can do in getting a good education. This is an example of the _____ process.
   a. multicultural
   b. feminist
   c. socialization
   d. hereditary

2. "Human development is fairly fixed." This statement is an example of which side of the nature/nurture debate?
   a. Nature side of the debate
   b. Nurture side of the debate
   c. Neither the nature nor nurture side of the debate
   d. Both the nature and nurture side of the debate

3. According to social learning theory, the greatest impact of social interaction occurs during
   a. childhood.
   b. adolescence.
   c. early adulthood.
   d. middle adulthood.

4. What are the three stages of Mead's role-taking theory?
   a. Prework stage, work stage, postwork stage
   b. Anal stage, phallic stage, postphallic stage
   c. Mirror stage, active stage, passive stage
   d. Preparatory stage, play stage, game stage

5. According to Mead, children learn how to perform a role they don't yet occupy. Mead referred to this process as
   a. a developing self.
   b. anticipatory socialization.
   c. role taking.
   d. the generalized other.

6. *True or False* Institutionalization is the process of learning cultural behaviors and expectations so deeply that we assume they are correct and accept them without question.

7. *True or False* Charles Horton Cooley proposed that the looking-glass self develops in five phases.

8. *True or False* According to Erving Goffman, social life mirrors theatrical performance.

9. *True or False* Healthy child development is most likely in authoritative homes.

10. *True or False* Resocialization is the process of reinforcing the established values and beliefs of an individual.

11. Think about your early socialization. Were social learning theories, symbolic interaction theories, or both important in your socialization? Explain and illustrate your answer using specific examples.

12. Have you engaged in impression management? Describe when, where, how, and why.

1. c  2. a  3. a  4. d  5. b  6. False  7. False  8. True  9. True  10. False

---

## TABLE 4.2

# Key Elements of Socialization Theories

| SOCIAL LEARNING THEORIES | SYMBOLIC INTERACTION THEORIES |
| --- | --- |
| • Social interaction is important in learning appropriate and inappropriate behavior. | • The self emerges through social interaction with significant others. |
| • Socialization relies on direct and indirect reinforcement. | • Socialization includes role taking and controlling the impression we give to others. |
| *Example:* Children learn how to behave when they are scolded or praised for specific behaviors. | *Example:* Children who are praised are more likely to develop a strong self-image than those who are always criticized. |

For practice tests, printable flash cards, and more, visit 4ltrpress.cengage.com/soc.

# Social Interaction and Social Structure

## CHAPTER 5 TOPICS

### 1 Social Structure

*Social interaction*, central to all social activity, affects people's behavior. Our interaction is part of the *social structure*, which guides our actions and gives us a feeling that life is orderly and predictable. Every society has a social structure that encompasses statuses and roles.

### 2 Status

A *status* is a social position that an individual occupies in a society. Every person has many statuses that form her or his *status set*, which include both ascribed and achieved statuses. An *ascribed status* is a social position that a person is born into and can't control, change, or choose (such as age, race, and being male or female). An *achieved status* is a social position that a person attains through personal effort or assumes voluntarily (such as college student or wife). Because we hold many statuses, some clash and result in *status inconsistency* because we occupy social positions that are ranked differently (such as being a low-paid college professor).

### 3 Role

A *role* defines how we are expected to behave in a particular status, but people vary considerably in fulfilling the responsibilities associated with their roles. These differences reflect *role performance*, the actual behavior of a person who occupies a status. A *role set* encompasses different roles attached to a single status (such as a parent who is a teacher, chauffeur, and PTA member). Playing many roles often leads to *role conflict* because it's difficult to meet the requirements of two or more statuses, and *role strain*, the stress that arises because of incompatible demands among roles within a single status.

Some common ways to resolve role conflict include compromising, negotiating, setting priorities, compartmentalizing, not taking on more roles, and exiting one or more current roles.

**Example**: Exiting a Marriage

Divorce is a good example of role exit, but it often involves a long process of five stages that may last several decades (Bohannon 1971). The "emotional divorce" begins when one or both partners feel disillusioned or unhappy. The "legal divorce" is the formal dissolution of the marriage during which the partner who does not want the divorce may try to stall the end of the marriage. During the "economic divorce" stage, the partners may argue about who should pay past debts, property taxes, and unforeseen expenses (such as moving costs). The "coparental divorce" stage involves parents' agreeing on issues such as child support and visitation rights. During the "community divorce" stage, partners inform friends, family, and others that they are no longer married. Finally, the couple goes through a "psychic divorce" in which the partners separate from each other emotionally. In many cases, one or both spouses never complete this stage because they can't let go of their pain, anger, and resentment—even if they remarry.

### 4 Explaining Social Interaction

See Table 5.2 on next page.

### 5 Nonverbal Communication

Our *nonverbal communication* includes gestures, facial expressions, eye contact, and silence. Touching and how we use space are also important forms of nonverbal communication because they send powerful messages about our feelings and power.

### 6 Online Interaction

Many people interact in *cyberspace*, an online world of computer networks. Internet usage varies by sex, age, ethnicity, and social class. Cyberspace can be impersonal and socially isolating, but it can also save time, foster closer relationships among family members and friends, and facilitate working from home.

## KEY TERMS

**social interaction** the process by which we act toward and react to people around us.

**social structure** an organized pattern of behavior that governs people's relationships.

**status** a social position that a person occupies in a society.

**status set** a collection of social statuses that a person occupies at a given time.

**ascribed status** a social position that a person is born into.

**achieved status** a social position that a person attains through personal effort or assumes voluntarily.

**master status** an ascribed or achieved status that determines a person's identity.

**status inconsistency** the conflict that arises from occupying social positions that are ranked differently.

**role** the behavior expected of a person who has a particular status.

**role performance** the actual behavior of a person who occupies a status.

**role set** the different roles attached to a single status.

**role conflict** the frustrations and uncertainties a person experiences when confronted with the requirements of two or more statuses.

**role strain** the stress that arises from incompatible demands among roles within a single status.

**self-fulfilling prophecy** a situation in which if we define something as real and act on it, it can, in fact, become real.

**ethnomethodology** the study of how people construct and learn to share definitions of reality that make everyday interactions possible.

**dramaturgical analysis** a research approach that examines social interaction as if occurring on a stage where people play different roles and act out scenes for the audiences with whom they interact.

**social exchange theory** the perspective whose fundamental premise is that social interaction is based on each person's trying to maximize rewards (or benefits) and minimize punishments (or costs).

**nonverbal communication** messages that are sent without using words.

# TEST YOUR LEARNING

1. Jenna is the youngest of three girls in her family. Being the youngest sister is an example of a(n)
   a. achieved status.
   b. ascribed status.
   c. problematic status.
   d. interchangeable status.

2. Jake has a full-time course load while also working 30 hours a week. Given the demands of both work and school, Jake might be likely to experience
   a. role set.
   b. role strain.
   c. role conflict.
   d. role exchange.

3. According to ____, social interaction is based on trying to maximize rewards for oneself while minimizing costs.
   a. ethnomethodology theory
   b. feminist theory
   c. conflict theory
   d. social exchange theory

4. Gestures, eye contact, and silence are all examples of
   a. emotional language.
   b. punishment.
   c. healthy relationship interaction.
   d. nonverbal communication.

5. Which one of the following statements is false?
   a. Almost equal numbers of women and men use the Internet.
   b. The higher a family's income, the greater the likelihood that its members are Internet users.
   c. A major disadvantage of cyberspace communication is that it's more impersonal than face-to-face interaction.
   d. White Americans are the most wired group in the United States.

6. *True or False* In sociology, status signifies prestige.

7. *True or False* A master status is based on one's achieved status.

8. *True or False* We can minimize role conflict and role strain by not setting priorities.

9. *True or False* If we define something as real and act on it, it can, in fact, become real. This is known as the self-fulfilling prophecy.

10. *True or False* Being an African American college student is an example of having two ascribed statuses.

11. List your social statuses. Which ones are ascribed, and which ones are achieved? Do you have one or more master statuses, and how do they affect you? After describing the difference between role conflict and role strain, choose an example from each and explain how you cope with or have resolved a role conflict or role strain.

12. Regarding online interaction, describe three benefits and three costs that you've encountered. Include any of the topics you read about on pages 95–97, but illustrate your answer with specific examples.

1. b  2. c  3. d  4. d  5. d  6. False  7. False  8. False  9. True  10. False

---

| TABLE 5.2 | |
|---|---|

## Sociological Explanations of Social Interaction

| PERSPECTIVE | KEY POINTS |
|---|---|
| Symbolic Interactionist | • People create and define their reality through social interaction. <br> • Our definitions of reality, which vary according to context, can lead to self-fulfilling prophecies. |
| Social Exchange | • Social interaction is based on a balancing of benefits and costs. <br> • Relationships involve trading a variety of resources, such as money, youth, and good looks. |
| Feminist | • The sexes act similarly in many interactions but often differ in communication styles and speech patterns. <br> • Men are more likely to use speech that's assertive (to achieve dominance and goals), while women are more likely to use language that connects with others. |

*For practice tests, printable flash cards, and more, visit 4ltrpress.cengage.com/soc.*

# Social Groups, Organizations, and Social Institutions

## CHAPTER 6 TOPICS

### 1 Social Groups

A *social group* (such as friends or work groups) gives us a common identity and a sense of belonging. Social groups include *primary groups* (such as family members) that shape our social and moral development and *secondary groups* (such as the students in your sociology class) that pursue a specific goal or activity.

Members of an *in-group* share a sense of identity and "we-ness," whereas *out-groups* are viewed and treated negatively because they are seen as having values, beliefs, and other characteristics different from those of the in-group. We also have *reference groups* that influence who we are, what we do, and who we'd like to be in the future. Groups often form a *social network*, a web of social ties that links an individual to others (such as members of a local hiking group).

**Example**: Secondary Groups Can Replace Primary Groups

In 1864, an alcoholic who had ruined a promising career on Wall Street because of his constant drunkenness cofounded Alcoholics Anonymous (AA), a program that would enable people to stop drinking by undergoing a spiritual awakening and seeking help from a buddy to stay sober. Initially, AA was a secondary group that tried to beat alcoholism by encouraging its members to attend regular meetings where alcoholics talked about their accomplishments in staying sober. Over the years, however, AA has become a primary group for many members because it offers a relatively small group of people who engage in face-to-face interaction over an extended period, especially when their family and friends have rejected them.

### 2 Formal Organizations

We depend on a variety of *formal organizations* to provide goods and services in a stable and predictable way. Two of the most widespread and important types of formal organizations in the United States are *voluntary associations* (such as charitable groups), whose members share a common set of interests and are not paid for their participation, and *bureaucracies* (such as your college), which are supposed to accomplish goals and tasks in the most efficient and rational way possible.

### 3 Sociological Perspectives on Social Groups and Organizations

See Table 6.4 on the next page.

### 4 Social Institutions

A *social institution* meets a society's basic survival needs. Functionalists identify five core social institutions—family, economy, political institutions, education, and religion—that are universal and interconnected.

## KEY TERMS

**social group**  two or more people who interact with one another and who share a common identity and a sense of belonging or "we-ness."

**primary group**  a relatively small group of people who engage in intimate face-to-face interaction over an extended period.

**secondary group**  a large, usually formal, impersonal, and temporary collection of people who pursue a specific goal or activity.

**ideal types**  general traits that describe a social phenomenon rather than every case.

**in-group**  people who share a sense of identity and "we-ness" that typically excludes and devalues outsiders.

**out-group**  people who are viewed and treated negatively because they are seen as having values, beliefs, and other characteristics different from those of an in-group.

**reference group**  a group of people who shape our behavior, values, and attitudes.

**groupthink**  a tendency of in-group members to conform without critically testing, analyzing, and evaluating ideas, which results in a narrow view of an issue.

**social network**  a web of social ties that links an individual to others.

**formal organization**  a complex and structured secondary group that has been deliberately created to achieve specific goals in an efficient manner.

**voluntary association**  a formal organization created by people who share common interests and who are not paid for their participation.

**bureaucracy**  a formal organization that is designed to accomplish goals and tasks through the efforts of a large number of people in the most efficient and rational way possible.

**goal displacement**  a preoccupation with rules and regulations rather than achieving the organization's objectives.

**alienation**  a feeling of isolation, meaninglessness, and powerlessness that may affect workers in a bureaucracy.

**iron law of oligarchy**  the tendency of a bureaucracy to become increasingly dominated by a small group of people.

**glass ceiling**  attitudes or organizational biases in the workplace that prevent women from advancing to leadership positions.

**social institution**  an organized and established social system that meets one or more of a society's basic needs.

## TEST YOUR LEARNING

1. A(n) ____ is a small group of people who engage in frequent and intimate face-to-face interaction.
   a. out-group
   b. reference group
   c. secondary group
   d. primary group

2. ____ conducted experiments that used "teachers" who administered electric shocks to "learners."
   a. Solomon Asch
   b. Stanley Milgram
   c. Philip Zimbardo
   d. Irving Janis

3. Max Weber outlined the characteristics of an efficient and productive bureaucracy. Which of the following was NOT one of those characteristics?
   a. High degree of specialization
   b. Explicit rules and regulations
   c. Qualifications-based employment
   d. Decentralized authority

4. Regarding social groups and organizations, symbolic interactionists maintain that
   a. some people benefit more than others.
   b. cooperation works.
   c. men benefit more than women.
   d. people define and shape their situations.

5. ____ contend that organizations are based on vast differences in power and control.
   a. Functionalists
   b. Conflict theorists
   c. Symbolic interactionists
   d. Exchange theorists

6. *True or False* A high school football team is a good example of a primary group.

7. *True or False* Groupthink occurs most often when in-group members discuss a diversity of ideas.

8. *True or False* The proper functioning of a bureaucracy might result in an alienation of individual workers.

9. *True or False* The iron law of oligarchy states that bureaucracies have a tendency to become increasingly dominated by a small group of people.

10. *True or False* The glass ceiling refers to organizational barriers, but not attitudes, in the workplace.

11. Describe yourself using the following concepts: primary group, secondary group, in-group, out-group, reference group, and social network. Be specific in illustrating your self-description with examples.

12. In the United States, we live in a "McDonaldized" society. What does this concept mean? What are some of the benefits and costs of a "McDonaldized" society? Illustrate your answer with specific examples from your everyday life.

1.d  2.b  3.d  4.d  5.b  6.False  7.False  8.True  9.True  10.False

---

### TABLE 6.4

# Sociological Perspectives on Groups and Organizations

| THEORETICAL PERSPECTIVE | LEVEL OF ANALYSIS | MAIN POINTS | KEY QUESTIONS |
|---|---|---|---|
| **Functionalist** | Macro | Organizations are made up of interrelated parts and rules and regulations that produce cooperation in meeting a common goal. | • Why are some organizations more effective than others?<br>• How do dysfunctions prevent organizations from being rational and effective? |
| **Conflict** | Macro | Organizations promote inequality that benefits elites, not workers. | • Who controls an organization's resources and decision making?<br>• How do those with power protect their interests and privileges? |
| **Feminist** | Macro and micro | Organizations tend not to recognize or reward talented women and regularly exclude them from decision-making processes. | • Why do many women hit a glass ceiling?<br>• How do gender stereotypes affect women in groups and organizations? |
| **Symbolic Interactionist** | Micro | People aren't puppets but can determine what goes on in a group or organization. | • Why do people ignore or change an organization's rules?<br>• How do members of social groups influence workplace behavior? |

*For practice tests, printable flash cards, and more, visit 4ltrpress.cengage.com/soc.*

# Deviance, Crime, and the Criminal Justice System

## CHAPTER 7 TOPICS

### 1  What Is Deviance?

*Deviance*, the violation of social norms, is usually punished by a *stigma*, a negative label. Perceptions of deviance vary across and within societies and can change over time. Those in authority or power decide what's right or wrong.

**Example**: Deviance and College Drinking

According to many college presidents, alcohol abuse is the most serious problem on campus. Alcohol abuse results in alcohol poisoning and blackouts and leads to sexual assault, violent behavior, injuries, and academic problems. Because drinking laws are rarely enforced, some college presidents have proposed that the drinking age be lowered from 21 to 18. Others argue that doing so would increase traffic fatalities and drinking problems. Young people can get a driver's license at 16 and vote and enlist in the military at 18. Should they be the ones, then, to decide whether drinking laws should be changed?

### 2  What Is Crime?

*Crime* violates societal norms and rules. *Criminologists* study the nature, extent, cause, and control of criminal behavior. Violent crimes are most likely to be covered by the media, but Americans are much more likely to be victimized by theft or burglary than to be murdered, raped, robbed, or assaulted with a deadly weapon. *Victimless crimes* violate laws, but the parties involved don't consider themselves victims.

### 3  Controlling Deviance and Crime

The purpose of *social control* is to eliminate, or at least reduce, deviance and crime. Formal social control is administered by those in authority or power. Informal social control is internalized from childhood. Most people conform because of positive and negative *sanctions*.

### 4–7  Sociological Explanations of Deviance and Crime

See Table 7.2 on the next page.

### 8  The Criminal Justice System and Social Control

The *criminal justice system* relies on three major approaches in controlling crime: prevention and intervention, punishment, and rehabilitation. A *crime control model* supports a tough approach toward criminals in sentencing, imprisonment, and capital punishment. In contrast, many people believe that *rehabilitation* can change offenders into productive and law-abiding citizens.

## KEY TERMS

**deviance**  traits or behavior that violate expected rules or norms.

**stigma**  a negative label that devalues a person and changes her or his self-concept and social identity.

**crime**  a violation of societal norms and rules for which punishment is specified by law.

**criminologists**  researchers who use scientific methods to study the nature, extent, cause, and control of criminal behavior.

**victimization survey**  interviewing people about their experiences as crime victims.

**victimless crimes**  acts that violate laws but involve individuals who don't consider themselves victims.

**social control**  the techniques and strategies that regulate people's behavior in society.

**sanctions**  rewards or punishments for obeying or violating a norm.

**anomie**  the condition in which people are unsure of how to behave because of absent, conflicting, or confusing social norms.

**strain theory**  the idea that people may engage in deviant behavior when they experience a conflict between goals and the means available to obtain the goals.

**white-collar crime**  illegal activities committed by high-status individuals in the course of their occupation.

**corporate crimes**  (also known as organizational crimes) illegal acts committed by executives to benefit themselves and their companies.

**cybercrime**  (also called computer crime) illegal activities that are conducted online.

**organized crime**  activities of individuals and groups that supply illegal goods and services for profit.

**patriarchy**  a hierarchical system in society in which cultural, political, and economic structures are controlled by men.

**differential association theory**  a perspective that asserts that people learn deviance through interaction, especially with significant others.

**labeling theory**  a perspective that holds that society's reaction to behavior is a major factor in defining oneself or others as deviant.

**primary deviance**  the initial act of breaking a rule.

**secondary deviance**  rule-breaking behavior that people adopt in response to the reactions of others.

**criminal justice system**  the government agencies—including the police, courts, and prisons—that are charged with enforcing laws, passing judgment on offenders, and changing criminal behavior.

**crime control model**  an approach that believes that crime rates increase when offenders don't fear apprehension or punishment.

**rehabilitation**  a social control approach that holds that appropriate treatment can change offenders into productive, law-abiding citizens.

## TEST YOUR LEARNING

1. Which of the following statements is false?
   a. Deviance can be a trait or a behavior.
   b. Informal deviance violates laws.
   c. Deviance is usually accompanied by social stigmas.
   d. Deviance varies across situations.

2. Who of the following is most likely to be the victim of a crime?
   a. A black man
   b. A black woman
   c. A white man
   d. A white woman

3. When 3-year-old Kyle colored the kitchen table blue with his new crayons, his mother frowned and scolded him. Kyle's mother used _____ to control his behavior?
   a. punishment
   b. negative sanctions
   c. behavior modification
   d. positive sanctions

4. Merton proposed four deviant modes by which people adapted to social strain. Which of the following was NOT one of Merton's modes?
   a. Innovation
   b. Retreatism
   c. Ritualism
   d. Recidivism

5. What is a fundamental question that a conflict theorist would ask regarding crime?
   a. "Why do some people commit crimes whereas others do not?"
   b. "Why are some acts defined as criminal whereas others are not?"
   c. "Why do men commit more violent crimes than women?"
   d. "How does one's social context impact deviant behavior?"

6. **True or False** Female crime rates have decreased.

7. **True or False** Differential association theory claims that people learn deviant behaviors through interaction with others.

8. **True or False** Labeling theory claims that society's reaction to a behavior is a major factor in defining oneself or others as deviant.

9. **True or False** Capital punishment decreases crime.

10. **True or False** The crime control model maintains that rehabilitation is the best way to decrease the frequency of crime.

11. What's the difference, if any, between deviance and crime? Include the positive and negative functions, if any, of deviance and crime in your answer.

12. Which sociological perspective do you think provides the best explanation of deviance and crime? Explain why, illustrating your answer with specific examples.

1.b  2.a  3.b  4.d  5.b  6.False  7.True  8.True  9.False  10.False

---

## TABLE 7.2
# Sociological Explanations of Deviance and Crime

| THEORETICAL PERSPECTIVE | LEVEL OF ANALYSIS | KEY POINTS |
|---|---|---|
| Functionalist | Macro | • Anomie increases the likelihood of deviance.<br>• Crime occurs when people experience blocked opportunities to achieve the culturally approved goal of economic success. |
| Conflict | Macro | • Laws protect the interests of the few (primarily those in the upper classes) rather than the rights of the many.<br>• Law enforcement is rarely directed at the illegal activities of the powerful. |
| Feminist | Macro and micro | • Crimes committed by women reflect their general oppression due to social, economic, and political inequality.<br>• Many women are criminal offenders or victims because of culturally organized beliefs and practices that are sexist and patriarchal. |
| Symbolic Interactionist | Micro | • People learn deviant and criminal behavior from others, such as parents and friends, who are important in their everyday lives.<br>• If people are labeled or stigmatized as deviant, they are likely to develop deviant self-concepts and engage in criminal behavior. |

*For practice tests, printable flash cards, and more, visit 4ltrpress.cengage.com/soc.*

# Social Stratification: United States and Global

## CHAPTER 8 TOPICS

### 1 What Is Social Stratification?

*Social stratification* is a hierarchical ranking of people who have different access to valued resources. A *closed stratification system* differs from an *open stratification system* because the latter allows movement from one *social class* to another.

### 2 Dimensions of Stratification

Stratification includes *wealth*, *prestige*, and *power*. People are more likely to experience status inconsistency if they rank differently on these three dimensions, such as a football player who has great wealth but little power.

### 3 Social Class in America

A good indicator of social class is *socioeconomic status (SES)*, an overall rank based on a person's income, education, and occupation. Using SES and other variables (such as values, power, and *conspicuous consumption*), most sociologists agree that there are at least four social classes in the United States: upper, middle, working, and lower. These groups can be divided further into upper-upper, lower-upper, upper-middle, lower-middle, and the working class. The lower class includes the *working poor* and the *underclass*. A major outcome of social stratification is *life chances*.

**Example**: Restaurant Menus and Stratification

Two sociologists—in Iowa and Virginia—asked their students in introductory sociology classes to do a content analysis (see Chapter 1) of 10 menus that represented a sampling of restaurants by social class. The students found that the restaurants that catered to upper-class clientele had higher than average entrée prices, described the entrées in foreign languages, used fancy sauces, recommended expensive wines, and had few illustrations. Middle-class menus emphasized "value for the dollar," presented photos of entrées with "bountiful plates overflowing with appetizing food," and popular items such as quesadillas. Menus at lower-class restaurants featured low prices ($3 to $10 entrées), the items were numbered, none of the entrées had "pretentious names," and the typesetting was simple (Wright and Ransom 2005). In effect, then, even menus denote social class and social status.

### 4 Poverty in America

*Absolute poverty* is a serious social problem compared with *relative poverty* because millions of Americans live below the *poverty line*. Explanations for poverty vary, but two perspectives propose that individual characteristics lead to poverty or that a society's organization creates and sustains poverty.

### 5 Social Mobility

*Social mobility* can be *horizontal*, *vertical*, *intragenerational*, or *intergenerational*. Structural, demographic, and individual factors affect a person's social mobility.

### 6 Global Inequality

Global inequality is widespread, but some societies are much wealthier than others. Sociologists use *modernization theory*, *dependency theory*, and *world-system theory* to explain why inequality is universal.

### 7 Sociological Explanations: Why There Are Haves and Have-Nots

See Table 8.3 on the next page.

## KEY TERMS

**social stratification** the hierarchical ranking of people in a society who have different access to valued resources, such as property, prestige, power, and status.

**open stratification system** a system that is based on individual achievement and allows movement up or down.

**closed stratification system** a system in which movement from one social position to another is limited by ascribed statuses such as one's sex, skin color, and family background.

**social class** a category of people who have a similar standing or rank in a society based on wealth, education, power, prestige, and other valued resources.

**wealth** abundance of economic assets and material possessions that a person or family owns, including property and income.

**prestige** respect, recognition, or regard attached to social positions.

**power** the ability of individuals or groups to achieve goals, control events, and maintain influence over others despite opposition.

**socioeconomic status (SES)** an overall ranking of a person's position in the class hierarchy based on income, education, and occupation.

**working poor** people who work at least 27 weeks a year but receive such low wages that they live in or near poverty.

**underclass** people who are persistently poor and seldom employed, residentially segregated, and relatively isolated from the rest of the population.

**life chances** the extent to which people have positive experiences and can secure the good things in life because they have economic resources.

**absolute poverty** not having enough money to afford the most basic necessities of life.

**relative poverty** not having enough money to maintain an average standard of living.

**poverty line** the minimal level of income that the federal government considers necessary for basic subsistence.

**feminization of poverty** the higher likelihood that female heads of households will be poor.

**social mobility** a person's ability to move up or down the social class hierarchy.

**horizontal mobility** moving from one position to another at the same class level.

**vertical mobility** moving up or down the class hierarchy.

**intragenerational mobility** moving up or down the class hierarchy over one's lifetime.

**intergenerational mobility** moving up or down the class hierarchy relative to the position of one's parents.

**Davis–Moore thesis** the functionalist view that social stratification benefits a society.

**meritocracy** a belief that individuals are rewarded for what they do and how well rather than on the basis of their ascribed status.

**bourgeoisie** those who own and control capital and the means of production.

**proletariat** workers who sell their labor for wages.

**corporate welfare** an array of subsidies, tax breaks, and assistance that the government has created for businesses.

## TEST YOUR LEARNING

1. Slavery and castes are ____ systems.
   a. capitalist
   b. open stratification
   c. closed stratification
   d. socialist

2. A person's socioeconomic status (SES) is based on
   a. income, education, and occupation.
   b. wealth, prestige, and power.
   c. wealth, prestige, and lifestyle.
   d. education, income, and age.

3. The economic gap between the wealthy and the poor
   a. fluctuates from year to year.
   b. is about the same as during the 1990s.
   c. is decreasing.
   d. is increasing.

4. Which of the following is NOT a major structural factor in social mobility?
   a. Changes in the economy
   b. Immigration patterns
   c. Consumer confidence
   d. Number of available positions in given occupations

5. ____ claim that social stratification ultimately benefits society.
   a. Functionalists
   b. Conflict theorists
   c. Feminist theorists
   d. Symbolic interactionists

6. ***True or False*** One of the criticisms of conflict theory is that it ignores structural factors in explaining stratification.

7. ***True or False*** Conspicuous consumption displays one's social status and enhances one's prestige.

8. ***True or False*** Intergenerational mobility refers to moving up or down the class hierarchy over one's lifetime.

9. ***True or False*** The Davis–Moore thesis is a symbolic interactionist perspective.

10. ***True or False*** According to world-system theory, inequality exists throughout the world because the global economic system helps richer countries stay rich while poorer countries remain poor.

11. Describe your current social class and explain why you place yourself in this class. Ten years from now, do you expect to experience upward or downward mobility? Again, explain why.

12. Should public assistance recipients be required to work, regardless of their education level? If no, why not? If yes, should the government subsidize preschool and afterschool care for their children while the parent(s) is/are at work? In your answer, incorporate the sociological perspectives on social stratification.

1. c   2. a   3. d   4. c   5. a   6. False   7. True   8. False   9. False   10. True

---

**TABLE 8.3**

# Sociological Explanations of Social Stratification

| PERSPECTIVE | LEVEL OF ANALYSIS | KEY POINTS |
|---|---|---|
| **Functionalist** | Macro | • Fills social positions that are necessary for a society's survival<br>• Motivates people to succeed and ensures that the most qualified people will fill the most important positions |
| **Conflict** | Macro | • Encourages workers' exploitation and promotes the interests of the rich and powerful<br>• Ignores a wealth of talent among the poor |
| **Feminist** | Macro and micro | • Constructs numerous barriers in patriarchal societies that limit women's achieving wealth, status, and prestige<br>• Requires most women, not men, to juggle domestic and employment responsibilities that impede upward mobility |
| **Symbolic Interactionist** | Micro | • Shapes stratification through socialization, everyday interaction, and group membership<br>• Reflects social class identification through symbols, especially products that signify social status |

*For practice tests, printable flash cards, and more, visit 4ltrpress.cengage.com/soc.*

# Gender and Sexuality

## CHAPTER 9 TOPICS

### 1 How Women and Men Are Similar and Different

*Sex* refers to biological characteristics, whereas *gender* refers to learned attitudes and behaviors. *Gender identity* and *gender roles* differ depending on whether people perceive themselves as masculine or feminine and because a society expects women and men to think and behave differently. Many Americans still have *gender stereotypes* about how people will look, act, think, and feel based on their sex.

### 2 Contemporary Gender Stratification and Inequality

*Sexism* is widespread because of *gender stratification*, which can lead to inequality in the family, education, the workplace (as in the case of a *gender pay gap*), and politics. *Sexual harassment* and *pregnancy discrimination* are also common in the workplace.

### 3 Sexuality

Our sexual identity incorporates a *sexual orientation* that can be *homosexual*, *heterosexual*, *bisexual*, or *asexual*. *Transgender* people include those living on the boundaries of the sexes. Many biological theories maintain that sexual orientation has a strong genetic basis, but social constructionists argue that culture, not biology, plays a large role in forming sexual identity.

### 4 Some Current Controversies About Sexuality

*Abortion* is controversial, with almost equal percentages of Americans supporting or condemning the practice. Those who favor *same-sex marriage* argue that people should have the same legal rights regardless of sexual orientation; those who oppose same-sex marriage contend that such unions are immoral and contrary to religious beliefs. Some people view *pornography* as erotic recreation, whereas others denounce it as obscene and as debasing women.

### 5 Gender and Sexuality Across Cultures

Worldwide, women have fewer rights and opportunities than men. People in some countries are more accepting of homosexuality than in the past, but *heterosexism* prevails.

### 6 Sociological Explanations of Gender Inequality and Sexuality

See Table 9.5 on the next page.

**Example**: Gender Roles and Hooking Up—Are Women the Losers?

*Hooking up* refers to physical encounters, no strings attached, and can mean anything from kissing and genital fondling to oral sex and sexual intercourse. At many high schools, hooking up is more common than dating. At some college campuses, 76 percent of the students have hooked up 5 times, on average, and 28 percent have had 10 or more such encounters (England et al. 2007). Hooking up has its advantages. It's much cheaper than dating. Also, it's assumed that hooking up requires no commitment of time or emotion. In addition, hookups remove the stigma from those who can't get dates but can experience sexual pleasure, and they make people feel sexy and desirable (Bogle 2008). Hooking up also has disadvantages, especially for women, because women who hook up generally get a bad reputation as being "easy" (England and Thomas 2009). In effect, then, and despite the advantages of hooking up, the sexual double standard persists.

## KEY TERMS

**sex** the biological characteristics with which we are born.

**gender** learned attitudes and behaviors that characterize women and men.

**gender identity** a perception of oneself as either masculine or feminine.

**gender roles** the characteristics, attitudes, feelings, and behaviors that society expects of females and males.

**gender stereotypes** expectations about how people will look, act, think, and feel based on their sex.

**sexism** an attitude or behavior that discriminates against one sex, usually females, based on the assumed superiority of the other sex.

**gender stratification** people's unequal access to wealth, power, status, prestige, and other valued resources because of their sex.

**gender pay gap** the overall income difference between women and men in the workplace (also called the wage gap, pay gap, and gender wage gap).

**sexual harassment** any unwanted sexual advance, request for sexual favors, or other conduct of a sexual nature that makes a person uncomfortable and interferes with her or his work.

**sexual identity** our awareness of ourselves as male or female and the ways that we express our sexual values, attitudes, feelings, and beliefs.

**sexual orientation** a preference for sexual partners of the same sex, of the opposite sex, of both sexes, or neither sex.

**homosexuals** those who are sexually attracted to people of the same sex.

**heterosexuals** those who are sexually attracted to people of the opposite sex.

**bisexuals** those who are sexually attracted to members of both sexes.

**asexuals** those who lack any interest in or desire for sex.

**transgender people** those who are transsexuals, intersexuals, or transvestites.

**sexual script** specifies the formal or informal norms for acceptable or unacceptable sexual activity, which individuals are eligible sexual partners, and the boundaries of sexual behavior.

**heterosexism** belief that heterosexuality is superior to and more natural than homosexuality or bisexuality.

**homophobia** the fear and hatred of homosexuality.

**abortion** the expulsion of an embryo or fetus from the uterus.

**pornography** the graphic depiction of images that may lead to sexual arousal.

# TEST YOUR LEARNING

1. _____ refers to learned attitudes and behaviors, whereas _____ refers to biological characteristics with which we are born.
   a. Sex; gender
   b. Gender; sex
   c. Sex; gender roles
   d. Sexual identity; gender stereotypes

2. Generally, the gender pay gap increases as the level of educational attainment
   a. increases.
   b. decreases.
   c. neither of the above because it depends on how hard a person works
   d. neither a nor b because such data are not available

3. Sean, age 16, lies to his friends about the frequency of his sexual activity and new sexual exploits every week. Sean is
   a. being a normal male.
   b. adhering to a sexual script.
   c. falling into a pathologic life course.
   d. challenging gender norms.

4. _____ is the belief that heterosexuality is superior to and more natural than homosexuality or bisexuality.
   a. Heterobiology
   b. Homosexism
   c. Heterosexism
   d. Bisexism

5. _____ posit that gender inequality and sexuality are socially constructed.
   a. Functionalists
   b. Conflict theorists
   c. Feminist theorists
   d. Symbolic interactionists

6. **True or False** Sexism can only be enacted by men against women.

7. **True or False** Sexual harassment is the fastest growing type of employment discrimination.

8. **True or False** U.S. abortion rates have been increasing.

9. **True or False** A majority of Americans support same-sex marriage.

10. **True or False** Conflict theorists believe that capitalism increases gender inequality.

11. How do sex and gender differ? How do they affect each other? "Our sexual behavior is spontaneous." Explain why you agree or disagree with this statement.

12. Describe how gender stratification affects women and men in family life, education, the workplace, and politics. Provide examples to illustrate your answer.

1.b  2.a  3.b  4.c  5.d  6.False  7.True  8.False  9.False  10.True

---

## TABLE 9.5

# Sociological Explanations of Gender Inequality and Sexuality

| THEORETICAL PERSPECTIVE | LEVEL OF ANALYSIS | KEY POINTS |
|---|---|---|
| **Functionalist** | Macro | • Gender roles are complementary, equally important for a society's survival, and affect human capital.<br>• Agreed-on sexual norms contribute to a society's order and stability. |
| **Conflict** | Macro | • Gender roles give men power to control women's lives instead of allowing the sexes to be complementary and equally important.<br>• Most societies regulate women's, but not men's, sexual behavior. |
| **Feminist** | Macro and micro | • Women's inequality reflects their historical and current domination by men, especially in the workplace.<br>• Many men use violence—including sexual harassment, rape, and global sex trafficking—to control women's sexuality. |
| **Symbolic Interactionist** | Micro | • Gender inequality is a social construction that emerges through day-to-day interactions and reflects people's gender role expectations.<br>• The social construction of sexuality varies across cultures because of societal norms and values. |

*For practice tests, printable flash cards, and more, visit 4ltrpress.cengage.com/soc.*

# Race and Ethnicity

## CHAPTER 10 TOPICS

### 1 U.S. Racial and Ethnic Diversity

Perhaps the most multicultural country in the world, the United States includes about 150 distinct ethnic or racial groups among more than 310 million inhabitants. By 2025, only 58 percent of the U.S. population will be white—down from 86 percent in 1950.

### 2 The Social Significance of Race and Ethnicity

*Race* refers to physical characteristics, whereas an *ethnic group* identifies with a common national origin or cultural heritage. A *racial-ethnic group* has both distinctive physical and cultural characteristics.

### 3 Our Changing Immigration Mosaic

In 1900, almost 85 percent of immigrants came from Europe; now immigrants come primarily from Asia and Latin America. Many Americans are ambivalent about immigrants, especially those who are in the country illegally, but many scholars argue that, in the long run, both legal and undocumented immigrants bring more benefits than costs.

### 4 Dominant and Minority Groups

A *dominant group* has more economic and political power than a *minority*. The latter may be larger in number than a dominant group but is often subject to differential and unequal treatment because of its physical, cultural, or other characteristics. Patterns of dominant-minority group relations include *genocide*, *segregation*, *assimilation*, and *pluralism*.

### 5 Some Sources of Racial-Ethnic Friction

*Racism* justifies and preserves the social, economic, and political interests of dominant groups. *Prejudice* is an attitude; *discrimination* is an act that occurs at both the individual and institutional level. All of us can be prejudiced, but minorities are typically targets of *stereotypes* and *ethnocentrism* that often lead to *scapegoating*.

### 6 Major Racial and Ethnic Groups in the United States

European Americans, who settled the first colonies, are declining in population, whereas Latinos now comprise 17 percent of the population. Other large racial and ethnic populations are African Americans (14 percent), Asian Americans (6 percent), and American Indians (almost 2 percent). Middle Eastern Americans, who comprise less than 0.05 percent of the population, come from more than 30 countries. All of these groups have experienced prejudice and discrimination, but they have enhanced U.S. society and culture.

**Example**: Stuff White People Like

The popular blog *Stuff White People Like* has generated clones (e.g., *Stuff Educated Black People Like* and *Stuff Asian People Like*). Why are these sites so popular? Many fans say that the descriptions are funny because they're true. According to some critics, however, by poking fun at privileged upper-middle-class white people, the sites fuel stereotypes instead of having painfully frank discussions about U.S. race and racism (Sternbergh 2008). Do you agree or disagree?

### 7 Sociological Explanations of Racial-Ethnic Inequality

See Table 10.4 on the next page.

### 8 Interracial and Interethnic Relationships

About 97 percent of Americans report being only one race, but the numbers of biracial children are increasing because of interracial dating and marriage. The rise of racial-ethnic intermarriage reflects many micro and macro factors that include greater acceptance of integration as well as interethnic and interracial contact.

## KEY TERMS

**race** a group of people who share physical characteristics, such as skin color and facial features, that are passed on through reproduction.

**ethnic group** a group of people who identify with a common national origin or cultural heritage that includes language, geographic roots, food, customs, traditions, and/or religion.

**racial-ethnic group** a group of people who have distinctive physical and cultural characteristics.

**dominant group** any physically or culturally distinctive group that has the most economic and political power, the greatest privileges, and the highest social status.

**apartheid** a formal system of racial segregation.

**minority** a group of people who may be subject to differential and unequal treatment because of their physical, cultural, or other characteristics, such as gender, sexual orientation, religion, ethnicity, or skin color.

**genocide** the systematic effort to kill all members of a particular ethnic, religious, political, racial, or national group.

**segregation** the physical and social separation of dominant and minority groups.

**assimilation** the process of conforming to the culture of the dominant group, adopting its language and values, and intermarrying with that group.

**pluralism** minority groups retain their culture but have equal social standing in a society.

**racism** a set of beliefs that one's own racial group is naturally superior to other groups.

**prejudice** an attitude, positive or negative, toward people because of their group membership.

**stereotype** an oversimplified or exaggerated generalization about a category of people.

**ethnocentrism** the belief that one's own culture, society, or group is inherently superior to others.

**scapegoats** individuals or groups whom people blame for their own problems or shortcomings.

**discrimination** any act that treats people unequally or unfairly because of their group membership.

**individual discrimination** harmful action on a one-to-one basis by a member of a dominant group against a member of a minority group.

**institutional discrimination**  unequal treatment and opportunities that members of minority groups experience as a result of the everyday operations of a society's laws, rules, policies, practices, and customs.

**gendered racism**  the overlapping and cumulative effects of inequality due to racism and sexism.

**contact hypothesis**  the idea that the more people get to know members of a minority group personally, the less likely they are to be prejudiced against that group.

**miscegenation**  marriage or sexual relations between a man and a woman of different races.

# TEST YOUR LEARNING

1. What is the difference between race and ethnicity?
   a. Race is a negative term that refers to difference; ethnicity is the politically correct term that refers to difference.
   b. Race refers to the physical characteristics of a group of people; ethnicity refers to the cultural heritage of a people.
   c. Race refers to people around the world; ethnicity refers only to a group of people within a specific nation.
   d. There is no difference between race and ethnicity; the terms are synonymous.

2. Today most immigrants to the United States come from which of the following countries?
   a. China and Mexico
   b. Mexico and Canada
   c. England and Canada
   d. Japan and Russia

3. _____ is the physical and social separation of dominant and minority groups.
   a. Racism
   b. Internal colonialism

c. Segregation
d. Institutional discrimination

4. Tyrone and Jeanette were up for the same job, which Jeanette ultimately won. Tyrone told his friends that he didn't get the job not because Jeanette was more qualified but because "women always get hired first now." This is an example of
   a. scapegoating.
   b. pluralism.
   c. stereotyping.
   d. prejudice.

5. Laws against miscegenation were overturned nationally in
   a. 1807.
   b. 1867.
   c. 1907.
   d. 1967.

6. ***True or False*** On the continuum of dominant versus minority group relations, genocide reflects the least tolerance, whereas pluralism illustrates the greatest tolerance.

7. ***True or False*** Ethnocentrism is the belief that one's own racial group is naturally inferior to others.

8. ***True or False*** The fastest growing ethnic minority in the United States is Asian Americans.

9. ***True or False*** American Indians are on the verge of vanishing from the American landscape.

10. ***True or False*** The lower the educational level, the greater the likelihood that a person will have an interracial marriage.

11. What's the difference between individual discrimination and institutional discrimination? Is one more powerful than the other in its effect on minority groups? If so, which one and why? If not, why not?

12. Choose any two minorities discussed in this chapter. Compare then in terms of factors such as population size, diversity, constraints, and economic success.

1.b  2.a  3.c  4.a  5.d  6.True  7.False  8.False  9.False  10.False

## TABLE 10.4

# Sociological Explanations of Racial-Ethnic Inequality

| THEORETICAL PERSPECTIVE | LEVEL OF ANALYSIS | KEY POINTS |
|---|---|---|
| **Functionalist** | Macro | Prejudice and discrimination can be dysfunctional, but they provide benefits for dominant groups and stabilize society. |
| **Conflict** | Macro | Powerful groups maintain their advantages and perpetuate racial-ethnic inequality primarily through economic exploitation. |
| **Feminist** | Macro and micro | Minority women suffer from the combined effects of racism and sexism. |
| **Symbolic Interactionist** | Micro | Hostile attitudes toward minorities, which are learned, can be reduced through cooperative interracial and interethnic contacts. |

*For practice tests, printable flash cards, and more, visit 4ltrpress.cengage.com/soc.*

# The Economy and Politics

## CHAPTER 11 TOPICS

### 1 Global Economic Systems

The *economy* determines how a society produces, distributes, and consumes goods and services. *Capitalism* frequently spawns *monopolies* and *oligopolies*, which dominate the market and discourage competition. Ideally, *socialism* emphasizes cooperation and a collective ownership of property and forbids private profits.

### 2 Corporations and the Economy

A *corporation*, usually created for profit, often forms a *conglomerate*. Both corporations and conglomerates are governed by *interlocking directorates* that have become more powerful than ever because of the growth of *transnational corporations* and *transnational conglomerates*.

### 3 Work in U.S. Society Today

*Work* produces goods or services. Many Americans have been casualties of *deindustrialization* and *globalization*. Others have lost their jobs to *offshoring*. Widespread *downsizing* has resulted in unemployment, *underemployment*, and discouraged workers.

### 4 Sociological Explanations of Work and the Economy

See Table 11.2 on the next page.

### 5 Global Political Systems

In a *democracy*, ideally, citizens have a high degree of control over the state. *Totalitarianism* controls people's lives; *authoritarianism* generally permits some degree of individual freedom. A *monarchy* is the oldest type of authoritarian regime.

### 6 Politics, Power, and Authority

Politics includes *power* and *authority*. Authority can be based on tradition, charisma, rational-legal power, or a combination of these sources.

### 7 Politics and Power in U.S. Society

A *political party* tries to influence and control government. The voting rate is much higher among some groups than others. Situational and structural factors, such as convenience, can also encourage or discourage voting.

### 8 Sociological Perspectives on Politics and Power

See Table 11.5 on the next page.

## KEY TERMS

**economy** a social institution that determines how a society produces, distributes, and consumes goods and services.

**politics** a social institution through which individuals and groups acquire and exercise power and authority and make decisions.

**capitalism** an economic system in which wealth is in private hands and is invested and reinvested to produce profits.

**monopoly** domination of a particular market or industry by one person or company.

**oligopoly** domination of a market by a few large producers or suppliers.

**socialism** an economic system based on the principle of the public ownership of the production of goods and services.

**communism** a political and economic system in which all members of a society are equal.

**corporation** a social entity that has legal rights, privileges, and liabilities apart from those of its members.

**conglomerate** a giant corporation that owns a collection of companies in different industries.

**interlocking directorate** a situation in which the same people serve on the boards of directors of several companies or corporations.

**transnational corporation** (sometimes called a multinational corporation or an international corporation) a large company that is based in one country but operates across international boundaries.

**transnational conglomerate** a corporation that owns a collection of different companies in various industries in a number of countries.

**work** a physical or mental activity that produces or provides either goods or services.

**deindustrialization** a process of social and economic change caused by the reduction of industrial activity, especially manufacturing.

**globalization** the growth and spread of investment, trade, production, communication, and new technology around the world.

**offshoring** sending work or jobs to another country to cut a company's costs at home.

**downsizing** a euphemism for firing large numbers of employees at once.

**underemployed** people who have part-time jobs but want full-time work or whose jobs are below their experience and education level.

**government** a formal organization that has the authority to make and enforce laws.

**democracy** a political system in which, ideally, citizens have control over the state and its actions.

**totalitarianism** a political system in which the government controls every aspect of people's lives.

**authoritarianism** a political system in which the state controls the lives of citizens but generally permits some degree of individual freedom.

**monarchy** a political system in which power is allocated solely on the basis of heredity and passes from generation to generation.

**power** the ability of a person or group to affect the behavior of others despite resistance and opposition.

**authority** the legitimate use of power.

**traditional authority** power based on customs that justify the position of the ruler.

**charismatic authority** power based on exceptional individual abilities and characteristics that inspire devotion, trust, and obedience.

**rational-legal authority** power based on the belief that laws and appointed or elected political leaders are legitimate.

**political party** an organization that tries to influence and control government by recruiting, nominating, and electing its members to public office.

**pluralism** a political system in which power is distributed among a variety of competing groups in a society.

**power elite** a small group of influential people who make the nation's major political decisions.

## TEST YOUR LEARNING

1. ____ is a market system dominated by a few large producers or suppliers.
   - **a.** Capitalism
   - **b.** Monopoly
   - **c.** Oligopoly
   - **d.** Socialism

2. Felipe serves on the board of directors of several different companies. He is part of
   - **a.** an interlocking directorate.
   - **b.** a transnational corporation.
   - **c.** a conglomerate.
   - **d.** a monopoly.

3. Which of the following is NOT a major reason for the surge in women's employment since the 1970s?
   - **a.** The falling wages of men
   - **b.** Women's increased education
   - **c.** Rising cost of home ownership
   - **d.** Decrease in sexism

4. According to Max Weber, which one of the following types of authority is most characteristic of the majority of U.S. presidents?
   - **a.** Charismatic
   - **b.** Absolute
   - **c.** Traditional
   - **d.** Rational-legal

5. ____ claim that the U.S. political structure is pluralistic.
   - **a.** Functionalists
   - **b.** Conflict theorists
   - **c.** Feminist theorists
   - **d.** Symbolic interactionists

6. *True or False* Compared with low-wage U.S. jobs, high-wage jobs are relatively safe from offshoring.

7. *True or False* Deindustrialization has spread because of globalization.

8. *True or False* Married people are more likely to vote than people who are not married.

9. *True or False* Some of the best-paid lobbyists are retired corporate executives (CEOs).

10. *True or False* In the United States, most Republicans believe that the federal government should provide social programs for its citizens.

11. Have you personally benefited or been harmed by globalization? Be specific in explaining your answer.

12. Suppose someone asked you to explain U.S. political power. Are you more likely to use a functionalist, conflict, or feminist perspective? A combination? Or none? Explain your answer.

1. c  2. a  3. d  4. d  5. a  6. False  7. True  8. True  9. False  10. False

---

### TABLE 11.2
# Sociological Explanations of Work and the Economy

| THEORETICAL PERSPECTIVE | KEY POINTS |
|---|---|
| Functionalist | Capitalism benefits society; work provides an income, structures people's lives, and gives them a sense of accomplishment. |
| Conflict | Capitalism enables the rich to exploit other groups; most jobs are low-paying, monotonous, and alienating; productivity isn't always rewarded. |
| Feminist | Gender roles structure women's and men's work experiences differently and inequitably. |
| Symbolic Interactionist | How people define and experience work in their everyday lives affects their workplace behavior and relationships with coworkers and employers. |

---

### TABLE 11.5
# Sociological Explanations of Political Power

| | FUNCTIONALISM: A PLURALIST MODEL | CONFLICT THEORY: A POWER ELITE MODEL | FEMINIST THEORIES: A PATRIARCHAL MODEL |
|---|---|---|---|
| **Who has political power?** | The people | Rich upper-class people—especially those at top levels in business, government, and the military | White men in Western countries; most men in traditional societies |
| **What is the source of political power?** | Citizens' participation | Wealthy people in government, business corporations, the military, and the media | Being white, male, and very rich |
| **Does one group dominate politics?** | No | Yes | Yes |
| **Do political leaders represent the average person?** | Yes, the leaders speak for a majority of the people. | No, the leaders are most concerned with keeping or increasing their personal wealth and power. | No, the leaders are rarely women who have decision-making power. |

# Families and Aging

## CHAPTER 12 TOPICS

### 1 What Is a Family?

Among other activities, the members of a *family* care for one another and any children. Worldwide, however, families vary in characteristics such as structure (a *nuclear* or an *extended* family), living arrangements (*patrilocal, matrilocal,* or *neolocal*), who has authority (*matriarchal, patriarchal,* or *egalitarian*), and how many marriage mates a person can have (*monogamy* or *polygamy*).

### 2 How U.S. Families Are Changing

*Divorce* is easier to obtain than in the past because all states now have *no-fault divorce* laws. The number of single people has risen greatly, primarily because many people are postponing marriage. There has also been a striking increase in *cohabitation* and out-of-wedlock births. In *dual-earner couples*, median income can be twice as high when both spouses work full-time, but domestic and employment responsibilities often conflict.

### 3 Diversity in American Families

Among many racial-ethnic groups, extended families are common and provide considerable emotional and economic support. However, family structures can vary widely depending on when people arrived in the United States and their socioeconomic status. Gay and lesbian families are very similar to heterosexual families but often lack the legal rights and benefits that married couples enjoy.

### 4 Family Conflict and Violence

We are more likely to experience *intimate family violence* than assault by a stranger. *Child maltreatment* is common, especially by parents. Similarly, in *elder abuse*, most of the offenders are adult children, spouses, or other family members.

### 5 Our Aging Society

How people define "old" varies across societies. *Gerontologists* emphasize that the aging population shouldn't be lumped into one group because, for example, there are significant differences between the young-old and the oldest-old. Our *old-age dependency* has increased, placing a larger burden on the working-age population.

**Example**: "He Gets Prettier; I Get Older"

When comparing her own public image with that of her actor husband, the late Paul Newman, actress Joanne Woodward once remarked, "He gets prettier; I get older." Was she right? If aging gracefully is acceptable, why do many companies tout antiaging products? And why are the products targeted primarily at women?

### 6 Sociological Explanations of Family and Aging

See Table 12.3 on the next page.

## KEY TERMS

**family** an intimate group consisting of two or more people who (1) live together in a committed relationship, (2) care for one another and any children, and (3) share close emotional ties and functions.

**incest** taboo cultural norms and laws that forbid sexual intercourse between close blood relatives.

**marriage** a socially approved mating relationship that people expect to be stable and enduring.

**endogamy** cultural practice of marrying within one's group.

**exogamy** cultural practice of marrying outside one's group.

**nuclear family** a family form composed of married parents and their biological or adopted children.

**extended family** a family form composed of parents and children, as well as other kin.

**patrilocal residence pattern** newly married couples live with the husband's family.

**matrilocal residence pattern** newly married couples live with the wife's family.

**neolocal residence pattern** a newly married couple sets up its own residence.

**boomerang generation** young adults who move back into their parents' home after living independently for a while or who never leave it in the first place.

**matriarchal family system** the oldest women control cultural, political, and economic resources and, consequently, have power over males.

**patriarchal family system** the oldest men (grandfathers, fathers, and uncles) control cultural, political, and economic resources and, consequently, have power over females.

**egalitarian family system** both partners share power and authority fairly equally.

**marriage market** a courtship process in which prospective spouses compare the assets and liabilities of eligible partners and choose the best available mate.

**homogamy** marrying someone with similar characteristics such as race, ethnicity, age, education, social class, or religion.

**monogamy** one person is married exclusively to another person.

**serial monogamy** individuals marry several people, but one at a time.

**polygamy** a marriage form in which a man or woman has two or more spouses.

**divorce** the legal dissolution of a marriage.

**no-fault divorce** state laws that do not require either partner to establish guilt or wrongdoing on the part of the other to get a divorce.

**stepfamily** a household in which two adults who are biological or adoptive parents, with a child from a prior relationship, marry or cohabit.

**cohabitation** an arrangement in which two unrelated people aren't married but live together and have a sexual relationship.

**dual-earner couples** both partners are employed outside the home (also called dual-income, two-income, two-earner, or dual-worker couples).

**fictive kin** nonrelatives who are accepted as part of a family.

**intimate partner violence (IPV)** abuse that occurs between people in a close relationship.

**child maltreatment** (also called child abuse) a broad range of behaviors that place a child at serious risk, including physical and sexual abuse, neglect, and emotional mistreatment.

**elder abuse** (sometimes called elder mistreatment) any knowing, intentional, or negligent act by a caregiver or other person that causes harm to people age 65 or older.

**gerontologists** scientists who study the biological, psychological, and social aspects of aging.

**sandwich generation** people in a middle generation who care for their own children and their aging parents.

**old-age dependency ratio** (sometimes called the elderly support ratio) the number of working-age adults ages 18 to 64 for every person age 65 and older who's not in the labor force.

**activity theory** proposes that many older people remain engaged in numerous roles and activities, including work.

**exchange theory** posits that people seek through their interactions with others to maximize their rewards and minimize their costs.

**ageism** discrimination against older people.

**continuity theory** posits that older adults can substitute satisfying new roles for those they've lost.

## TEST YOUR LEARNING

1. This chapter defines a *family* as two or more people who adhere to three criteria. Which of the following is NOT one of those criteria?
   a. Are legally sanctioned by the state
   b. Live together in a committed relationship
   c. Care for one another and any children
   d. Share close emotional ties and bonds

2. The family serves many important functions in a society. Which of the following functions does the family NOT serve?
   a. Sexual regulation    b. Social placement
   c. Peer approval        d. Economic security

3. Around the world, the most common residence pattern is
   a. patriarchal.    b. matriarchal.
   c. neolocal.       d. egalitarian.

4. Which of the following is NOT a common factor in intimate partner violence (IPV)?
   a. Income level
   b. Employment status
   c. Drug abuse
   d. Number of children

5. ____ use both a macro and a micro approach in examining families and aging.
   a. Functionalists
   b. Conflict theorists
   c. Feminist theorists
   d. Symbolic interactionists

6. *True or False* Jack marries several people, but one at a time. Jack is a serial monogamist.

7. *True or False* U.S. divorce rates have decreased since 1980.

8. *True or False* Couples who live together before marriage have lower divorce rates than those who don't cohabit before marriage.

9. *True or False* Most offenders of child maltreatment are relatives and unrelated caregivers such as foster parents and boyfriends.

10. *True or False* The world's aging population characterizes industrialized and not developing countries.

11. "U.S. families are falling apart." Explain whether you agree or disagree with this statement, using functionalist, conflict, feminist, and symbolic interactionist perspectives.

12. What do you think your life will be like at age 85? How might your options be different because of your sex, social class, race, and ethnicity? Support your answer with specific examples.

1.a  2.c  3.a  4.d  5.c  6.True  7.False  8.False  9.False  10.False

| TABLE 12.3 |
|---|

## Sociological Perspectives on Families and Aging

| THEORETICAL PERSPECTIVE | LEVEL OF ANALYSIS | KEY POINTS |
|---|---|---|
| **Functionalist** | Macro | • Families are important in maintaining societal stability and meeting family members' needs.<br>• Older people who are active and engaged are more satisfied with life. |
| **Conflict** | Macro | • Families promote social inequality because of social class differences.<br>• Many corporations view older workers as disposable. |
| **Feminist** | Macro and Micro | • Families both mirror and perpetuate patriarchy and gender inequality.<br>• Women have an unequal burden in caring for children as well as older family members and relatives. |
| **Symbolic Interactionist** | Micro | • Families construct their everyday lives through interaction and subjective interpretations of family roles.<br>• Many older family members adapt to aging and often maintain previous activities. |

# Education and Religion

## CHAPTER 13 TOPICS

### 1 What Is Education?

U.S. *education* and *schooling* have changed in four important ways: Universal education has expanded, community colleges have flourished, public higher education has burgeoned, and student diversity has greatly increased.

### 2 Sociological Perspectives on Education

See Table 13.1 on the next page.

### 3 Some Problems With U.S. Education

Compared with those in many other countries, many U.S. students are performing poorly in elementary and high schools—especially in mathematics and the sciences. American teachers' salaries are low, many teachers are out of field, and teachers have less control over curricula than ever before. Despite grade inflation, high school and college dropout rates are high, and cheating is widespread.

### 4 What Is Religion?

*Religion* unites believers into a community. Every known society distinguishes between *sacred* and *secular* activities. Religion, *religiosity*, and spirituality differ; for example, people who describe themselves as religious may not attend services.

**Example**: Sacred vs. Secular

The corporate mission of Chick-fil-A (a franchise that prepares sandwiches), as stated on a plaque at company headquarters, is "to glorify God." Chick-fil-A is the only national fast-food chain that closes on Sunday so employees can go to church, and prospective employees are asked about their religious activities. Some franchise operators are delighted with the religious emphasis; others believe that a business should stay out of its workers' personal lives (Schmall 2007). What do *you* think?

### 5 Types of Religious Organization and Some Major World Religions

People express their religious beliefs most commonly through organized groups, including *cults* (also called *new religious movements [NRMs]*), *sects*, *denominations*, and *churches*. Some NRMs, which usually organize around a *charismatic leader*, have been short-lived, whereas others have become established religions with highly organized bureaucracies. Worldwide the largest religious group is Christians, followed by Muslims, but no religious group comes close to being a global majority. The third largest group is nonreligious persons. Five religions—Christianity, Islam, Hinduism, Buddhism, and Judaism—have an impact on economic, political, and social issues.

### 6 Religion in the United States

Religion in the United States is complex and diverse. About half of U.S. adults have changed their religion since childhood, many opting for no religion at all. Mainline Protestant groups have declined whereas evangelicals have surged. Religious participation varies by sex, age, race, ethnicity, and social class. Some sociologists maintain that *secularization* is increasing rapidly in the United States; others contend that this claim is greatly exaggerated, especially as witnessed by the growth of *fundamentalism* and the prevalence of *civil religion*.

### 7 Sociological Perspectives on Religion

See Table 13.4 on the next page.

## KEY TERMS

**education** a social institution that transmits attitudes, knowledge, beliefs, values, norms, and skills to its members through formal, systematic training.

**schooling** formal training and instruction provided in a classroom setting.

**intelligence quotient (IQ)** an index of an individual's performance on a standardized test relative to the performance of others of the same age.

**hidden curriculum** school practices that transmit nonacademic knowledge, values, attitudes, norms, and beliefs that legitimize economic inequality and fill unequal work roles.

**credentialism** an emphasis on certificates or degrees to show that people have certain skills, educational attainment levels, or job qualifications.

**tracking** (also called streaming or ability grouping) assigning students to specific educational programs and classes on the basis of test scores, previous grades, or perceived ability.

**religion** a social institution that involves shared beliefs, values, and practices related to the supernatural that unites believers into a community.

**sacred** anything that people see as mysterious, awe-inspiring, extraordinary and powerful, holy, and not part of the natural world.

**secular** anything that is not related to religion.

**religiosity** the ways people demonstrate their religious beliefs.

**cult** a religious group that is devoted to beliefs and practices that are outside of those accepted in mainstream society.

**new religious movement (NRM)** term used instead of cult by most sociologists.

**charismatic leader** a religious leader whom followers see as having exceptional or superhuman powers and qualities.

**sect** a religious group that has broken away from an established religion.

**denomination** a subgroup within a religion that shares its name and traditions and is generally on good terms with the main group.

**church** a large established religious group that has strong ties to mainstream society.

**secularization** a process of removing institutions such as education and government from the dominance or influence of religion.

**fundamentalism** the belief in the literal meaning of a sacred text.

**civil religion** (sometimes called secular religion) practices in which citizenship takes on religious aspects.

**Protestant ethic** a belief that hard work, diligence, self-denial, frugality, and economic success will lead to salvation in the afterlife.

**false consciousness** an acceptance of a system of beliefs that prevents people from protesting oppression.

**ritual** (sometimes called a rite) a formal and repeated behavior in which the members of a group regularly engage.

## TEST YOUR LEARNING

1. ____ maintain that the education system creates and perpetuates social inequality.
   a. Functionalists
   b. Conflict theorists
   c. Feminist theorists
   d. Symbolic interactionists

2. An emphasis on certificates or degrees is called
   a. meritocracy.
   b. tracking.
   c. the hidden curriculum.
   d. credentialism.

3. In examining educational outcomes, ____ are the most likely to focus on process.
   a. functionalists
   b. conflict theorists
   c. feminist theorists
   d. symbolic interactionists

4. A ____ is a religious group that is devoted to beliefs and practices that are outside of those accepted in mainstream society.
   a. cult          b. sect
   c. denomination  d. church

5. According to ____, religion is taught and not innate.
   a. functionalists
   b. conflict theorists
   c. feminist theorists
   d. symbolic interactionists

6. **True or False** Cultural innovation is one of the latent functions of education.

7. **True or False** U.S. high school dropout rates have been increasing.

8. **True or False** Both religion and religiosity are social institutions.

9. **True or False** The Protestant work ethic is a belief that hard work, diligence, self-denial, frugality, and economic success will lead to salvation in the afterlife.

10. **True or False** One of the criticisms of functionalist theories on religion is that they often ignore the role that religion plays in creating social cohesion and harmony.

11. How does education reinforce existing inequalities in the United States based on sex, social class, and race or ethnicity? Illustrate your answer with specific examples.

12. Explain how religious and secular rituals differ and overlap. Be specific. Also, why do people who say that they have no religion practice civil religion?

1.b  2.d  3.d  4.a  5.d  6.False  7.False  8.False  9.True  10.False

---

### TABLE 13.1

## Major Sociological Perspectives on Education

| THEORETICAL PERSPECTIVE | VIEW OF EDUCATION | SOME MAJOR QUESTIONS |
|---|---|---|
| Functionalist | Contributes to society's stability, solidarity, and cohesion and provides opportunities for upward mobility | What are the manifest and latent functions of education? |
| Conflict | Reproduces and reinforces inequality and maintains a rigid social class structure | How does education limit equal opportunity? |
| Feminist | Produces inequality based on gender | What are the gender gaps in education? |
| Symbolic Interactionist | Teaches roles and values through everyday face-to-face interaction and practices | How do tracking, labeling, self-fulfilling prophecies, and engagement affect students' educational experiences? |

### TABLE 13.4

## Sociological Perspectives on Religion

| THEORETICAL PERSPECTIVE | VIEW OF RELIGION | SOME MAJOR QUESTIONS |
|---|---|---|
| Functionalist | Religion benefits society by providing a sense of belonging, identity, meaning, emotional comfort, and social control over deviant behavior. | How does religion contribute to social cohesion? |
| Conflict | Religion promotes and legitimates social inequality, condones strife and violence between groups, and justifies oppression of poor people. | How does religion control and oppress people, especially those at lower socioeconomic levels? |
| Feminist | Religion subordinates women, excludes them from decision-making positions, and legitimizes patriarchal control of society. | How is religion patriarchal and sexist? |
| Symbolic Interactionist | Religion provides meaning and sustenance in everyday life through symbols, rituals, and beliefs and binds people together in a physical and spiritual community. | How does religion differ within and across societies? |

# Health and Medicine

## CHAPTER 14 TOPICS

### 1 Health and Illness in the United States

*Health* varies among individuals and societies, but all people experience *disease*. *Medicine* is a vital part of *health care* in diagnosing, treating, and preventing illness, injury, and other health impairments. Many people experience *disability*; others die at an earlier age than expected for a variety of macro- and micro-level reasons.

In addressing the issue of why some people are healthier than others, health and medical practitioners, researchers, and sociologists examine *epidemiology*. Epidemiology includes both the incidence and prevalence of illness or health problems within a population during a specific time period.

The three most important types of interlocking reasons for illness and early death are environmental, demographic, and lifestyle factors. Environmental variables include access to health care, preventable medical mistakes, and medication errors. Regarding demographic factors, after age 65, people are more likely to experience *chronic diseases* rather than *acute diseases*. Women tend to live longer than men but have more chronic diseases, such as arthritis, asthma, cancer, and mental illness, and experience higher depression rates than men. Among U.S. racial-ethnic groups, black babies have the *highest infant mortality rate*, and Latinos are the least likely to die from illicit drugs and prescription drug abuse. Regarding social class, people with a higher SES have an array of resources, such as money and knowledge, that increase good health and provide greater access to health care.

Our lifestyle choices also improve or impair health. The top three preventable health hazards, in order of priority, are smoking, obesity, and drug abuse. Sexually transmitted diseases are another important source of preventable health hazards.

### 2 Health Care: The United States and Around the World

The United States is one the richest nations in the world, but only the very wealthy don't have to worry about receiving and paying for the best medical care available. About 17 percent of Americans have no health insurance. Since 2000, employer-based health insurance coverage has deteriorated. Since 2006, workers at both large and small firms have been making higher contributions to premiums and paying higher deductibles or copayments.

*Medicare* pays many of the medical costs of Americans age 65 and over, regardless of income. *Medicaid*, another government program, provides medical care for those living below the poverty level. Whether their health insurance is private or government sponsored, many Americans belong to a *health maintenance organization*.

The United States spends more on health care than any other nation in the world, but ranks only 37th worldwide in life expectancy. Compared with other high-income countries, we cover a smaller percentage of the total population; have poorer health outcomes, fewer and shorter doctor visits, and fewer days of inpatient hospital care; and pay considerably more for prescription drugs and routine medical services (such as doctor's office visits and normal birth deliveries).

In 2010, Congress passed health care reform legislation—the Patient Protection and Affordable Care Act (which goes by the acronym ACA)—that is designed to be implemented fully by 2020. Some key features of the ACA that took effect immediately prohibit health care insurers from denying coverage to children with a preexisting condition and imposing lifetime spending limits. ACA also requires insurers to provide free preventive health services, such as tests for diabetes, mammograms, colonoscopies, and routine vaccinations; allows young adults to be insured under their parents' policies until age 26; permits people to choose a primary care doctor outside of the health plan's provider network; and allows people to visit the nearest emergency room outside of a plan's network without penalty.

### 3 Sociological Perspectives on Health and Medicine

See Table 14.3 on the next page.

## KEY TERMS

**health** the state of physical, mental, and social well-being.

**disease** an alteration of the normal physical and/or mental structures of the body or mind.

**health care** any activity that improves a person's well-being.

**medicine** a system of individuals, organizations, and institutions that provide scientific diagnosis, treatment, and prevention of illness, injury, and other health impairments.

**disability** physical or mental impairments that limit a person's ability to perform an important activity.

**epidemiology** the study of the causes and distribution of disease within a population.

**chronic diseases** long-term or lifelong illnesses that develop gradually or are present from birth.

**acute diseases** illnesses that strike suddenly and often disappear rapidly but can cause incapacitation and sometimes death.

**infant mortality rate** the number of deaths of infants (younger than 1 year) per 1,000 live births in a population.

**health maintenance organization (HMO)** a business organization that provides medical care to subscribers for a fixed fee.

**sick role** a pattern of behavior accepted as appropriate for people who are ill.

**medical-industrial complex** a network of business enterprises that influences medicine and health care.

**medicalization** a process that defines and treats a nonmedical condition or behavior as an illness, disorder, or disease that requires a medical solution.

## TEST YOUR LEARNING

1. The sick role is most closely associated with which theoretical perspective?
   a. Functionalist
   b. Conflict
   c. Feminist
   d. Symbolic interactionist

2. ____ is the top preventable lifestyle health hazard.
   a. Obesity
   b. Drug abuse
   c. Sexually transmitted diseases
   d. Smoking

3. Some have hailed the ____ as an important milestone in U.S. health care reform.
   a. AMA
   b. DSM
   c. ACA
   d. HMO

4. Which one of the following statements about U.S. health and medicine is *false*?
   a. Illness and death rates vary considerably, especially by social class.
   b. According to functionalists, the medical profession and health care industry maintain the status quo because of the medical-industrial complex.
   c. Women use more health care services than men do.
   d. According to symbolic interactionists, physicians, mental rights advocates, and parents all benefit from medicalization.

5. One of the most common criticisms of ____ theories is that they often ignore the contributions of medical and health care systems.
   a. functionalist
   b. conflict
   c. feminist
   d. symbolic interaction

6. *True or False* Acute diseases increase as people age.

7. *True or False* Texas has the largest number of uninsured residents.

8. *True or False* Gender rating is the practice of charging men more than women for identical health care plans.

9. *True or False* All states have obesity rates equal to or greater than 30 percent.

10. *True or False* From a functionalist perspective, physicians play a key role as gatekeepers for the sick role.

11. Is government-paid health care a basic human right? Or should be limited to taxpayers? Explain your answer. Be as specific as possible, and incorporate the material in this chapter such as health care in other countries and the ACA.

12. Provide any two limitations of functionalist, conflict, feminist, and symbolic interactionist theories in explaining health and medicine. (Extra points if you add a third limitation for each theory that's not discussed in this chapter.)

1.a  2.d  3.c  4.b  5.b  6.False  7.False  8.False  9.False  10.True

### TABLE 14.3

# Sociological Perspectives on Health and Medicine

| PERSPECTIVE | LEVEL OF ANALYSIS | KEY POINTS |
|---|---|---|
| Functionalist | Macro | • Health and medicine are critical in ensuring a society's survival and preserving social order.<br>• Illness is dysfunctional because it prevents people from performing expected roles.<br>• Sick people are expected to seek professional help and get well. |
| Conflict | Macro | • There are gross inequities in the health care system.<br>• The medical establishment is a powerful social control agent.<br>• A drive for profit ignores people's health needs. |
| Feminist | Macro and Micro | • Women are less likely than men to receive high-quality health care.<br>• Gender stratification in medicine and the health care industry reduces women's earnings.<br>• Men control women's health. |
| Symbolic Interactionist | Micro | • Illness and disease are socially constructed.<br>• Labeling people as ill increases the likelihood of being stigmatized.<br>• Medicalization has increased the power of medical associations, parents, and mental health advocates and the profits of pharmaceutical companies. |

*For practice tests, printable flash cards, and more, visit 4ltrpress.cengage.com/soc.*

# Population, Urbanization, and the Environment

## CHAPTER 15 TOPICS

### 1 Population Dynamics

*Demography* examines the interplay among *fertility*, *mortality*, and *migration*. The *crude death rate* and the infant mortality rate measure a population's life expectancy and health. Push and pull factors affect international migration and internal migration. Demographers also use *sex ratios* and *population pyramids* to understand a population's composition and structure.

Demographers who believe that population growth is a ticking bomb subscribe to *Malthusian theory*, which argues that the world's food supply will not keep up with population growth. *Demographic transition theory*, in contrast, maintains that population growth is kept in check and stabilizes as countries experience greater economic and technological development.

### 2 Urbanization

Globally and in the United States, *cities* and *urbanization* mushroomed during the twentieth century and are expected to increase. As more people move from rural to urban areas, many of the world's largest cities are becoming *megacities*. In the United States, urban growth has led to **s**uburbanization, *edge cities*, *exurbs*, *gentrification*, and *urban sprawl*.

Sociologists offer several perspectives on how and why cities change, and how these changes affect people. See Table 15.3 on the next page.

### 3 Environmental Issues

Population growth and urbanization are changing the planet's *ecosystem* and, many argue, endangering plants, animals, and humans. Water and air pollution and global warming are good examples of threats to the ecosystem in the United States and globally. Clean water has been depleted for many reasons, including pollution, privatization, and mismanagement.

Four of the most common sources and causes of air pollution are the burning of fossil fuels, manufacturing plants that spew pollutants into the air, winds that carry contaminants across borders and oceans, and lax governmental policies. Air pollution, which can lead to the *greenhouse effect*, is a major cause of *climate change* and *global warming*.

The rise of environmental problems has sparked a concern about *sustainable development*. Those who are pessimistic about achieving sustainable development show that, worldwide, the United States has one of the worst records on environmental performance, largely because of the close ties between government officials and corporations. Others are optimistic about achieving sustainable development and point to examples such as decreases in the emission of major air pollutants and some large U.S. corporations' switching to practices that decrease pollution and energy consumption.

## KEY TERMS

**demography**  the scientific study of human populations.

**population**  a group of people who share a geographic territory.

**fertility**  the number of babies born during a specified period in a particular society.

**crude birth rate**  (also known as the birth rate) the number of live births per 1,000 people in a population in a given year.

**mortality**  the number of deaths during a specified period in a population.

**crude death rate**  (also called the death rate) the number of deaths per 1,000 people in a population in a given year.

**migration**  the movement of people into or out of a specific geographic area.

**sex ratio**  the proportion of men to women in a population.

**population pyramid**  a visual representation of the makeup of a population in terms of the age and sex of its members at a given point in time.

**Malthusian theory**  the idea that the population is growing faster than the food supply needed to sustain it.

**demographic transition theory** the idea that population growth is kept in check and stabilizes as countries experience economic and technological development.

**zero population growth (ZPG)** a stable population level that occurs when each woman has no more than two children.

**city**  a geographic area where a large number of people live relatively permanently and make a living primarily through nonagricultural activities.

**urbanization**  population movement from rural to urban areas.

**megacities**  metropolitan areas with at least 10 million inhabitants.

**suburbanization**  population movement from cities to the areas surrounding them.

**edge cities**  business centers that are within or close to suburban residential areas.

**exurbs**  areas of new development beyond the suburbs that are more rural but on the fringe of urbanized areas.

**urban sprawl**  the rapid, unplanned, and uncontrolled spread of development into regions adjacent to cities.

**gentrification**  a process in which middle-class and affluent people buy and renovate houses and stores in downtown urban neighborhoods.

**urban ecology**  the study of the relationships between people and urban environments.

**new urban sociology**  the view that urban changes are largely the result of decisions made by powerful capitalists and high-income groups.

**ecosystem**  an area in which all forms of life live in relation to one another and a shared physical environment.

**global warming**  the increase in the average temperature of Earth's atmosphere.

**greenhouse effect**  the heating of Earth's atmosphere because of the presence of certain atmospheric gases.

**climate change**  a change in overall temperatures and weather conditions over time.

**sustainable development**  economic activities that meet the needs of the present without threatening the environmental legacy of future generations.

# TEST YOUR LEARNING

1. ____ is the study of human populations.
   a. Sociology
   b. Demography
   c. Ecology
   d. Zero population growth (GPG)

2. Emma is comparing the number of deaths per 1,000 people in Spain and Russia for 2011. She is measuring
   a. fertility.
   b. mortality.
   c. the crude birth rate.
   d. the crude death rate.

3. Which one of the following is NOT one of the phases of demographic transition theory?
   a. Preindustrial society
   b. Early industrial society
   c. Advanced industrial society
   d. Declining industrial society

4. Maleek and Shonda live in a newly developed area that's rural but on the fringe of an urbanized region. They live in
   a. an exurb.
   b. a suburb.
   c. a gentrified area.
   d. an edge city.

5. Why do many feminist scholars maintain that women suffer more problems than men when living in urban areas?
   a. Women don't have the natural aggressive instinct that men have.
   b. Husbands tend to maintain tighter controls over their wives in urban areas.
   c. Urban areas have typically been designed by men and for men.
   d. All of the above explain why women suffer more problems in urban areas.

6. **True or False** Malthusian theory claims that population growth is kept in check and stabilizes as developing countries experience economic and technological development.

7. **True or False** Population pyramids are visual representations of a population's makeup in terms of the age and sex of its members at a given point in time.

8. **True or False** Concentric zone theory emphasizes the development of suburbs around a city but away from its center.

9. **True or False** A sex ratio of 115 means that there are 115 women for every 100 men in a population.

10. **True or False** Global warming begins with the greenhouse effect.

11. Does Malthusian theory or demographic transition theory come closer to your perspective on population growth? Why?

12. How do functionalist, conflict, feminist, and symbolic interactionist theories differ in their explanations of urbanization? Be specific in illustrating your answer.

1.b  2.d  3.d  4.a  5.c  6.False  7.True  8.False  9.False  10.True

---

## TABLE 15.3

# Sociological Perspectives on Urbanization

| PERSPECTIVE | LEVEL OF ANALYSIS | KEY POINTS |
|---|---|---|
| Functionalist | Macro | People create urban growth by moving to cities to find jobs and to suburbs to enhance their quality of life. |
| Conflict | Macro | Driven by greed and profit, large corporations, banks, developers, and other capitalist groups determine the growth of cities and suburbs. |
| Feminist | Macro and micro | Whether they live in cities or suburbs, women generally experience fewer choices and more constraints than men. |
| Symbolic Interactionist | Micro | City people are more tolerant of different lifestyles, but they tend to interact superficially and are generally socially isolated. |

*For practice tests, printable flash cards, and more, visit 4ltrpress.cengage.com/soc.*

# Social Change: Collective Behavior, Social Movements, and Technology

## CHAPTER 16 TOPICS

### 1 Collective Behavior

According to an influential sociological theory, six macro-level conditions can encourage or discourage *collective behavior*: structural conduciveness, structural change, the growth and spread of a generalized belief, precipitating factors, mobilization, and social control. There are many types of collective behavior, some more short-lived or harmful than others. *Rumors*, *gossip*, and *urban legends* are typically untrue, but many people believe and pass them on for a number of reasons, such as anxiety or to reinforce a community's moral standards. In contrast, *panic* and *mass hysteria* can have dire consequences, including death.

*Fashions*, *fads*, and *crazes* are harmless because they usually last only a short time and change over time. People choose to participate in fashions, fads, and crazes. In contrast, a *disaster* is an unexpected event due to social, technological, or natural causes.

*Publics*, *public opinion*, and *propaganda* also affect large numbers of people, and some of these types of collective behavior are more harmful than others. *Crowds* vary in their motives, interests, and emotional level. A casual crowd, for example, has little, if any, interaction, the gathering is temporary, and there is little emotion. On the other hand, protest crowds, especially *mobs* and those involved in a *riot*, can wreak considerable havoc on property and result in death.

**Example**: Crowds Can Be Deadly

On Thanksgiving Day, 2008, crowds started gathering at 9:00 P.M. outside the Wal-Mart store in Valley Stream, New York, for a bargain-hunting ritual known as Black Friday, the day after Thanksgiving. By 4:55 A.M., the crowd had grown to more than 2,000 people and could no longer be held back. Suddenly, according to witnesses, the glass doors shattered and "the shrieking mob surged through in a blind rush for holiday bargains." A 34-year-old male temporary worker, who had been hired for the holiday season, was trampled to death, and four other people, including a 28-year-old woman who was eight months pregnant, were treated for injuries

(McFadden and Macropoulos 2008). Review the types of crowds in this chapter. Which type of crowd do you think this Wal-Mart incident most nearly represents?

### 2 Social Movements

Unlike collective behavior, *social movements* are typically organized and have long-lasting effects. Some of the most common social movements are alternative, redemptive, reformative, resistance, and revolutionary (see Table 16.1 on the next page). Sociologists have offered several explanations for the emergence of social movements, including mass society theory, relative deprivation theory, resource mobilization theory, and new social movements theory. Each theory has strengths and weaknesses in helping us understand social movements.

Social movements generally go through four stages: emergence, organization, institutionalization, and decline. Decline occurs when a social movement is successful and becomes a part of society's fabric; when the members become distracted because the group loses sight of its original goals and/or their enthusiasm diminishes; when the membership fragments because the participants disagree about goals, strategies, or tactics; or when a government quashes dissent. Social movements are important because they can create or resist change at the individual, institutional, and societal levels.

### 3 Technology and Social Change

*Technology* also generates changes. Some of the most important technological advances have included computer technology, biotechnology, and nanotechnology—all of which have changed our lives dramatically. Technology has both benefits and costs, however. For example, the Internet and other forms of telecommunication technology can bring people together but can also intrude on our privacy. In addition, technological advances raise numerous ethical questions, such as their greater availability to the wealthy and educated.

## KEY TERMS

**social change** the transformations of societies and social institutions over time.

**collective behavior** the spontaneous and unstructured behavior of a large number of people.

**rumor** unfounded information that people spread quickly.

**gossip** rumors, often negative, about other people's personal lives.

**urban legends** (also called contemporary legends and modern legends) a type of rumor consisting of stories that supposedly happened.

**panic** a collective flight, typically irrational, from a real or perceived danger.

**mass hysteria** an intense, fearful, and anxious reaction to a real or imagined threat by large numbers of people.

**fashion** a standard of appearance that enjoys widespread but temporary acceptance within a society.

**fad** a form of collective behavior that spreads rapidly and enthusiastically but lasts only a short time.

**craze** a fad that becomes an all-consuming passion for many people for a short time.

**disaster** an unexpected event that causes widespread damage, destruction, distress, and loss.

**public** a collection of people, not necessarily in direct contact with each other, who are interested in a particular issue.

**public opinion** widespread attitudes on a particular issue.

**propaganda** the presentation of information to influence people's opinions or actions.

**crowd** a temporary gathering of people who share a common interest or participate in a particular event.

**mob** a highly emotional and disorderly crowd that uses the threat of force, actual force, or violence against a specific target.

**riot** a violent crowd that directs its hostility at a wide and shifting range of targets.

**social movement** a large and organized activity to promote or resist a particular social change.

**relative deprivation** a gap between what people have and what they think they should have compared with others in a society.

**technology** the application of scientific knowledge for practical purposes.

# TEST YOUR LEARNING

1. According to sociologist Neil Smelser, which one of the following does NOT encourage collective behavior?
   a. Structural strain
   b. Growth and spread of a generalized belief
   c. Social control
   d. Personal dissatisfaction at work or home

2. A major difference between panic and mass hysteria is that
   a. panics are typically not as severe as mass hysteria.
   b. panics don't usually last as long as mass hysteria.
   c. panics stem from imagined events, whereas mass hysteria arises from real events.
   d. panics tend to occur less today, whereas mass hysteria occurs more frequently.

3. The "amazing ball" was a toy that became incredibly popular and sold for a very high price, but its popularity dropped off very quickly. This is an example of a
   a. fashion.
   b. fad.
   c. craze.
   d. disaster.

4. ____ maintains that social movements emerge because there's a gap between what people have and what they think they should have compared with other people.
   a. Mass society theory
   b. Relative deprivation theory
   c. Resource mobilization theory
   d. New social movements theory

5. Which of the following is NOT one of four stages of a social movement?
   a. Emergence
   b. Conflict
   c. Institutionalization
   d. Decline

6. **True or False** During the organization stage of social movements, the movement becomes more bureaucratic.

7. **True or False** Gossip and urban legends are types of rumors.

8. **True or False** Mobs typically last longer than riots.

9. **True or False** Environmentalism is an example of the new social movements theory.

10. **True or False** A positive trend in technology is that the advances are now available to most low-income people.

11. What conditions are necessary to produce social movements? What draws people into a social movement? If you're involved in a social movement now, which one and why?

12. What, personally, have been some of the benefits and costs of technology in your life? Offer specific examples to illustrate your answer.

1.d  2.b  3.c  4.b  5.b  6.False  7.True  8.False  9.True  10.False

---

TABLE 16.1

# Five Types of Social Movements

| MOVEMENT | GOAL | EXAMPLES |
| --- | --- | --- |
| Alternative | Change some people in a specific way | Alcoholics Anonymous, transcendental meditation |
| Redemptive | Change some people, but completely | Jehovah's Witnesses, born-again Christians |
| Reformative | Change everyone, but in specific ways | Gay rights advocates, Mothers Against Drunk Driving (MADD) |
| Resistance | Preserve status quo by blocking or undoing change | Antiabortion groups, white supremacists |
| Revolutionary | Change everyone completely | Right-wing militia groups, Communism |

*For practice tests, printable flash cards, and more, visit 4ltrpress.cengage.com/soc.*